The Role of Pollinators in Tropical Landscapes

A. K. Chakravarthy
Editor

The Role of Pollinators in Tropical Landscapes

Conservation and Sustainability across Wild and Managed Systems

Springer

Editor
A. K. Chakravarthy
Society for Science and Technology
Applications (SSTA)
Bangalore, Karnataka, India

ISBN 978-981-95-9138-1 ISBN 978-981-95-9139-8 (eBook)
https://doi.org/10.1007/978-981-95-9139-8

This Springer imprint is published by the registered company Springer Nature Singapore Pte Ltd.
The registered company address is: 152 Beach Road, #21-01/04 Gateway East, Singapore 189721, Singapore

Foreword

The world is witnessing severe declines in both the diversity and populations of pollinators. This is occurring across a wide range of habitats at unprecedented and alarming rates. Sustaining pollinator populations and their contributions to food security on Planet Earth has become a pressing issue, particularly in the tropics. This is because pollinators perform vital functions in ecosystems and their interactions impact organisms at multitrophic levels. In particular, they provide pollination services to an increasing number of crops in the tropics, yielding flowers, fruits, nuts, and vegetables—many that contribute to human health and well-being. Conservation strategies within wild and managed landscapes should include pollinators that are at high risk. Despite this situation, there has been limited action "on the ground" to ensure survival of pollinators and their associated economic and ecological services. In particular, there have been few action plans developed for pollinator conservation in many regions, in part because of a paucity of knowledge on pollinator biology and ecology, as well as lack of incorporation of pollinator safety into pest management practices. These above-mentioned concerns have provided the impetus for this book, *The Role of Pollinators in Tropical Landscapes: Conservation and Sustainability Across Wild and Managed Systems*. It is comprehensive, comprising 21 chapters under four sections: Biology and Behavior of Pollinators, Pollinators in Wild and Managed Ecosystems, Conservation Strategies, and Beekeeping as a Sustainable Enterprise for Livelihood Security. In the first section, the chapters illustrate how changes in pollinator bioecology bring about changes in pollination-plant interactions. The second section addresses how rising temperatures and changes in precipitation can bring about changes in flower-visitation by pollinators at species, population, community and ecosystem levels, and their resultant impacts on crop yield and quality. The chapters in the third section focus on Conservation Strategies. Threats and their impact on key component species in pollination systems and the important role of coordination among key stakeholders to execute conservation plans are emphasized. Such conservation measures include monitoring pollinator populations and provision of suitable habitats (especially floral and nesting resources). The above three sections provide the context for the important issue of supporting livelihoods. Therefore, the fourth section in this book focuses on managed pollinators and livelihoods. "Meliponiculture" and "Village concept" are two concepts included in this section.

I believe this book will contribute to the development of broad conservation plans for pollinators as well as other biodiversity components in south and south-east Asia, and support deaccelerating of pollinator population declines in both natural and agroecosystems. This will therefore be a useful reference and guide for not only pollination biologists but also for policymakers, zoologists, landscape designers, ecologists, entomologists, conservation biologists and practitioners, researchers, teachers, and students alike.

I therefore congratulate the editor, and all the authors for their valuable contributions.

Hawkesbury Institute for the Environment
and School of Science, Western Sydney University
Penrith, NSW, Australia

Scientific Adviser in Sustainable Pest Management
Australian Olive Association
Allambie Heights, NSW, Australia

Member, Scientific Task Force, Bee-Friendly Farming
Wheen Bee Foundation
Yendon, VIC, Australia

NATA Examiner, Life Sciences
Sydney, NSW, Australia

Visiting Professor, TNAU
Coimbatore, India April 2025

Robert Spooner-Hart

Preface

As teenage children in the 1950s, we use to hear a lot about honeybees and honey. In meetings, discourses, and chats, pollination by honeybees and conservation was emphasized. Then came in the 1960s the catastrophic events like climate change, urbanization, pollution, habitat deterioration, and fragmentation that impacted honeybee populations. Coupled with this, the dramatic changes in agriculture and intensification, land-use changes, globalization, and other factors such as epidemic diseases further hit apiculture. Since then, the populations of honeybees and other pollinators are decimating continually, and the role and development of apiculture as a sustainable enterprise, providing livelihood security, is still haunting! Currently, the trend is for researchers to teach teenagers on-farm honey bees to non-honeybees; from bees to non-bee pollinators, day to night pollinators, and terrestrial to pollinators in aquatic ecosystems. Apiculture to Meliponiculture, backyard to Village/Landscape apiculture, rural to urban apiculture and so on. It looks like researchers and policymakers seem to have a master plan to comprehend the entire pollinator system holistically and act on ground. This is crucially important in tropics and sub-tropics where population of pollinators is steadily declining. In this context, a compilation on pollinators providing insights for the present and the future generations was thought imperative and valuable.

This book, *The Role of Pollinators in Tropical Landscapes: Conservation and Sustainability Across Wild and Managed Systems*, is planned to address emerging issues and technologies, conveying ideas and scientific opinions of 80 contributors on pollinators from tropics, mainly from India. This book is divided into four sections with 212 illustrations and several tables: Biology and Behavior of Pollinators; Pollinators in Wild and Managed Ecosystems; Conservation Strategies; and Beekeeping as a Sustainable Enterprise for Livelihood Diversity, populations, impact assessment and management. How crop productivity is enhanced with pollinators, why and how pollinators can be conserved, and how and why pollinators should form an integral part of farming systems and support livelihoods are also addressed. Hopefully, the collective efforts of the teams of contributors and the editor will be useful to a wider section of human society culminating in saving the pollinators, ecosystems, and the planet Earth! We are already closing on the irreversible global crisis.

Bangalore, Karnataka, India A. K. Chakravarthy

Acknowledgments

The planning and execution of this book would not have been possible without the cooperation and coordination of several scientists, specialists, and scholars. Sustained and concerted efforts of over two years resulted in the book manuscript. Dr. K.T. Vijayakumar, Dept. Of Apiculture, UAS, GKVK, Bangalore, was instrumental in pursuing workers to contribute for the book. He also has contributed chapters in this book, besides helping me in planning the book contents. I owe a debt of gratitude to Dr. Vijayakumar. Drs. Narayanappa and Bettaiah contributed a chapter on Gaint Honeybee, *Apis dorsata*, while Pritha Ghosh and team organized and wrote chapters on pollinator-plant relationships and behavior, Drs. Raj Kumar Thakur and team members from Himachal Pradesh, India, contributed chapters on village concept, and Bumble bee pollination, Dr. Pavithra Naik provided an updated review on the impact of climate change on insect pollinators. Dr. Shahnawaz and coworkers dealt with foraging and home range of *Apis* bees. Dr. Rani and fellow-entomologists developed an understanding of the pollinators on Avocado, Rambhutan, and Litchi in South India. A community of pollinators in Mangrove ecosystems in west coast of India has been documented by Drs. Remadevi and D. Chatterjee, while in forests by Drs. Tombat and others. Management of pollinators in cultivated crops like Onion, Pomegranate, fruits, vegetables, and flower crops have also been included in this compilation. Dr. Abu Hassan Jalil, Malaysia, shared his life-long experience in Meliponiculture in this book. The contributors have also dealt with pollinator conservation strategies and beekeeping as a subsidiary enterprise for livelihoods. The editor profoundly thanks all the contributors for their untiring efforts and contributions, despite their hectic schedules. The editors thank immensely the Springer Nature Group, New Delhi and Singapore for their interests and support for this book.

A. K. Chakravarthy

Contents

About the Editor

A. K. Chakravarthy, Head and Principal Scientist (Rtd.), Division of Entomology and Nematology, is an author of many books and over 424 scientific papers and 80 chapters in Entomology and Natural history. His interest includes insects, birds, bats, rodents, and mammals. With nearly four decades of experience in teaching, research and extension, he has been the investigator for over 35 research projects and has guided more than 28 students for post- graduation. Recently, he has published books titled, *Innovative Pest Management Approaches for the 21st Century* (Foreword by Prof. Dr. Remzi ATLIHAN, Turkey), *New Horizons in Insect Science: Towards Sustainable Pest Management* (Foreword by Prof. Hiroshi Honda, Japan), *Economic and Ecological Significance of Arthropods in Diversified Ecosystems: Sustaining Regulatory Mechanisms* (Foreword by Fani Hatijina, Greece), *Arthropod Diversity and Conservation in the Tropics and Sub-tropics* (Foreword by Prof. Madhav Gadgil, India), *The Black Spotted, Yellow Borer, Conogethes punctiferalis Guenee and Allied Species* (Foreword by Prof. Alma Solis, USA), and *Experimental Techniques in Host Plant Resistance* (Foreword by Prof. Michael C. Smith, USA) by Springer, *Animal Behaviour in the Tropics-Vertebrates* (Foreword by Dr. Mewa Singh, Mysore University, India) by Springer.

Contributors

U. Amala ICAR-National Bureau of Agricultural Insect Resources, Bengaluru, Karnataka, India

Mangla Ram Bajiya Department of Entomology, Dr. YSP UHF Nauni, Solan, Himachal Pradesh, India

G. N. Chaithra Department of Biotechnology, College of Agriculture, UASM, Hassan, Karnataka, India

A. K. Chakravarthy Society for Science and Technology Applications (SSTA), Bangalore, Karnataka, India

Debajyoti Chatterjee Division of Entomology, ICAR - Indian Agricultural Research Institute, New Delhi, India

Nilanjana Chaudhuri Agricultural Entomology, Regional Research Station, Terai Zone, Uttar Banga Krishi Viswavidyalaya, Pundibari, Cooch Behar, West Bengal, India

N. Deeksha Raj Department of Forestry and Environmental Science, UAS, GKVK, Bangalore, Karnataka, India

Diksha Devi Department of Entomology, Dr YS Parmar University of Horticulture and Forestry, Nauni, Himachal Pradesh, Solan, India

Pritha Ghosh Department of Entomology, School of Agriculture, Lovely Professional University, Phagwara, Punjab, India

K. S. Jagadish Department of Entomology, GKVK, UAS, Bengaluru, Karnataka, India

Abu Hassan Jalil Agronomist, Meliponiculturist, Stamford College, Petaling Jaya, Malaysia

B. R. Jayanthi Mala ICAR-Indian Institute of Horticultural Research, Hessaraghatta, Bengaluru, Karnataka, India

Manoj Kumar Jena Department of Plant Protection, Institute of Horticultural Sciences, Warsaw University of Life Sciences, Warsaw, Poland

S. I. Kazmi Plant Protection Division, Aligarh Muslim University, Aligarh, Uttar Pradesh, India

Shivani Khokhar Maharana Pratap Horticultural University, Karnal, Haryana, India

Rajiv Kumar Division of Horticulture and Medicinal plants, ICAR-Indian Institute of Horticultural Research, Bengaluru, Karnataka, India

G. S. Madhu ICAR-IIHR-Central Horticultural Experiment Station, Chettalli, Kodagu, Karnataka, India

B. Majumdar Zoological Survey of India, Prani Vigyan Bhawan, New Alipore, Kolkata, West Bengal, India

Rathna Kumari Bettaiah Mallamma Department of Botany, Government First Grade College, Vijayanagara, Bengaluru, Karnataka, India

Ajinkya Markad Department of Entomology, School of Agriculture, Lovely Professional University, Phagwara, Punjab, India

Bikash Mohapatra The Cotton Corporation of India Ltd, Sirsa, Haryana, India

Rajeeb Lochan Moharana Department of Seed Science and Technology, College of Agriculture, OUAT, Bhawanipatna, Odisha, India

Monika Department of Entomology, Dr. YSP UHF, Nauni, Solan, Himachal Pradesh, India

B. M. Muralidhara ICAR-IIHR-Central Horticultural Experiment Station, Chettalli, Kodagu, Karnataka, India

Narayanappa Nagaraja UGC-Malaviya Mission Teacher Training Centre, Bangalore University, Jnana Bharathi Campus, Bengaluru, Karnataka, India

Pavithra Prakash Nayak Department of Climate Change, Environmental Management & Policy Research Institute, Doresanipalya Forest Campus, Bangalore, India

Nayan Deepak ICAR-IIHR-Central Horticultural Experiment Station, Chettalli, Kodagu, Karnataka, India

H. L. Nithin Kumar AICRP on Honey Bees and Pollinators, University of Agricultural Sciences, GKVK, Bangalore, India

Ahmad Pervez Department of Zoology, Pt. L. M. S. Campus, Sri Dev Suman Uttarakhand University, Rishikesh, India

M. Gopi Prasad Department of Entomology, Sri Kinjarapu Yerramnaidu College of Agricultural Sciences (SKYCAS), Srikakulam, Andra Pradesh, India

D. Rajanna AINP on Soil Arthropod Pests, UAS, GKVK, Bangalore, Karnataka, India

S. Rajendiran ICAR-IIHR-Central Horticultural Experiment Station, Chettalli, Kodagu, Karnataka, India

V. Rakshith ICAR-IIHR-Central Horticultural Experiment Station, Chettalli, Kodagu, Karnataka, India

T. N. Rakshitha AICRP on Honey Bees and Pollinators, University of Agricultural Sciences, GKVK, Bangalore, India

Bijitha Ramesh Department of Climate Change, Environmental Management & Policy Research Institute, Doresanipalya Forest Campus, Bangalore, India

A. Rameshkumar Zoological Survey of India, Prani Vigyan Bhawan, New Alipore, Kolkata, West Bengal, India

Kiran Rana Department of Entomology, Dr YS Parmar University of Horticulture and Forestry, Nauni, Himachal Pradesh, Solan, India

A. T. Rani ICAR-IIHR-Central Horticultural Experiment Station, Chettalli, Kodagu, Karnataka, India

D. Devika Rani Department of Entomology, Sri Kinjarapu Yerramnaidu College of Agricultural Sciences (SKYCAS), Srikakulam, Andra Pradesh, India

Department of Entomology, University of Agricultural Sciences, GKVK, Bangalore, Karnataka, India

Priyanka Rani Department of Entomology, Dr. YSP UHF, Nauni, Solan, Himachal Pradesh, India

K. M. Ravi Kiran Department of Studies and Research in Zoology, Tumkur University, Tumkur, Karnataka, India

P. V. R. Reddy ICAR-Indian Institute of Horticultural Research, Hessaraghatta, Bengaluru, Karnataka, India

P. V. Rami Reddy ICAR-Indian Institute of Horticultural Research, Bengaluru, Karnataka, India

O. K. Remadevi Environmental Management Policy Research Institute, Bangalore, Karnataka, India

S. Sardar High Altitude Regional Centre, Zoological Survey of India, Saproon, Solan, Himachal Pradesh, India

A. Shahnawaz Department of Studies and Research in Zoology, Tumkur University, Tumkur, Karnataka, India

Ruchi Sharma Senior Research Fellow, Department of Entomology, Dr. YSP UHF, Nauni, Solan, Himachal Pradesh, India

M. Shashikant Department of Studies in Zoology, Vijayanagar Sri Krishnadevaraya University, Ballari, Karnataka, India

Sindhu Sheoran Hawkesbury Institute for the Environment, Western Sydney University, Penrith, NSW, Australia

Shivakumara Department of Entomology, College of Agriculture, V. C. Farm, Mandya, Karnataka, India

T. M. Shivalingaswamy ICAR-National Bureau of Agricultural Insect Resources, Bengaluru, Karnataka, India

H. S. Shwetha Department of Studies and Research in Zoology, Tumkur University, Tumkur, Karnataka, India

A. N. Shylesha ICAR-National Bureau of Agricultural Insect Resources, Bengaluru, Karnataka, India

Gaurav Singh Maharana Pratap Horticultural University, Karnal, Haryana, India

Sawraj Jit Singh Department of Entomology, Dr YS Parmar University of Horticulture and Forestry, Nauni, Himachal Pradesh, Solan, India

P. Sree Chandana ICAR-Indian Institute of Rice Research, Hyderabad, Telangana, India

Kolla Sreedevi ICAR-National Bureau of Agricultural Insect Resources, Bengaluru, Karnataka, India

S. M. Sreelakshmi Department of Studies and Research in Zoology, Tumkur University, Tumkur, Karnataka, India

A. N. Sringeswara Department of Forestry and Environmental Science, UAS, GKVK, Bangalore, Karnataka, India

B. Tambat Department of Forestry and Environmental Science, UAS, GKVK, Bangalore, Karnataka, India

Nayimabanu Taredahalli AICRP on Honey Bees and Pollinators, University of Agricultural Sciences, GKVK, Bangalore, India

Meena Thakur Department of Entomology, Dr. YSP UHF, Nauni, Solan, Himachal Pradesh, India

Priyanka Thakur Department of Entomology, Dr. YSP UHF, Nauni, Solan, Himachal Pradesh, India

Raj Kumar Thakur Department of Entomology, Dr. YSP UHF, Nauni, Solan, Himachal Pradesh, India

L. Vani Department of Studies and Research in Zoology, Tumkur University, Tumkur, Karnataka, India

L. Veena Department of Studies and Research in Zoology, Tumkur University, Tumkur, Karnataka, India

H. S. Venu ICAR-National Bureau of Agricultural Insect Resources, Bengaluru, Karnataka, India

K. T. Vijayakumar AICRP on Honey Bees and Pollinators, University of Agricultural Sciences, GKVK, Bangalore, India

Meena Yadav Department of Zoology, Maitreyi College, University of Delhi, Delhi, India

Part I

Bioecology and Behavior of Pollinators

Bio-ecology and Utilization of Nomia for Pollination

1

K. T. Vijayakumar, T. N. Rakshitha, Nayimabanu Taredahalli, and H. L. Nithin Kumar

Abstract

Nomia bees are a group of native bee species that are important pollinators of crops and wild plants. *Nomia* bee species are a genus of sweat bees in the family Halictidae with over 130 named species, which are found in most parts of the world. *Hoplonomia westwoodi* (Gribodo 1894) is a common species of southern India, recorded as an important pollinator of many cultivated crops. The maximum number of nests of alkali bees is seen in soil with higher pH. The conservation and utilization of *Nomia* bees need to be worked out.

Keywords

Bio-ecology · *Nomia* · Pollination · Utilization

1 Introduction

Pollinators provide an essential ecosystem service by enabling plants to produce fruits and seeds. These pollinators come from diverse groups within the animal kingdom, including insects, birds, bats, reptiles, and others. Among them, insects play the most dominant role in pollination, supporting a vast range of plant species worldwide. Within insect pollinators, several taxonomic orders contribute significantly, such as Hymenoptera, Lepidoptera, Coleoptera, Diptera, and Hemiptera. Of these, members of the order Hymenoptera stand out as particularly advanced and effective pollinators. The most critical group within Hymenoptera is the

K. T. Vijayakumar (✉) · T. N. Rakshitha · N. Taredahalli · H. L. Nithin Kumar
AICRP on Honey Bees and Pollinators, University of Agricultural Sciences, GKVK, Bangalore, India

A. K. Chakravarthy (ed.), *The Role of Pollinators in Tropical Landscapes*,
https://doi.org/10.1007/978-981-95-9139-8_1

superfamily Apoidea, which comprises around 25,000 described species and includes bees, as well as related insects like wasps, ants, ichneumons, chalcids, and sawflies, distributed across roughly 250 genera and 11 families (Teale 1957). Beyond the commonly recognized honeybees, 20,000 other bee species occur worldwide that contribute significantly to pollination. These include numerous solitary bee species such as *Megachile*, *Nomia*, *Osmia*, *Halictus*, *Andrena*, and *Ceratina,* which play a vital role in pollinating fruit crops and many other flowering plants.

Plant–pollinator relationships are highly intricate and dynamic (Mitchell et al. 2009). It is estimated that nearly three-quarters of angiosperm species depend on animal pollinators for effective pollen transfer between flowers (Cresswell 2005). Among flowering plants, anthers in about 72 families and 574 genera open through specialized pores, and in roughly 54 families and 357 genera, pollen release is further restricted to bees capable of a behavior known as buzz pollination (Buchmann 1983). This unique process involves bees landing on a flower (Larson and Barrett 1999), gripping or curling around the "anther cone" (Buchmann 1985), and then activating their indirect flight muscles at high frequency. These rapid vibrations shake the anthers, causing pollen grains to be forcibly expelled (Proenca 1992). The process produces a characteristic buzzing sound, hence the term "buzz pollination" (Buchmann 1983). A wide range of bees, including solitary and social species, as well as generalists and specialists, use this sonicating behavior to collect pollen. Sonicating bees are distributed across at least seven bee families and more than 50 genera globally (Symon 1979). In Sri Lanka, bees belonging to genera such as *Hoplonomia*, *Lasioglossum*, *Patellapis* (family Halictidae), and *Amegilla* and *Xylocopa* (family Apidae) have been recorded buzzing flowers having poricidal anthers (Karunaratne et al. 2005). Recognizing the important ecological service provided by these non-*Apis* bees, this chapter focuses on *Nomia*—commonly referred to as sweat bees or alkali bees. It covers their biology, nesting behavior, habitats, and their significance in pollinating various crops.

Nomia bees are native bee species that pollinate both cultivated crops and wild plants. These bees are generally small to medium in size, measuring about 7–12 mm in length. They appear robust and hairy, with coloration that changes, and are often recognized by the presence of yellow or metallic green iridescent bands across the abdomen. Slightly smaller than honeybees, *Nomia* females carry pollen on their hind legs and typically nest by digging burrows in the soil. Despite having a sting, *Nomia* bees are usually less aggressive, and their sting is considered milder and less painful compared to that of honeybees. Morphologically, they can be identified by a distinct, strongly curved basal vein on their forewings. Additionally, members of the family Halictidae, including *Nomia,* tend to have hind legs that are positioned more widely apart than in many other bees. In India, research focusing on these non-*Apis* bee pollinators remains limited. Halictid bees, to which *Nomia* belongs, represent one of the six existing bee families, and globally there are about 3500 described species in this family (Michener 2007). These bees are commonly called "sweat bees," a name that comes from the belief that they are attracted to human sweat to meet their salt requirements, though conclusively not proven.

The subfamily Nomiinae (family Halictidae, order Hymenoptera) consists of around 600 species distributed globally, with the notable exception of South America (Astafurova and Pesenko 2005). The Orient subfamily comprises approximately 154 valid species (Pannure and Belavadi 2017). In India alone, Nomiinae is represented by about 15 genera and 72 species (Saini and Rathor 2012), of which 48 species of 13 genera have been documented specifically from South India (Pannure and Belavadi 2017). The genus *Nomia* itself comprises six subgenera and about 111 described species distributed worldwide (Michener 2007). While many Nomiinae are solitary in nesting habits, communal nesting has been observed in several genera and subgenera (Batra 1966; Wcislo and Engel 1996; Michener 2007). One notable species, *Nomia (Acunomia) chalybeata* Smith, previously known under the synonym *N. pavonula* Cockerell in Japan, is a multivoltine, medium-sized bee. Females typically reach around 12.0 mm in length, and males are slightly larger, about 12.5 mm, and both sexes feature characteristic emerald green stripes on the metasomal terga. The first documented nesting of *N. chalybeata* was reported from Iriomote Island; however, the species is already known to inhabit other subtropical islands of Japan, including Amami, Ryukyu, and Sakishima islands (Hirashima 1989). Beyond Japan, its distribution extends to the Korean Peninsula, Taiwan, China, Myanmar, and India (Ascher and Pickering 2012).

Four species belonging to the genus *Nomia* Latreille, 1804, specifically *Nomia (Nomia) crassipes* (Fabricius 1798), *Nomia (Nomia) curvipes* (Fabricius 1793), *Nomia (Hoplonomia) elliotii* (Smith 1875), and *Nomia (Hoplonomia) westwoodii* Gribodo, 1894, are newly documented from Pakistan (Bodlah et al. 2016). The genus *Hoplonomia* Ashmead, which has a predominantly Oriental distribution (Ascher and Pickering 2022), is currently represented by at least five known species in India. Among these, *Hoplonomia westwoodi* (Gribodo 1894) is particularly noteworthy as it is frequently observed in southern India, where it serves as a key pollinator for several cultivated crops (Udayakumar and Shivalingaswamy 2018). Despite these records, many species of *Nomia* and *Hoplonomia* likely remain undiscovered or unreported. This highlights the importance of systematic surveys, conservation, and active promotion of native pollinators like sweat bees to support sustainable crop pollination.

2 Taxonomic Details

Kingdom: Animalia, **Phylum**: Arthropoda, **Class**: Insecta, **Order**: Hymenoptera, **Family:** Halictidae, **Subfamily:** Nomiinae, **Genus**: *Nomia* Latreille, 1804

Nomia is a genus of sweat bees of the family Halictidae, encompassing over 130 described species distributed world over. Members of this genus can be recognized by several distinctive morphological features. Notably, they have a rounded facial profile where the compound eyes blend smoothly into the lower edge of the clypeus, giving the appearance that the clypeus does not extend sharply below the face. The abdominal segments of *Nomia* bees often display striking bands in shades of pearl, ivory, yellow, white, green, or even metallic blue, contributing to their distinctive appearance. Body length generally ranges between 9 and 13 mm. Other characteristic features include lamelliform projections on the metanotum in some species, and in males of the subgenus *Nomia*, which have one or two preapical teeth in the hind femur. Additional diagnostic traits include filiform (thread-like) antennae, a labial palpus divided into four segments, and a maxillary palpus with six segments. The forewings typically exhibit three submarginal cells, with the radial cell rounded at its apex (Bodlah et al. 2016). The marginal cell is also rounded, and notably, the second submarginal cell is shorter compared to the first and third. Unlike many other halictid bees, *Nomia* species have a basal vein that is less sharply curved. Ecologically, species within this genus construct their nests in compacted or hard soils, on banks, flat ground, and occasionally even near human dwellings. Bees bring to nest pollen stores to feed their developing larvae.

2.1 *Nomia crassipes* (Fabricius, 1798)

The head and abdomen of this species show fine yet dense punctation, whereas the thorax is more coarsely and densely punctured. The clypeus is slightly convex in shape. The antennae are mostly testaceous brown, except for the scape (first segment), which is pale yellow. The abdominal tergites display bands that appear yellowish green, and the apical sections of the forewings are slightly smoky in color. A distinct morphological feature of the males is the hind femur, which is notably enlarged and curved toward the distal end; the basal half of the hind femur is dark, while the distal half is yellowish. The hind tibiae have small apical teeth. Additionally, the metanotum lacks any projections, and the abdomen is approximately as broad as the thorax. In terms of size, males of this species have a body length of around 13 mm, with the forewings measuring about 9 mm in length and 3 mm in width (Fig. 1.1). Geographically, the species occurs in a broad range extending from China, Thailand, India, and Pakistan, to parts of Madagascar and Africa, spanning from Gambia and Sudan southwards to Angola, Transvaal, and Mozambique. Regarding host plants, *Nomia crassipes* is known to visit and pollinate species such as wild safflower (*Carthamus oxycanthus*), common bean (*Phaseolus vulgaris*), and milk thistle (*Silybum marianum*).

2.2 *Nomia curvipes* (Fabricius, 1793)

This species closely resembles *N. crassipes* in general appearance, but can be distinguished by specific tibial modifications. The abdominal terga bear yellow transverse bands, while forewings appear slightly less smoky; the hind wings are entirely clear. The hind femora are brown and transition to a pale yellow color toward the apex. In males, the hind tibiae are characteristically broadened on the inner side, forming a distinct acute angle. The thorax is more densely punctured compared to the head and abdomen, providing a textural difference. Males typically measure about 11 mm in body length, with forewings approximately 7 mm long and

Fig. 1.1 *Nomia crassipes* F. (Source: Pannure 2014)

Fig. 1.2 *Nomia curvipes* F. (Source: Pauly 2009)

2.5 mm wide. *Nomia curvipes* is reported from India, Pakistan, Bengal, and upper Burma, where it is associated with host plants including wild safflower (*Carthamus oxycanthus*), common bean (*Phaseolus vulgaris*), and groundnut (*Arachis hypogaea*) (Fig. 1.2).

2.3 *Nomia strigata* (Fabricius, 1793)

The species is among the common, abundant solitary bees native to Asia, with its distribution primarily concentrated in Southeast Asia. Females typically build underground nests and are capable of buzz pollination—using sonication to release pollen of plants with poricidal anthers. These bees forage on a variety of floral resources, including native wild plants such as *Melastoma malabathricum* and *Ardisia elliptica*, as well as cultivated crops like brinjal (eggplant), chili, and tomato. Morphologically, individuals are relatively small, measuring about 9 mm in body length. They have black eyes, a dark thorax covered with yellow hairs along the rear

Fig. 1.3 *Nomia strigata* F. (Source: Anon 2020)

margin, and a black abdomen marked by smooth, iridescent bands whose color can range from yellow-green to bright blue depending on the individual (Fig. 1.3).

2.4 *Nomia* (*Hoplonomia*) *elliotii* (Smith, 1875)

The head and thorax show fine punctuations, while the clypeus is notably flat. The scutellum is more coarsely punctured and bears a distinct feature—either a notch or a pair of small tubercles on its posterior margin. On the metanotum, there are two prominent lamelliform projections located centrally. The anterior abdominal segments have greenish-yellow apical bands, each marked with a transverse impressed line at the middle. In males, the hind femora are distinctly swollen, and the tibiae possess a blunt, testaceous-colored process. The typical body size of males is around 9 mm in length, with forewings measuring about 6 mm long and 3 mm wide. The subgenus *Hoplonomia* has a broad distribution across South and Southeast Asia, the Philippines, Indonesia, New Guinea, the Bismarck Archipelago, the Solomon Islands, Australia (extending south to southern Queensland), and Madagascar. Regarding host associations and habitat, these bees utilize a variety of weeds and ornamental flowers for forage and excavate nests directly in soil (Fig. 1.4).

2.5 *Nomia* (*Hoplonomia*) *westwoodii* Gribodo, 1894

This species closely resembles *N. elliotii*, particularly in having bifid (two-pronged) projections on the metanotum. However, it can be distinguished by differences in coloration patterns and body size. The hind tibiae and femora are not as prominently swollen as in related species and lack the apical, sharply truncated tibial process. The legs and abdominal sternites show a pale rufo-testaceous (reddish-brown) coloration. In terms of size, females typically measure about 9 mm in body length,

Fig. 1.4 *Nomia (Hoplonomia) elliotii* S. (Source: Pauly 2009)

Fig. 1.5 *Nomia (Hoplonomia) westwoodii* (Source: Pauly 2009)

with forewings approximately 6 mm long and 3 mm wide. Males are slightly larger, reaching around 10 mm in body length, with forewings about 7 mm long and 4 mm wide. Geographically, bees occur in India in the Bengal region (Fig. 1.5).

2.6 *Nomia melanderi* Cockerell, 1906

The species *N. melanderi* has a body length of approximately 12.5 mm. Abdominal segments 2–4 each feature broad, light emerald-green tegumentary bands along their hind margins, while the first abdominal segment shows only a small patch of green at the posterior corners. On the mesothorax, the anterior portion bears pale grayish-ochreous hairs interspersed with a few black bristles. In contrast, the posterior part of the mesothorax is largely exposed, smooth, and shiny, appearing almost impunctate apart from some larger punctures near the hairy regions and a few faint, scattered puncture traces. The scutellum itself is similarly smooth and virtually impunctate. Another notable feature is found on the first abdominal segment, where the apical depression is distinctly bounded above by a strong ridge that curves inward toward

Fig. 1.6 *Nomia melanderi* C. (Source: Tokach and Wu-Smart 2022)

the depression. This ridge is more pronounced than in *N. foxii*, where the boundary around the depression is typically less distinct (Fig. 1.6).

3 Phylogeny of Nomiinae Bees

The family Halictidae is recognized as a monophyletic group, a classification that is consistently supported by five identified synapomorphies. Although phylogenetic analyses have placed the subfamily Nomiinae in varying positions within the family tree, its two representative genera consistently emerge as sister groups across different analyses. This consistent relationship supports the conclusion that Nomiinae itself is also monophyletic (Alexander and Michener 1995) (Fig. 1.7).

The monophyly of these groups is supported by multiple synapomorphies: specifically, by 7, 9, and 3 synapomorphies, respectively. Within the family, the subfamily Rophitinae is identified as the sister group to the clade formed by Nomiinae and Halictinae combined. All three tribes within the subfamily Halictinae, namely, Nomioidini, Augochlorini, and Halictini, are likewise considered strictly monophyletic (holophyletic). Their monophyly is supported by 10, 1, and 1 synapomorphies, respectively (Pesenko 2000).

4 Keys to Subfamily of Halictidae (Saini and Rathor 2012; Michener 2007)

Antenna arising near mid-length of eyes, or, if below that level, then it is separated from upper clypeal margin by much more than the diameter of antennal socket; hind basitarsus of female, except in some parasitic forms, with apical process bearing penicillus; trochanteral and femoral scopa well developed except in parasitic forms.

Apex of marginal cell broadly rounded; when there are three sub-marginal cells, the third is usually about as long as the first, or, if shorter, then usually twice as long

Fig. 1.7 Phylogenetic tree of families of Halictidae (Danforth et al. 2008)

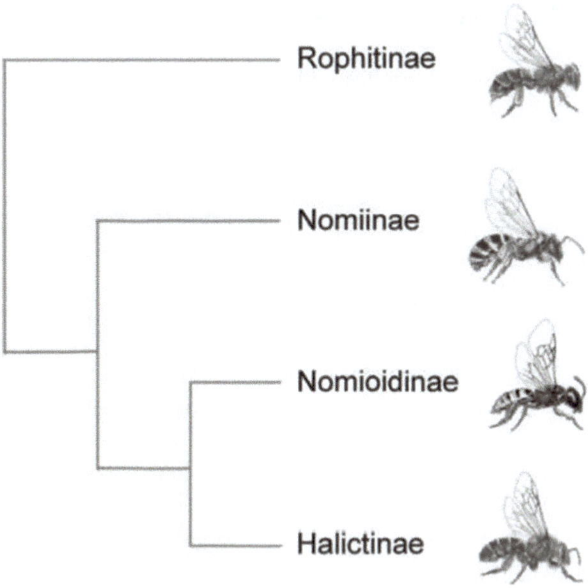

as the second; episternal groove below scrobal groove absent or a weak depression, sometimes directed strongly forward; prepygidial fimbria of female not divided. . . . . **Nomiinae**

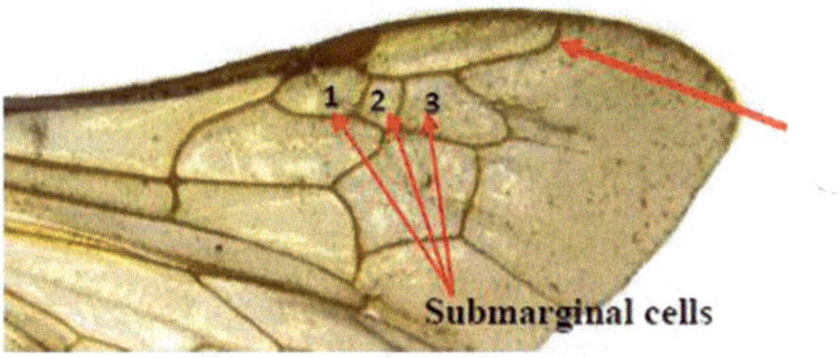

Apex of marginal cell is truncated (sometimes rounded); prepygidial fimbria of female not divided medially... . . . . **Nomioidinae**

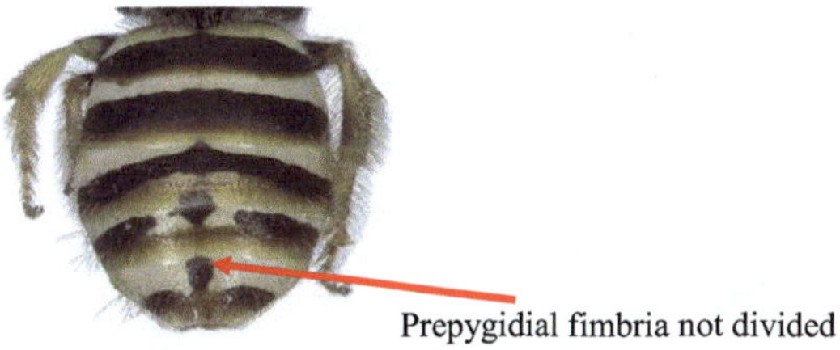

4.1 Key to the Genera of Nomiinae

1a. Metasomal tergites with apical integumental bands. **2**

1b. Metasomal tergites without apical integumentary bands (only sometimes pubescent bands). **6**

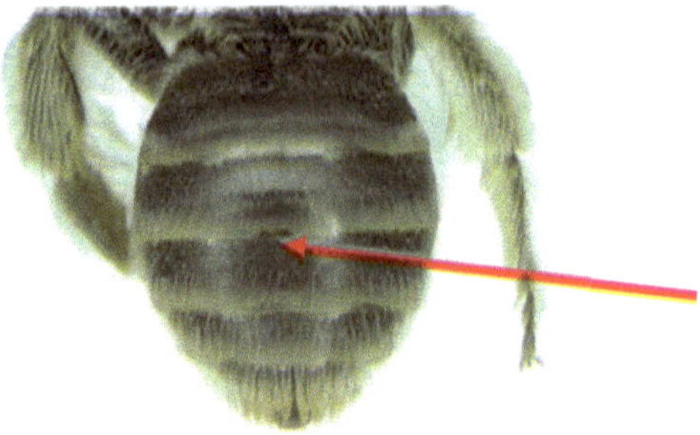

2a. *(1a)* Metanotum with a bifid projection . . . . ***Hoplonomia*** Ashmead (in part)

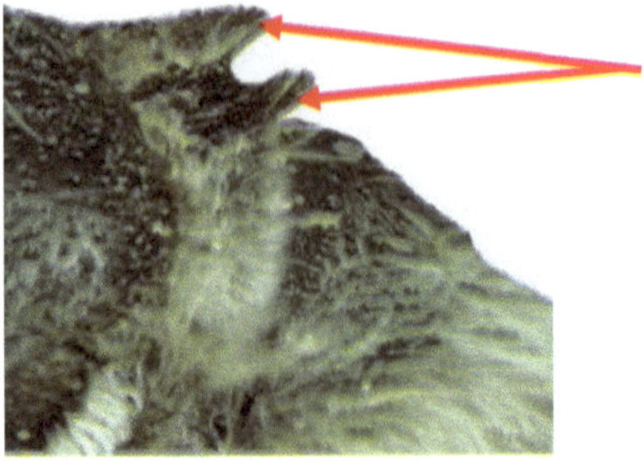

2b. Metanotum without bifid projection. **3**

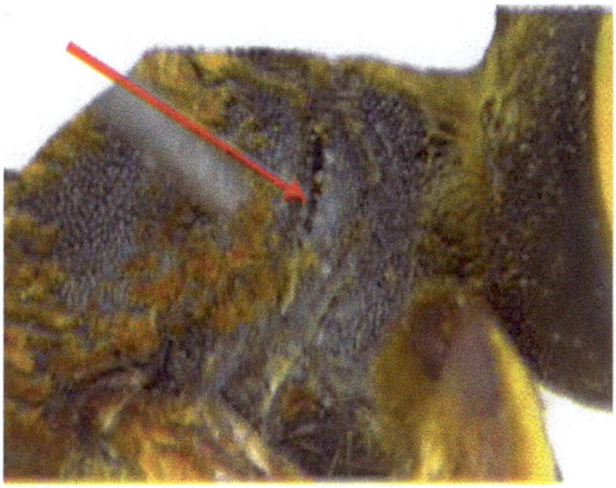

3a. *(2b)* Females with the complete basal plate of hind tibia. **4**

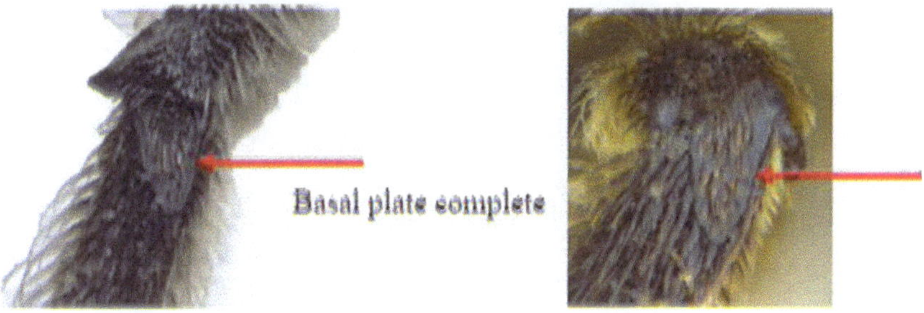

3b. Females with incomplete basal plate of hind tibia. **5**

4a. *(3a)* Tegulae oval, smaller; males with well-developed mandibles often with one or two hook on the inside edge . . . . ***Gnathonomia*** Pauly.

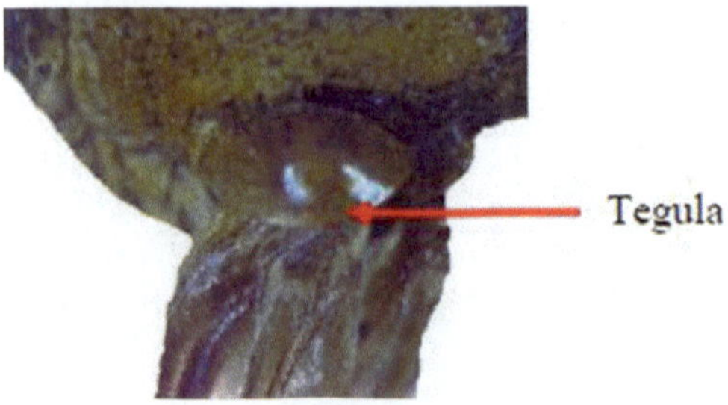

4b. Tegula "auriforme" (earlike), the apex being narrowed, larger; male mandibles simple . . . . ***Maculonomia*** Wu

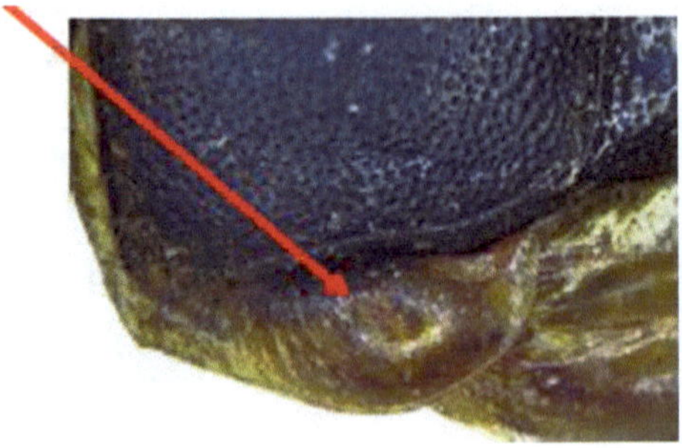

5a. *(3b)* Metasomal tergite I with an enamel band; male with a preapical tooth underside of femurs . . . . ***Nomia* s.str** Latreille

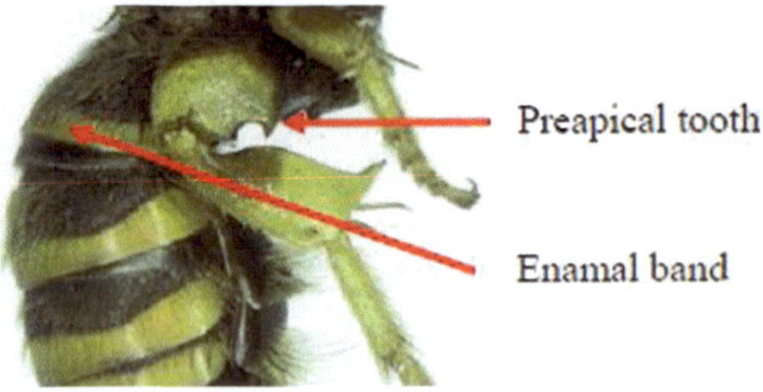

5b. Metasomal tergite I without enamel band; male without preapical tooth underside of femurs . . . . **_Leuconomia_** Pauly

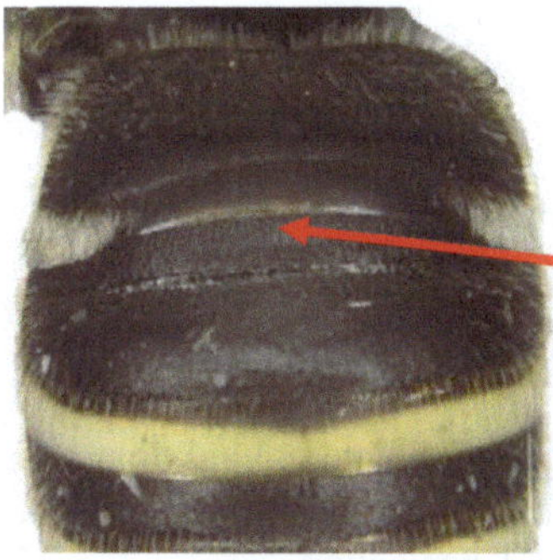

6a. *(1b)* Tegulae very large, strongly developed, posteriorly extending behind the level of scutoscutellar suture **7**

6b. Tegulae normal, much smaller **9**

7a. *(6a)* Forewing with two cells submarginal cells; subocular (below the eye) carina absent . . . . **_Steganomus_** Ritsema

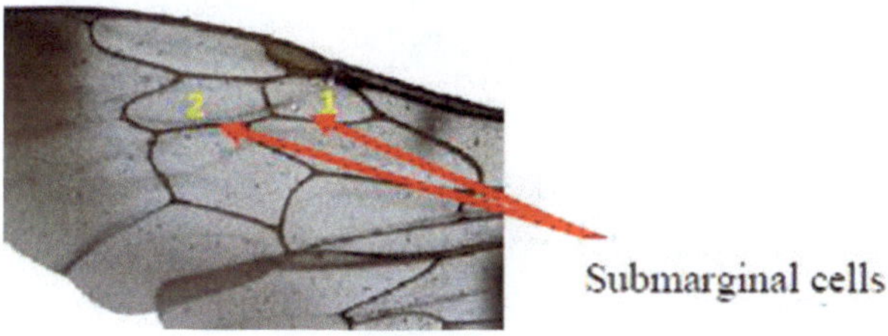

7b. Forewing with three submarginal cells; subocular carina present **8**

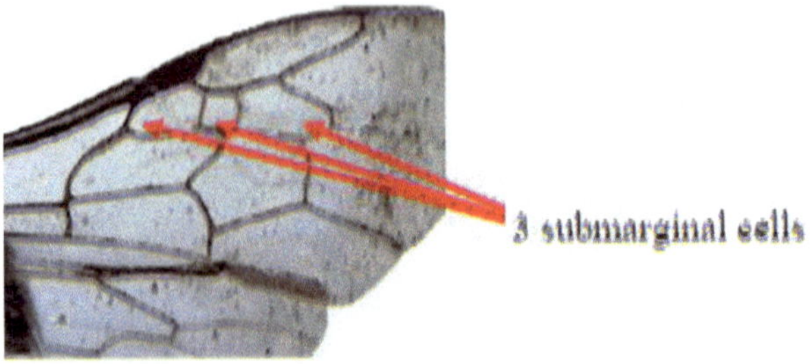

8a. (7b) Females with incomplete basal plate in hind tibia........... ***Pachynomia*** Pauly
8b. Females with the complete basal plate of hind tibia................ ***Pseudapis*** Kirby
9a. *(6b)* Metanotum with a bifid projection.......................***Hoplonomia*** **Ashmead** (in part)

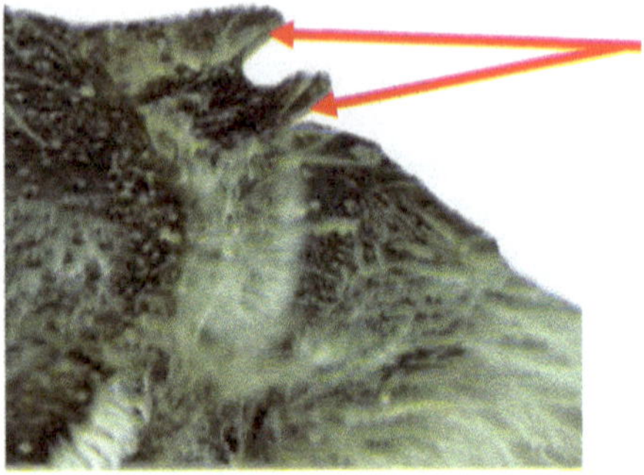

9b. Metanotum simple, without any projection. **10**
10a. *(9b)* Pronotum with a carina along its anterior edge; female with the basal plate of the tibia posteriorly incomplete, linear . . . .*Lipotriches* Gerstaecker

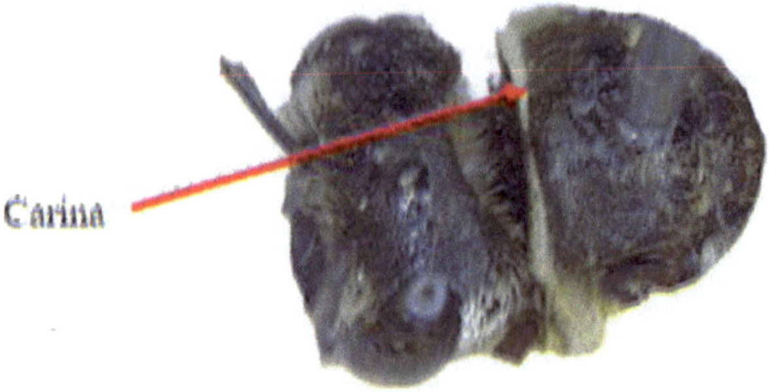

10b. Pronotal carina interrupted medially, laterally developed; females with the basal plate of hind tibiae posteriorly carinate or not **11**

11a. *(10b)* Antero-apical lobe of the wing smoky; females with the basal plate of hind tibiae carinate and triangular. . . . . ***Maculonomia*** Wu

11b. Antero-apical lobe of wing not smoky; females with basal plate of hind tibiae carinate. **12**

12a. *(11b)* Basal plate of hind tibiae in females complete. . . . . ***Austronomia*** Michener

12b. Basal plate hind tibiae in females incomplete . . . . ***Macronomia*** Cockerell

5 Biology of Nomia

The various life stages of *H. westwoodi* are described below.

Egg Eggs are found exclusively in brood cells that contain an adequate pollen provision, known as bee bread. This bee bread, with a creamy yellow color and semisolid consistency, accumulates at the bottom of each brood cell. In observations, the average weight of the bee bread per cell was about 0.067 g ($n = 1$). The egg itself is creamy white, cylindrical, and slightly curved, laid singly atop the brood food. The caudal (rear) end of the egg is bent and anchored into the brood provision.

Larva Upon hatching, the larva initially appears straight, resting on the surface of the brood food. As it develops, the larval body becomes robust and assumes a characteristic 'C'-shaped posture. In the second instar, the mandibles become faintly pigmented. The mature larva has a swollen body, and by this stage, only a small quantity of unused brood food typically remains in the cell. During the final instar, the larva consumes not only the remaining pollen provisions but also gnaws at the inner cell linings. The larval head capsule is well-formed, bearing two prominent mandibles. As is typical for bee larvae, these are apodous, lacking legs on both thoracic and abdominal segments.

Prepupa and Pupa After completing feeding, the fully developed larva contracts slightly, forming the prepupal stage. At this stage, the brood cell is left without leftover brood food or wax covering. The prepupa then undergoes a molt to become an exarate pupa, which remains uncovered within the brood cell. Initially, the pupa appears yellow with yellow compound eyes; as development proceeds, the eye color transitions through orange and brown, eventually becoming black when the pupa is close to emergence. Under controlled laboratory (in vitro) conditions, the transformation from pupa to adult bee occurs within about 2 weeks.

Adult After emergence, the young adult bee typically remains inside its brood cell for about two to 3 days. During this period, it engages in cleaning behaviors: wiping its compound eyes using the forelegs, rubbing its gaster (abdomen) with the hind legs, and exercising its wings by flapping. The recorded weight of newly emerged adult bees ranged between 0.037 and 0.044 grams ($n = 4$) (Fig. 1.8).

The developmental timeline of *Hoplonomia westwoodi* comprises an egg stage lasting approximately 5.6 ± 1.14 days, a larval stage of about 23.40 ± 3.58 days, and a pupal stage lasting roughly 12.80 ± 2.17 days. Altogether, the complete life cycle spans around 41.8 days. At the onset of pupation, pupae exhibit white eyes, which gradually darken, changing first to pale brown and eventually to dark brown as they mature. The pre-defecating larvae are typically curved and whitish, surrounded by the residual pollen provision. In contrast, the post-defecating larvae appear fully grown, curved, yellowish-white in color, and are usually found in brood cells that lack pollen but contain scattered fecal pellets. Notably, examination of excavated

Fig. 1.8 Life stages of *H. westwoodi*. (**a**) Adult male, (**b**) egg with cell provisions, (**c**) and (**d**) larvae at different stages, (**e**) pupae

nests revealed no evidence of brood parasitism, as all pupae successfully developed into adult bees, resulting in a 100% emergence rate. Additionally, there were no visible signs of predation or parasitism in brood cells containing larvae. The presence of four adult females within the nesting shaft may have contributed to this lack of brood parasitism or predation inside the nest (Udayakumar and Shivalingaswamy 2018).

6 Nesting Behavior and Nest Structure

All Halictid bees construct their nests underground, but exhibit a wide range of nesting strategies. These range from widely scattered solitary nests, to more closely packed nests where individual bees may share common entrance tunnels, and even to primitive forms of social organization. In these social arrangements, lateral tunnels typically terminate in a single brood cell. A notable example is the alkali bee (*Nomia melanderi*), a solitary, ground-nesting species native to western North America. True to its name, this bee nests in alkali soils, favoring bare soil conditions that remain moist below the surface yet dry on top. Such conditions naturally occur in landscapes where a hard pan layer helps retain subsurface moisture. The alkali salts in the soil act to seal the surface, effectively maintaining the necessary moisture conditions for nesting (Hannan et al. 2013).

The alkali bee, *Hoplonomia westwoodi*, is a solitary ground-nesting species distributed across India, Afghanistan, Pakistan, and Sri Lanka. Studies have provided artificial nesting substrates, and these bees exhibited the highest nesting preference for a mixture of red soil and farmyard manure (FYM) in a 1:1 ratio, followed by a blend of red soil, sand, and FYM (2:1:1) (Vijayakumar et al. 2022). Key factors influencing key parameters affecting nest included adequate shade, well-drained soil, and the absence of excess roots within containers like polypropylene bags and pots (Udayakumar and Shivalingaswamy 2018). Although each female constructs a nest, solitary ground-nesting bees often build nests in close proximity, leading to the formation of dense aggregations (Michener et al. 1958). This behavior is linked to philopatry, where females are drawn to sites already containing holes or partial nests created by other females.

Nesting structure varies depending on substrate conditions: typically, horizontal and vertical burrows are created in well-drained soils (Vijayakumar et al. 2022). Bees often use drainage holes at the bottom of polypropylene bags as horizontal entrances or exits. Interestingly, nests tend to be concentrated around the peripheral edges of these bags, as the sidewalls provide structural support for the tunnels. Such nest congregations are thought to enhance reproductive success by creating a dilution effect against natural enemies: the higher density of nests makes it harder for predators and parasites to target individual brood cells (Sandeep and Muthuraman 2019).

6.1 Relationship Between Nests and Soil Characteristics

Certain fundamental requirements define a suitable nesting bed for alkali bees. One critical factor is the presence of a subsurface moisture source that can move upward toward the soil surface. Ideally, a hardpan layer located about a foot or more beneath the surface helps retain this moisture by preventing deeper percolation. Above this, the porous upper soil layer should effectively hold moisture and enable its upward capillary movement. Additionally, the site must allow rapid drainage of surface water to prevent flooding of brood cells. The texture of the underlying soil layer is also important: soils varying from silt loam to sandy loam, with no more than seven percent clay-sized particles, are typically preferred. Observations have shown that maximum nest occupancy occurs when the soil's clay content is relatively low, even when mixed into clay loam (Vijayakumar et al. 2022).

Soil texture directly influences the bees' ability to excavate nests and supports a continuous capillary flow that replaces evaporated water in the upper soil layers. Maintaining a balance where capillary flow meets or slightly exceeds evaporation is crucial (Cane 1991). Excessive clay content greater than 25% can hinder capillary rise, while soils with sand (45–80%) are difficult to seal, leading to increased water loss and rapid drying. Research indicates that silt loams containing 2–6% fine silt, 42–68% coarse silt, 13–24% clay, and 10–40% sand offer the best nesting conditions for alkali bees (Kaushik and Yadav 2012). In contrast, for other halictid bees like *Lasioglossum* species, preferred soils typically have a high silt content (44–53.8%) and lower clay percentages (Cane 1991). Interestingly, *Hoplonomia westwoodi* showed a nesting preference for soils with lower silt and higher clay proportions than those favored by other halictids, suggesting that soil preference can vary based on species and local environmental conditions (Vijayakumar et al. 2022). Moreover, only certain patches within a landscape might meet these specific soil and moisture conditions, which is why nest aggregations form in localized clusters (Polidori et al. 2010).

6.2 Nest Architecture of *Hoplonomia westwoodi*

The nest structure typically features a vertical main burrow that extends to an average depth of 13.5 ± 0.97 cm. From this main shaft, a series of horizontal and vertical lateral tunnels branch out, each terminating in clusters of brood cells arranged horizontally (Vijayakumar et al. 2022). The nest entrance itself is circular, measuring about 0.5 cm in diameter, and often features a small turret approximately 1.2 cm high at the opening. In some areas, parts of this turret can become hardened, likely due to soil compaction or moisture conditions (Fig. 1.9).

Female *Hoplonomia westwoodi* often utilize multiple nest entrances, as the subterranean tunnels can branch and interconnect. The main tunnel, which reaches an average depth of 70.1 cm with a diameter of 0.5 cm ($n = 10$), typically gives rise to horizontal laterals originating around 41.7 cm below the surface. These laterals end in individual brood cells or clusters of cells, with cells largely determined by

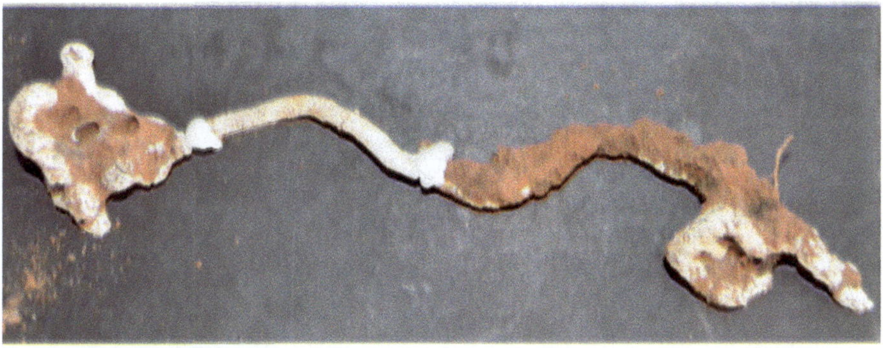

Fig. 1.9 Nest architecture of *Hoplonomia westwoodi*

food availability and quality nesting substrate (Abrol 2012). Brood cell clusters are commonly located at depths ranging from 62.8 to 70.0 cm, likely an adaptation that helps protect developing brood from flooding during heavy rainfall (Udayakumar and Shivalingaswamy 2018). The inner walls of the tunnels are smooth and form a continuous hollow passage. Clusters themselves are usually round or oval, reinforced by mud pillars and/or plant roots, and are separated from the surrounding cavity by a space of approximately 5.0 ± 1.4 mm. Within these clusters, brood cells are oriented upward and inclined at an angle of about 70–80° inward. The size of each cluster varies, reflecting the number of brood cells it contains. The brood cells are semi-oval in shape and, after egg-laying, are securely sealed with soil. The inner surface of each brood cell is polished smooth and coated with a secretion produced by the Dufour's gland (Batra 1970). This secretion, rich in triglycerides, likely acts as a waterproofing layer (Cane and Carlson 1984).

Occasionally, some brood cells become partially filled with loose soil—possibly the spoil excavated during the creation of new cells nearby. The average distance between adjacent brood cells is about 2.8 ± 0.3 mm. The pollen provision, weighing roughly 0.067 g (Yogesh 2012), is formed into spherical, pill-shaped pollen balls placed horizontally at the cell's base. A translucent egg, approximately 4 mm long, is laid on top of this pollen ball, anchored at both ends. Once the larva hatches, it consumes the pollen ball. As the larva matures, it defecates, depositing fecal material at the base of the cell. Brood cells containing prepupae and pupae are often lined with a delicate silken web on the rear surface of the brood plug. This web is secreted by the final (5th) instar larva and may help reinforce or protect the developing pupa. Comparable nest architectures are explained for other halictid bees such as *Nomioides divisus*, *Nomioides variegata* (Batra 1966), and *N. melanderi* (Batra 1970). As a solitary, mass-provisioning bee, *H. westwoodii* completes its brood care by fully provisioning each cell with pollen before sealing it (Table 1.1).

The number of brood cells varies notably between solitary and communal nests. In solitary nests, the brood count generally ranges from 5 to 20, while in communal nests, it can increase significantly, from 12 up to 101 brood cells. Within these communal nests, each brood cell cluster may contain anywhere from 0 to 32

Table 1.1 Nest characteristics of *Hoplonomia westwoodi*

Nest parameters	Mean ± SD	Range
Diameter of entrance (cm)	0.88 ± 0.14 ($n = 11$)	0.7–1.2
Closeness of nest burrow (cm)	5.83 ± 3.49 ($n = 11$)	2 –12
Depth of the nest (cm)	13.58 ± 0.97 ($n = 6$)	12–14.5
Shaft depth (cm)	4.17 ± 1.33 ($n = 6$)	2.5–6.5
Total length of the nest (cm)	23 ± 3.89 ($n = 6$)	18–28
Height of the tumuli (cm)	0.9 ± 0.15 ($n = 7$)	0.6–1
Number of brood cells	2.67 ± 1.03 ($n = 6$)	2–4
Height of the brood cell (cm)	1.8 ± 0.44 ($n = 16$)	1.2–2.5
Diameter of brood cell (cm)	0.8 ± 0.25 ($n = 16$)	0.5–1
Weight of the bee bread (gm)	0.067 ($n = 1$)	–
Weight of the just emerged bee (gm)	0.04 ± 0.002 ($n = 4$)	0.037–0.044

individual brood cells. Interestingly, in *Nomia incerta*, there are documented cases of both the main burrow and brood cells being reused by subsequent generations, with reported reuse rates of up to 47.9% (Masuda 1943; Hirashima 1961; Okumura 1966). The tendency to reuse maternal nests was recorded in other species like *Lipotriches yasumatsui*, where it appears closely linked to the formation of communal nesting behavior (Maeta and Sakagami 2011). For example, the closely related solitary species *N. melanderi* typically builds brood clusters averaging 15–20 cells per nest (Bohart and Cross 1955). However, even in communal arrangements, it seems that female constructs brood cell clusters independently. These clusters become quite large and even form double layers, as observed in one cluster that contained 32 brood cells.

The facultative communal bee *N. chalybeata* also shows a pattern where each foundress builds her own single brood cell cluster within the shared nest. Although direct in-nest interactions among cohabiting females were not observed, each female foraged for the requirements of her own progeny. Similar communal nesting is recorded in *N. incerta* (Masuda 1943; Okumura 1966; Iwata 1975), where each founding bee appears to construct its brood cells separately. In related nomiine bees, like *Lipotriches (Austronomia) australica* (previously described as *Nomia (Austronomia) australica*), communal nests typically house one to three females. Studies revealed that all nestmates actively forage for the brood (Vogel and Kukuk 1994). Meanwhile, in *L. (Lipotriches) yasumatsui*, there is evidence of some division of labor among nestmates, suggesting it may be more than simply communal (Maeta and Sakagami 2011). Furthermore, semisocial behaviors which include cooperation beyond mere cohabitation have been documented in Indian nomiine bees like *N. capitata* and *N. oxybeloides* (Batra 1966). Despite these observations across related species, the detailed nesting behavior of *H. westwoodi* remains relatively unexplored and warrants further research to fully understand its social and nesting dynamics.

Fig. 1.10 *Eupetersia* sp.

7 Natural Enemies

Inside nests, earthworms are only occasionally encountered, but when present, they can damage brood cells. In addition, intrusions by certain ant species can be particularly harmful, as ants often destroy developing pupae by feeding on them (Hannan et al. 2013). A notable natural enemy of *Hoplonomia westwoodii* is a cleptoparasitic wasp species of the genus *Eupetersia*, recorded parasitizing *H. westwoodii* for the first time (Vijayakumar et al. 2022). Cleptoparasitism by *Eupetersia* species has been reported in other halictid bees, including species of *Lasioglossum* and *Zonalictus* (Michener 1978). At nesting sites, parasite females typically seek out unguarded nests to lay their eggs in freshly provisioned brood cells (Legewie 1925; Knerer and Plateaux-Qunu 1967; Knerer 1973; Sick et al. 1991). In solitary bee hosts, the nests are left unattended while the female bees are away foraging, creating an opportunity for parasites to enter. In contrast, for primitively eusocial host species, this window of vulnerability mainly occurs during the early solitary stage of the nest cycle. Once the first worker brood emerges, nest entrances usually become guarded by worker bees, which helps to deter both parasites and conspecific intruders from nearby nests (Knerer 1969) (Fig. 1.10).

8 Conservation of Nomia

Buzz-pollinating bees are generally most active during early morning, which is compatible with the timing of anther dehiscence in many buzz-pollinated plants. The handling time, the duration a bee spends on each flower fluctuate considerably in bee species (Shelly and Villalobos 2000). Research by Buchmann and Cane (1989) demonstrated a positive correlation between the quantity of pollen available and the handling time per floral visit. Bees are capable of immediately evaluating pollen rewards, spending more time on flowers that offer higher pollen returns, as observed in *Solanum elaeagnifolium*. Additionally, bees tend to prefer younger flowers, which typically contain more pollen, over older ones. Plant families that often rely on buzz pollination include Solanaceae, Melastomataceae, Bixaceae, Cochlospermaceae, Fabaceae, and Dilleniaceae (Harder and Barclay 1994). These families produce flowers whose poricidal anthers require sonication by bees to efficiently release pollen.

At the study site in Kegalle, two buzzing bee species *Hoplonomia westwoodi* and *Patellapis kaluterae* were commonly observed, with *H. westwoodi* being the larger of the two (Wanigasekara and Karunaratne 2012). *P. kaluterae* was noted to vibrate each anther cone of a flower individually, thereby spending more time per flower, whereas *H. westwoodi* vibrated all the anther cones of a single flower simultaneously, resulting in quicker visits. As flowers of *Solanum* species lack nectar (Anderson and Symon 1988), bee visits to *S. violaceum* were solely for pollen collection. The most frequent visitors were buzzing bees from the family Halictidae, which includes the highest number of bee species documented in Sri Lanka (Karunaratne et al. 2005). Genera such as *Hoplonomia*, *Patellapis*, and *Amegilla* are ground-nesting bees and, like domesticated honeybees and leafcutter bees, can be managed for crop pollination by offering suitable nesting sites. An experiment successfully domiciled *H. westwoodi* by providing artificial nesting media (Vijayakumar et al. 2022). Among the tested nesting structures, 25 × 40 cm polypropylene bags received the highest nesting preference. Regarding substrates, the most favored was Red soil mixed with manure (FYM) at a 1:1 ratio, followed by a mix of Red soil + Sand + FYM (2:1:1). These findings emphasize the importance of conserving wild buzzing bee populations in and around crop fields, even during the off-season. Buzzing bees also showed a tendency to visit freshly opened flowers more often than older, senescent ones with faded white petals and browned anthers (Weiss 1991; Jayasinghe and Karunaratne 2009). The change in floral color and contrast as flowers age may serve as a visual cue, helping bees identify pollen-rich flowers from a distance, thereby improving foraging efficiency and reducing energy wasted on depleted flowers (Fandino 2007).

In India, there is still limited information on the direct benefits of native buzzing bees to local communities. However, targeted research initiatives are now underway to establish best practices for conserving and managing these bees, aiming to boost agricultural productivity. In principle, conservation and management of bees are cost-effective interventions that can also enhance the visual and ecological value of agricultural landscapes. Key practices include designating parts of farmland to provide year-round floral resources, as well as secure nesting, mating, and resting sites, which also serve as refuges from predators and adverse conditions. During crop flowering periods, it is essential for farmers to carefully regulate pesticide use to minimize harm to visiting pollinators. Steps should also be taken to reduce pesticide drift into neighboring habitats. Furthermore, national regulations on the registration and application of plant protection products contribute significantly to pollinator safety. To protect nesting sites of soil-nesting bees, farmers should control trampling by livestock and people and reduce frequent tillage. The All India Coordinated Research Project (AICRP) on honeybees and pollinators is currently working to develop mass-rearing protocols for various solitary bee species. Outcomes of this research will be shared openly to support broader conservation efforts. Ultimately, the success of conserving native bee populations will depend heavily on creating public awareness of their critical role in sustaining crop yields and maintaining biodiversity.

9 Utilization of Nomia for Pollination of Crops

Several crops benefit significantly from pollination by non-Apis bee species (Parker et al. 1987; Klein et al. 2007), many of which are ground-nesting bees (Cane 1997). These ground-nesting bees frequently make up the majority of wild bee communities in various regions, for example, about 85% of bee species in eastern North America construct nests underground (Cane 2003). Despite their abundance, these bees are rarely managed for agricultural pollination due to challenges like crop rotation and difficulties in preparing suitable nesting sites. An important exception is the alkali bee (*Nomia melanderi* Cockerell), native to arid desert basins of western North America. This bee, though a floral generalist, is valued for efficiency in pollinating alfalfa (Cane 2002). Alkali bees prefer to nest gregariously in moist, salt-crusted basin soils (Johansen et al. 1978). To establish new nesting sites within alfalfa seed fields, farmers transplant small soil blocks typically measuring 1–3 feet cut or cored from dense natural nesting aggregations (Stephen 1960).

Nomia bees are active visitors to Solanaceous crops, with noticeable foraging activity throughout the day. Their foraging peaks between 8:00 and 13:00 h, as bees typically emerge during the morning. The duration of a single foraging trip varies from 2 to 60 min, averaging about 18.41 min. In the morning, bees spend less time inside their nests between trips compared to the afternoon. Each bee makes around 9–22 trips daily. During morning hours, flowers are more receptive, containing abundant nectar and viable pollen, which attracts a higher number of solitary bees, especially pollen gatherers. This increased visitation indirectly boosts pollination efficiency. Handling time, the period a bee spends on a flower, is a critical factor. Upon landing, Nomia bees begin vibrating the anthers, which helps pollen grains transfer to the stigma. Initial vibrations on newly opened flowers release large amounts of pollen, but as the flowers age, the number of pollen grains per anther declines, reducing the pollen available during subsequent visits. This depletion is closely linked to handling time (Shelly and Villalobos 2000) observed in species such as *Hoplonomia westwoodi* and *Pseudapis kaluterae*. Interestingly, in *H. westwoodi*, body size is inversely related to handling time (Anderson and Symon 1988). Symon (1979) noted that only larger insects are capable of effective buzzing, enabling them to extract and transfer pollen, thus functioning as primary pollen vectors. When Nomia bees perform buzz pollination on open flowers, they may facilitate self-pollination, cross-pollination, or both. Vibrations can cause pollen to land directly on the stigma of the same flower (self-pollination). Alternatively, bees already carrying pollen from prior visits may deposit it on the stigma, enabling cross-pollination. In *Solanum melongena*, seed set is generally higher when cross-pollinated by bees compared to artificial pollination (Aizen et al. 2002). Cross-pollination tends to produce a larger number of seeds, whereas artificial pollination results in fewer seeds. Buzz pollination has also proven highly effective in enhancing seed formation. For example, the presence of bees increased the germination rate of *Brassica napus* seeds from 83% to 96% (Kevan and Eisikowitch 1990).

The alkali bee, *Hoplonomia westwoodi*, is recognized as an efficient buzz pollinator that actively forages on a wide variety of medicinal, aromatic, Solanaceous, and ornamental crops. Its host plant spectrum includes Guava (*Psidium guajava*; Myrtaceae), Leucas (*Leucas aspera*; Lamiaceae), Holy basil (*Ocimum tenuiflorum*; Lamiaceae), Portulaca (*Portulaca oleracea*; Portulacaceae), Jacquemontia (*Jacquemontia violacea*; Convolvulaceae), Cuphea (*Cuphea hyssopifolia*; Lythraceae), and Neem (*Azadirachta indica*; Meliaceae). Additionally, *H. westwoodi* visits Fire bush (*Hamelia patens*; Rubiaceae), Brinjal (*Solanum melongena*; Solanaceae), Tomato (*Lycopersicon esculentum*; Solanaceae), Chili (*Capsicum annum*; Solanaceae), Cucumber (*Cucumis sativus*; Cucurbitaceae), Ridge Gourd (*Luffa acutangula*; Cucurbitaceae), Sunflower (*Helianthus annuus*; Asteraceae), Red Gram (*Cajanus cajan*; Fabaceae), Pongamia (*Pongamia pinnata*; Fabaceae), and Okra (*Abelmoschus esculentus*; Malvaceae). The extensive diversity of floral resources visited by *H. westwoodi* demonstrates its polylectic nature (Vijayakumar et al. 2022).

Alkali bees (*Nomia melanderi*) exhibit remarkable adaptability to climatic conditions that are typically unsuitable for honeybees (*Apis mellifera*). They preferentially establish nests in soils with high salinity, which are generally considered marginal for agricultural use. Of particular ecological and economic importance is their effectiveness as pollinators of leguminous crops, notably alfalfa (*Medicago sativa*). The pollination of alfalfa flowers presents a unique mechanical challenge. The floral reproductive organs, the anthers and stigma, are concealed within the keel petal, which must be mechanically "tripped" to facilitate pollen transfer. This process requires the bee to land on the keel and apply targeted pressure, thereby snapping it open and exposing the pollen-bearing anthers and receptive stigma. Flowers that have undergone this process are identifiable by their visibly exposed reproductive structures. Honeybees generally avoid tripping alfalfa flowers due to the mechanical effort required and the recoil action of the keel. In contrast, alkali bees have evolved to efficiently perform this task, making them highly effective pollinators. Notably, *N. melanderi* individuals are capable of visiting and successfully pollinating over a thousand alfalfa flowers per day (James 2002).

Solanaceous crops such as tomato (*Solanum lycopersicum*), chilli (*Capsicum annuum*), and brinjal (*Solanum melongena*) exhibit a characteristic floral morphology marked by poricidal anthers. In these flowers, the anthers are fused into a tubular structure with a small apical pore through which pollen is released (Buchmann 1985). Effective pollen extraction from these poricidal anthers necessitates a vibratory mechanism commonly termed "buzz pollination," which is typically performed by the wing vibrations of certain wild native bees, including *Amegilla zonata* (Anthophorinae: Apidae), *Hoplonomia westwoodi* (Nominae: Halictidae), and bumble bees (*Bombus* spp.) (King and Buchmann 2003). The targeted use of these

Table 1.2 Host plant and reward collected by *H. westwoodi*

Family	Plant species	Reward pollen	Collected nectar
Acanthaceae	*Asystasia gangetica* (L.) T. Anderson		++
Asteraceae	*Gaillardia pulchella* Foug. Jacquemontia sp.		++
Convlvulaceae	Argyreia cuneata *Ker Gawl.*		++
Fabaceae	*Crotolaria retusa* L. *Ocimum basilicum* L. Ocimum gratissimum L.	–	++
Lamiaceae	*Strobilanthus barbatus* Nees	–	++
Malpighiaceae	*Tristellateia australasiae* A. Rich		++
Polygonaceae	*Antigonon leptopus* Hook. & Arn.		++
Rubiaceae	*Hamelia patens* Jacq.		++
Solanaceae	*Solanum lycopersicon* L Solanum melongena L.	++	–

solitary buzz-pollinating bees, particularly *A. zonata* and *H. westwoodi*, has been shown to significantly enhance fruit set in tomato crops (Udayakumar and Shivalingaswamy 2018) (Table 1.2).

Pollination within greenhouse environments presents unique challenges, particularly because abiotic factors such as wind, which can aid pollination in open fields, are absent. In this context, the potential application of solitary native bees for effective crop pollination remains an underexplored yet promising area of research, especially against the backdrop of global pollinator declines threatening sustainable agricultural production. Native bees have an important part in enhancing crop yields, as they are naturally abundant and adapted to diverse ecological habitats. Notably, for crops requiring buzz pollination due to specialized floral morphologies (e.g., poricidal anthers), the conventional use of honeybees (*Apis mellifera*) proves inadequate. Such crops demand the vibratory mechanism provided by buzz-pollinating bees to achieve efficient pollen release and fruit set. Therefore, under greenhouse conditions, rather than relying on the introduction of exotic pollinators, the targeted use of native alkali bees and other solitary buzz-pollinating species represents a more sustainable and ecologically compatible strategy to ensure effective pollination and optimal crop production.

References

Abrol (2012) Conservation, augmentation and utilization of wild bees in crop pollination. In: Sharma SK, Yogesh K (eds) Advances in bio ecology and management of insect pollinators of crops. CCS. HAU, Hisar, p 68

Aizen MA, Ashworth L, Galetto L (2002) Reproductive success in fragmented habitats: do compatibility systems and pollination specialization matter? J Veg Sci 13(6):885–892

Alexander BA, Michener CD (1995) Phylogenetic studies of the families of short-tongued bees (Hymenoptera: Apoidea). Univ Kans Sci Bull 55:377–424

Anderson J, Symon D (1988) Insect foragers on Solanum flowers in Australia. Ann Missouri Bot Garden 75(3):842–852

Anonymous (2020) Pearl banded bee- *Nomia strigata.* Singapore geographic https://www.singaporegeographic.com/insects/bee-wasp-hornet/pearly-banded-bee-nomia-strigata

Ascher JS, Pickering J (2012) Discover life bee species and world checklist (Hymenoptera: Apoidea: Anthophila). <http://discoverlife. org/mp/20q?guide=Apoidea species>

Ascher J, Pickering J (2022) Discover life bee species guide and world checklist (Hymenoptera: Apoidea: Anthophila). http://www.discoverlife.org/mp/20q?guide=Apoidea_species. Accessed date: 15 June 2021

Astafurova V, Pesenko A (2005) Contributions to the Halictid fauna of the eastern Palaearctic region: subfamily Nomiinae (Hymenoptera: Halictidae). Far East Ent 154:1–16

Batra SWT (1966) Social behavior and nests of some nomiine bees in India (Hymenoptera, Halictidae). Insect Soc 13:145–154

Batra SWT (1970) Behavior of the alkali bee, *Nomia melanderi*, within the nest (Hymenoptera: Halicitdae). Ann Entomol Soc Am 63:400–406

Bodlah I, Amjad M, Bodlah MA, Saeed M (2016) Record of sweet bees, genus Nomia Latreille, 1804 (Halictidae: Apoidea) from Pothwar tract. Pakistan J Entomol Zool Stud 4(3):178–182

Bohart GE, Cross EA (1955) Time relationships in the nest contribution and life cycle of the alkali bee. Ann Entomol Soc Am 48:403–406

Buchmann SL (1983) Buzz pollination in angiosperms. In: Jones CE, Little RJ (eds) Handbook of experimental pollination biology. Van Nostrand Reinhold, New York, pp 73–133

Buchmann SL (1985) Bees use vibration to aid pollen collection from non-poricidal flowers. J Kansas Entomol Soc 58(3):517–525

Buchmann SL, Cane JH (1989) Bees assess pollen returns while sonicating Solanum flowers. Oecologia 81(3):289–294

Cane JH (1991) Soils of ground-nesting bees (Hymenoptera: Apoidea): texture, moisture, cell depth, and climate. J Kansas Entomol Soc 64:406–413

Cane JH (1997) Ground-nesting bees: the neglected pollinator resource for agriculture. In: Richards KW (ed) Pollination: from theory to practice. Acta Hortic Leiden, pp 309–324

Cane JH (2002) Pollinating bees (Hymenoptera: Apiformes) of U.S. alfalfa compared for rates of pod and seed set. J Econ Entomol 95:22–27

Cane JH (2003) Exotic non-social bees (Hymenoptera: Apoidea) in North America: ecological implications. In: Strickler, K.V. AND J. H. Cane (eds.), For non-native crops, whence pollinators of the future? Thomas say publications in entomology, Entomol Soc America Lanham, pp 113–126

Cane JH, Carlson RG (1984) Dufour's gland triglycerides from Anthophora, Emphoropsis (Anthophoridae) and Megachile (Megachilidae) bees (Hymenoptera: Apoidea). Comp Biochem Physiol 78:769–770

Cresswell JE (2005) Accurate theoretical prediction of pollinator mediated gene dispersal. Ecology 86(3):574–578

Danforth BN, Eardley C, Packer L, Walker K, Pauly A, Randrianambinintsoa FJ (2008) Phylogeny of Halictidae with an emphasis on endemic African Halictinae. Apidologie 39:86–101

Fandino MTA (2007) Management of *Bombus atratus* bumblebees to pollinate lulo (S*olanum quitoense* L), a native fruit from the Andes of Colombia. In: Vlek PLG, Denich M, Martius C, Rodgers C, van de Giesen N (eds) Ecology and development series, p 50

Hannan A, Yasuo M, Ryoichi M (2013) Nesting biology and life cycle of *Nomia* (Acunomia) *chalybeata* Smith on Iriomote Island, southernmost Archipelago of Japan, with notes on the simultaneous occurrence of diapausing and non-diapausing prepupae within the same nests (Hymenoptera: Halictidae). J Saudi Soc Agric Sci 12:91–99

Harder LD, Barclay RMR (1994) The functional significance of poricidal anthers and buzz pollination: controlled pollen removal from Dodecatheon. Funct Ecol 8(4):509–517

Hirashima Y (1961) Monographic study of the subfamily Nomiinae of Japan (Hymenoptera, Apoidea). Acta Hymenopterologica 1:241–303

Hirashima Y (1989) A check list of Japanese insect II. Entomological Laboratory, Kyushu Univ Japan Wild Life Center, p 1088
Iwata K (1975) Note of naturalist. Asahi News Paper Co. Ltd., Tokyo, p 565
James HC (2002) Pollinating bees (Hymenoptera: Apiformes) of U.S. alfalfa compared for rates of pod and seed set. J Econ Entomol 95(1):22–27
Jayasinghe UJMSR, Karunaratne WAIP (2009) Buzz pollination of *Solanum melongena* by bees. Proceedings of the Peradeniya University Research Sessions, 14:290–292
Johansen CA, Mayer DF, Eves JD (1978) Biology and management of the alkali bee, *Nomia melanderi* Cockerell (Hymenoptera: Halictidae). Melanderia 28:25–46
Karunaratne WAIP, Edirisinghe JP, Gunatilleke CVS (2005) Floral relationships of bees in selected areas of Sri Lanka. Ceylon J Sci 34:27–45
Kaushik HD, Yadav S (2012) Bioecology and management of alkali bee. In: Sharma SK, Yogesh K (eds) Advances in bio ecology and management of insect pollinators of crops. HAU, Hisar, p 55. compendium
Kevan PG, Eisikowitch D (1990) The effects of insect pollination on canola (Brassica napus L. cv. O.A.C. Triton) seed germination. Euphytica 45(1):39–41
King MJ, Buchmann SL (2003) Floral sonication by bees: mesosomal vibration by Bombus and Xylocopa, but not Apis (Hymenoptera: Apidae), ejects pollen from poricidal anthers. J Kansas Entomol Soc 76:295–305
Klein AM, Vaissiere B, Cane JH, Steffan-Dewenter I, Cunningham SA, Kremen C, Tscharntke T (2007) Importance of crop pollinators in changing landscapes for world crops. Proc R Soc Lond B Biol Sci 274:303–313
Knerer G (1969) Stones, cement and guards in halictine nest architecture and defense. Entomol News 80:141–147
Knerer G (1973) Periodizitit und Strategie der Schmarotzer einer sozialen Schmalbiene, Ellaeus malachurus (K.) (Apoidea: Halictidae). Zool Anz 190:41–63
Knerer G, Plateaux-Qunu C (1967) Usurpation des nids etrangers et parasitisme facultatif chez Halictus scabiosae (Rossi). Insect Soc 14:47–50
Larson BMH, Barrett SCH (1999) The pollination ecology of buzz-pollinated Rhexia virginica (Melastomataceae). Am J Bot 86(4):502–511
Legewie H (1925) Zum Problem des tierischen Parasitismus I. Teil: Die Lebensweise der Schmarotzerbiene Sphecodes monilicornis K. (= subquadratus SM.) (Hymenoptera: Apidae). Z Morphol Okol Tiere 4:430–464
Maeta Y, Sakagami SF (2011) Nesting biology and communal nests of the nomiine bee, *Lipotriches* (Lipotriches) *yasumatsui* (Hirashima) in northern Honshu, Japan (Hymenoptera, Halictidae, Nomiinae). Chugoku Kontyu 25:33–56
Masuda H (1943) Bionomics of *Nomia punctata* Westwood. Mushi 15:16–23
Michener CD (1978) The parasitic groups of Halictidae (Hymenoptera, Apoidea). Univ Kansas Sci Bull 51:291–339
Michener CD (2007) The bees of the world. The Johns Hopkins University Press, Baltimore, pp 334–341
Michener CD, Lange RB, Bigarella J, Salamuni R (1958) Factors influencing the distribution of bees nests in earth banks. Ecology 39:207–217
Mitchell RJ, Irwin RE, Flanagan RJ, Karron JD (2009) Ecology and evolution of plant-pollinator interactions. Ann bot 103(9):1355–1363
Okumura T (1966) Studies on the biology of *Nomia punctata* Westwood (Hym., Apoidea). Life Study, Fukui 10:17–22
Pannure A (2014) *Nomia crassipes* (Fabricius, 1789). India Biodiversity portal, https://indiabiodiversity.org/observation/show/16182590
Pannure A, Belavadi VV (2017) An updated distributional checklist of bees of the subfamily Nomiinae (Hymenoptera: Apoidea: Halictidae) with new records from South India. Entomon 42:311–328

Parker FD, Griswold TL, Botsford JH (1987) Biological notes on Nomia heteropoda Say (Hymenoptera: Halictidae). Pan Pac Entomol 62:91–94

Pauly A (2009) Classification des Nomiinae de la Région Orientale, de Nouvelle-Guinée et des îles de l'Océan Pacifique (Hymenoptera: Apoidea: Halictidae). Bull Inst R Sci Nat Belg 79:151–229

Pesenko YA (2000) Phylogeny and classification of the family Halictidae revised (Hymenoptera: Apoidea). J Kansas Entomol Soc 72:104–123

Polidori C, Rubichi A, Barbieri V, Trombino L, Donegana M (2010) Floral resources and nesting requirements of the ground-nesting social bee, *Lasioglossum malachurum* (Hymenoptera: Halictidae), in a Mediterranean semi agricultural landscape. Psyche 2010(1):851947. https://doi.org/10.1155/2010/851947

Proenca CEB (1992) Buzz pollination-older and more widespread than we think. J Trop Ecol 8(1): 115–120

Saini MS, Rathor VSA (2012) Species checklist of family Halictidae (Hymenoptera: Apoidea) along with keys toits subfamilies, genera & subgenera from India. Int J Environ Sci 3:134–166

Sandeep KJ, Muthuraman M (2019) Bio-ecology of blue banded bees, *Amegilla zonata* L. (Apidae: Hymenoptera). J Entomol Zool Stud 7:428–434

Shelly TE, Villalobos E (2000) Buzzing bees on Solanum (Solanaceae): floral choice and handling time track pollen availability. Fla Entomol 83(2):180–187

Sick M, Ayasse M, Tenge J, Liibke G, Francke W (1991) Spieten Duftstoffe eine Rolle bei der Fortpflanzung yon Kuckucksbienen der Gattung Sphecodes (Hymenoptera: Halictidae). Verhandlungen der deutschen Zoologischen Gesellschaft 84:393

Stephen WP (1960) Maintaining alkali bees for alfalfa seed production. Oregon State College Agricultural Experiment Station, Station Bulletin, p 568

Symon DE (1979) Fruit diversity and dispersal in Solanum in Australia. J Adelaide Bot Gard 1: 321–331

Teale EW (1957) Insect friends. Dodd Mead and Co, New York, p 96

Tokach R, Wu-Smart J (2022) Alkali bee (*Nomia melanderi*) pests. Pacific Northwest Pest Management handbook. https://pnwhandbooks.org/insect/bee-protection/alkali-bee

Udayakumar A, Shivalingaswamy TM (2018) Nesting biology, seasonality and host range of sweat bee, *Hoplonomia westwoodi* (Gribodo) (Hymenoptera: Halictidae: Nomiinae). Sociobiology 65: 491–496

Vijayakumar KT, Nayimabanu T, Arati P (2022) Bioecology and domiciliation of the alkali bee, *Hoplonomia westwoodi* (Gribodo, 1894) (Hymenoptera: Halictidae: Nomiinae) from India. Sociobiol 69(2):e7339

Vogel ME, Kukuk PF (1994) Individual effort in the facultatively social halictid bees, *Nomia* (*Austronomia*) *australica* (Smith). J Kansas Entomol Soc 67:225–235

Wanigasekara RWMUM, Karunaratne WAIP (2012) Efficiency of buzzing bees in fruit set and seed set of *Solanum violaceum* in Sri Lanka. Psyche 231638:1–7. https://doi.org/10.1155/2012/231638

Wcislo WT, Engel MS (1996) Social behavior and nest architecture of nomiine bees (Hymenoptera: Halictidae; Nomiinae). J Kansas Entomol Soc 69:158–167

Weiss MR (1991) Floral colour changes as cues for pollinators. Nature 354(6350):227–229

Yogesh K (2012) Bioecology and management of leafcutter bees. In: Sharma SK, Yogesh K (eds) Advances in bio ecology and management of insect pollinators of crops. HAU, Hisar, p 50. compendium

Bio-Ecology, Threats, and Conservation of Honeybee, *Apis dorsata* in Tropical Wild and Cultivated Ecosystems

2

Narayanappa Nagaraja and Rathna Kumari Bettaiah Mallamma

Abstract

Apis dorsata is successfully established in Southeast Asia, producing large amounts of honey and beeswax, while also pollinating both wild and cultivated crops. The nests of *A. dorsata* are normally sheltered on both arboreal and terrestrial structures, especially at an inaccessible elevation. They build their colonies rapidly during pollen-rich periods and produce huge amounts of honey during winter and summer seasons. Reproductive swarming of these species most commonly occurs during summer season. *A. dorsata* exhibits colony migration during floral dearth periods. Nonetheless, the population of *A. dorsata* is currently under threat primarily due to anthropogenic interventions, including traditional methods of honey harvesting, insecticide poisoning, deforestation, and habitat deterioration. These interventions led to the mortality of several hundred giant honeybee colonies annually. Accordingly, conservation of these bee species is vital through novel methods of honey harvesting. Apparently, creating awareness on scientific methods of honey harvesting together with afforestation is required for the conservation of *A. dorsata* colonies.

Keywords

Apis dorsata · Anthropogenic threats · Rafter technology · Conservation

N. Nagaraja (✉)
UGC-Malaviya Mission Teacher Training Centre, Bangalore University, Jnana Bharathi Campus, Bengaluru, Karnataka, India
e-mail: nagaraja@bub.ernet.in

R. K. B. Mallamma
Department of Botany, Government First Grade College, Vijayanagara, Bengaluru, Karnataka, India

A. K. Chakravarthy (ed.), *The Role of Pollinators in Tropical Landscapes*,
https://doi.org/10.1007/978-981-95-9139-8_2

1 Introduction

Honeybees are the most familiar and productive social insects, exhibiting fascinating and spectacular ways of life (Moritz and Southwick 1992; Page et al. 2016). They are social insects living in societies that succeed in dealing with varied challenges posed by social life, including communication, swarming, nest architecture, etc. (Oldroyd and Pratt 2015; Koeniger et al. 2017). The Asian continent is blessed with nine honeybee species, including the exotic honeybee, *Apis mellifera* (Abrol 2020). Among them, five species build their nests naturally in cavities and are domesticated in respective bee hives. These honeybee species build multiple parallel combs by maintaining species-specific bee space.

Honeybees mainly produce honey, the first biological sweet, mostly from the nectar of blossoms, which they collect, combine with their secretions, transform, and store in combs (Nagaraja and Rajagopal 2009; Giampieri et al. 2022). The colonies of *A. dorsata* annually produce 20 to 30 kgs of honey in Indian conditions. Furthermore, about 60–70% of honey sold in the Indian market is produced from *A. dorsata* colonies. Honey has been used continuously through the ages, both as food and a medicine in curing various ailments. Beeswax secreted by young worker bees to build their combs is used in pharmaceuticals, cosmetics, waterproofing materials, etc. Royal jelly is a secretion of the hypopharyngeal glands of young worker honeybees. Royal jelly is a highly nutritious food fed to early larval stages and the queen bee, and is found to be antitumor, anti-inflammatory, and antiallergic in human beings (Baitala et al. 2010). Propolis has a variable composition with pharmacological and nutritional applications (Marcucci 1995), and bee venom is used in curing arthritis and other ailments (Sharaf et al. 2022).

Insects are the principal pollinators of most cross-pollinated plants and are efficient in maintaining natural ecosystems through gene flow (Kevan 1999). According to an estimate, about 60% to 70% of human food crops are pollinated by many species of bees, including honeybees (Klein et al. 2007). Honeybees are diligent pollinators of natural and cultivated plant species (Hung et al. 2018; Requier et al. 2019). They forage on the plants within a vicinity of 2 to 3 km from the nests. Long foraging hours coupled with floral fidelity made them efficient pollinators of most of the crop plants. Honeybee pollination increases fruit production (Mahanta and Rahman 1997), vegetable production, and seed yield (Rajagopal et al. 1999; Nagaraja and Reddy 2002), and is an efficient pollinator of hundreds of cultivated and natural plants in the tropics (Chantawannakul et al. 2018; Sihag 2025). Furthermore, *A. dorsata* not only pollinates crop plants but is also pivotal in the pollination of thousands of tree species on a large scale for its wide foraging range.

The population of honeybees is gradually declining primarily due to colony collapse disorder, parasitic mites, and varieties of pests, including anthropogenic activities (De Jong and Lester 2023). Reports strongly stipulate that natural ecosystems are deteriorating due to many types of anthropogenic activity. Despite *A. dorsata* being free from most diseases and tolerating pests effectively, its population is declining on a large scale due to threats such as traditional honey hunting, habitat loss, deforestation, excessive use of agricultural chemicals, pesticide poisoning, and climate change (Nagaraja 2012, 2022). Declining bee pollination may result in adverse effects on crop production, floral diversity, and human health (Oldroyd and Wongsiri 2006). Nevertheless, there is an immediate need to develop sustainable conservation strategies for protecting these natural pollinators from anthropogenic threats.

2 Bio-Ecology

2.1 Nesting Biology

Knowledge of the biology and behavior of honeybees is crucial for their utilization in honey production and pollination services (Hepburn 2011). On the basis of geographic and morphological variations, *Apis dorsata* is grouped into three subspecies. The colonies of *A.d. dorsata* are abundant in India and southern Asia, in contrast to *A.d. breviligula* found only in Luzon of the Philippines. However, other subspecies such as *A.d. binghami* are distributed in Sulawesi and its surrounding islands of Indonesia (Hepburn and Radloff 2011). *A.d. breviligula* and *A.d. binghami* exhibit aggregation of three to five colonies on a suitable nesting structure compared to aggregation of 250–300 *A.d. dorsata* colonies found on a banyan tree near Bengaluru, India (Nagaraja 2017).

The nests of *A. dorsata* are comparatively larger in size than those of other *Apis* species and are generally covered by 3–6 layers of worker bees (Koeniger et al. 2017). The larger body size of worker bees corresponds with variation in the length of their tongue (Cariveau et al. 2016), foraging region (Greenleaf et al. 2007), and sensory functions (Streinzer et al. 2013). The workers and drones of *A. dorsata* are reared in similar-sized hexagonal cells, in contrast to small-sized cells for workers and large-sized cells for drones in other honeybee species. The worker bees of these species perform periodic mass flights according to changing environmental conditions and nutrient demands of colonies (Woyke et al. 2005). Nonetheless, the drones perform dusk mass flights, mostly a few minutes after sunset (Woyke et al. 2007). *A. dorsata* foragers perform dances on the comb surface from where they can view the location of the sun. The definite lifestyle, together with moderate to high aggressive behavior, has limited the domestication of *A. dorsata* (Koeniger et al. 2010).

2.2 Nesting Ecology

The nests are a prerequisite for routine colony activities, including colony growth and reproduction in honeybees (Hepburn and Radloff 2011). The combs furnish an effective niche for bees to rear brood and store pollen and honey. *A. dorsata* constructs medium to large-sized semicircular combs on tall trees, multistoried apartments, rock cliffs, metro flyovers, etc. (Fig. 2.1). They usually build nests at heights ranging between 3 and 60 m above the ground. The congregation of

Fig. 2.1 Nesting of *Apis dorsata* colonies on a building, tree, and rock cliff

A. dorsata nests is quite popular, and the number may vary from 10 to 250 on a suitable nesting structure, occasionally up to 500 colonies, as on a banyan tree near a village about 30 km from Bengaluru, India, during honey flow seasons (Reddy and Peterson 2014).

The colonies *of A. dorsata* are highly migratory and reoccupy the same nest sites for several years (Oldroyd et al. 2000) despite workers mostly living for 40 to 50 days. *A. dorsata* colonies generally consist of a large-sized comb with brood, pollen, and honey, and a protective cover of many layers of adult bees. Nonetheless, worker bees build a waxy structure called wax specks on the top portion of the comb, and these wax specks are useful for bees in hanging on their nest curtain.

Orientation of nests toward the northeast direction was known to minimize the exposure to stable wind and dazzling sunlight in and around Bengaluru, India (Reddy and Reddy 1993). Furthermore, honeybees use the Earth's magnetic field as a cue in deciding the comb orientation (De Jong 1982). However, Lindauer (1971) observed that bees decide the orientation of combs in the direction of the wind.

2.3 Nest Distribution

Apis dorsata colonies are abundantly distributed in rural regions, possibly due to uninterrupted supply of pollen and nectar from cultivated and forest plants. The agricultural lands are seasonally cultivated with crops. Similarly, the forest plants around the cultivated ecosystem, such as pongamia, neem, jamun, and eucalyptus species, supply ample quantities of nectar seasonally. The existence of ideal nesting structures in urban regions attracts hundreds of giant honeybee colonies. The shielding from wind, rain, and direct sunlight was responsible for the selection of huge buildings for nesting in metropolis. Itioka et al. (2001) reported variation in the population of *A. dorsata* nests in the plains of tropical forests, and these were known to be associated with the flowering season. However, *A. dorsata* preferred buildings and huge trees in suburban sites for nesting. Roy et al. (2011) reported that the rich bee flora, coupled with suitable nesting sites, influences the spatial spacing of giant honeybee colonies. *A. dorsata* colonies were found in greater numbers in the regions of agro-forestry than in forest patches (Pavageau et al. 2018). Basavarajappa and Raghunandan (Basavarajappa and Raghunandan 2003) found a maximum number of *A. dorsata* colonies in winter in the regions of Mysuru and Kodagu, India. The population of *A. dorsata* colonies varies seasonally, and a huge number of colonies were found in winter and summer in the plains of Karnataka (Nagaraja 2019).

2.4 Colony Defense

Protection of nests through various defensive strategies is well developed in *A. dorsata*. These bee species build nests on terrestrial and arboreal structures beyond the reach of their natural enemies at higher heights. The workers of *A. dorsata* also exhibit shimmering behavior for colony defense against flying predators (Woyke et al. 2008). This behavior reaches the entire nest within a few seconds in the form of a wavelike pattern on the nest surface. The worker bees caution the predators by visual signals by twisting their bodies in the vertical direction on their nest surface (Kastberger et al. 1998). Bees also perform quick runs over the nest curtain by forming clusters at the lower fringe of the colony. The workers of nearby giant bee colonies also join in defensive activity and emit defense waves on the nest surface (Seeley et al. 1982). The shimmering behavior of workers serves as an alerting signal with low-frequency comb vibrations and finally releases the Nasonov pheromone to attack the predators.

2.5 Colony Growth and Development

The progress of honeybee colonies is characterized by an increase in worker population, followed by colony reproduction. After settling in a suitable nest site, the young workers build up a small-sized new comb which grows gradually. Venkatesh and Reddy (1989) found medium-sized combs in *A. dorsata* colonies after one month of settlement. *A. dorsata* combs are mostly semicircular and rarely irregular in shape.

The colonies of *A. dorsata* rear brood continuously throughout the year with seasonal fluctuations (Nagaraja 2020), and it depends on the presence of floral resources in the vicinity of bee colonies (Neupane and Thapa 2005). The brood rearing activity of *A. dorsata* was found greater from October to December months of the year in the plain regions of Karnataka, South India. However, a limited amount of brood was found from April to July. The better pollen sources, a protein-rich food in the vicinity of bee colonies, enhance brood production (Rajagopal and Nagaraja 1999). However, the honey flow season was recorded from December to February for the blooming of nectar-producing flowers. *A. dorsata* colonies rear drone brood during the breeding season and are scattered on the comb (Chuttong et al. 2019). Koeniger and Wijayagunasekera (1976) reported drone bees generally in old giant honeybee colonies with larger comb sizes. Woyke et al. (2007) recorded drone bees even in newly built colonies.

2.6 Reproductive Swarming

Polyandry, mating males with only females, but females mating with many males, is rare among insects. The virgin queens of *A. dorsata* mate with many successful drones during the flight, a few meters away from the colony. During the mating process, the drones generally assemble under an identified canopy of tall emergent trees at heights between 20 and 25 m. The virgin queens fly to identified mating yards called "drone congregation areas" (DCAs) and return to nests after mating, after 15–30 min (Tan et al. 1999). Moritz et al. (1995) reported mating of *A. dorsata* queens with 30 to 35 drones. The drones of these bee species may either transfer sperm into the oviduct or directly into the spermatheca of the queen during mating (Nagaraja 2023). After successful mating, the successful queen establishes a new nest along with a group of worker bees (Fig. 2.2). Venkatesh and Reddy (1989) found one reproductive swarm per colony in a nesting cycle of *A. dorsata*.

The scouts of *A. dorsata* perform a few dances during the swarming period. The quality of the newly found nesting sites is determined by greater number of circuits per dance. The swarming was generally noticed during December, followed by March, and absconding was observed during April and May in the colonies in Bengaluru, India.

2.7 Colony Migration

Migration is a seasonal movement of animals across regions, usually to exploit resources. Honeybee colonies generally migrate to exploit the bloom of flowering plants that supply floral resources. Short-distance migrations occur in response to internal disturbances in the colonies. However, migrations for long distances are due to a shortage of floral resources and unbearable weather conditions. *A. dorsata* colonies migrate seasonally between floral dearth to floral rich areas for 200 km in search of new colonies and food (Crane et al. 1993; Dyer and Seeley 1994). The colonies of *A. dorsata* arrive in the plains of South Karnataka during July, develop

Fig. 2.2 A swarm of *Apis dorsata* settling on the rooftop of a building

from October to December, and depart these regions during the onset of summer. The colonies of *A. dorsata* initiate migration by leaving barren combs behind. The dancers signal the direction of migration (Dyer and Seeley 1994), and the chemical cues left out in combs also constitute a navigational map in coordination.

3 Floral Resources

The floral resources are a prerequisite for routine activities of honeybee colonies (Loper and Berdel 1980). Honeybees visit the flowers with accessible moist pollen and nectar with a suitable sugar concentration (Kevan and Baker 1983). The quality of flora resources is evaluated primarily on long flowering period, nectar sugar content, size of the flower, etc. Nectar is an important carbohydrate source produced by the flowers to fascinate different pollinators. Continuous supply of pollen and nectar is a prerequisite for healthy development of bee colonies (Michener 2007). The ratio of pollen to nectar that is collected largely depends on the nutritional demands of the nests (Somme et al. 2016). Deficiency of food leads to a decline in colony populations, followed by the absconding of bee colonies (Williams and Kremen 2007).

3.1 Bee Flora in Wild Ecosystem

Honeybees gather pollen and nectar from wild flora found in natural and semi-natural habitats (Joshi et al. 2016). Studies report that the plants of the families such as Fabaceae and Rutaceae were among the most frequently visited by *A. dorsata* for pollen and nectar. The major plant species visited by *A. dorsata* in the wild are *Citrus* spp., *Dalbergia sisso*, *Litchi* spp., *Pongamia pinnata*, *Peltophoram pterocarpum*, Eucalyptus, etc. (Table 2.1). Raffiudin et al. (2024) found *A. dorsata* as a prime pollinator of tropical forests in the Asian continent. Forest patches show lower anthropogenic interference and have a high number and variety of suitable large nesting trees (Bhagwat et al. 2015). Availability of abundant resources from these forest patches determines the settlement and migration of bee colonies from forest patches (Krishnan 2011). Establishing forest patches that supply better pollen and nectar for bees could be a part of a strategy to restore urban pollinators.

During winter, *A. dorsata* foragers collect huge amounts of pollen and nectar from agricultural, horticultural, and plantation crops such as *Allium cepa*, *Brassica* species, *Raphanus sativus*, *Helianthus annuus*, *Brassica juncea*, *Cajanus cajan*, *Cucurbits*, *Musa paradisiaca*, *Cocos nucifera*, etc. (Table 2.1).

Guava is a major nectar-producing plant and is pollinated by different species of honeybees. (Fig. 2.3). In an area where there is equal distribution of resource patches, bees forage near nesting sites to decrease energy and increase foraging efficiency (Rahimi et al. 2021).

Table 2.1 Major bee flora of *Apis dorsata* in wild and cultivated ecosystems

Sl. No.	Scientific name	Common name	Family	Uses
1	*Allium cepa*	Onion	Liliaceae	Vegetable
2	*Allium sativum*	Garlic	Liliaceae	Spice
3	*Amomum compactum*	Large cardamom	Zingiberaceae	Spice
4	*Anacardium occidentale*	Cashew nut	Anacardiaceae	Fruit
5	*Antigonon leptopus*	Coral creeper	Polygonaceae	Ornamental
6	*Azadirachta indica*	Neem	Meliaceae	Insect repellent
7	*Brassica compestris*	Field mustard	Brassicaceae	Oil
8	*Brassica juncea*	Indian mustard	Brassicaceae	Oil
9	*Brassica nigra*	Black mustard	Brassicaceae	Spice
10	*Brassica oleraceae var. botrytis*	Cauliflower	Brassicaceae	Vegetable
11	*Cajanus cajan*	Pegeon pea	Leguminaceae	Pulse
12	*Carica papaya*	Papaya	Caricaceae	Fruit
13	*Carthamus tinctorius*	Safflower	Asteraceae	Oil
14	*Cichorium intybus*	Chicory	Asteraceae	Beverage
15	*Citrullus vulgaris*	Watermelon	Cucurbitaceae	Fruit
16	*Citrus aurantifolia*	Lime	Rutaceae	Fruit
17	*Citrus aurantium*	Sour orange	Rutaceae	Fruit
18	*Citrus limetta*	Sweet lemon	Rutaceae	Fruit
19	*Citrus limon*	Lemon	Rutaceae	Fruit
20	*Citrus sinensis*	Sweet orange	Rutaceae	Fruit
21	*Cocos nucifera*	Coconut	Araceae	Oil /beverage
22	*Coffea arabica*	Arabian coffee	Rubiaceae	Beverage
23	*Coriandrum sativum*	Coriander	Apiaceae	Pulse
24	*Cucumis anguria*	Gherkin	Cucurbitaceae	Vegetable
25	*Cucumis sativus*	Cucumber	Cucurbitaceae	Vegetable
26	*Cuminum cyminum*	Cumin	Apiaceae	Spice
27	*Cucumis melo*	Melon	Cucurbitaceae	Fruit
28	*Cucurbita pepo*	Pumpkin	Cucurbitaceae	Vegetable
29	*Dalbergia sissoo*	Shisham	Fabaceae	Rosewood
30	*Daucus carota*	Carrot	Apiaceae	Vegetable
31	*Ellettaria cardamomum*	Cardamom	Zingiberaceae	Spice
32	*Eucalyptus* spp.	Eucalyptus	Myrtaceae	Oil
33	*Guizotia abyssinica*	Niger	Asteraceae	Oil
34	*Helianthus annuus*	Sunflower	Asteraceae	Pulse
35	*Linum usitatissimum*	Flax/Linseed	Linaceae	Fiber
36	*Mangifera indica*	Mango	Anacardiaceae	Fruit
37	*Malus domestica*	Apple	Rosaceae	Fruit
38	*Medicago sativa*	Alfalfa	Leguminaceae	Forage
39	*Momordica charantia*	Bitter gourd	Cucurbitaceae	Vegetable

(continued)

Table 2.1 (continued)

Sl. No.	Scientific name	Common name	Family	Uses
40	*Moringa oleifera*	Drumstick	Moringaceae	Vegetable
41	*Musa paradisiaca*	Banana	Musaceae	Fruit
42	*Litchi chinensis*	Litchi	Sapindaceae	Fruit
43	*Ocimum basilicum*	Basil	Laminaceae	Essence
44	*Peltophorum pterocarpum*	Peltophoram	Fabaceae	Ornamental tree
45	*Phyllanthus indica*	Gale of the wind	Phyllanthaceae	Amla
46	*Pongamia pinnata*	Karanj	Fabaceae	Oil
47	*Punica granatum*	Pomegranate	Punicaceae	Fruit
48	*Psidium guajava*	Guava	Myrtaceae	Fruit
49	*Prunus armeniaca*	Apricot	Rosaceae	Fruit
50	*Prunus amygdalus*	Almond	Rosaceae	Fruit
51	*Prunus domestica*	Plum	Rosaceae	Fruit
52	*Prunus communis*	Peach	Rosaceae	Fruit
53	*Pyrus persica*	Pear	Rosaceae	Fruit
54	*Raphanus sativus*	Radish	Cruciferae	Vegetable
55	*Rubus fruticosus*	Black berry	Rosaceae	Forage
56	*Sesamum indicum*	Sesame	Pedaliaceae	Oil
57	*Solanum melongena*	Brinjal	Solanaceeae	Vegetable
58	*Tamarindus indica*	Tamarind	Leguminosae	Multiple
59	*Trifolium fragiferum*	Strawberry clover	Fabaceae	forage
60	*Trifolium pratense*	Red clover	Fabaceae	Fruit

Fig. 2.3 A mango tree in full bloom—a major source of nectar to *Apis dorsata*

3.2 Foraging Ecology

Knowledge on foraging ecology of pollinators is crucial for their effective utilization in pollination of crops in tropical ecosystems (Dainese et al. 2019; Thimmegowda et al. 2020). Bees have evolved a mutual relationship with flowering plants for pollination (Michener 2007). The floral resources are a prerequisite for the growth and development of honeybee colonies (Dimou et al. 2006). Honeybees communicate the quality of the food to nest-mates through physical dances during foraging activity (von Frisch 1967). This mechanism has made bees to gather floral resources faster by spending minimal energy. *A. dorsata* is a pollinator of wild and cultivated crops for its wide foraging activity and seasonal migration from the regions of floral-dearth to floral-rich areas by increasing population in new areas (Stewart et al. 2018; Mukherjee et al. 2019). The ability of bees to visit a large number of flowers is governed by its large sized colonies with thousands of foraging bees.

Normally, *A. dorsata* foragers gather pollen and nectar continuously throughout the day in the plains of Karnataka. A greater number of pollen collectors was documented between 8.00 and 10.00 h, followed by 16.00 to 18.00 h. Similarly, the nectar collectors were recorded in maximum numbers between 9.00 and 11.00 h, followed by 16.00 and 17.00 h of the day (Nagaraja and Rajagopal 2000). The supply of plenty of moist sticky pollen during the blooming of flowers prompts the bees to collect the pollen. *A. dorsata* foragers gather pollen and nectar from a distance of 2 to 3 km. The higher nectar collectors in late morning hours were due to the abundant supply of good-quality nectar. Similarly, foragers of *A. dorsata* gathered larger amounts of pollen and nectar from 8.00 to 11.00 h of the day.

The maximum pollen and nectar loads carried by *A. dorsata* foragers were 36.71 mg and 42.51 μl at 9.00 and 10.00 h of the day, respectively. The foragers of *A. dorsata* exhibit foraging activity for long hours even on bright moonlight (Wongsiri et al. 1996). The high levels of visual pigment in the reticular cells of bees ommatidia have made gathering of floral resources even in dark moonlight (Suwannapong and Wongsiri 1999; Young et al. 2021).

4 Threats

Honeybees have been threatened by natural and anthropogenic threats since time immemorial in their distribution regions. Modern agricultural practices, with the creation of agricultural landscapes, have severely affected most honeybee species (De Jong and Lester 2023). The major factors inducing stress in bees are the intensification of land use, habitat deterioration through agriculture and urban development, and application of pesticides on crops (Goulson et al. 2015). Furthermore, the traditional methods of honey harvesting, bee poisoning, pests and diseases, deforestation, and habitat destruction are the major threats faced by *A. dorsata* colonies in India.

4.1 Honey Harvesting

Honey is the major product produced by honeybees and is extracted scientifically without damaging both brood and super combs through honey extractors from

Fig. 2.4 The comb of *Apis dorsata* damaged due to traditional methods of honey harvesting

colonies of domesticated bees like *A. cerana* and *A. mellifera*. However, no such extractors have been developed so far to extract honey from feral honeybee species such as *A. florea*, *A. dorsata*, and *A. laboriosa*. Furthermore, the nests of *A. dorsata* are hunted through traditional methods for their honey and beeswax since ancient times (Crane 1999). The honey hunters are experienced personnel, mainly from forest regions where plenty of *A. dorsata* build colonies. Honey hunters primarily depend on honey, beeswax, and remaining forest products for their livelihoods.

Honey from *A. dorsata* colonies is being harvested by traditional methods. The nests of *A. dorsata* with surplus quantities of honey during the honey season are easily identified by the honey hunters. The honey hunters mostly collect honey during the nighttime due to the mass stinging behavior of *A. dorsata*. In the process of honey harvesting, the hunters generate huge amounts of smoke, especially from wet leaves available in those sites, and cloth after reaching the colonies. Normally, they destroy brood combs followed by adult bees, including queens, with flames of the fire (Matias et al. 2018) (Fig. 2.4).

4.2 Bee Poisoning

Bee poisoning is the accidental killing of bees by the use of insecticides on the crops in pest control practices. Pesticides used for plant protection are generally hazardous to honeybees (Nagaraja 2016). Most of the cultivated crops in the tropics and subtropics are sprayed with less toxic to highly toxic insecticides in insect pest control practices (Klein et al. 2007). Bee poisoning occurs when insecticides are applied to the crop during blooming. Foragers are exposed to insecticide residues on flowers and get poisoned either by taking contaminated dew, pollen, or nectar. It has been quite apparent that systemic insecticides sprayed on plants can penetrate floral nectar in sufficient quantity to kill honeybees. Despite a few insecticides being less toxic, most of them are highly toxic. These insecticides even kill the brood as well as the adult population and contaminate the hive products (De La Rua et al. 2009). Some insecticides may even persist on the pollen and nectar, causing long-term mortality.

Mortality of thousands of *A. dorsata* colonies is quite common in most metropolitan cities in India. For example, *A. dorsata* colonies rarely attack school children and the general public in a metropolitan city like Bengaluru by constructing nests on the school and public buildings. These mass attacks cause major injuries and even death of people occasionally. Consequently, *A. dorsata* colonies on school/public buildings are destroyed by the application of toxic insecticides. This method of killing is responsible for large-scale mortality of *A. dorsata* colonies in urban regions.

4.3 Pests and Predators

Asian honeybees are tolerant of most pests and diseases. However, a few diseases and pests are catastrophic to both wild and hive honeybee colonies. The brood mite, *Tropilaelaps clareae*, infests both the worker and drone brood of *A. dorsata* colonies. However, *A. dorsata* colonies tolerate these parasitic mites by migrating after the brood-rearing season, leaving the mite-infested brood in the deserted combs. Furthermore, these species do not open the mite-infested pupae in the capped cells. Therefore, the pressure that kills the pupae does not reach the adult bees or other brood cells (Woyke et al. 2004). The combs of *A. dorsata* are most frequently attacked by insect pests during the dearth season. *Gallaria mellonella* (greater wax moth) is a seasonal pest inflicting damage, especially in weaker bee colonies in tropical Asia. Mahindre (1983) found about 90% wax moth damage in the combs of *A. dorsata*. On severe wax moth damage, the infested colonies desert their nests and reconstruct them elsewhere.

4.4 Deforestation

Deforestation is quite common throughout the world; however, the Asian continent is significantly impacted by deforestation. The urbanization, gathering of firewood, production of charcoal, livestock grazing, and expansion of agricultural land are the major causes of deforestation. Furthermore, large-scale mechanical removal of vegetation reduces the bee population due to a dearth of floral resources (Kremen et al. 2004). Removal of trees often causes a dearth of pollen and nectar for honeybees. For example, the University of Agricultural Sciences, Bengaluru, India, found about 250 *A. dorsata* colonies in a banyan tree (*Ficus* spp.), especially during the winter season, during the 1990s, primarily for a rich source of nectar from Eucalyptus plantations. However, uprooting of Eucalyptus trees during later periods made the university campus almost free from *A. dorsata* colonies (Nagaraja 2012). The removal of trees for the widening of roads is also responsible for the decline in colonies in rural and urban regions.

4.5 Habitat Destruction

Natural landscapes provide suitable habitats and floral resources for honeybee species. *A. dorsata* nests are mostly found on huge trees in forests and rural regions. Removal of trees for timber apparently destroys the natural habitat of bees. Furthermore, removal of huge trees in the forest regions as well as on roadsides where they most frequently build nests usually destroys the habitats of *A. dorsata*. The forest clearance and canopy loss occur due to logging and clear-cutting for food production and agriculture (Imai et al. 2018). Estimates predict that habitat loss is severe in terms of biodiversity loss in Southeast Asia (Sodhi et al. 2010). The level of deforestation in India is the second highest in the world and was found to be about 668,000 hectares per year during 2015–2020 (FAO 2020). Forest fire is a common problem in the tropics, wherein the wildfire spreads over thousands of kilometers, destroying nesting habitats. It leads to the death of millions of feral honeybee colonies (Taylor et al. 1999).

5 Conservation of *Apis dorsata* Colonies

Honeybees contribute a lot to human health and well-being in addition to maintaining biodiversity (Potts et al. 2016). Recent data on global analysis of bee occurrence showed that a quarter of bee species are disappearing in view of threats (Zattara and Aizen 2021). Apparently, there is a need to safeguard honeybees from threats. Among them, a few are as follows.

5.1 Scientific Methods of Honey Harvesting

Harvesting of honey through crude methods severely reduces the population of *A. dorsata*. Nonetheless, in a few parts of Southeast Asia, honey is harvested by exposing colonies to sufficient quantities of smoke with medium-sized bee smokers along with protective clothing (Waring and Jump 2004). Bee smokers developed to be lightweight and easy to handle produce an ample amount of smoke required for silencing bees during honey harvesting. Development and distribution of such bee smokers among the traditional hunters reduces the destruction of *A. dorsata* nests. Application of *Amomum aculeatum* sap on the body while harvesting honey significantly reduced the number of stings. Nevertheless, creating awareness on the usage of bee smokers and sap of the *Amomum* among honey hunters reduces the mortality of giant honeybee colonies.

5.2 Application of Rafter Technology

Rafters (bamboo cliffs) are used to attract migratory swarms of *A. dorsata* during swarming seasons by mimicking tree branches. These rafters are successfully used in a few Asian countries, including India, for attracting giant honeybees (Tan et al. 1997). Giant honeybees successfully build their nests on the rafters. The wooden rafters, along with *A. dorsata* nests, are pulled down using ropes during honey harvesting safely. The honey portion is collected without much damage to the comb. Such colonies could develop rapidly and store honey for a few weeks during the honey flow season. This technique facilitates many honey harvests compared to a single harvest through traditional methods of harvesting.

5.3 Application of Bee-Friendly Pesticides

The application of toxic insecticides on crop plants causes the mortality of bees through poisoning. However, if necessary, crops are sprayed with low-toxic, safer insecticides only during early morning or late evening hours while bees are not actively foraging. Application of pesticides on crops during evening hours is generally less hazardous to bees than during early morning hours. Plant-based organic pesticides are available in the market under different trade names. Such pesticides may be used in pest control practices in place of inorganic toxic pesticides. The application of eco-friendly microbial/biopesticides is advisable on the crops in areas rich with honeybee species.

5.4 Habitat Protection

Suitable nesting structures, such as huge trees and rock cliffs, are crucial for the successful establishment of *A. dorsata* colonies. Nonetheless, preservation of huge, branched trees and rock cliffs is essential for attracting giant honeybee colonies. Therefore, steps should be initiated to plant thousands of bee flora that supply pollen, nectar, and both pollen and nectar continuously throughout the year. Furthermore, bee flora should not be sprayed with toxic insecticides during the blooming of the flowers.

5.5 Use of Insect Repellents

Thousands of giant honeybee colonies are being killed in metropolitan cities by the application of pesticides. Instead of removing *A. dorsata* colonies using toxic pesticides, steps should be taken to vacate bee colonies from buildings by application of smoke during day time. Since sap of the *Amomum* plant shows bee repellent properties, usage of its purified active compounds would greatly help in handling bees during honey harvesting, and also for protecting bees on the flowers from pesticides. Pasting suitable long-lasting bee repellents such as neem oil and motor grease on the substratum of the deserted nest site was found to be effective in preventing reoccupation of giant honeybee colonies. There is a scope to initiate research on screening and developing eco-friendly bee repellents to repel the nests of

A. dorsata to safer zones, where they are under severe threat (Qasim and Nagaraja 2024).

6 Conclusion

The colonies of *A. dorsata* are threatened by both natural and anthropogenic risks. If such threatening continues, there is every possibility of a significant decline in the population of *A. dorsatsa*. Harvesting of honey during daytime by wearing a bee suit to shield from bee stings, followed by using a sufficient amount of smoke to silence the aggressive behavior of bees, reduces the mortality of bee colonies. Thorough training is required for honey hunters on bee-friendly and scientific techniques of honey collection. Creating awareness on the impact of traditional honey collection, deforestation, habitat destruction, and bee poisoning is essential among honey hunters and the general public to save the bee colonies. Providing lucrative sources of income to honey producers can reduce their traditional honey-hunting activities.

Acknowledgments This study was supported by the research grants from University Grants Commission (UGC) (No. 42-623/2013 (SR), New Delhi, and financial assistance from Bangalore University (No. DEV: D2a: BU-RP: 2020-21), Bengaluru, India.

References

Abrol DP (2020) The future role of dwarf honeybee in natural and agricultural systems. CRC Press (Taylor and Francis Group), Boca Raton, London, New York, p 322

Baitala TV, Faquinello P, de Toledo VAR, Mangolin CA, Martins EN, Ruvolo-Takasusuki MCC (2010) Potential use of major royal jelly proteins as molecular markers for royal jelly production in Africanized honeybee colonies. Apidologie 41:160–168

Basavarajappa S, Raghunandan KS (2003) Colony status of Asian giant honeybee, *Apis dorsata* Fabricius in Southern Karnataka, India. Afr J Agric Res 8:680–689

Bhagwat SA, Kushalappa CG, Williams PH, Brown ND (2015) The role of informal protected areas in maintaining biodiversity in the Western Ghats of India. Ecol Soc 10(1):8

Cariveau DP, Nayak GK, Bartomeus I, Zientek J, Ascher JS, Gibbs J, Winfree R (2016) The allometry of bee proboscis length and its uses in ecology. PLoS One 11(3):e0151482

Chantawannakul P, Williams G, Neumann P (2018) Asian beekeeping in the 21st century. Springer Nature, Singapore, p 325

Chuttong B, Somana W, Burgett M (2019) Giant honey bee (*Apis dorsata* F.) rafter beekeeping in southern Thailand. Bee World 96:66–68

Crane E (1999) The world history of beekeeping and honey hunting. Routledge, New York, p 704

Crane E, van Luyen V, Mulder V, Ta TC (1993) Traditional management system for *Apis dorsata* in submerged forests in Southern Vietnam and Central Kalimantan. Bee World 74:27–40

Dainese M, Martin EA, Aizen MA, Albrecht M, Bartomeus I, Bommarco R, Steffan-Dewenter IA et al (2019) A global synthesis reveals biodiversity-mediated benefits for crop production. Sci Adv 5:eaax0121

De Jong D (1982) Orientation of comb building by honeybees. J Comp Physiol A 147(495):501

De Jong D, Lester PJ (2023) The global challenge of improving bee protection and health. Front Bee Sc 1:1118292. https://doi.org/10.3389/frbee.2023.1118292

De La Rua P, Jaffe R, Dall Olio R, Munozi I, Serrana J (2009) Biodiversity, conservation and current threats of European honeybees. Apidologie 40:263–284

Dimou M, Thrasyvoulou A, Tsirakoglou V (2006) Efficient use of pollen traps to determine the pollen flora used by honey bees. J Apic Res Bee World 45:42–46

Dyer FC, Seeley TD (1994) Colony migration in the tropical honeybee, *Apis dorsata* F. (Hymenoptera: Apidae). Insect Soc 41:129–140

FAO (2020) Global Forest Resources Assessment 2020: Main report. Rome

Giampieri F, Quiles JL, Cianciosi D, Forbes-Hernandiz TY, Onantes-Bermejo FJ, Alvarez-Squarez JM, Battino M (2022) Bee products: an emblematic example of underutilized sources of bioactive compounds. J Agric Food Chem 70(23):6833–6848

Goulson D, Nicholls F, Botias C, Rotheray EL (2015) Bee declines driven by combined stress from parasites, pesticides and lack of flowers. Science 347(6229):1255957

Greenleaf SS, Williams NM, Winfree R, Kremen C (2007) Bee foraging ranges and their relationship to body size. Oecologia 153(3):589–596

Hepburn HR (2011) Absconding, migration and swarming. In: Hepburn R, Radloff SE (eds) Honeybees of Asia. Springer-Verlag, Berlin, pp 133–158

Hepburn HR, Radloff SE (2011) Biogeography. In: Hepburn R, Radloff SE (eds) Honeybees of Asia. Springer-Verlag, Berlin, pp 62–63

Hung KJ, Kingston JM, Albrecht M, Holway DA, Kohn JR (2018) The worldwide importance of honey bees as pollinators in natural habitats. Proc R Soc B 285:20172140

Imai N, Furulawa T, Tsujino R, Kitamura S, Yumota T (2018) Factors affecting forest area change in Southeast Asia during 1980-2010. PLoS One 13(6):e0199908

Itioka T, Inoue T, Kaliang H, Kato M, Nagamitsu T, Momose K, Sakai S et al (2001) Six-year population fluctuation of the giant honey bee *Apis dorsata* (Hymenoptera: Apidae) in a tropical lowland dipterocarp forest in Sarawak. Ann Entomol Soc Am 94:545–549

Joshi NK, Otieno M, Rajotte EG, Fleischer SJ, Biddinger DJ (2016) Proximity to woodland and landscape structure drives pollinator visitation in apple orchard ecosystem. Front Ecol Evol 4:38

Kastberger G, Raspotnig G, Biswas S, Winder O (1998) Evidence of Nasonov scenting in colony defence of the giant honeybee *Apis dorsata*. Ethology 104:27–37

Kevan PG (1999) Pollinators as bioindicators of the state of the environment: species, activity and diversity. Agric Ecosyst Environ 74:73–393

Kevan PG, Baker HG (1983) Insects as flower visitors and pollinators. Annu Rev Entomol 28:407–453

Klein AM, Vaissiere BE, Cane JH, Steffan-Dewenter I, Cunningham SA, Kremen C, Tscharntke T (2007) Importance of pollinators in changing landscapes for world crops. Proc R Soc B Biol Scie 274:303–313

Koeniger N, Wijayagunasekera HNP (1976) Time of drone flight in the three Asian honeybee species (*Apis cerana, Apis florea, Apis dorsata*). J Apic Res 15:67–71

Koeniger N, Koeniger G, Tingek S (2010) Honeybees of Borneo: exploring the centre of *Apis* diversities. Natural History Publications (Borneo), Kota Kinabalu, Sabah, Malaysia, p 262

Koeniger N, Kurze C, Phiancharoen M, Koeniger G (2017) Up or down that makes the difference. How giant honeybees (*Apis dorsata*) see the world. PLoS One 12(11):e0185325

Kremen C, Williams NM, Brigg RL, Fay JP, Thorp RW (2004) The area requirements of an ecosystem service: crop pollination by native bee communities in California. Ecol Lett 7:1109–1119

Krishnan S (2011) Pollinator services and coffee production in a forested landscape mosaic. Ph. D Thesis. ETH, Zurich

Lindauer M (1971) Communication among social bees. Harvard Uni. Press, Cambridge M.A., p 61

Loper GM, Berdel RLA (1980) Nutritional bioassay of honeybee brood-rearing potential. Apidologie 11:181–189

Mahanta M, Rahman A (1997) Studies on pollination of Litchi (*Litchi chinensis*) by honeybees (*Apis cerana* F.). Indian Bee J 59(2):74–75

Mahindre DB (1983) Handling rock bee colonies. Indian Bee J 45:72–73

Marcucci MC (1995) Propolis: chemical composition, biological properties and therapeutic activity. Apidologie 26:83–99

Matias DMS, Borgemeister C, von Wehrden H (2018) Ecological changes and local knowledge in a giant honey bee (Apis dorsata F.) hunting community in Palawan, Philippines. Ambio 247(8): 924–934

Michener CD (2007) The bees of the world, 2nd edn. The Johns Hopkins Univ. Press, Baltimore and London

Moritz RFA, Southwick EE (1992) Bees as super organisms. Springer, Berlin, Heidelberg, New York

Moritz RFA, Kryger P, Koeniger N, Estoup A, Tingek S (1995) High degree of polyandry in *Apis dorsata* queens detected by DNA microsatellite variability. Behav Ecol Sociobiol 37:357–363

Mukherjee R, Deb R, Devy SM (2019) Diversity matters: effects of density compensation in pollination service during rainfall shift. Ecol Evol 9(17):9701–9711

Nagaraja N (2012) Asian honeybees: biology, threats and their conservation. In: Florio RM (ed) Biology, threats and colonies. Nova Publishers Inc., USA, pp 99–123

Nagaraja N, Rajagopal D (2009) Honeybees: diseases, parasites, pests, predators and their management. MJP Publishers, Chennai, India, p 210

Nagaraja N (2016) Effect of insecticide poisoning on mortality of giant honeybee, *Apis dorsata* colonies. Pesquisa 2(1):96–100

Nagaraja N (2017) Population fluctuation of giant honeybee, *Apis dorsata* in plains of Karnataka. J Ent Res 43(4):503–508

Nagaraja N (2019) Nesting patterns of giant honeybee, *Apis dorsata* in plains of Karnataka, India. J Ent Res 41(3):307–310

Nagaraja N (2020) Biology of dwarf honeybee, *Apis florea* Fabricius (Hymenoptera: Apidae). In: Abrol DP (ed) The future role of dwarf honeybee in natural and agricultural systems. CRC Press, Taylor and Francis Group, USA, pp 13–23

Nagaraja N (2022) Conservation of Asian giant honeybee *Apis dorsata* Fab (Hymenoptera: Apidae). Int J Zool Appl Biosci 7(3):29–33

Nagaraja N (2023) Reproductive biology of Asian giant honeybee, Apis dorsata Fabricius (Hymenoptera: Apidae). In: Abrol DP (ed) The role of giant honeybees in natural and agricultural systems. CRC Press, Taylor and Francis Group, USA, pp 27–36

Nagaraja N, Rajagopal D (2000) Foraging and brood rearing activity of rock bee, *Apis dorsata* F. (Hymenoptera: Apidae). J Ent Res 24(3):243–248

Nagaraja N, Reddy CC (2002) Influence of bee pollinators on the quality of seed production in Sunflower (*Helianthus annuus*) and coconut (*Cocas nucifera*). Asian Bee J 4(2):1–5

Neupane KR, Thapa RB (2005) Pollen collection and brood production by honeybees (*Apis mellifera* L.) under Chitwan condition of Nepal. J Inst Agric Anim Scie 26:143–148

Oldroyd BP, Pratt SC (2015) Comb architecture of the eusocial bees arises from simple rules during cell building. Adv Insect Physiol 49:101–121

Oldroyd BP, Wongsiri S (2006) Asian honey bees. Biology, conservation and human interactions. Harvard University Press, Cambridge, MA, p 360

Oldroyd BP, Osborne KE, Mardan M (2000) Colony relatedness in aggregations of *Apis dorsata* Fabricius - (Hymenoptera, Apidae). Insect Soc 47:94–95

Page P, Lin Z, Buawangpong N et al (2016) Social apoptosis in honeybee super organisms. Sci Rep 6:27210

Pavageau C, Gaucherel C, Garcia C, Guazoul J (2018) Nesting sites of giant honeybees modulated by landscape patterns. J Appl Ecol 55:1230–1240

Potts SG, Imperatriz-Fonseca V, Ngo H et al (2016) Safeguarding pollinators and their values in human-welbeing. Nature 540:220–229

Qasim SS, Nagaraja N (2024) Repellent activity of citronella and clove oils against workers of the giant honeybee, *Apis dorsata* (Hymenoptera: Apidae). Appl Biol Res 26(3):351–358

Raffiudin R, Dyahastuti M, Nugraha R, Sayusti T, Djuita NR, Suwananda E, Allvioningrum V, Mardhony R, Biagioni S, Setyaningsi CA, Prasetyo LB, Priawandiputra W, Atmowidi T,

Saad A, Behling H (2024) The effect of land cover on the foraging behavior and pollen in the honey of the giant bee *Apis dorsata* in Sumatra. Front Bee Scie 2:1366287
Rahimi E, Barghjelveh S, Dong P (2021) Estimating landscape structure effects on pollination for management of agricultural landscapes. Ecol Process 10:59. https://doi.org/10.1186/s13717-021-00331-3
Rajagopal D, Nagaraja N (1999) Beekeeping status in Karnataka, India. Asian Bee J 1:50–59
Rajagopal D, Veeresh GK, Chikkadevaiah NN, Kencharaddi RN (1999) Potentiality of honeybees in hybrid seed production of sunflower (*Helianthus annuus* L). Indian J Agric Sci 69:40–43
Reddy MS, Peterson S (2014) Nominating the bee trees of Ramagovindapura as a world heritage site. J Apic 29(3):207–212
Reddy CC, Reddy MS (1993) Studies on the distribution of nests of giant honeybee (*Apis dorsata* F). Indian Bee J 55:36–39
Requier F, Jowanowitsch KK, Kallnik K, Steffan-Dewenter I (2019) Limitation of complementary resources affects colony growth, foraging behavior and reproduction in bumble bees. Eco 101(3):e02946
Roy P, Leo R, Thomas SG, Varghese A, Sharma K, Prasad S, Bradbear N, Roberts S, Potts SG, Davida P (2011) Nesting behaviour of rock bee, *Apis dorsata* in the Nilgiri biosphere reserve in India. Trop Ecol 52:285–291
Seeley TD, Seeley RH, Aktratanakul P (1982) Colony defense strategies of the honeybees in Thailand. Ecol Monogr 52:43–63
Sharaf SE, Alsanosi S, Alzahrani AR, Al-Hhamdi SS, Sharaf SE, Ayoub N (2022) Knowledge, attitude and practice of bee venom acupuncture therapy on rheumatoid arthritis among patients in Saudi Arabia. Int J Gen Med 15:1171–1183
Sihag RC (2025) Conservation of giant honeybee (*Apis dorsata*) for honey and beeswax production and sustainable pollination. Insects 16(6):560
Sodhi NS, Koh LP, Clemdents R et al (2010) Conserving southeast Asian forest diversity in human modified landscapes. Biol Conserv 143:2375–2384
Somme L, Moquet L, Quinet M et al (2016) Food all in row: urban trees offer valuable floral resources to pollinating insects. Urban Ecosyst 19:1149–1161
Stewart AB, Sritongchuay T, Teartisup P, Kaewsomboon S, Bumrungsri S (2018) Habitat and landscape factors influence pollinators in a tropical megacity, Bangkok, Thailand. Peer J 6: e5335
Streinzer M, Brockmann A, Nagaraja N, Spaethe J (2013) Sex and caste specific variation in compound eye morphology of five honeybee species. PLoS One 8(2):e57702
Suwannapong G, Wongsiri S (1999) Ultra structure of the compound eyes of the giant honey bee queens, *Apis dorsata* Fabricius, 1793. J STREC 7(1–2):60–68
Tan NQ, Chinh PH, Thai PH, Muler V (1997) Rafter beekeeping with *Apis dorsata*: some factors affecting occupation of rafters by bees. J Apic Res 36:49–54
Tan NQ, Mardan M, Thai PH, Chinh PH (1999) Observations on multiple mating flights of *Apis dorsata* queens. Apidologie 30:339–346
Taylor D, Saksena P, Sanderson PG, Kucera K (1999) Environmental change and rain forests on the Sunda shelf of Southeast Asia: drought, fire and biological cooling of biodiversity by hotspots. Biodivers Conserv 8:1159–1177
Thimmegowda GG, Mullen S, Sottilare K, Sharma A, Mohanta SS, Brockmann A, Dhandapany PS, Olsson SB (2020) A field-based quantitative analysis of sublethal effects of air pollution on pollinators. Proc Natl Acad Sci USA 117(34):20653–20661
Venkatesh G, Reddy CC (1989) Rates of swarming and absconding in the giant honey bee, *Apis dorsata* F. Proc Indian Acad Sci (Anim Sci) 98(6):425–430
von Frisch K (1967) The dance language and orientation of bees. Harvard Univ. Press, Cambridge, MA
Waring C, Jump DR (2004) Rafter beekeeping in Cambodia with *Apis dorsata*. Bee World 84:14–18

Williams NM, Kremen C (2007) Resource distributions among habitats determine solitary bee offspring production in a mosaic landscape. Ecol Appl 17:910–921

Wongsiri S, Thapa R, Oldroyd B, Burgett DM (1996) A magic bee tree: home to *Apis dorsata*. Am Bee J 136:796–799

Woyke J, Kruk C, Wilde J, Wilde M (2004) Periodic mass flights of the giant honey bee *Apis dorsata*. J Apic Res 43:180–185

Woyke J, Wilde J, Reddy CC, Nagaraja N (2005) Periodic mass flights of giant honeybee, *Apis dorsata* performed in successive days at two environmental conditions. J Apic Res 44(4): 180–189

Woyke J, Wilde J, Wilde M, Reddy MS, Nagaraja N, Sivaram V (2007) Presence or absence of drones in drone dusk mass flights performed by *Apis dorsata breviligula* worker bees. J Apic Res 46(1):40–49

Woyke J, Wilde J, Wilde M, Sivaram V, Cernancia C, Nagaraja N, Reddy MS (2008) Comparison of defense body movements of *Apis laboriosa*, *Apis dorsata dorsata* and *Apis dorsata breviligula* honeybees. J Insect Behav 21:481–494

Young AM, Kodabalagi S, Brockmann A, Dyer FC (2021) A hard day's night: patterns in the diurnal and nocturnal foraging behaviour of *Apis dorsata* across lunar cycles and seasons. PLoS One 16(10):e0258604

Zattara EE, Aizen MA (2021) Worldwide occurrence records suggest a global decline in bee species richness. One Earth 4:114–123

Plant Pollinator: Relationship Based on Morphology and Physiology

3

Pritha Ghosh and Ajinkya Markad

Abstract

Plant-pollinator interactions are fundamental to ecosystem health and biodiversity, driven by intricate adaptations in the morphology and physiology of both plants and pollinators. Flower morphology, including size, shape, color, and nectar production, evolves to attract specific pollinators, enhancing reproductive success. Similarly, pollinators exhibit physiological and morphological traits, such as specialized mouthparts, sensory adaptations, and flight capabilities, that optimize foraging efficiency and facilitate pollination. This mutualistic relationship promotes co-evolution, with certain plant species becoming highly dependent on specialized pollinators. However, disruptions such as habitat loss, climate change, and the introduction of invasive species threaten these relationships, jeopardizing pollination services critical to food security and ecological balance. Understanding the interplay of morphology and physiology in plant-pollinator systems is critical for conserving biodiversity and mitigating the impacts of environmental change.

Keywords

Pollination · Adaptation · Biodiversity · Climate change

P. Ghosh (✉) · A. Markad
Department of Entomology, School of Agriculture, Lovely Professional University, Phagwara, Punjab, India

A. K. Chakravarthy (ed.), *The Role of Pollinators in Tropical Landscapes*,
https://doi.org/10.1007/978-981-95-9139-8_3

1 Introduction

1.1 Pollination

Pollination is the transfer of pollen from an anther of a plant to the stigma of a plant, later enabling fertilization and the production of seeds, most often by an animal or by wind (Barrows 2011). Pollination is a phenomenon that evolved over millions of years, involving a mutually beneficial relationship between plants and pollinators (Kaur and Singh Kaleka 2022). Pollination can be facilitated by various agents, including animals like insects, birds, and bats, as well as water, wind, and even self-pollination within a flower. Typically, pollination takes place within the same species. However, when pollination happens between different species, it can produce hybrid offspring in natural environments and in plant breeding efforts. Over 90% of flowering plants in hot and humid regions depend on pollination (Imran 2020).

2 Role of Pollination in Ecosystems

Pollination is crucial for keeping ecosystems in balance and essential for crop production, connecting agriculture to the cycle of life. Pollination has economic significance as it enhances the quality and quantity of agricultural output (Breeze et al. 2011; Gill et al. 2016; Hristov et al. 2020). Pollination plays a critical role in ecosystem functioning and biodiversity.

Plant-pollinator interactions are important mutualisms in agricultural food production because they provide essential ecosystem services helping global biodiversity (Ollerton 2017). Animal pollinators are responsible for the reproduction of 87.5% of flowering plants (Ollerton et al. 2011). Animal pollination is required for 87 of the world's leading food crops and 35% of crop production volumes (Klein et al. 2007). Reports of declining pollinators from around the world pose an urgent ecological challenge (Potts et al. 2010; Christmann 2019).

2.1 Plant Reproduction

Pollination is essential for the sexual reproduction of flowering plants. It enables the transfer of pollen grains from the male reproductive organs (stamens) to the female reproductive organs (pistils) of flowers, leading to fertilization and production of seeds. Fertilization process leads to the formation of seeds (Proctor et al. 1996a, b). The successful pollination process is vital for the reproductive success of plants. It guarantees the transfer of compatible pollen to the female reproductive organs, significantly enhancing the likelihood of successful fertilization and subsequent seed production (Johnson and Steiner 2000). Cross-pollination, the transfer of pollen between flowers of the same species, enhances genetic diversity within plant populations. This variation in genetic material is crucial for plant species to adapt and thrive in dynamic and evolving environments (Benoit and Kalisz 2020).

2.2 Biodiversity Conservation

Pollination relies on the crucial role of pollinators, including bees, butterflies, birds, bats, and insects (Ollerton et al. 2011). These pollinators play a vital role in promoting the diversity of plants through cross-pollination, wherein pollen is transferred from one flower to another flower of the same species. This process fosters genetic advancements in plant populations and contributes to the overall maintenance of biodiversity in the ecosystems (Valido et al. 2019). Pollination contributes to plant diversity within ecosystems. Cross-pollination facilitated by various pollinators enhances genetic diversity in plant populations (Ollerton et al. 2011). This genetic variation is vital for the adaptation and survival of plants (Eckert et al. 2010). The conservation of endangered plant species heavily depends on pollination by specific pollinators. Safeguarding pollinators and their interactions with endangered plants is of utmost importance to ensure the long-term survival and conservation of plant species.

2.3 Food Production

Pollination plays a critical role in ensuring agricultural productivity and food security. Crops such as fruits, vegetables, nuts, and oilseeds rely on pollinators for effective pollination (Klein et al. 2007). It is estimated that about 75% of globally significant crops depend on animal pollinators, and approximately 35% of global food production is reliant on pollination (Gallai et al. 2009). It enhances fruit set, quality, and yield of several crops, including fruits, vegetables, nuts, and oilseeds (Klein et al. 2007).

2.4 Habitat Creation

Plants that rely on pollinators for reproduction serve as important habitats and food sources for animal species, including insects, birds, and mammals (Kremen et al. 2007). These habitats play a crucial role in supporting biodiversity within ecosystems, contributing to the ecological balance and functioning of natural communities.

2.5 Ecosystem Services

The ecological services provided by pollination are invaluable. Pollinators aid in the maintenance of ecosystem services such as nutrient cycling, soil fertility, and the provision of habitat for other organisms (Potts et al. 2016). They also support the reproduction of wild plants, which form the foundation of many terrestrial ecosystems.

2.6 Decline Pollinator

A significant decline in pollinators, including the well-known phenomenon called colony collapse disorder, has been witnessed recently. This decline harmed the early stages of plant regeneration, specifically seed dispersal and pollination. The successful regeneration of plants relies heavily on interactions between plants and animals, and the interruption of these interactions poses a threat to biodiversity and the functioning of ecosystems (Neuschulz et al. 2016). Because of the economic importance of insect pollinators, particularly honeybees, honeybees are transported and introduced to most parts of the world. Recent dramatic declines in pollinator abundances, however, have raised concerns about the sustainability of pollination services. Pollinators perform multiple roles in enhancing the ecosystem.

3 Role of Pollination in Agriculture

Pollination is an important method for agriculture. In most agronomical crops, horticultural crops rely upon pollination. Crops such as wheat, maize, rice, soybeans, and sorghum are primarily wind-pollinated or capable of self-pollination. Out of the top 15 crops that play a significant role in the global human diet in 2013, over 10% of the total calorie intake from plant crops (211 out of 1916 kcal/person/day) relies on insect pollination (FAO 2017; Aizen et al. 2019).

The significance of natural insect pollination in agriculture, both ecologically and economically, is increasingly recognized, leading to emerging financial opportunities. The presence of native pollinators, facilitated by nearby forests or wild grasslands, can have a positive impact on crops like apples, almonds, or coffee. Studies have shown that the proximity of such natural habitats enhances crop yields by approximately 20%, thereby improving the quantity and quality of the produce. (IPCC 2007). This highlights the importance of preserving and promoting native pollinators and their habitats to capitalize on the benefits they bring to agricultural systems (Munyuli 2010).

4 Methods in Pollination

Pollination occurs through biotic or abiotic means (Fig. 3.1). Biotic pollination involves the living pollinators, such as insects, birds, bats, and other animals, to transfer pollen between flowers (Ollerton et al. 2011). Abiotic pollination relies on nonliving factors for pollen transfer. Wind pollination, or anemophily, occurs when pollen is dispersed by the wind to reach other flowers (Bawa 1990). Biotic and

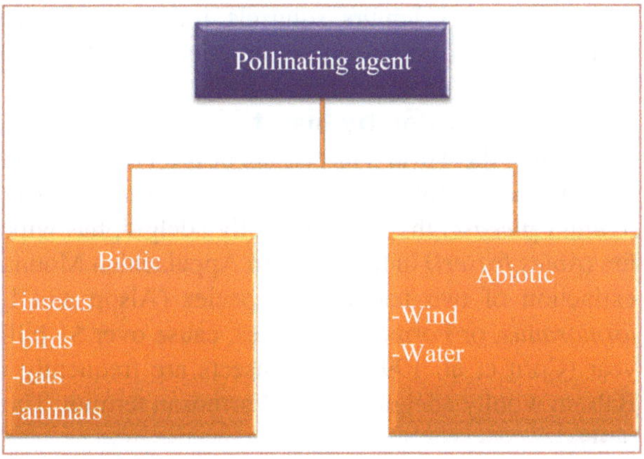

Fig. 3.1 Methods of pollination

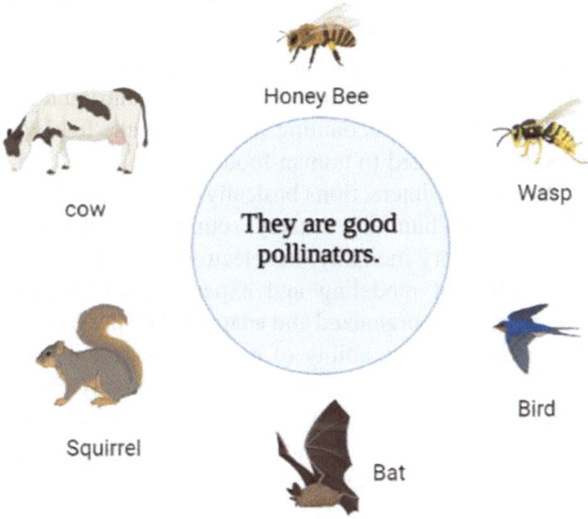

Fig. 3.2 Examples of pollinators present in nature

abiotic pollination mechanisms are essential and are observed in plant species, depending on their characteristics and environmental conditions.

4.1 Biotic Pollination

Biotic pollination is relied upon by about 80% of angiosperms, indicating that the majority of flowering plants depend on living pollinators for successful reproduction (Ackerman 2000). An estimated range of 100,000–200,000 animal species serve as pollinators for the world's 250,000 species of flowering plants. While the majority of pollinators are insects, there are also approximately 1500 species of birds and mammals that visit flowers and contribute to pollen transfer between them. Animals such as monkeys, lemurs, squirrels, rodents, and possums (Fig. 3.2) are among the frequent visitors involved in pollination (Abrol and Abrol 2012).

4.1.1 Pollination by Insect

Only 1% of the world's insects are pests (Triplehorn and Johnson 2005), but they are responsible for 13% of crop and 9% of forest production losses (Pimental 2000). *Adelges picease*, the balsam woolly adelgid, has wiped out nearly all of the Fraser firs (*Abies fraseri)* in the southern Appalachian Mountains and is responsible for the extinction of two native bird species (Alsop and Laughlin 1999). *Coptotermes formosanus*, or Formosan termites, cause over $1 billion in structural damage each year (Corn et al. 1999). Pest insects are frequently invasive species, such as the Balsam woolly adelgid and the Formosan termite. These insects are less likely to be pests.

Entomophily, which refers to pollination by insects, commonly takes place on plants that have evolved colored petals and emit a strong scent to attract specific insects such as bees, wasps, ants (Hymenoptera), beetles (Coleoptera), moths and butterflies (Lepidoptera), and flies (Diptera) (Dar et al. 2017). According to Gallai et al. (2009), insect-mediated pollination contributed a significant economic value of EUR 153 billion, accounting for 9.5% of the total value of agricultural production specifically related to human food.

Bee-flower interactions basically depend on sensory cues, viz., olfaction, vision, chemical and humidity sensing, sound, and touch. Clarke et al. (2017) unveiled yet another sensory modality, the electrostatic field arising between a bee and a flower. Finite-element modeling and experiments offered new insights into how these interactions are organized and adapted. The three ingredients that enable electrostatic interactions are the ability of bees to accumulate positive charge, the atmospheric electric field, and the plants that are negatively charged. Further investigations are required to understand the mechanisms of electrostatic interactions and aerial electroreception. The particles including the nanoparticles that surround the bee foraging zone may have effects.

Cananthrophily

Beetle pollination is characterized by the association between flowers and beetles, which is assumed to occur in more than 90% of pollinated plant species. While beetles may not always be the primary pollinators, their widespread presence makes them significant contributors to pollination. Flowers that rely on beetle pollination or possess characteristics to attract beetles typically exhibit certain traits. These flowers are often large in size and can be pale white, cream, or green. They emit a strong scent, which can sometimes resemble carrion. Beetle-pollinated flowers are usually flat or disc-shaped, accommodating the robust bodies of beetles. The pollen in these flowers is easily accessible, as beetles have limited ability to handle and manipulate it. Although beetles have been referred to as destructive pollinators, as they may feed on floral parts or directly consume pollen, many beetle-pollinated plants have developed extra protective structures around their ovaries. The main reward offered by beetle-pollinated flowers is pollen (Korotkova and Barthlott 2009; Wojcik 2021).

Melittophily

Bees play a crucial role as the primary pollinators of plants. Pollination by bees is a critical process for the reproduction of numerous plant species (Free 1993). Certain flowers that attract bees tend to have yellow or blue coloration, sometimes with ultraviolet nectar guides, and emit a scent. These flowers offer rewards in the form of nectar, pollen, or both, albeit in varying quantities (Fig. 3.3). The nectar of bee flowers typically contains sucrose as the dominant sugar. Additionally, a small number of bee species collect oil from specialized glands on the flower (Martins et al. 2014).

Social bees such as honeybees, bumblebees, and stingless bees are highly important and efficient pollinators for both wildflowers and agricultural crops (Kleijn et al. 2015). Their abundance and effectiveness in pollination make them a prominent focus of research on insect pollinators. The domestication of honeybees has provided humans with the opportunity to utilize pollination services by relocating colonies in man-made hives to areas with high pollination needs (Breeze et al. 2011; Wright et al. 2013; Potts et al. 2010).

Psychophily

Butterfly pollination involves flowers that are typically disc-shaped or compound in structure, providing a convenient landing pad for butterflies. Asteraceae family flowers are frequently visited by butterflies. These flowers are often vibrant and eye-catching, displaying shades of pink, red, or purple, and emitting strong scents. Butterflies perceive the color red, which explains why many of the flowers visited by butterflies exhibit this color. Butterflies feed on nectar using proboscis and are attracted to flowers with abundant nectar reserves. Nectar sources in butterfly-pollinated flowers are in tubes or spurs accessible to their long proboscis. The anthers of flowers are usually positioned in a way that deposits pollen on the heads or undersides of butterflies when they come into contact with them. It is common for butterflies to land on compound flowers and feed from multiple sites within the flower structure (Wojcik 2021).

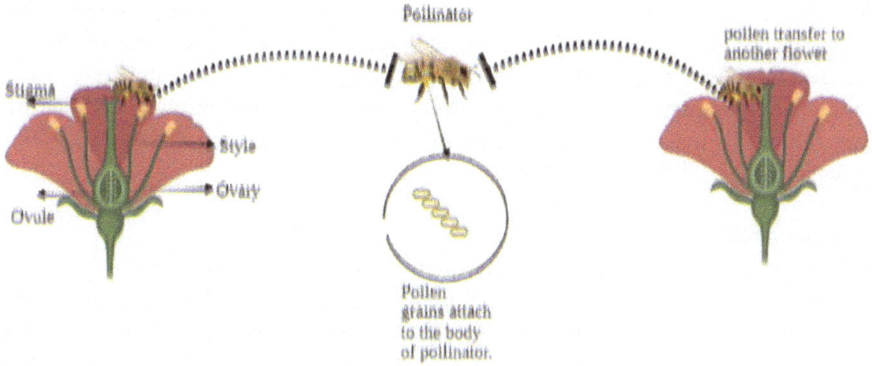

Fig. 3.3 Pollination by bees

Phalaenophily

Hawk moths (Sphingidae) are significant pollinators within the moth family. Similar to hummingbirds, they exhibit a hovering behavior with rapid wing beats when approaching flowers. Most hawk moths are active during the night or twilight, making them nocturnal or crepuscular pollinators. Consequently, flowers pollinated by moths possess distinct characteristics. These flowers are typically white, aiding visibility in low-light conditions. They are often large and showy, featuring tubular corollas permitting access for the proboscis of hawk moths. Moth-pollinated flowers emit a strong, sweet scent, typically produced in the evening, night, or early morning to attract moths. Flowers produce abundant nectar to fuel the high metabolic rates required for the fast-flying nature of hawk moths. Other moth species, such as Noctuids, Geometrids, and Pyralids, have a slower flight pattern and settle on flowers. These moths do not require as much nectar as their fast-flying counterparts. Consequently, the flowers they pollinate are generally smaller in size, though they may cluster together in inflorescences or flower heads (Oliveira et al. 2004).

Myophily

Flies that participate in pollination are attracted to a similar range of flower types as bees, as they also seek nectar as a food source. Flowers pollinated by flies exhibit a diversity of colors and shapes, although they are generally less intricate compared to bee-pollinated flowers. Flowers often emit a strong and fragrant scent to assist flies in locating them. Flies primarily feed on nectar and flowers pollinated by flies typically offer abundant nectar reserves.

Sapromyophily

Flower pollination by flies attracted to carrion or dung is less common than pollination by nectar-feeding flies. These flowers have evolved to mimic the appearance and scent of animal flesh or dung, as they serve as egg-laying substrates for flies. They emit a strong and foul odor to resemble the scent of decay. The flowers are small in size, colorless, or dull purple/pink, often secreting a sticky substance. Trap blossoms, which deceive the insects into visiting without providing any rewards, are also prevalent. An extraordinary example of this type of pollination is seen in the giant corpse plant, also known as titan arum (*Amorphophallus titanum*). This enormous flower reaches heights of over 2 meters and emits a powerful, putrid odor resembling rotting flesh. Flesh flies and carrion beetles are attracted to this plant, and as they attempt to find food or lay eggs, they become trapped inside the giant flower till it withers, releasing them after they have been covered in pollen (Jones and Jones 2001).

Myrmecophily

Despite the abundance of ants and frequent visits to flowering plants, the actual number of plants that rely on ants for pollination is surprisingly low. In general, ants are often regarded as nectar robbers rather than effective pollinators. This means that they extract nectar from flowers without actively contributing to the pollination process (Rostás et al. 2018). Ant-pollinated plants are primarily found in arid or alpine environments. These plant species have developed specific adaptations to attract and utilize ants as pollinators. Examples of ant-pollinated plants include Small's stonecrop (*Diamorpha smallii*), alpine nailwort (*Paronychia pulvinata*), Cascade knotweed (*Polygonum cascadense*), as well as two alpine orchid species, *Chamorchis alpine* and *Dactylorhiza viridis*. The plants have developed exclusive structures and traits that make them attractive to ants for pollination (Wojcik 2021).

Pollination by Birds (Ornithophily)

Pollination by birds is called ornithophily. While many bird species occasionally visit flowers in search of food, particularly to find hidden insects within inflorescences, they may also consume nectar. Reports indicate that flower visitation has been observed in as many as 50 families of birds (Proctor and Yeo 1973; Proctor et al. 1996a, b). However, three specific families of birds have evolved as significant groups specializing in flower interactions. These groups are hummingbirds (Trochilidae), sunbirds (Nectariniidae), and honeyeaters (Meliphagidae). These birds have developed unique adaptations and behaviors that enable them to efficiently interact with flowers and extract nectar as their primary food source. According to available research, there is a solitary instance of a bird-pollinated native plant called *Anagyris foetida* L. (Leguminosae). In Spain, this particular plant species is pollinated by three warbler species: *Phylloscopus collybita* Vieillot, *Sylvia atricapilla* L., *and Sylvia. melanocephala* Gmelin (Ortega-Olivencia et al. 2005).

4.1.2 Pollination by Bat (Chiropterophily)

There are striking differences between bat pollination in the New World and the Old World. In the Old World, pollinating bats primarily consist of large fruit bats from the family Pteropodidae. Bats lack the ability to hover and instead must perch on the plant to access and consume the nectar. The Old-World bats also do not possess echolocation capabilities (Fleming et al. 2009). In lowland tropical forests, pollination by birds and bats is limited to a relatively small percentage of plant species, ranging from 3% to 11% (Devy and Davidar 2003). Bird pollination is generally more prevalent than bat pollination, at local and global scales. Bird pollination is observed in nearly 500 plant genera, while bat pollination is found in approximately 250 genera (Sekercioglu 2006).

4.2 Pollination by Mammals (Therophily)

The characteristics of the pollination syndrome associated with mammal pollination, excluding bats, include the following: a yeasty scent; inconspicuous, dull-colored, ground-level flowers or inflorescences often hidden from view; large and robust flowers, or clustered together in multi-flowered inflorescences; flowers or inflorescences that are either sessile or supported by short and sturdy stalks; bowl-shaped flowers or inflorescences; abundant nectar that is rich in sucrose and typically produced during the night; sturdy and wiry styles; an appropriate distance between the stigma and nectar to accommodate the pollinator's snout or proboscis; and potentially a flowering period during winter or spring (Wiens et al. 1983; Melidonis and Peter 2015).

Nocturnal non-flying mammals possess a keen sense of smell and are often attracted to plants that do not exhibit bright, showy colors but instead emit strong odors. These plants have adapted to the preferences of mammals by producing a significant amount of pollen, as mammals tend to be larger compared to other pollinators. Unlike smaller pollinators, mammals may not have the same level of precision in transferring pollen, necessitating the production of larger pollen quantities to increase the chances of successful pollination (Carthew and Goldingay 1997).

4.3 Abiotic Pollination

These plants do not specifically attract animal pollinators; they do exhibit shared characteristics.

4.3.1 Pollination by Wind (Anemophily)

Around 98% of abiotic pollination, which refers to pollination by wind, is attributed to anemophily. This method of pollination is believed to have originated from insect pollination (entomophily) and is likely a result of environmental changes or the limited availability of pollinators (Whitehead 1969; Culley et al. 2002; Faegri and Van Der Pijl 2013).

Pollination by wind is a mechanism of pollination where pollen is dispersed by air currents. Unlike biotic pollination, wind-pollinated plants do not rely on animals for pollen transfer. Instead, they produce large quantities of lightweight and small-sized pollen grains that are easily carried by the wind (Friedman and Barrett 2009). Wind-pollinated flowers often possess inconspicuous characteristics, including small size, lack of showy colors, and a green appearance. These flowers produce an immense quantity of relatively small pollen grains to increase the chances of successful wind dispersal.

Wind pollination is a prevalent strategy among conifers and approximately 12% of flowering plants worldwide. This mode of pollination is particularly common in grasses, cereal crops, various tree species, and notorious allergenic plants like ragweeds. Wind-pollinated plants disperse vast quantities of pollen grains into the air, relying on a chance for a small percentage of these grains to successfully reach their intended targets.

5 Characteristics of Pollinator

5.1 Vision

Insects, such as bees and butterflies, have visual variations that extend from ultraviolet light (around 300 nm) to yellow-orange light (approximately 650 nm). Heliconius butterflies, known for their long lifespan, possess a well-developed visual system and learning ability, which is evident in their strong preference for male

Anguria plants (Gilbert 1972). Bee pollinators have spherical compound eyes composed of around 6300 cone-shaped facets. Their visual system is sensitive to blue, yellow, and blue-green colors, as well as ultraviolet and polarized light (Anonymous 2003). Pollinating bats have large eyes, providing them with good vision. They also have a keen sense of smell (olfaction), a long, absorbent tongue, and digestive modifications to efficiently feed on nectar (Anonymous 2000).

5.2 Olfaction (Sense of Smell)

Honeybees have a highly developed olfactory capability, estimated to be 40 times better than that of humans. This strong sense of smell plays a crucial role in locating food sources and communication within and outside the hive (Anonymous 2003).

Trap lining is a foraging behavior in certain pollinators, such as euglossine bees, *Heliconius ethilla* butterflies, and sphinx moths. This behavior involves visiting a series of floral resources in a linear pattern, which helps the pollinators efficiently locate and exploit food sources (Janzen 1970; Ehrlich and Gilbert 1973; Linhart and Mendenhall 1977).

At low plant densities, the significance of specialized floral morphology and floral visibility for outcrossing (the transfer of pollen between different individuals) is reduced. Other factors, such as trap lining behavior or other mechanisms, may play a more important role in ensuring successful pollination (Grant 1971).

5.3 Foraging Behavior

The evolution of changes in floral characteristics is closely associated with modifications in the behavior of insects and pollinators. Several factors influence foraging behavior and determine the profitability of a food source, including weather conditions, distance to the food source, and the quantity and quality of food available.

Pollinators have varying foraging ranges, typically ranging from 3 to 12 km (Anonymous 2003). Additionally, the foraging rate can differ depending on the type of flowers and the specific pollinator. For example, bumblebees have been found to have a foraging rate twice as fast as honeybees (Free 1968), while solitary bees may have a slower foraging rate that can vary based on size (Wood 1965). In comparison, sphinx moths tend to forage fairly rapidly, and their foraging behavior requires lower energy compared to crawling insects like ants (Hickman 1974).

Several studies have demonstrated short-distance movements among pollinators, including honeybees, syrphid flies, bees, flies, butterflies, hawk moths, and others (Beattie 1978; Levin 1978; Waddington and Holden 1979). Turning behavior during foraging has also been recorded in honeybees, bumblebees, and butterflies, with bees exhibiting more frequent turns when foraging in rich patches with abundant nectar (Pyke 1979).

The ability to return to the same location over time has been observed in landmarking honeybees and bumblebees (Manning 1956; Heinrich 1979). In trap-lining euglossine bees and territorial anophila like Syrphids and bumblebees, individuals learn to recognize specific plant positions and visit them in a particular sequence, although they repeat parts of the same path multiple times in a single foraging trip (Janzen 1970; Grant 1971; Manning 1956; Heinrich 1979).

Honeybees are known to recognize rewards based on color and scent, and their learning process begins a couple of seconds before feeding and diminishes quickly (Von Frisch 1967; Menzel 1967). Bumblebees have innate flower-probing behavior, but when they deal with complex flowers, bumble bees learn how to successfully exploit them (Laverty 1980). The attraction of native butterflies and other nectar feeders to exotic plant species leads to reduced visits to native plant species, resulting in a decline in native plant populations (Simonson et al. 2000). Broad-tailed and Rufous hummingbirds avoid plants and flowers that have recently been visited by nectar-robbing bumblebees, using nectar as a proximate cue in combination with spatial and visual cues (Irwin 2000).

6 Factors Attracting Pollinators

Flower traits like shape, size, color, scent production (Fig. 3.4), electric fields, and movement have been measured and acknowledged as factors that contribute to the recognition and attraction of pollinators to flowers (Alcorn et al. 2012; Balamurali et al. 2015).

6.1 Flower Color

The world of flowering plants displays a remarkable array of flower colors and patterns, encompassing the full spectrum of colors visible to both humans and pollinators (Menzel and Shmida 1993). Certain colors tend to attract particular

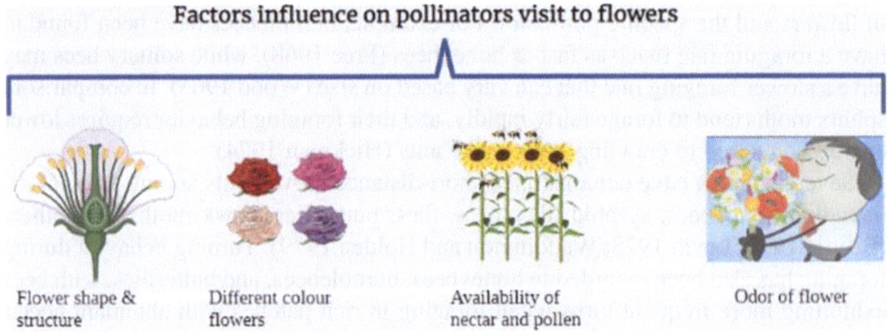

Fig. 3.4 Factors influencing attraction of pollinators to the flower

pollinators, such as bees and butterflies. Bees are commonly drawn to flowers with blue, purple, and yellow hues, whereas butterflies are more inclined toward vibrant colors like red, orange, and pink (Chittka and Menzel 1992). Pollinators have the ability to detect and respond to flower colors at varying distances, and different color attributes come into play depending on the proximity of the pollinator to the flowers. Bees, for instance, utilize achromatic (non-color) cues to detect flowers from a distance. As they get closer or when the flowers are large-sized, chromatic (color) cues become significant in their interaction with the flowers (de Ibarra et al. 2015).

6.2 Flower Shape and Size

Flowers exhibit a wide range of diverse forms, encompassing different structures and arrangements, evolved to fulfill specific pollination functions. The diversity in flower forms is primarily shaped by natural selection operating in conjunction with the pollinators of each flowering plant species. Consequently, the specific form of a flower is influenced by both the characteristics of the plant species itself and the outcomes of convergent evolution (Ghosh et al. 2017; Hodges 1997). The shape of a flower enhances its appeal to pollinators, facilitates the deposition of pollen, affects how the pollinators interact with the flower, and influences the level of specialization in pollinator interactions. Additionally, flower shape impacts the electrostatic properties of pollen deposition (Valenta et al. 2017). Flower size plays a unique role in attracting pollinators, as larger floral displays are preferred by bumble bees (*Bombus diversus* Smith) due to their specific preferences (Makino et al. 2007; Giuliani et al. 2020).

6.3 Flower Scent

Olfactory signals play a crucial role in attracting pollinators by advertising the presence of rewarding resources, often working in conjunction with visual cues. In comparison to colors and color patterns, olfactory signals have a higher adaptability and are more precisely selected, making them more resilient (Primante 2015). The scent is a multifaceted aspect of the floral phenotype that plays a significant role in communication between flowering plants and pollinators. It promotes specialization in plant-pollinator relationships through secretive channels and uncommon connections. The scent also facilitates outcrossing and reproductive isolation through the persistence of floral traits (Raguso 2008).

6.4 Nectar Production

The quantity and quality of nectar produced by flowers significantly influenced pollinator visitation. Higher nectar volumes often attract more pollinators and increase the likelihood of successful pollination (Nicolson and Thornburg 2007).

7 Threats to Pollinator

In recent years, there has been a significant decline in pollinator populations, commonly referred to as pollinator decline. The loss of pollinators has a noticeable impact on early plant regeneration processes, including seed dispersal and pollination. These critical processes heavily rely on interactions between plants and animals. When interactions are interrupted due to the decline of pollinators, it poses a threat to biodiversity and the functioning of ecosystems. As a result, the overall health and sustainability of ecosystems are compromised (Neuschulz et al. 2016).

According to researchers, the decline in bee and pollinator populations is attributed to a combination of factors. These factors include the loss of habitat or flowers that provide pollen and nectar, pesticide exposure, and the presence of parasites and pathogens. While each of these factors individually affects bees negatively, there is evidence that the combination of stresses has an even more detrimental impact (Potts et al. 2010; Vanbergen and the Insect Pollinators Initiative 2013; Goulson et al. 2015). Bees and other pollinators rely on flowers for food source. Nectar from flowers provides them with carbohydrates, while pollen serves as their protein source, necessary for rearing their young. Landscapes lacking flowers, such as mowed lawns with strict weed control or heavily paved areas in cities, as well as fields with limited plant diversity, offer little food for bees. This results in poor nutrition and compromises the immune systems of the bees. When bees are nutritionally weakened, they become more susceptible to diseases and the harmful effects of pesticides (Smitley 2019).

7.1 Habitat Loss and Fragmentation

Pollinators require diverse habitats for foraging, nesting, and reproduction. The destruction and fragmentation of natural habitats due to urbanization, agriculture, and land-use changes limit their access to essential resources (Potts et al. 2016).

7.2 Exposure to Pesticides

According to Woodcock et al. (2017), exposure to pesticides is a major threat to pollinators, with different types of pesticides like insecticides, herbicides, and fungicides having harmful effects on their overall health. The impact of pesticide exposure on pollinators includes:

7.2.1 Reducing Foraging Activity

Pesticides hinder the ability of pollinators to efficiently locate and gather nectar and pollen from flowers, leading to a decline in foraging efficiency. This reduced efficiency results in insufficient nutrition, having adverse effects on survival and reproductive success.

7.2.2 Direct Mortality

Some pesticides have high toxicity levels for pollinators, causing direct mortality when they come into contact with these chemicals. This can result in immediate or subsequent death, depending on the concentration and duration of exposure.

7.3 Invasive Pest

The presence of invasive species poses a threat to pollinators in multiple ways. Invasive plant species have the ability to outcompete native plants, diminishing the availability of suitable food sources for pollinators. Invasive insect species, including certain bee parasites, introduce diseases and pathogens that have detrimental effects on native pollinators (Goulson et al. 2015).

7.4 Modern Agricultural Practices

Alterations in land management practices have an impact on pollinators. Changes in how land is managed influence the availability of suitable habitats and food resources for pollinators. For instance, practices such as the removal of hedgerows, monoculture farming, and the decline of wildflower-rich meadows lead to a reduction in floral resources and nesting sites for pollinators. Similarly, modifications in forest management affect the presence of appropriate habitats for pollinators. These shifts in land use practices have significant implications for pollinator populations and their overall well-being. Contemporary agricultural practices, including the disappearance of hedgerows, the adoption of monoculture farming, and the decline of meadows abundant in wildflowers, have resulted in a decrease in floral resources and nesting environments for pollinators. Likewise, alterations in how forests are managed affect the accessibility of suitable habitats for pollinators. These modifications in land use practices had a substantial effect on the availability of resources and habitats necessary for the well-being of pollinator populations (Carvell et al. 2006).

7.5 Effect of Diseases

Diseases can significantly contribute to the decline of pollinator populations. Several diseases and pathogens affect bees and other pollinators, leading to adverse effects on their health and overall population dynamics, e.g., varroa mites and bee viruses (Martin 2001), Colony Collapse Disorder (CCD) in Honeybees (VanEngelsdorp et al. 2009).

7.6 Climate Change

Climate change can change the timing of flowering in plants, which, in turn, affects the synchronization between plants and pollinators. These alterations in phenology have adverse consequences for pollinators. Extreme weather events like droughts and heat waves can disrupt the availability of essential floral resources and nesting sites for pollinators. These disruptions lead to a decline in pollinator populations. In summary, climate change affects the timing of flowering, disrupts resource availability, and ultimately leads to a reduction in pollinator numbers. Details on the climate change impacts on pollinators are found in other chapters.

References

Abrol DP, Abrol DP (2012) Non bee pollinators-plant interaction. In: Pollination biology: biodiversity conservation and agricultural production, pp 265–310

Ackerman JD (2000) Abiotic pollen and pollination: ecological, functional, and evolutionary perspectives. In: Pollen and pollination, pp 167–185

Aizen MA, Aguiar S, Biesmeijer JC, Garibaldi LA, Inouye DW, Jung C et al (2019) Global agricultural productivity is threatened by increasing pollinator dependence without a parallel increase in crop diversification. Glob Chang Biol 25(10):3516–3527

Alcorn K, Whitney H, Glover B (2012) Flower movement increases pollinator preference for flowers with better grip. Funct Ecol 26(4):941–947

Alsop FJ, Laughlin TF (1999) Changes in the spruce-fir avifuana of Mt. Guyot, Tennessee, (1967–1985). J Tenn Acad Sci 66:207–209

Anonymous (2000) Ecology of pollination. http://www.sci.uidaho.edu/biosci/lecture/314%20 Austad/Microsoft%20 PowerPoint%20-%20Lecture%2020.pdf

Anonymous (2003) "Bees behavior during foraging", Apiary Fact Sheet # 111. http://agf.gov.bc.ca/apiculture/factsheets/111_forage.html

Balamurali GS, Krishna S, Somanathan H (2015) Senses and signals: evolution of floral signals, pollinator sensory systems and the structure of plant–pollinator interactions. Curr Sci 108(10): 1852–1861

Barrows EM (2011) Animal behavior desk reference. A dictionary of animal behavior, ecology, and evolution, 3rd edn. CRC Press LCC, Boca Raton, p 794

Bawa KS (1990) Plant-pollinator interactions in tropical rain forests. Annu Rev Ecol Syst 21(1): 399–422

Beattie A (1978) Plant-animal interactions affecting gene flow in Viola

Benoit AD, Kalisz S (2020) Predator effects on plant-pollinator interactions, plant reproduction, mating systems, and evolution. Annu Rev Ecol Evol Syst 51:319–340

Breeze TD, Bailey AP, Balcombe KG, Potts SG (2011) Pollination services in the UK: how important are honeybees? Agric Ecosyst Environ 142(3–4):137–143

Carthew SM, Goldingay RL (1997) Non-flying mammals as pollinators. Trends Ecol Evol 12(3): 104–108

Carvell C, Roy DB, Smart SM, Pywell RF, Preston CD, Goulson D (2006) Declines in forage availability for bumblebees at a national scale. Biol Conserv 132(4):481–489

Chittka L, Menzel R (1992) The evolutionary adaptation of flower colours and the insect pollinators' colour vision. J Comp Physiol A 171:171–181

Christmann S (2019) Do we realize the full impact of pollinator loss on other ecosystem services and the challenges for any restoration in terrestrial areas. Restor Ecol 27:720–725. https://doi.org/10.1111/rec.12950

Clarke D, Erica M, Daniel R (2017) The bee, the flower and the electrica field: electric ecology and aerial electro reception. J Comp Physiol A Neuroethol Sens Neural Behav Physiol 203(9): 737–748. https://doi.org/10.1007/s00359-017-1176-6

Corn ML et al (1999) Harmful non-native species: issues for congress. Congressional Research Service, Library of Congress, Washington, DC

Culley TM, Weller SG, Sakai AK (2002) The evolution of wind pollination in angiosperms. Trends Ecol Evol 17(8):361–369
Dar SA, Hassan GI, Padder BA, Wani AR, Parey SH (2017) Pollination and evolution of plant and insect interaction. J Pharmacogn Phytochem 6(3):304–311
de Ibarra NH, Langridge KV, Vorobyev M (2015) More than colour attraction: behavioural functions of flower patterns. Curr Opin Insect Sci 12:64–70
Devy MS, Davidar P (2003) Pollination systems of trees in Kakachi, a mid-elevation wet evergreen forest in Western Ghats, India. Am J Bot 90(4):650–657
Eckert CG, Kalisz S, Geber MA, Sargent R, Elle E, Cheptou PO, Goodwillie C, Johnston MO, Kelly JK, Moeller DA, Porcher E, Ree RH, Vallejo-Marín M, Winn AA (2010) Plant mating systems in a changing world. Trends Ecol Evol 25(1):35–43. https://doi.org/10.1016/j.tree.2009.06.013
Ehrlich PR, Gilbert LE (1973) Population structure and dynamics of the tropical butterfly Heliconius ethilla. Biotropica 5:69–82
Faegri, K., and Van Der Pijl, L (2013) Principles of pollination ecology
FAO (2017) Food and Agriculture Organization of the United Nations, Statistics Division. "FAOstats Food Supply - Crops Primary Equivalent"
Fleming TH, Geiselman C, Kress WJ (2009) The evolution of bat pollination: a phylogenetic perspective. Ann Bot 104(6):1017–1043
Free JB (1968) Dandelion as a competitor to fruit trees for bee visits. J Appl Ecol 5:169–178
Free JB (1993) Insect pollination of crops. Academic, London/New York, pp 172–180
Friedman J, Barrett SC (2009) Wind of change: new insights on the ecology and evolution of pollination and mating in wind-pollinated plants. Ann Bot 103(9):1515–1527
Gallai N, Salles JM, Settele J, Vaissière BE (2009) Economic valuation of the vulnerability of world agriculture confronted with pollinator decline. Ecol Econ 68(3):810–821
Ghosh S, Ganga M, Manimaran P (2017) Pollination mechanisms and adaptations in flower and ornamental crops-a review. J Pharmacogn Phytochem 6(5):662–665
Gilbert LE (1972) Pollen feeding and reproductive biology of Heliconius butterflies. Proc Natl Acad Sci 69(6):1403–1407
Gill RJ, Baldock KC, Brown MJ, Cresswell JE, Dicks LV, Fountain MT et al (2016) Protecting an ecosystem service: approaches to understanding and mitigating threats to wild insect pollinators. In: Advances in ecological research, vol 54. Academic, pp 135–206
Giuliani C, Giovanetti M, Lupi D, Mesiano MP, Barilli R, Ascrizzi R et al (2020) Tools to tie: flower characteristics, voc emission profile, and glandular trichomes of two mexican salvia species to attract bees. Plants 9(12):1645
Goulson D, Nicholls E, Botías C, Rotheray EL (2015) Bee declines driven by combined stress from parasites, pesticides, and lack of flowers. Science 347(6229):1255957
Grant V (1971) Euglossine bees as long-distance pollinators of tropical plants. Science 171:203–205
Heinrich, B. (1979) "Bumblebee economics", Harvard Univ. Press, Cambridge, MA
Hickman JC (1974) Pollination by ants: a low-energy system. Science 184(4143):1290–1292
Hodges SA (1997) Floral nectar spurs and diversification. Int J Plant Sci 158(S6):S81–S88
Hristov P, Neov B, Shumkova R, Palova N (2020) Significance of apoidea as main pollinators. Ecological and economic impact and implications for human nutrition. Diversity 12(7):280
Imran M (2020) Neonicotinoid insecticides: a threat to pollinators. IntechOpen. https://doi.org/10.5772/intechopen.88814
IPCC (2007) Intergovernmental panel on climate change, fourth assessment report, climate change 2007: syntheses report. UNEP, Gene've
Irwin RE (2000) Effects of nectar-robbing bumblebees on hummingbird-pollinator behavior and plant reproductive success in *Ipomopsis Aggregata* (Polemoniaceae) (Doctoral dissertation, University of Vermont)
Janzen DH (1970) Herbivores and the number of tree species in tropical forests. Am Nat 104(940): 501–528

Johnson SD, Steiner KE (2000) Generalization versus specialization in plant pollination systems. Trends Ecol Evol 15(4):140–143. https://doi.org/10.1016/s0169-5347(99)01811-x

Jones GD, Jones SD (2001) The uses of pollen and its implication for entomology. Neotrop Entomol 30:314–349

Kaur N, Singh Kaleka A (2022) Diversity, importance and decline of pollinating insects in present era. IntechOpen. https://doi.org/10.5772/intechopen.100316

Kleijn D, Winfree R, Bartomeus I, Carvalheiro LG, Henry M, Isaacs R et al (2015) Delivery of crop pollination services is an insufficient argument for wild pollinator conservation. Nat Commun 6(1):7414

Klein AM, Vaissière BE, Cane JH (2007) Importance of pollinators in changing landscapes for world crops. Proc R Soc B Biol Sci 274(1608):303–313. https://doi.org/10.1098/rspb.2006.3721

Korotkova N, Barthlott W (2009) On the thermogenesis of the Titan arum (Amorphophallus titanum). Plant Signal Behav 4(11):1096–1098

Kremen C, Williams NM, Aizen MA, Gemmill-Herren B, LeBuhn G, Minckley R et al (2007) Pollination and other ecosystem services produced by mobile organisms: a conceptual framework for the effects of land-use change. Ecol Lett 10(4):299–314

Laverty T (1980) The flower visiting behavior of bumblebees; learning and flower complexity. Can J Zool 58:1324–1335

Levin DA (1978) Pollinator behaviour and the breeding structure of plant populations

Linhart YB, Mendenhall JA (1977) Pollen dispersal by hawk moths. Biotropica 9:143

Makino TT, Ohashi K, Sakai S (2007) How do floral display size and the density of surrounding flowers influence the likelihood of bumble bee revisitation to a plant? Funct Ecol 21(1):87–95

Manning A (1956) Some aspects of the foraging behaviour of bumble-bees. Behaviour 9(1):164–200

Martin SJ (2001) Varroa destructor reproduction during the winter in *Apis mellifera* colonies in UK. Exp Appl Acarol 25(4):321–325

Martins AC, Melo GA, Renner SS (2014) The corbiculate bees arose from New World oil-collecting bees: implications for the origin of pollen baskets. Mol Phylogenet Evol 80:88–94

Melidonis CA, Peter CI (2015) Diurnal pollination, primarily by a single species of rodent, documented in Protea foliosa using modified camera traps. S Afr J Bot 97:9–15

Menzel R (1967) Untersuchungen zum Erlernen von Spektralfarben durch die Honigbiene (Apis mellifera). Z Vergl Physiol 56:22

Menzel R, Shmida AVI (1993) The ecology of flower colours and the natural colour vision of insect pollinators: the Israeli flora as a study case. Biol Rev 68(1):81–120

Munyuli T (2010) Pollinator biodiversity and economic value of pollination services in Uganda. (Doctoral dissertation, Makerere University)

Neuschulz EL, Mueller T, Schleuning M, Böhning-Gaese K (2016) Pollination and seed dispersal are the most threatened processes of plant regeneration. Sci Rep 6(1):29839

Nicolson SW, Thornburg RW (2007) Nectar chemistry. In: Nectaries and nectar. Springer Netherlands, Dordrecht, pp 215–264

Oliveira PE, Gibbs PE, Barbosa AA (2004) Moth pollination of woody species in the Cerrados of Central Brazil: a case of so much owed to so few? Plant Syst Evol 245(1–2):41–54

Ollerton J (2017) Pollinator diversity: distribution, ecological function, and conservation. Annu Rev Ecol Evol Syst 48:353–376. https://doi.org/10.1146/annurev-ecolsys-110316-022919

Ollerton J, Winfree R, Tarrant S (2011) How many flowering plants are pollinated by animals? Oikos 120(3):321–326

Ortega-Olivencia A, Rodriguez-Riano T, Valtuena FJ, Lopez J, Devesa JA (2005) First confirmation of a native bird-pollinated plant in Europe. Oikos 110:578

Pimental D (2000) Environmental and economic costs of nonindigenous species in the United States. Bioscience 50:53–65

Potts SG, Biesmeijer JC, Kremen C, Neumann P, Schweiger O, Kunin WE (2010) Global pollinator declines: trends, impacts and drivers. Trends Ecol Evol 25(6):345–353

Potts SG, Imperatriz-Fonseca VL, Ngo HT, Biesmeijer JC, Breeze TD, Dicks LV et al. (2016) IPBES (2016): summary for policymakers of the assessment report of the intergovernmental science-policy platform on biodiversity and ecosystem services on pollinators, pollination and food production. Secretariat of the Intergovernmental Science-Policy Platform on Biodiversity and Ecosystem Services, Bonn, Germany, 36

Primante C (2015) The role of floral traits in structuring plant-pollinator interactions. Universitat Autònoma de Barcelona

Proctor M, Yeo P (1973) The pollination of flowers

Proctor M, Yeo P, Lack A (1996a) The natural history of pollination. Timber Press

Proctor M, Yeo P, Lack A (1996b) The natural history of pollination. HarperCollins Publishers

Pyke GH (1979) Optimal foraging in bumblebees: rule of movement between flowers within inflorescences. Anim Behav 27:1167–1181

Raguso RA (2008) Wake up and smell the roses: the ecology and evolution of floral scent. Annu Rev Ecol Evol Syst 39:549–569

Rostás M, Bollmann F, Saville D, Riedel M (2018) Ants contribute to pollination but not to reproduction in a rare calcareous grassland forb. PeerJ 6:e4369. https://doi.org/10.7717/peerj.4369

Sekercioglu CH (2006) Increasing awareness of avian ecological function. Trends Ecol Evol 21(8): 464–471

Simonson S, Chong GW, Stohlgren TJ (2000) Exotic plant species and native pollinators: implications for native biodiversity. http://abstracts.allenpress.com/esa/web2000/abstracts/SAR-3-44- 11.html

Smitley D (2019) Elsner, Michigan State University Extension; Joy N. Landis, Michigan State University IPM; Paula M. Shrewsbury, University of Maryland Department of Entomology; Daniel A. Herms, The Davey Tree Expert Company, Kent, Ohio; and Cristi L Palmer, IR-4 Project Rutgers University

Triplehorn CA, Johnson NF (2005) Borror and DeLong's introduction to the study of insects, 7th edn. Brooks/Cole, Thomson Learning, Belmont

Valenta K, Nevo O, Martel C, Chapman CA (2017) Plant attractants: integrating insights from pollination and seed dispersal ecology. Evol Ecol 31:249–267

Valido A, Rodríguez-Rodríguez MC, Jordano P (2019) Diversity and importance of animal pollinators. In: Encyclopedia of ecology, 2nd edn. Academic, pp 352–359

Vanbergen AJ, Initiative T I P (2013) Threats to an ecosystem service: pressures on pollinators. Front Ecol Environ 11(5):251–259

VanEngelsdorp D, Evans JD, Saegerman C, Mullin C, Haubruge E, Nguyen BK et al (2009) Colony collapse disorder: a descriptive study. PLoS One 4(8):e6481

Von Frisch K (1967) The dance and orienation of bees (tr. Chadwick LE)

Waddington KD, Holden LR (1979) Optimal foraging: on flower selection by bees. Am Nat 114(2): 179–196

Whitehead DR (1969) Wind pollination in the angiosperms: evolutionary and environmental considerations. Evolution 23:28–35

Wiens D, Rourke JP, Casper BB, Rickart EA, LaPine TR, Peterson CJ, Channing A (1983) Nonflying mammal pollination of southern African Proteas: a non-coevolved system. Ann Mo Bot Gard 70:1–31

Wojcik V (2021) Pollinators: their evolution, ecology, management, and conservation. In: Arthropods: are they beneficial for mankind. IntechOpen, pp 1–22

Wood GW (1965) Note on the activity of native pollinators in relation to the bloom period of lowbush blueberry. J Econ Entomol 58(4):777–777

Woodcock BA, Bullock JM, Shore RF, Heard MS, Pereira MG, Redhead J et al (2017) Country-specific effects of neonicotinoid pesticides on honey bees and wild bees. Science 356(6345): 1393–1395

Wright SI, Kalisz S, Slotte T (2013) Evolutionary consequences of self-fertilization in plants. Proc Biol Sci 280(1760):20130133. https://doi.org/10.1098/rspb.2013.0133

Behavior and Ecology of Pollinators: Focus on Honeybees

4

M. Gopi Prasad, A. K. Chakravarthy, and D. Devika Rani

Abstract

Global hunger and malnutrition demand high crop yields. In this scenario, pollination, especially by honeybees, play crucial role in agriculture and horticultural crops. Modern crop cultivation enhances yields by nearly threefold. Simultaneously, hybrids and high-yielding varieties will be prone to the attack of arthropod pests and diseases, and require thorough chemical applications, harming the ecosystem balancers like honeybees. Conservation of these yield-promoting bees is a prime concern worldwide. For better conservation practices, it is important to study every aspect of them. In this chapter evolution of bees, communication, bee flora, foraging behavior, economic value of pollinators, reasons for the decline of pollinators, and their conservation are highlighted.

Keywords

Honeybees · Behavior · Ecology · Pollination

M. G. Prasad
Department of Entomology, Sri Kinjarapu Yerramnaidu College of Agricultural Sciences (SKYCAS), Srikakulam, Andra Pradesh, India

A. K. Chakravarthy
Society for Science and Technology Applications (SSTA), Bangalore, Karnataka, India

D. D. Rani (✉)
Department of Entomology, Sri Kinjarapu Yerramnaidu College of Agricultural Sciences (SKYCAS), Srikakulam, Andra Pradesh, India

Department of Entomology, University of Agricultural Sciences, GKVK, Bangalore, Karnataka, India

A. K. Chakravarthy (ed.), *The Role of Pollinators in Tropical Landscapes*,
https://doi.org/10.1007/978-981-95-9139-8_4

1 Introduction

A total of 2500 species of plants are currently being utilized as food by the world population. The world can feed more than 80% of its population with just 250 crops, of which only 10–12 are widely grown (Thurston 1969; Khan et al. 2021). In the process of developing fruits and seeds, pollination is essential (Bommarco et al. 2013; Aslan et al. 2016). Indeed, pollination is the first step, followed by fertilization, which is an important step for seed development. The pollination process transfers pollen from the anther (male) to the stigma (female) of the same or another flower. Abiotic and biotic processes are involved in transforming anthers into the stigma. Wind, water, and gravity are all forms of abiotic pollination. The pollination of biological plants is affected by animals. On Earth, 3,52,000 plant species are flowering type; out of these, nearly 3,08,000 plant species (87.5%) are pollinated by animals, including insects, birds, bats, etc. (Kumar et al. 2022). Certain flowering plants reproduce without pollen vectors, using mechanisms such as vegetative reproduction, apomixis, or automatic selfing. About 40 million decades before plant-pollinator relationship co-evolved by developing adaptations like size, color, odor, nectar, pollen, resins, oils, nutrient contents and materials for building nests, and morphology, positioning, and development of the floral parts (Kevan 1999; Percival 1965; Baker and Hurd Jr. 1968; Faegri and van der Pijl 1971; Leppik 1972; Tepedino 1979).

Crop acreage dependent on pollinators has increased by over 300% in the last 50 years. Animal pollination is required for many important crop species. Animal pollination increases the quantity and quality of fruit production. Important pollinators include honeybees, solitary bees and bumblebees, pollen wasps, ants, fly species such as bee flies, hoverflies, and others. Of these, honeybees are the species of insects that give honey and food for human beings. Honey contains enzymes, vitamins, minerals, water, and pinocembrin (Baskar et al. 2017). A typical bee colony population can make four million trips each year. Each trip covers nearly 100 flowers, explaining that the role-playing by honeybees is maintaining plant diversity through pollination. Bees are incredibly vital to humans; as Einstein comments, "If the bees disappear from the earth's surface, man will have just 4 years left to survive." (Prasad and Rani 2023) These biological transporters are inextricably connected to humans through pollination services, crop production, honey production and other beekeeping products, ecosystem function, food security, and as bio-indicators of pollution (Mauricio da Rocha et al. 2010).

2 Entomophily

Pollen grains are transmitted to mature ovaries by insects such as moths, butterflies, wasps, bees, beetles, etc (Table 4.1).

Beetles Coleoptera is the largest insect order, pollinating certain primitive plant species. The many hairs on the small tongue and maxillae of foraging beetles aid

Table 4.1 Pollination types

Pollination type	Insects	Pollinated crops
Myophily	Flies	Orchids, mango, other fruit crops
Cantharophily	Beetles	Magnolias, pond lilies, spicebush, goldenrods, Spirea, other flower crops
Phalaenophily	Moths	Morning glory, tobacco, yucca, and gardenia
Psychophily	Butterflies	Floricultural crops
Sphecophily	Wasps	Fig, goldenrod
Myrmecophily	Ants	Small's stonecrop, alpine nailwort and Cascade knotweed
Melittophily	Bees	Many horticultural and agricultural crops and forest plants

Fig. 4.1 Beetle pollination

pollination. These coleopterans are weak, with poorly developed sensory organs and horny, robust elytra wings. Beetle-pollinated flowers are flat and disc-shaped, in accordance with beetle structure. Beetles consume floral parts, so flowers possess defensive structures; well, beetles are called destructive pollinators. Beetle pollination is called Cantherophily (Mishra 1995; Wojcik 2021) (Fig. 4.1).

Flies Dipterans are important in areas where specialized hymenopterans are absent, i.e., Arctic regions. Dipterans flies are nectar feeders and contain sucking and lapping (or) sponging type of mouthparts. Usually, they prefer yellow to other colors, having ample copious amounts of nectar, and they also show flower constancy. Pollination aided by flies is called myophily, and in the case of carrion or dung seekers, it is Sapromyophily. In orchid and mango plantations, house flies perform high rates of pollination (Mishra 1995; Wojcik 2021).

Butterflies Butterfly pollination is termed Psychophily. To provide easy landing, flowers develop into compound or disc-shaped structures. Butterflies strongly prefer plants belonging to the family Asteraceae. While sucking nectar from flowers, adult Lepidopterans use long coiled proboscis, pollen attached to the body, and the proboscis is transferred to another flower. The Hawk moth is the most important pollinator in Lepidoptera. Phalaenophily is the pollination aided by moths (Mishra 1995; Wojcik 2021) (Fig. 4.2).

Fig. 4.2 Butterfly pollination

Fig. 4.3 Ant pollination

Ant Pollination (Myrmecophily) These are minor pollinators, mostly pollinating the flowers in clumps. Ant-pollinated plants occur in arid or alpine environments. Small's stonecrop, alpine nailwort, and Cascade knotweed are important ant-pollinated plants (Mishra 1995; Wojcik 2021) (Fig. 4.3).

Bee Pollination This is also called Melittophily. Although melittophily predominantly refers to bees, pollen wasps are attracted to flowers of similar morphological and anatomical characteristics. Bees pollinate a diverse range of plant species. Bees gained importance due to their ability to pollinate the typical floral forms (Peas) and are found in various habitats. Bees are the most important pollinators, visiting more than 90% of the leading 107 global crop types (Mishra 1995; Wojcik 2021; Klein et al. 2007) (Fig. 4.4).

Fig. 4.4 Wasp pollination

3 Honeybees

3.1 Evolution of Honeybees

Although fossil evidence is limited, bees most likely appeared in the Cretaceous era, 146–74 million years ago, at the same time as flowering plants. *Trigona prisca*, a stingless bee discovered in the Upper Cretaceous of New Jersey, USA, is the oldest known fossil bee (Gupta 2014). Bee evolution is intimately related to a shift in feeding from insect prey to pollen and nectar collected from angiosperm blossoms (Wojcik 2021; Linksvayer et al. 2012; Gupta 2014).

Bees are considered to have developed from hunting wasps that became vegetarians after developing a liking for honey. The availability of nectar and pollen caused certain wasps' feeding patterns to transition from a predatory life on larvae (Sphecoidea) to one of nectar and pollen collection. Over 120 million years ago, the honeybee evolved morphologies designed for pollen and nectar collection, such as dense hairs, pollen baskets, long tongues, and nests to store supplies. It has no resemblance to modern *Trigona* spp.

The *Apis* genus is considered to have originated in India. Ten species of honeybees in the genus *Apis* are well known (Engel 1999; Arias and Sheppard 2005). Of them, six species are found in India: *A. cerana indica*, *A. florae*, *A. laboriosa*, *A. dorsata*, *A. andreniforms*, and *A.koschevnikovi*. *Apis mellifera* L. is considered to have existed on Earth for 10–20 million years earlier than humans. Honeybees are the earliest living creatures that have survived since the Neolithic Era. This honeybee species is assumed to have evolved in southern Asia and spread to other countries. Another hypothesis states that the origin of western honeybees is outside of Asia (Han et al. 2012). It was brought to North America by Dutch colonists in 1638. Because bees visit a wide variety of host plants and can preserve heat by aggregating, they have spread widely and presently occur on planet Earth. Globally, beekeeping has grown by about 50% over the past five decades

(Valido et al. 2019). The migration of bee hives (e.g., *Apis mellifera* hives) seasonally introduces many bees into the cultivating and wild areas for pollination and honey gathering (Aizen and Harder 2009; Mu et al. 2014).

3.2 Biology

Honeybees undergo complete metamorphosis. Bees pass through four developmental stages, i.e., egg, grub, pupa, and adult.

Egg Even though the hive's queen and workers are both females, only the queen produces eggs. It lays fertilized and unfertilized eggs vertically at the bottom of the comb cell @ one egg per cell. Both ends are rounded, and one pole seems larger than the other. Grubs hatch on third or fourth day from eggs. Fertilized eggs turn into females (workers and queen), whereas unfertilized eggs turn into males (drones). A typical colony's brood (eggs and young) includes 5000–7000 eggs (Fig. 4.5).

Grub Drone and bee-worker larvae are given royal jelly for the first three to four days after hatching. After that, pollen and nectar are fed to the larvae. Royal jelly is continually given to a queen larva until the eighth day, when the queen cells are further stuffed with royal jelly and sealed. Before pupating, the worker bee larvae consume 1300 times daily and attain 300 times their original weight (Malik et al. 2020; Erickson Jr et al. 1986). In a larva, fat makes up about three-fourths of its weight. Before pupating, the capped royal queen will continue to eat. Larvae are initially loop-shaped, but when the cell cap approaches, they become stretched in the cell with their heads upright. Larvae transition into the pupal stage after four moults. Between 7000 and 11,000 larvae will be in the brood.

Pupa Larvae consume much nourishment before entering the pupa and develop and build a silky cocoon around itself. When the larva transforms into a pupa, worker bees restrict cell entry by capping, flat for workers and concave for drones. At this phase, it is easy to differentiate between the head, thorax, and abdomen. In sealed

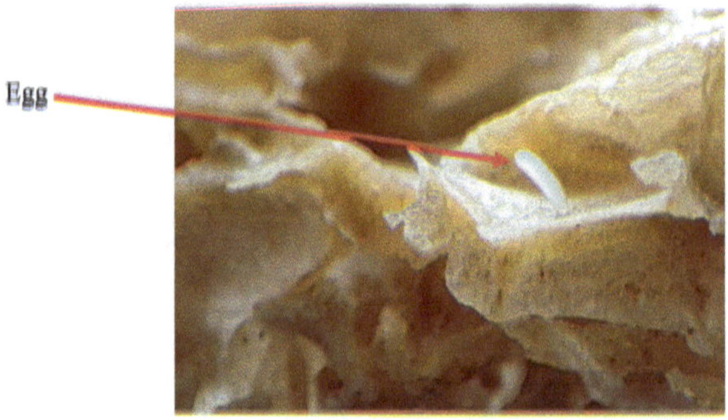

Fig. 4.5 Honeybee egg

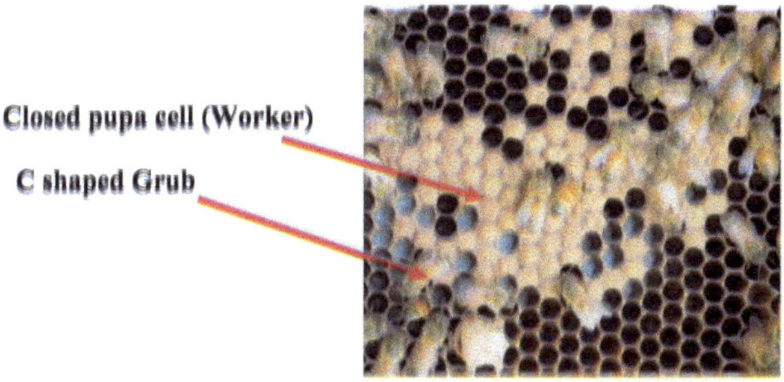

Fig. 4.6 Bee brood

Fig. 4.7 Honeybee life cycle

compartments, the normal brood keeps 16,000–24,000 pupae. The compound eyes of the drone pupa meet over the head, with the worker and queen being far apart (Fig. 4.6).

Adult Insects undergo metamorphosis to become adults and finally emerge from sealed cells as adults by gnawing. The proportionately longer abdomen of a queen makes her wings appear smaller than a worker's (Fig. 4.7).

Wax Glands Four pairs of wax glands are found on the ventral side of abdominal segments 4–7 in worker bees and become active between 14 and 18 days old. On their sterna, underneath these glands, are wax mirrors or plates. The basitarsus collects the secretions of these glands, from where they settle on the abdomen. Afterwards, the mandibles remove the secretions, depositing them into the comb construction.

Sting The ovipositor, a modified egg-laying device, causes honeybee stings. Strong muscles are linked to the poison gland surrounding the sting's three lengthy stylets, each containing two lancets at the end. Workers and the queen have stings; drones do not. While the queen's sting lacks curved barbs, her sting's tip is densely barbed. When a worker bee pierces an adversary while attacking in self-defense or hive defense, she typically loses her sting; shortly after, she dies because of the substantial loss of vital tissues and organs. The worker cannot restock the venom after emptying the poison sac if she retains her sting after impaling. Venom accumulates in the glands once the bee is 15 days old. Those who exhibit potentially fatal hypersensitivity to bee venom are cautioned not to approach bee boxes.

Stingless Bee *Trigona prisca*, a stingless bee, is the earliest fossilized bee species. It cannot be distinguished from the current *Trigona*. The biggest genus of stingless bees is *Trigona*. There are now 32 stingless *Trigona* species documented in Thailand under the genus *Trigona* and *Hypotrigona*,

Scent Glands The final abdominal segments include a thin membrane containing these glands. By bending the segments of these glands, the bee presses the glands and exhales the aroma outside. These aromas come from specific cells' metabolic waste.

4 Caste System

Castration is a form of reproduction in which the development of the reproductive organs plays a major role in separating the individuals.

Queen The colony's mother is the mature female insect with fully developed ovaries. The amount of food (royal jelly), juvenile hormones, and water larvae consumed is thought to affect caste. The entire process of development from the egg takes 16 days. The pheromones produced by the queen preserve the hive's integrity. Queen performs a series of mating flights after reaching maturity. Throughout that flight time, a queen mates with a drone, collecting 5–12 billion dormant sperm from a single partner, which she stores in a pouch-like sac known as the spermatheca. In certain situations, polyandry may occur (a queen mates with many drones), enhancing the colony's diversity. While over 5-year life spans have been observed, queens typically survive for 1–2 years. A queen accurately deposits unfertilized eggs in drone cells and fertilized eggs in worker cells by determining the sex of offspring. Her entire life is spent working as an egg-laying machine in a chemical factory (Gadagkar 1996) (Fig. 4.8).

Fig. 4.8 Queen bee

Fig. 4.9 Drone bee

Fig. 4.10 Worker bee

Drones Insect males grow testicles that are fully developed. Drones account for 15% of the hive's population (Malik et al. 2020). Their large compound eyes, about twice as large as the workers' and queen's, make them identifiable. Their physique falls in the middle of other castes. Their life span is approximately 2 months, and development from egg to adult takes 24 days. Worker bees can develop functional ovaries and lay drone-producing unfertilized eggs. These drones will be fully grown if raised in drone-sized cells, but if raised in worker cells, they will be considerably smaller. Drones are all sexually capable. Drones' sole goal is to mate with fresh queen. Bees copulate in midflight 200–300 feet above the colony, away from the hive (Malik et al. 2020) (Fig. 4.9).

Workers/Mason Bees The female insects' workers grow from undeveloped ovaries. Because of a restricted diet that causes hormonal imbalances, workers cannot reproduce. Occasionally, a few selected workers mature into ideal females and lay (fertilized) eggs through which drones form. Certain situations, such as colonies without queens, require workers to produce diploid eggs to create the queen parthenogenetically. They take up most of the population. Compared to drones and queens, they are smaller. To finish development from the egg, it takes 21 days. Bee workers are the comb's labor force. They clean the comb, remove dead brood, feed and care for growing larvae, attend to the queen, collect nectar and pollen, regulate the hive's humidity and temperature, make beeswax, and supervise the hive (Gadagkar 1996). Worker bees live for 4–6 months in winter but only for 28–35 days in the spring and summer. Mature *A. mellifera* colonies can range from 20,000 to 70,000 worker bees (Lee and Winston 1987) (Fig. 4.10).

5 Types of Honeybees

Ten species of honeybees from the genus *Apis* are now recognized as existing (Engel 1999). Nuclear DNA and mitochondrial (mtDNA) marker-based phylogenetic studies strongly support categorizing them into three groups: cavity-nesting bees

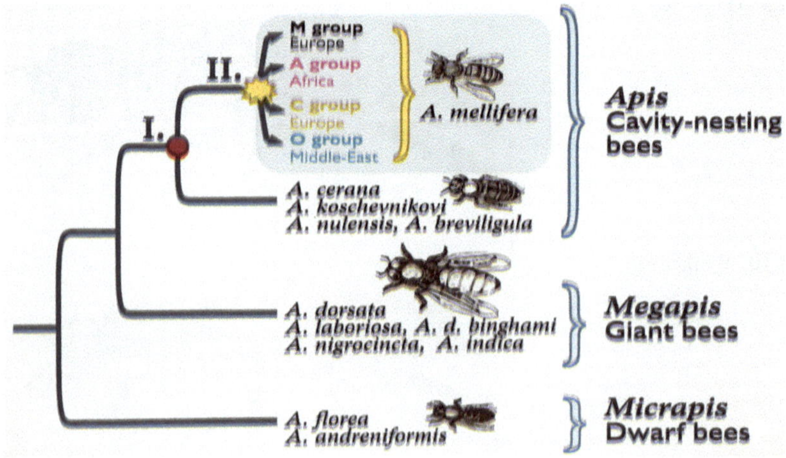

Fig. 4.11 Phylogenetic relationships in honeybees

(*A. mellifera*, *A. cerana*, *A. koschevnikovi*, *A. nulensis*, and *A. breviligula*), giant bees (*A. dorsata*, *A. laboriosa*, *A. binghami*, *A. nigrocincta*, and *A. indica*), and open nesting the dwarf bees (*A. florea* and *A. andreniformis*) (Arias and Sheppard 2005; Raffiudin and Crozier 2007; Tanaka et al. 2001; Willis et al. 1992) (Fig. 4.11).

5.1 The Wild Bee (*Apis dorsata*)

Large-sized bees (Gupta 2014) found in sub-mountainous areas of India up to 2700 metres are *Apis dorsata*. Pakistan, Sri Lanka, Malaysia, Indo-China, the Philippines, China, and Indonesia constitute part of the geographical distribution. A "bee tree" is typical when 50–100 colonies are concentrated on a Tree. They often shift the colony's location. The phenomenon of returning to a similar or the same niche generation after generation is attributed to swarming leaves to aroma behind. They construct a single, 1–2 m long comb in an open area. Rock bees are difficult to grow and have a fierce attitude. Behavior and biology of *A. dorsata* are comparable to those of domesticated *Apis* species. If enough moonlight is available, *A. dorsata* may feed at night. Each comb produces roughly 36 kilograms of honey annually.

5.2 Little Bee (*Apis florae*)

This species is dispersed in India, Indonesia, Malaysia, the Philippines, Pakistan, Iran, Sri Lanka, and the Indo-China area in the East and up to Iran and Oman in the west. They create solitary vertical combs similar to the size of a palm. They nest in the open among branches, shrubs, hedges, caverns, empty cases, etc. Drone bees are 1.55 times bigger than worker bees. Very often, swarms of these bees occur.

Between 300 and 900 grams of honey are produced annually by each hive. Their honey is thin in consistency. Among the four recognized *Apis* species, they are the smallest, even smaller than the Indian bee. Plains and slopes less than 300 ams will be suitable for nesting. When disruptions occur, they abandon the comb, leaving the honey, pollen, and brood collections to comb; they return to gather any leftover collections.

5.3 The Indian Hive Bee (*Apis cerana indica*)

They are a subdued species that produces 20–25 kg of honey annually in Kashmir but only 3–5 kg annually in the lowlands. They built several multiple combs. Their place of origin is from Asia or India. This species is found in Pakistan, Sri Lanka, Malaysia, the Philippines, Indo-China, China, the USSR, Japan, and Indonesia. It flourishes up to 2500 m above sea level; they surpass *Apis florae* in size but falls short of *Apis mellifera*. They undergo swarming five to six times a year. *A. mellifera*, imported decades ago, has currently replaced/supplanted the native domesticated *A. cerana indica* in China and Japan. *A. cerana indica* based bee keeping hardly exists in Haryana, Punjab, and the plains of Uttar Pradesh. These bees have a pleasant temperament and react to smoking. Since its behaviors are modest, a shortage of flora rapidly manifests as absconding.

5.4 European Bee/Italian Bee (*Apis mellifera*)

European bees behave similarly to the Indian bees, which construct parallel combs in parallel design in tree hollows, along walls, or in shaded areas. They are widespread throughout Europe and have reached every other nation. Apart from *A. dorsata*, they are bigger than any other honeybees. The colony's average annual production is 25–40 kg. They frequently swarm and abscond. *A. mellifera mellifera*, *Apis mellifera linguistica*, *Apis mellifera carnica*, and *Apis mellifera caucasia* are four significant European races for Beekeeping. *Apis mellifera* was found to have the finest language among these races and is widespread throughout other nations. It keeps the efficient queen, does not swarm as much, and is an excellent honey gatherer with a calm disposition. This race shows superior performance in India compared to *Apis cerana indica*.

6 Distribution

These bees reside in chilly, harsh winters and tropical locations where winters never happen, and summer temperatures are typically greater. The capacity of bees to adapt to various climatic conditions and environments has proved genuinely amazing. During their evolutionary history, honeybees evolved into different breeds due to unique climatic circumstances and characteristics of nectariferous plants. *Apis mellifera* is found in Europe, Africa, northwestern Asia, etc., and sporadically in America and Australia (Ruttner 1988, 1992; Ruttner and Kauhausen 1985; Sheppard and Meixner 2003; Kerr 1957; Sheppard 1989; Engel 1999; Moritz et al. 2005). Except for *Apis florea*, also known from Jordan, the eastern Arabian Peninsula, and northern Africa, the remaining Honeybees were primarily limited to Asia (Radloff et al. 2011).

7 Communication

The language and communication of the honeybee, which has been studied in great detail, can serve as an excellent example of nonhuman language (von Frisch 1967, 1971; Lindauer 1975, 1981; Seeley 1982, 1992). Worker bees that are experienced forage for pollen and nectar, and they communicate via performing dances between nest individuals, which boosts foraging effectiveness. Bees perform dances to inform colony individuals of the locations of food sources (Wray et al. 2012). These dances convey information regarding food source direction, distance from the nest, quantity, etc., in general; these are called bee languages (Gadagkar 1996). It was described as the only known form of "'symbolic communication"' in insects (Duangphakdee et al. 2011). Without sign-use, there would be no coordination processes. Many honeybee species perform these dances on the vertical surface of the comb, i.e., in the dark nest (*A. cerana indica* and *A. mellifera*). Hence, the bee uses gravity and not the Sun as the reference point while dancing.

Round Dance Bees perform this dance to convey message about the presence of pollen and nectar, including its profitability. Profitability is defined by its nectar energy content, mainly sugars (Munz 2005; Riley et al. 2005). A bee forager resorts to round dance when the nectar/pollen source is less than 50 meters from hive (Wolf et al. 2016; Bukovac et al. 2017).

Sickle Dance It is performed when a flower or nectar source is 50–150 m from the hive. This dance is performed in a crescent and a transitional shape or outline (Riley et al. 2005; Seeley 2009).

Waggle Dance Bees perform the waggle dance if food is higher than 100 m from the nest. The behavior conveys the scope and direction of food source. The bee forager performing the cargo dance continues dancing for a short period and then returns to the point of departure in a circuit, forming a full figure-eight circuit. When running down the dance's straights, the shaking body/body vibrating leads to a waggle movement (13 times per second); the bee's body, particularly the abdomen, moves powerfully side by side to each other. The angle of the waggling phase of dance with regard to the vertical line of the nest represents the direction at which the nectar source related to sun azimuth is located. Suppose food is along the direction of the Sun. If the food source is in the direction of the sun, the bee represents a waggle run pointing staraight up or, if food direction against the Sun, with a waggle run pointing straight down or else food located 60° to the Sun with a waggle run 60° from the right of the vertical (Gadagkar 1996; Singla 2020). The bee conveys a reacting alert formed by 250–300 hertz or cycles per second, at a low sound frequency (Abrol 2013).

Bees, like *Apis mellifera* gets rewarded through dance performed on the vertical surface of the bee hive inside the nest cavity. The orientation of the waggle phase relative to vertical on the comb correlates with the direction of the resource relative to the solar azimuth on departure from the hive; hence, the angle of the waggle run relative to vertical is considered a signal of direction for *A. mellifera* (Von Frisch 1967). The waggle dance is performed more than 100 times (Bukovac et al. 2017; Kumari et al. 2020). However, how do these bees gain the dance information? Several signs, like landmarks, can lead to the direction in which food is located (Barron and Plath 2017). Sun's location, skylight (Le Moel and Wystrach 2020), polarization pattern, and magnetic fields are detected. Bees can prefer one mechanism over another, based on the environmental situations. The dancer delivers a food sample via trophallaxis because these dances appear to represent quantitative information about the location of food sources, called "symbolic communication" (Couvillon 2012; Dyer 2002; Preece and Beekman 2014; Von Frisch 1967). Wenner discovered that the food distance from the hive correlated to the sounds of the dancing bee (Donovan 2000).

8 Bee Flora

Many agricultural and horticultural plants serve as food resources for bees (McGregor 1976). Over 40 plant species are visited by *A. florea. A. dorsata* depends on fewer plants than *A. florea. A. dorsata* uses only 38 species (Maksong 2008; Suwannapong et al. 2011). Thailand has over 68 species of plants for *A. cerana*. However, more than 100 plant species in the USA are generalist forager *Apis mellifera* bee plants due to the bee's behavior and morphology (Giannini et al. 2015). The pollination effectiveness is not the same for all crops, and it varies enormously from species to species and system (Aslan et al. 2016). Population size, climatic conditions, pollen quality and quantity, nectar, etc., drive efficiency. Effectiveness relates the overall visits of a pollinator to the seed set (Motten et al. 1981). Honeybees visit many crops for their food. Some bee foraging crops are provided below (Fig. 4.12).

9 Flower Visitation and Foraging Behavior

Foragers are worker bees and can search, collect, and carry food resources like pollen, nectar, and propolis (Marcucci 1995; Bankova et al. 1983, 2000) to their nests. This activity is called foraging. Honeybees have advanced foraging coordination and communication (von Frisch 1971; Suwannapong et al. 2012). Apis dorsata

Fig. 4.12 Foraging bee

recorded the highest foraging rate among all species, averaging 19.63 flower visits per minute. (Sujatha et al. 2024)

This reward, due to their olfactory abilities, includes social communication within the hive, olfactory recognition when foraging for food, distinguishing between different types of flowers, and determining whether a flower is carrying pollen or nectar from several meters away using nearly 170 olfactory receptors (Baskar et al. 2017). Bees carrying only pollen are pollen foragers, only nectar are nectar foragers, and both pollen and nectar are generalist foragers. The olfactory ability of honeybees is approximately 40 times higher than that of humans. Variations in floral features are intimately linked to changes in insect/pollinator behavior. They have an amazing ability to learn, memorize things, travel great distances, and search for food. Many insects, particularly Apidae (larvae), beetles, flies, thrips, and butterflies, rely on pollen for survival. Pollen contains 16–30% protein, 1–7% starch, 0–15% sugar, and 3–10% fat (Anonymous 2002). Weather, quality and quantity of available food, and distance from the food source all influence foraging behavior and determine profitability. According to Barker (1977), some carbohydrates in pollen and pollen substitutes may prove toxic.

Pollinators rely heavily on nectar as a source of nourishment and energy. It is a compound of sugars (Anonymous 2000), proteins (Baker and Baker 1975; Anonymous 2000), amino acids (House 1974), lipids (Downer 1978; Cobert et al. 1979), alkaloids (Baker and Baker 1975), antioxidants (Anonymous 2000), vitamins, organic acids, and inorganic materials such as minerals (House 1974; Waddington and Holden 1979) are all present. They all have pollination implications (Baker and Baker 1973; Baker 1978). Certain sugars are poisonous to insects; for example, galactose, lactose, and raffinose are toxic to honeybees but not to other insects, such as Diptera (Sotavalta et al. 1962).

Pollinators, particularly bees, are significantly impacted by floral shape, outline form, and length (Leppik 1956; Kevan and Baker 1983), color, scent, nectar, pollen, and other rewards. As a result, the efficacy of pollination is determined by floral features. Blooming inflorescences with vertical orientation are beneficial to bees since the amount of nectar decreases and the sugar content increases in higher inflorescences (Hocking 1968), and bees begin feeding at the bottom and work their way up (Pyke 1979; Waddington and Heinrich 1979).

Bees benefit from floral constancy because they learn to harvest nectar and pollen while visiting the same flower type (Grant 1950). The consistency of bees may be determined by examining the sorts of pollen on their bodies after foraging. *Apis cerana indica* has better floral fidelity, carrying a 5200 pollen load (Choudary 1978) than other foragers feeding on alfalfa, such as *A. florae*, *A. dorsata*, and *A. mellifera* (Dhaliwal and Atwal 1986). A forager of *A. mellifera* carries as much as 70–80 mg of nectar per load (Partap and Partap 1997). The color and aroma recognize the reward (Frisch 1967), and learning begins 2 s before feeding and fades quickly (Menzel 1967).

When foraging in a dense plant area with copious nectar, bees turn around more frequently (Pyke 1979). The ability of honeybees and bumblebees to return to the same nest after foraging is due to the orientation of land marking. Rock bees may forage at night on full moon days with enough light (Gupta 2014). Bees strongly prefer blooms that provide much food and visit them until this food is depleted. This behavior is referred to as floral fidelity.

10 Influence of Environmental Factors

Altitude The initiation of foraging activity of bees, *Apis mellifera* and *Apis cerana*, was delayed and ceased earlier with increasing altitude (Verma and Dulta 1986).

Temperature The hive bee, *Apis cerana*, forages at a lower temperature than *Apis mellifera* (Verma 1968). Foraging of bees is also greatly affected by changing temperatures, e.g., no foraging occurs below 8 °C, some activity between 8 and 16 °C, optimum activity between 16 and 32 °C, and reduced foraging above 32 °C (Roberts 1979; Anonymous 2003), but under certain circumstances, foraging continues up to 42–48 °C (Lundie 1925). Bees keep their flight mechanism at flight temperature by maintaining thoracic temperature at 32 °C (Heinrich 1979).

Light Honeybees forage when light intensity is about 500 lux or more; below that, their activity reduces and ceases at 10 lux (Levchenko 1961). In the morning, honeybees work at a lower light intensity than when they cease in the evening (Schricker 1965).

Wind Wind velocity of 24–34 km/hr. adversely affects bees' foraging behavior (Lundie 1925; Verma 1990).

Rains Heavy or continuous rains hinder the foraging efficiency of pollinators. Sunny days with limited wind and moderate temperatures are more convenient for foraging (Tandon et al. 2016).

Extreme Climatic Conditions Crops and pollinators suffer detrimental effects when any situation deviates from the normal range. Elevated temperatures, sudden and continuous flash rain, and late frost may harm pollinators by reducing insect numbers (Tandon et al. 2016).

11 Symptoms of Bites

Honeybee stings are most typically encountered in the rural population in fields. Itching, swelling, burning, and pain are all experienced at the bite site. Apart from these broad symptoms, Acharya Sushruta reported Kashayee and Sthalika symptoms such as greyish black pustules or furuncles and fever similar to upadrava (Malik et al. 2020). The vast majority of human beings are allergic to bee venom. Massive honeybee envenomation occurs when more than 50 stings occur at the same time (Betten et al. 2006). In mammals, the fatal dosage is believed to be around 20 stings/kg body weight. Three allergic components in bee venom are Phospholipase A (allergic to many people), hyaluronidase (few people are sensitive), and mellitin (a small number are allergic) (Betten et al. 2006; Habermann 1972). Honeybee venom is analgesic as well as anti-inflammatory. Honeybee keepers allergic to bee venom should carry epinephrine or undergo immunotherapy with pure venom extract.

Uses It is used against pain syndrome, herniated nucleus pulpous, cervical disc protrusion, and progressive muscle atrophy.

12 The Economic Value of Pollinators

Shreds of evidence from history say that honeybees support life on this Earth in many ways, from maintaining natural diversity to bee venom used to cure arthritis. Due to their cumulative benefits, it takes work to estimate the monetary value of functions (Costanza et al. 1997).

1. Around 90% of angiosperms depend on animal pollinators for reproduction (Buchmann and Nabhan 1996). The value of pollination services offered by all insects (including honeybees) is estimated to be $40 billion annually (Pimental et al. 1997; Dar et al. 2013).
2. Jammu & Kashmir also faces a pollination problem in the horticulture industry; according to Tej Partap's (Honorable V.C., Sher-e-Kashmir University of Agricultural Sciences and Technology (SKUAST)) 2011 unpublished statistics, Jammu & Kashmir loses Rs. 1600 crores per year due to a lack of pollination in apple orchards (Dar et al. 2017).
3. The worldwide economic value of pollination services was anticipated $215 billion (in 2005 US dollars), or 9.5% of global food production value (Gallai et al. 2009; Vanbergen et al. 2012).

4. Fruits and vegetables that are not correctly pollinated will not develop fully and will not reach levels to meet international demand (Christmann 2022).
5. Honeybee and bumblebee pollinators are valued at 172 million Euros for outdoor crops (rapa, beans, tree, and soft fruits) and 30 million Euros for glasshouse crops (tomatoes and sweet peppers) (Carreck and Williams 1998).
6. The pollinators involved in plant production have an economic value of about 153 billion euros worldwide (Baskar et al. 2017).
7. Kevan et al. (1997) found that keeping one honeybee colony per field yielded approximately one additional seed per apple. Compared to an orchard without honeybees, these superior apples were expected to fetch $200/ha.
8. The decline in pollinator populations significantly increases preventable diseases such as ischemic heart disease, potentially resulting in 1.4 million additional deaths per year (Potts et al. 2016).
9. Non-*Apis* bee pollinators contribute to successfully producing $6 billion worth of agricultural goods in the USA (Morse and Calderone 2000).

13 Population Decline

1. At the global level, studies recorded a decline in pollinators, especially honeybees. Dioecious species suffer more pollination failure due to the absence of transport of pollinators (Wilcock and Neiland 2002). Raising human demand for bees illustrates the bee population's deterioration (Williams 1982; Biesmeijer et al. 2006; Fitzpatrick et al. 2007). A study by (VanEngelsdorp 2011) reported a far reduction in bee population than normal rates. According to a 2016 research conducted by the Intergovernmental Science-Policy Platform on Biodiversity and Ecosystem Services, around 40% of insect pollinators, primarily moths and butterflies, are on the edge of extinction. Six million honeybee colonies were documented in 1947, but this number has reduced by more than half, with an annual loss of 0.9%. The decline is due to not a single cause. There are many factors responsible bee number reduction, accepted by many researchers (Oldroyd 2007) consists of global warming, agricultural pesticide use, bee pests and diseases, habitat loss, and fragmentation, and colony collapse disorder is associated with pollinator loss (Kovac and Crailsheim 1988; Kearns et al. 1998; Kremen and Ricketts 2000; Martin-Hernandez et al. 2007).
2. Loss of habitat appears to be the primary cause of decrease in bees, as it is of biodiversity decline in general (Foley et al. 2005). Habitat fragmentation, a direct outcome of habitat loss, will influence surviving populations via genetic isolation and subsequent inbreeding (Zayed 2009) or simply the inability of smaller habitat islands to maintain healthy bee populations (e.g., Ellis et al. 2006). The conversion of forest and landscape regions into urban areas for human use leads to a shortage of flowering plants and, thus, to a shortage of basic supplies for bees (e.g., pollen, nectar, and oil supplies, as well as various nesting sites).

Deforestation causes a loss of nesting sites and flora for foraging, along with an increase in temperature, making it more difficult for bees to survive. Trampling and grazing of grasslands by livestock limit nesting opportunities on the ground and in cavities. Gess and Gess (1993) discovered that changes in the availability of nesting places in grasslands harm bees and wasps in African grasslands. Forest fires burn huge plant biomass, i.e., live plants and dead wood, which destroys the nesting sites for bees and oil-yielding plants, resulting in decreased provisions for bee development.

3. Pollution is suggested as the key element of pollinators' deterioration after habitat loss; its proportion is almost 25.8%. It is shown that inorganic selenium is more toxic for honeybees than organic forms, leading to high mortality rates and negative effects on growth. The effects of Flupyradifurone (FPF) on honeybees predicted that its effect is more upon worker bees performing foraging activities. The effect was more dominant in summer than in the spring season (Sponsler et al. 2019; Riaz et al. 2020). Due to global warming, the synchrony between plants' flowering time and pollinators' peak availability is lacking, leading to pollination failure. In some cases, the flowering period of a plant may not correspond to the peak or emergence of its corresponding pollinator (Peñuelas et al. 2002). Agricultural intensification, of which increased use of pesticides is an integral part, is often argued to be a key driver of pollinator decline (Baskar et al. 2017; Simon-Delso et al. 2014). The application of pesticides in agricultural fields directly affects bees, as low-fecundity organisms are highly susceptible to pesticides. Research states that bees feed on neonicotinoids, coumaphos could affect the bee brains, and they produce few offspring only (Baskar et al. 2017; Simon-Delso et al. 2014).
4. Species-specific pollination-dependent plants suffer pollination failure due to pollinator limitation due to environmental change or disturbance. (The best-documented example comes from South Africa. Here, the absence of a specialized oil-collecting Rediviva bee to pollinate the rare shrub *Ixianthes retzioides* was hypothesized to cause the observed pollination failure (Wilcock and Neiland 2002)).
5. Some sugars are even toxic to certain insects, e.g., galactose, lactose, and raffinose are toxic to honeybees (Baker 1997) but not to other insects, e.g., Diptera (Sotavalta et al. 1962) (Faheem et al. 2004).
6. In managed bee hives, there is a threat from the bee mites, *Acarapis woodi* and *Varroa jacobsoni.* These are Endoparasitic mites of adult bees that pierce soft intersegmental tissues and feed on haemolymph, resulting in the short life of bees. European *A. mellifera* across most of the world is nowadays plagued by the exotic parasitic mite, *Varroa destructor*, that transmits a range of emergent viral pathogens, usually leading to colony demise (Cox-Foster et al. 2007; Evans 2003; Gary Jr et al. 1990; Kevan et al. 1997; Nitharwal et al. 2021).
7. Several insect pests harm honeybees and cause significant damage. Some major pests are the Greater wax moth (*Gallaria melonella*), lesser wax moth (*Achroia*

Fig. 4.13 Greater wax moth damage

grisella), black ants (*Componotus compresessus*), bee-hunter wasps (*Palaras orientalis*), black bee-eater (*Dicrurus macrocercus*), etc., and mammals (bears, skunks, badgers, and baboons) (Erickson Jr et al. 1986).

8. In addition to these insect pests, honeybee brood is afflicted by various diseases. The diseases of the adult bees are less hazardous than those of the brood.
   A. Bacterial diseases: American foulbrood (*Bacillus larvae*) and European foulbrood (*Melissococcus plutonious*).
   B. Viral Diseases: Thai sac brood and sacbrood.
   C. Fungal diseases: Chalkbrood (*Ascosphaera apis*), Stone brood (*Aspergillus flavus*).
9. In recent years, a new threat, Colony collapse disorder (CCD), first identified in the USA in 2006, caused a reduction in pollination rates due to considerable damage to bee colonies. After the detection of CCD, concern toward the previous bee decline cases was raised due to the CCD (Fig. 4.13).

14 Conservation

In the event of widespread population declines among native and non-native pollinators, there will be a negative impact on food security and native communities. (Meffe 1998;; Biesmeijer et al. 2006; Potts et al. 2010). Conservation of honeybees and other pollinators is an important issue in the global context of agriculture and natural, sustainable productivity (Dar et al. 2013). Honeybees' locally adapted strains, subspecies, and ecotypes suffer less from elevated losses than non-native bees. Therefore, their conservation as a genetic resource for breeding disease and stress-resistant strains is essential (Gupta 2014).

Maintaining non-cropping lands as a model in an agri-ecosystem, as suggested by agri-environment schemes for conserving pollinators in agricultural landscapes, can naturally enhance the pollinator number (Morandin and Winston 2006; Nitharwal et al. 2021). To increase agricultural productivity and reduce the need for agrochemicals, ecological intensification involves increasing the intensity of ecosystem services in field and landscape management. Examples include biotic pest regulation, nutrient cycling, and pollination.

Agro-forestry; managing forests or home gardens; creating, restoring, or maintaining native wildflower habitats; and intercropping are expected to foster diverse pollinator communities and pollination. Pesticide poisoning can be reduced by considering the impact of the toxic nature of a particular chemical on pollinators while applying it in fields. Choose the one having a negligible effect on pollinators. Labeling should include the toxicity levels against pollinators on pesticide containers. Awareness camps should be conducted regarding the role of pollinators in crop diversity.

Relevant research and agricultural policy should be made to improve techniques for the conservation of pollinators. Promoting the conservation and management of naturally occurring insect pollinators is important for sustainable agricultural productivity in the long run. Lack of awareness at all levels, be it farmers, extension workers, and professionals at the policy and planning level, is one of the main problems in promoting managed pollination. Pollination must be addressed in agricultural development strategies and included as a technical input in agricultural development packages. Lack of knowledge among farmers about the pollination behaviour of honeybees is a constraint hindering the use of honeybees for crop pollination. There is also a need to develop human resources and build capacity to initiate activities in conservation, rearing, and using pollinators to improve pollination and, thus, agricultural productivity (Dar et al. 2013). The first major global conference on pollinator declines was held in São Paulo, Brazil, in 1998, where more than 60 pollinator experts from different countries raised many questions about pollinators (Frankie and Thorp 2009).

References

Abrol D (2013) Beekeeping: A compressive guide to bees and beekeeping. Scientific Publishers

Aizen MA, Harder LD (2009) The global stock of domesticated honey bees is growing slower than agricultural demand for pollination. Curr Biol 19:951–951

Anonymous (2000) "Ecology of pollination", http://www.sci.uidaho.edu/biosci/lecture/314%20 Austad/Microsoft%20 PowerPoint%20-%20Lecture%2020.pdf

Anonymous (2002) "Mutualism lecture outline", Plant ecology EVE/PLB 117. http://www-eve. ucdavis.edu/pearcy/Mutualism%20Lecture%20 outline.pdf

Anonymous (2003) "Bees behavior during foraging", Apiary Fact Sheet # 111. http://agf.gov.bc.ca/ apiculture/factsheets/111_forage.html

Arias MC, Sheppard WS (2005) Phylogenetic relationships of honeybees (Hymenoptera: Apinae: Apini) inferred from nuclear and mitochondrial DNA sequence data. Mol Phylogenet Evol 37(1):25–35

Aslan CE, Liang CT, Galindo B, Kimberly H, Topete W (2016) The role of honey bees as pollinators in natural areas. Nat Areas J 36(4):478–479

Baker HG (1997) Pollen viability related to type of pollination in six angiosperm species. Ann Bot 80(1):83–87

Baker HG (1978) Chemical aspects of the pollination of woody plants in the tropics. In: Tomlinson PB, Zimmerman M (eds) Tropical trees as living systems. Cambridge Univ. Press, New York, pp 57–82
Barker RJ (1977) Some carbohydrates found in pollen and pollen substitutes are toxic to honey bees. J Nutr 107(10):1859–1862
Baker HG, Baker I (1973) Some anthecological aspects of the evolution of nectar-producing flowers, particularly amino acid production in nectar. In: Heywood VH (ed) Taxonomy and ecology. Academic, New York/London, pp 243–264
Baker HG, Baker I (1975) Studies on nectar constitution and pollinator- plant coevolution. In: Gilbert LE, Raven PH (eds) Coevolution of animals and plants. Univ. Texas Press, Austin, pp 100–140
Baker HG, Hurd PD Jr (1968) Intrafloral ecology. Annu Rev Entomol 1(3):385–414
Bankova VS, Popov SS, Marekov NL (1983) A study on flavonoids of propolis. J Nat Prod 46:471–474
Bankova V, De Castro SL, Marcucci MC (2000) Propolis: recent advances in chemistry and plant origin. Apidologie 31:3–15
Barron AB, Plath JA (2017) The evolution of honey bee dance communication: a mechanistic perspective. J Exp Biol 220(23):4339–4346
Baskar K, Sudha V, Jayakumar M (2017) Effect of pesticides on pollinators. MOJ Ecol Environ Sci 2(8):00052
Betten DP, Richardson WH, Tong TC, Clark RF (2006) Massive honey bee envenomation induced rhabdomyolysis in an adolescent. Paediatrics 117:231–235
Biesmeijer JC, Roberts SP, Reemer M, Ohlemueller R, Edwards M, Peeters T, Schaffers A, Potts SG, Kleukers R, Thomas CD, Settele J, Kunin WE (2006) Parallel declines in pollinators and insect-pollinated plants in Britain and The Netherlands. Science 313:351–354. https://blog.umd.edu/agronomynews/2020/06/29/beetles-and-pollination/
Bommarco R, Kleijn D, Potts SG (2013) Ecological intensification: harnessing ecosystem services for food security. Trends Ecol Evol 28:230–238
Buchmann SL, Nabhan GP (1996) The forgotten pollinators. Island Press, Washington, DC
Bukovac Z, Shrestha M, Garcia JE, Burd M, Dorin A, Dyer AG (2017) Why background colour matters to bees and flowers. J Comp Physiol A 203(5):369–380
Carreck N, Williams I (1998) The economic values of bees in the U.K. Bee World 79(3):115–123
Choudary RK (1978) Floral fidelity in the Indian honey bee (Apis cerana indica). Indian Bee J 40(2):33–35
Christmann S (2022) Flowers, pollinators and pollination
Cobert SA, Unwin DM, PrysJones OE (1979) Humidity, nectar and insect visits to flowers, with special reference to Crataegus, Tilia and Echium. Ecol Entomol 4:9–22
Costanza R, d'Arge R, De Groot R, Farber S, Grasso M, Hannon, B,..., Van Den Belt M (1997) The value of the world's ecosystem services and natural capital. nature 387(6630):253–260
Couvillon MJ (2012) The dance legacy of Karl von Frisch. Insect Soc 59:297–306
Cox-Foster DL, Conlan S, Holmes EC, Palacios G, Evans JD, Moran NA, Quan P-L, Briese T, Hornig M, Geiser DM, Martinson V, vanEngelsdorp D, Kalkstein AL, Drysdale A, Hui J, Zhai J, Cui L, Hutchison SK, Simons JF, Egholm M, Pettis JS, Lipkin WI (2007) A metagenomic survey of microbes in honey bee colony collapse disorder. Science 318:283–287
Dar SA, Khan ZH, Khan AA, Ahmad SB (2013) Pollinator biodiversity and their conservation. In: Animal diversity, natural history, and conservation. Daya Publishing House
Dar SA, Lone GM, Parey SH, Hassan Gh I, Rather BA (2017) Insect pollinators and their conservation. J Entomol Zool Stud 5(3):1121
Dhaliwal JS, Atwal AS (1986) Factors influencing tripping efficiency of bees on alfalfa (Medicago sativa L.). In: Kapil RP (ed) Pollination biology: an analysis. Inter. India Publ., New Dehli, pp 79–90
Donovan BJ (2000) A comprehensive honey bee dance and odour 'language'hypothesis. Bee World 81(1):5–10

Downer RGH (1978) Functional role of lipids in insects. In: Rockstein M (ed) Biochemistry of insects. Academic, New York/ London, pp 58–92

Duangphakdee O, Hepburn HR, Tautz J (2011) The dance language. In Honeybees of Asia, Springer, Berlin, Heidelberg, pp 313–332

Dyer FC (2002) The biology of the dance language. Annu Rev Entomol 47:917–949

Ellis JS, Knight ME, Darvill B, Goulson D (2006) Extremely low effective population sizes, genetic structuring and reduced genetic diversity in a threatened bumblebee species, Bombus sylvarum (Hymenoptera: Apidae). Mol Ecol 15:4375–4386

Erickson EH Jr, Carlson SD, Garment MB (1986) A scanning electron microscope atlas of the honey bee. Iowa State University Press

Engel MS (1999) The taxonomy of recent and fossil honeybees (Hymenoptera: Apidae; Apis). J Hymen Res 8(2):165–196

Evans JD (2003) Diverse origins of tetracycline resistance in the honey bee bacterial pathogen Paenibacillus larvae. J Invertebr Pathol 83:46–50

Faegri K, van der Pijl L (1971) The principles of pollination ecology. Pergamon Press, London

Faheem M, Aslam M, Razaq M (2004) Pollination ecology with special reference to insects a review. J Res Sci 4(1):395–409

Fitzpatrick Ú, Murray TE, Paxton RJ, Breen J, Cotton D, Santorum V, Brown MJF (2007) Rarity and decline in bumblebees—a test of causes 416 M.J.F. Brown, R.J. Paxton and correlates in the Irish fauna. Biol Conserv 136:185–194

Foley JA, DeFries R, Asner GP, Barford C, Bonan G, Carpenter SR, Chapin FS, Coe MT, Daily GC, Gibbs HK, Helkowski JH, Holloway T, Howard EA, Kucharik CJ, Monfreda C, Patz JA, Prentice IC, Ramankutty N, Snyder PK (2005) Global consequences of land use. Science 309: 570–574

Frankie GW, Thorp RW (2009) Pollination and pollinators. In: Encyclopedia of insects. Academic, pp 813–819

Frisch KV (1967) Honeybees: do they use direction and distance information provided by their dancers? Science 158(3804):1072–1076

Gadagkar R (1996) The honeybee dance-language controversy: robot bee comes to the rescue. Resonance 1(1):63–70

Gallai N et al (2009) Economic valuation of the vulnerability of world agriculture confronted with pollinator decline. Ecol Econ 68(3):810–821

Gary NE Jr, Page RE, Morse RA, Henderson CE, Nasr ME, Lorenzen K (1990) Comparative resistance of honey bees (Apismellifera L.) from Great Britain and United States to infestation by tracheal mites (Acarapiswoodi). Am Bee J 130:667–669

Gess FW, Gess SK (1993) Effects of increasing land utilization on species representation and diversity of aculeate wasps and bees in the semi-arid areas of southern Africa. In: Lasalle J, Gauld ID (eds) Hymenoptera and biodiversity. CAB International, Wallingford, pp 83–114

Giannini TC, Garibaldi LA, Acosta AL, Silva JS, Maia KP, Saraiva AM, Guimarães PR Jr, Kleinert AMP (2015) Native and non-native supergeneralist bee species have different effects on plant-bee networks. PLoS One 10:e0137198

Grant V (1950) The flower constancy of bees. Bot Rev 16:379–398

Gupta RK (2014) Taxonomy and distribution of different honeybee species. In: Beekeeping for poverty alleviation and livelihood security. Springer, Dordrecht, pp 63–103

Habermann E (1972) Bee and wasp venoms. Science 177:314–322

Han F, Wallberg A, Webster MT (2012) From where did the western honeybee (Apis mellifera) originate? Ecol Evol 2(8):1949–1957

Heinrich B (1979) Keeping a cool head: honeybee (Apis mellifera) thermal regulation. Science 205: 1269–1271

Hocking B (1968) Insect-flower associations in the high Arctic with special reference to nectar Oikos: 359–387

House HL (1974) Nutrition. In: Rockstein M (ed) The physiology of insecta, vol 5, 2nd edn. Academic, pp 1–62

Kearns CA, Inouye DW, Waser NM (1998) Endangered mutualisms: the conservation of plant-pollinator interactions. Annu Rev Ecol Syst 29:83–112

Kerr WE (1957) Introdução de abelhas africanas no Brasil. Brasil Apicola 3:211–213

Kevan PG (1999) Pollinators as bioindicators of the state of the environment: species, activity and diversity. In: Invertebrate biodiversity as bioindicators of sustainable landscapes. Elsevier, pp 373–393

Kevan PG, Baker HG (1983) Insects as flower visitors and pollinators. Annu Rev Entomol 28:407–453

Kevan PG, Greco CF, Belaoussoff S (1997) Log-normality of biodiversity and abundance in diagnosis and measuring of ecosystemic health: pesticide stress on pollinators on blueberry heaths. J Appl Ecol 34:1122–1136

Khan MA, Sulehri NI, Talha M, Nazar A, Ahmed HGMD (2021) Honey bees show dance pattern to communicate–a review. World J Biol Biotechnol 6(2):15–19

Klein AM et al (2007) Importance of pollinators in changing landscapes for world crops. Proc R Soc B 274:303–313

Kovac H, Crailsheim K (1988) Lifespan of Apis mellifera carnica Pollm. infested by Varroa jacobsoni Oud. in relation to season and extent of infestation. J Apic Res 27:230–238

Kremen C, Ricketts T (2000) Global perspectives on pollination disruptions. Conserv Biol 14: 1226–1228

Kumar V, Sharma J, Kumar A (2022) Importance of bee flora and their relative utility to honey bees in Bihar

Kumari M, Kumar B Kumari R, Kishor DR, (2020) Honeybee communication-a signal for their food. Agriculture & food e-Newsletter: 35

Lee PC, Winston ML (1987) Effects of reproductive timing and colony size on the survival, offspring colony size and drone production in the honey bee (Apis mellifera). Ecol Entomol 12:187–195

Leppik EE (1956) The form and function of numeral patterns in flowers. Amer J Bot 43:445–455

Leppik EE (1972) Origin and evolution of bilateral symmetry in flowers. Evol Biol 5:49–85

Levchenko IA (1961) Twilight vision of bees. Pchelovodstvo 38(1):5–6

Lindauer M (1975) Verständigung im Bienenstaat. Gustav Fischer Verlag, Stuttgart

Le Moël F, Wystrach A (2020) Towards a multi-level understanding in insect navigation. Curr Opin Insect Sci 42:110–117

Lindauer M (1981) Kommunikation im Bienenstaat—Arbeitsteilung, Nahrungssuche, Wohnungssuche. In: Scharf JH, Kämmerer W (eds) Nova Acta Leopoldina. Leipzig

Linksvayer TA, Fewell JH, Gadau J, Laubichler MD (2012) Developmental evolution in social insects: regulatory networks from genes to societies. J Exp Zool (Mol Dev Evol) 318:159–169

Lundie AE (1925) Flight activities of the honeybees. US Depreciation. Agric Bull No. 1328. pp. 37

Maksong S (2008) Identification of bee flora from the midgut of honeybees of Thailand. Burapha University, Chon Buri, p 104

Malik A, Malik S, Kaushish S, Kaur R, Goyal C (2020) A comprehensive review on Bee w.s.r to Honey Bee. 13.350–359

Marcucci MC (1995) Propolis: chemical composition, biological properties and therapeutic activity. Apidologie 26:83–99

Martin-Hernandez R, Meana A, Prieto L, Salvador AM, GarridoBailon E, Higes M (2007) Outcome of colonization of Apis mellifera by Nosema ceranae. Appl Environ Microbiol 73:6331–6338

Mauricio da Rocha JR, De Almeida JR, Lins GA, Durval A (2010) Insects as indicators of environmental changing and pollution: a review of appropriate species and their monitoring. Holos Environ 10(2):250

McGregor SE (1976) Insect pollination of cultivated crop plants (No. 496). US Department of Agriculture, Agricultural Research Service

Meffe GK (1998) The potential consequences of pollinator declines on the conservation of biodiversity and stability of food crop yields. Conserv Biol 12:8–17

Menzel R (1967) Untersuchungen zum Erlernen von Spektralfarben durch die Honigbiene (Apis mellifera). Z Vergl Physiol 56:22

Mishra RC (1995) Honeybees and their management in India

Morandin LA, Winston ML (2006) Pollinators provide economic incentive to preserve natural land in agroecosystems. Agric Ecosyst Environ 116:289–292

Moritz RFA, Hartel S, Neumann P (2005) Global invasions of the western honeybee (Apis mellifera) and the consequences for biodiversity. Ecoscience 12(3):289–301

Morse R, Calderone NW (2000) The value of honeybees as pollinators of U.S. crops in 2000. Bee culture

Motten AF, Campbell DR, Alexander DE, Miller HL (1981) Pollination effectiveness of specialist and generalist visitors to a North Carolina population of Claytonia virginica. Ecology 62(5): 1278–1287

Mu J, Peng Y, Xi X, Wu X, Griffin JN, Niklas KJ, Sun S (2014) Domesticated honey bees evolutionarily reduce flower nectar volume in a Tibetan lotus. Ecology 95:3161–3172

Munz T (2005) The bee battles: Karl von frisch, Adrian wenner and the honey bee dance language controversy. J Hist Biol 38(3):535–570

Nitharwal M, Rolania R, Jatav HS, Chandra K, Khan MA, Kumawat S et al (2021) Pollinators: their relevance in conservation and sustainable agrosystems. In: A Rustagi, B Chaudhry (ed) Plant reproductive ecology-recent advances. London, UK

Oldroyd BP (2007) What's killing American honey bees? PLoS Biol 5(6):e168

Partap U, Partap T (1997) Managed crop pollination: the missing dimension of mountain agricultural productivity. Mountain Farming Systems' Discussion Paper Series No. MFS 97/1. Kathmandu: International Centre for Integrated Mountain Development (ICIMOD)

Percival MS (1965) Floral biology. Pergamon Press, London

Peñuelas J, Filella I, Comas P (2002) Changed plant and animal life cycles from 1952 to 2000 in the Mediterranean region. Glob Chang Biol 8(6):531–544

Pimental D, Wilson C, McCullum C, Huang R, Dwen P, Flack J, Tran Q, Saltman T, Cliff B (1997) Economic and environmental benefits of biodiversity. Bioscience 47(11):747–757

Potts SG, Biesmeijer JC, Kremen C, Neumann P, Schweiger O, Kunin WE (2010) Global pollinator declines: trends, impacts and drivers. Trends in Ecology and Evolution 25:345–353

Potts SG, Imperatriz-Fonseca V, Ngo HT, Aizen MA, Biesmeijer JC, Breeze TD et al (2016) Safeguarding pollinators and their values to human well-being. Nature 540(7632):220–229

Prasad MG, Rani DD (2023) Voice of honey bees: a threat to humans. Vigyan Varta 4(6):161–164

Preece K, Beekman M (2014) Honeybee waggle dance error: adaption or constraint? Unravelling the complex dance language of honeybees. Anim Behav 94:19–26

Pyke GH (1979) Optimal foraging in bumblebees: rule of movement between flowers within inflorescences. Anim Behav 27:1167–1181

Radloff SE, Hepburn HR, Engel MS (2011) The asian species of Apis. In Honeybees of Asia Springer, Berlin, Heidelberg, pp 1–22

Raffiudin R, Crozier RH (2007) Phylogenetic analysis of honeybee behavioral evolution. Mol Phylogenet Evol 43(2):543–552

Riaz J, Razzaq H, Amjad T, Anjum A, Afzal T, Fatima F et al (2020) Pollution and pollinators: a review. Pure Appl Biol (PAB) 9(3):2049–2058

Riley JR, Greggers U, Smith AD, Reynolds DR, Menzel R (2005) The flight paths of honeybees recruited by the waggle dance. Nature 435(7039):205–207

Roberts RB (1979) "Energetics of cranberry pollination", Proc. 4th Int. Symp. Pollination, Md Agric Expenditure Stn Spec Misc Publ No. 1, pp 431–440

Ruttner F (1988) Biogeography and taxonomy of honeybees. Springer, Berlin, p xii+284

Ruttner F (1992) Naturgeschichte der Honigbienen. Ehrenwirth, Munich, p 357

Ruttner F, Kauhausen D (1985) Honeybees of tropical Africa: ecological diversifi cation and isolation. In: proceedings of the third international conference on apiculture in tropical climates, Nairobi, Kenya, 5–9 Nov 1984, pp 45–51

Schricker B (1965) Die Orientierung der Honigbiene in der Dammerung, zugleich ein Bietrag zur Frage der Ocellenfunktion bei Bienen. Z Vgl Physiol 49:420–458
Seeley TD (1982) How honeybees find a home. Sci Am 247(4):158–169
Seeley TD (1992) The tremble dance of the honey bee: message and meanings. Behav Ecol Sociobiol 31:375–383
Seeley TD (2009) The wisdom of the hive: the social physiology of honey bee colonies. Harvard University Press
Sheppard WS (1989) A history of the introduction of honeybee races into the United States. Am Bee J 129:664–667
Sheppard WS, Meixner MD (2003) Apis mellifera pomonella, a new honeybee subspecies from Central Asia. Apidologie 34(4):367–175
Simon-Delso N, Martin GS, Bruneau E, Minsart LA, Mouret C, Hautier L (2014) Honeybee colony disorder in crop areas: the role of pesticides and viruses. PLoS One 9:e103073
Singla A (2020) Dancing bees speak in a code-a review. Emerg Life Sci Res 6:44–53
Sotavalta O, Lajasmo L, Tapana M, Tuomisto J (1962) On the preferences shown by insects for certain food-sugar solution concentrations in feeding. Ann Acad Sci Fenn Ser A 63:1–40
Sponsler DB, Grozinger CM, Hitaj C, Rundlöf M, Botías C, Code A, Douglas MR (2019) Pesticides and pollinators: a socioecological synthesis. Sci Total Environ 662:1012–1027
Sujatha GS, Manoj K, Mandar VT, Mareedu GP, Ranjitha SM, Shashank DU, Gadde AK, KattaSubramanya ST (2024) Foraging behaviour of honeybee species on fenugreek (Kasuri Methi). Int J Adv Biochem Res 8(8S):1365–1368
Suwannapong G, Yemor T, Boonpakdee C, Benbow ME (2011) Nosema ceranae, a new parasite in Thai honeybees. J Invertebr Pathol 106:236–241
Suwannapong G, Eiri DM, Benbow ME (2012) Honeybee communication and pollination. In: New perspectives in plant protection, pp 39–62
Tanaka H, Roubik DW, Kato M, Liew F, Gunsalam G (2001) Phylogenetic position of Apis nuluensis of northern Borneo and phylogeography of A. cerana as inferred from mitochondrial DNA sequences. Insect Soc 48(1):44–51
Tandon V, Kumar A, Rana C (2016) Pollination—its type, threats and role in environment conservation. Int J Curr Res 8(8):35744–35751
Tepedino VJ (1979) The importance of bees and other insect pollinators in maintaining floral species composition. Great Basin Natur Mem 3:139–150
Thurston HD (1969) Tropical agriculture- a key to the world food crisis. Bioscience 19:29–34
Valido A, Rodríguez-Rodríguez MC, Pedro J (2019) Honeybees disrupt the structure and functionality of Plant Pollinator networks. Sci Rep 9:4711. https://doi.org/10.1038/s41598-019-41271-5
VanEngelsdorp D (2011) Morbidity and Mortality in the Bee Yard
Vanbergen AJ, Ambrose N, Aston D, Biesmeijer JC, Bourke A, Breeze T et al (2012) Insect pollinators: linking research and policy. Church House Conference Centre, Dean's Yard
Verma LR (1968) Biological and economic characters of Apis cerana indica F. Honeybee Sci 7: 151–157
Verma LR (1990) Beekeeping in integrated mountain development. Oxford & IBH Pub. Co. Pvt. Ltd., New Dehli/Calcutta, p 23
Verma LR, Dulta DK (1986) Foraging behavior of Apis cerana indica and Apis mellifera in pollinating apple flowers. J Apic Res 25:197–201
Von Frisch K (1967) The dance language and orientation of honeybees. Harvard University Press, Cambridge
von Frisch K (1971) Bees - their vision, chemical senses, and language. Revised Edition. CorneD University Press, Ithaca and London
Waddington KD, Heinrich B (1979) The foraging movements of bumblebees on vertical "inflorescences": an experimental analysis. J Comp Physiol 134:113–117
Waddington KD, Holden LR (1979) Optimal foraging on flowers selection by bees. Am Nat 114: 179–199

Wilcock C, Neiland R (2002) Pollination failure in plants: why it happens and when it matters. Trends Plant Sci 7(6):272

Williams PH (1982) The distribution and decline of British bumble bees (Bombus Latr.). J Apic Res 21:236–245

Willis LG, Winston ML, Honda BM (1992) Phylogenetic relationships in the honeybee (genus Apis) as determined by the sequence of the cytochrome oxidase II region of mitochondrial DNA. Mol Phylogenet Evol 1(3):169–178

Wojcik V (2021) Pollinators: their evolution, ecology, management, and conservation. In: Arthropods: are they beneficial for mankind. IntechOpen, pp 1–22

Wolf S, Nicholls E, Reynolds AM, Wells P, Lim KS, Paxton RJ, Osborne JL (2016) Optimal search patterns in honeybee orientation flights are robust against emerging infectious diseases. Sci Rep 6(1):1–10

Wray MK, Klein BA, Seeley TD (2012) Honey bees use social information in waggle dances more fully when foraging errors are more costly. Behav Ecol 23(1):125–131

Zayed A (2009) Bee genetics and conservation. Apidologie 40:237–262

Nest Variation, Foraging Preference and Host Plant Range of Genus *Apis* Bees from Karnataka, South India

5

A. Shahnawaz, M. Shashikant, L. Vani, L. Veena, S. M. Sreelakshmi, H. S. Shwetha, and K. M. Ravi Kiran

Abstract

During the evolutionary journey, honey bees have shown an intimate relationship with host or flowering plants and have become the most important pollinators of wild and crop plants of the terrestrial ecosystem. So, the information on nest variation, nectar or pollen preferences of foraging honey bees and host plant range is, therefore, imperative to increase the honey production to keep pace with public demand. This chapter was conducted to determine the parameters of paramount importance of genus *Apis* from Tumakuru, Karnataka, India. This chapter presents an account of three species of genus Apis viz. *Apis dorsata*, *Apis florea* and *Apis mellifera* with reference to nest variation, foraging preference and host plant range.

Keywords

Genus *Apis* · Nest variation · Foraging · Host plant range · Karnataka, India

1 Introduction

All the species on earth are classified in a taxonomic system that organizes the evolutionary relationship among the species. Taxonomy is hierarchical, with the highest categories as the most inclusive and the lower categories as the most restricted (Linksvayer et al. 2012; Gupta et al. 2014). During the evolutionary

A. Shahnawaz (✉) · L. Vani · L. Veena · S. M. Sreelakshmi · H. S. Shwetha · K. M. Ravi Kiran
Department of Studies and Research in Zoology, Tumkur University, Tumkur, Karnataka, India

M. Shashikant
Department of Studies in Zoology, Vijayanagar Sri Krishnadevaraya University, Ballari, Karnataka, India

A. K. Chakravarthy (ed.), *The Role of Pollinators in Tropical Landscapes*,
https://doi.org/10.1007/978-981-95-9139-8_5

history, honey bees are having a close association with flowering plants and have become the most important pollinators of wild and crop plants of terrestrial ecosystem (Michener 2007). On earth, there are bees on every continent except Antarctica and 17,500 species are identified (Michener 2000).

Honey bee groups widely differ in their host preference and exploit them (Cane and Sipes 2006; Muller and Kuhlmann 2008; Carvalho and Schlindwein 2011). Pollination of crops is mostly done by honey bees that depend on natural habitats (Greenleaf and Kremen 2006; Kremen et al. 2004). So, the host plants and bees show a mutualistic relationship, where host plants provide bees with food and other by-products, and the bees bring about pollination, providing the way to sexual reproduction and ensuring the survival of hosts (Simpson and Neff 1981; Claus et al. 2020). This plant-animal relationship is also important from the ecosystem perspective. Globally, about 75% of main crops depend on animal pollination for increased yield (Klein et al. 2007). Besides the importance of agricultural productivity (Gallai et al. 2009), crop yields reach their full productivity rarely due to inadequate pollination (Garibaldi et al. 2016). Additionally, this mutualistic relationship in the form of pollination improves crop quality, size and shelf life, which plays an important role in increasing the market price (Klatt et al. 2014). Hence, with this brief account, we provide information on honey bee diversity, nest variation, preference and host plant range to generate interest in these ecological specializations and to highlight the importance of honey bees in terrestrial ecosystems.

2 Species Diversity, Distribution and Occurrence

The present study of taxonomy and distribution of honey bees from seven sites of Tumakuru city, Karnataka, was conducted during 4 months, from January 2022 to April 2022. The current investigation was conducted in Tumakuru (Tumkur) city and seven study sites, viz. Dabasppette, Hirehalli, Tumkur University Campus, Siddaganga Mutt, Sree Siddaganga College of Arts, Science and Commerce for women (SSCW), Hosakere and Harenalli were selected based on vegetation cover and availability of food crops. During the current study, three species of honey bees viz. *Apis dorsata*, *Apis florea* and *Apis mellifera* (Fig. 5.1) were recorded. These

Fig. 5.1 Three species of honey bees from genus *Apis*

recorded few species of honey bees belonging to the genus are generally recognised (Engel 1999; Arias and Sheppered 2005). Even phylogenetic analysis studies based on molecular markers, i.e. nuclear and mitochondrial (mt DNA), revealed that species constitute three different clustering groups, habituating and exploiting diverse habitats (Arias and Sheppered 2005; Raffiudin and Crozier 2007; Tanaka et al. 2001; Willis et al. 1992).

All three species, i.e. *Apis dorsata*, *Apis florea* and *Apis mellifera*, are reported from Asia, which show a type of evolutionary relationship among the species. The Western honey bee, *Apis mellifera*, basically originated in Europe, Africa and the Middle East, though now found almost throughout the world because of its economic, agricultural and environmental importance.

The Karnataka state has all four species of honey bees, i.e. *A. dorsata*, *A. florea*, *A. mellifera* and *Apis cerana indica*, and during the present study, all the species except *A. cerana indica* are reported. However, *A. mellifera* was recently introduced in the state. In Karnataka, since prehistorical times, beekeeping has been practiced with *A. cerana indica*. During 1992, the state experienced an epidemic of Thai Sac Brood Disease (TSBD), which led to a great loss to beekeepers, and they lost most of the honey bee colonies. This pushed the state government to introduce *A. mellifera* with the aim of rehabilitating the people who actually depend on the beekeeping occupation (Sivaram 2001). But the beekeepers could not get satisfied with the overall results of crop production from this exotic species. The Department of Industries and Commerce of the Karnataka State Government has embarked on initiatives to revive the *Apis cerana indica* beekeeping to strengthen the conditions of the poor beekeepers and enhance pollination process in Karnataka.

3 Morphological Specialization

It was observed that all three collected honey bee species exploited food resources either from the same plant or different plants. So, all honey bee species showed month-wise variation with morphological features like body, size, proboscis, fore and hind limbs and wing characteristics. This is because honey bee showed complex morphological and behavioural adaptations to their host plant (Thorp 2000). Even variations with regard to habitat, shape and orientation of the hives of all three species are summarized in Tables 5.1, 5.2, and 5.3. Such morphological adaptations (Tables 5.4, 5.5, and 5.6) are related to pollen and nectar gathering, as some of the bees do function as specialists in pollen collecting rather than nectar collection exhibited by general types of honey bees (Ferrari 2015; Engel 2002). This morphological specialization for extracting and transporting pollen involves the modification of various body structures, including long and hooked hairs, proboscis size, wing pattern, fore and hindwing features (Neff 2004). These important morphometric characters show geographical variations among honey bees.

Proboscis is another morphometric feature in *Apis* for nectar collection and is a type of phenotypic adaptation depending on partitioning and availability of food resources (Hadisoesilo et al. 1995). It also shows the geographical and ecological

Table 5.1 Morphological characteristics of honeybee nests/hives of *Apis dorsata*

Months	No. of hives/ nests	Shape of the hives/nests			Location of hives/ nests		Surrounding of hives/nests		Orientation of hives/nests	
		U	V	A	RS	I S	Open	Covered	NE-SW	N-S
January	20	18	2	0	12	8	14	6	7	13
February	65	58	4	3	52	13	57	8	49	16
March	23	20	1	2	19	4	5	18	19	4
April	24	22	2	0	14	10	13	11	15	9

Note: *U* U-shaped, *V* V-shaped, *A* asymmetrical, *RS* road side, *IS* interior side, *NE-SW* north east to south west, N-S north to south

Table 5.2 Morphological characteristics of honey bee nests/combs of *Apis florea*

Months	No. of hives	Shape of the hives			Location of the hives		Hive surroundings		Hive orientation	
		U	V	A	RS	I	Open	Covered	NE-SW	N-S
January	1	0	0	0	0	1	0	1	1	0
February	11	9	2	0	6	5	0	11	6	5
March	0	0	0	0	0	0	0	0	0	0
April	1	1	0	0	0	1	0	1	0	1

Note: *U* U-shaped, *V* V-shaped, *A* asymmetrical, *RS* road side, *IS* interior side, *NE-SW* north east to south west, *N-S* north to south

Table 5.3 Morphological characteristics of honey bee nests/combs of *Apis mellifera*

Months	No. of hives	Shape of the hives			Location of the hives		Hive surroundings		Hive orientation	
		U	V	A	RS	IS	Open	covered	NE-SW	N-S
January	0	0	0	0	0	0	0	0	0	0
February	1	0	0	1	0	1	0	1	1	0
March	0	0	0	0	0	0	0	0	0	0
April	0	0	0	0	0	0	0	0	0	0

Note: *U* U-shaped, *V* V-shaped, *A* asymmetrical, *RS* road side, *IS* interior side, *NE-SW* north east to south west, *N-S* north to south

Table 5.4 Species-specific morphological characters of *Apis dorsata*

Sl. No.	Characters	Sample 1	Sample 2	Sample 3	Average
1	Proboscis (mm)	0.3	0.27	0.3	0.29 mm
2	Hind limb (mm)	1.5	1.4	1.5	1.46 mm
3	Forelimb (mm)	0.8	0.8	0.8	0.8 mm
4	Hind wing (mm)	0.9	0.8	0.8	0.83 mm
5	Forewing (mm)	1.3	1.3	1.3	1.3 mm

Table 5.5 Species-specific morphological characters of *Apis florea*

Sl. no	Characters	Sample 1	Sample 2	Sample 3	Average
1	Proboscis (mm)	0.29	0.2	0.2	0.23 (mm)
2	Hindlimb (mm)	0.7	0.7	0.7	0.7 (mm)
3	Forelimb (mm)	0.51	0.5	0.5	0.5 (mm)
4	Hindwing (mm)	0.5	0.4	0.4	0.4 (mm)
5	Forewing (mm)	0.7	0.6	0.7	0.66 (mm)

Table 5.6 Species-specific morphological characters of *Apis mellifera*

Sl. no	Characters	Sample 1	Sample 2	Sample 3	Average
1	Proboscis (mm)	0.30	0.3	0.3	0.3 (mm)
2	Hindlimb (mm)	0.8	0.79	0.8	0.79 (mm)
3	Forelimb (mm)	0.61	0.6	0.6	0.6 (mm)
4	Hindwing (mm)	0.62	0.62	0.62	0.62 (mm)
5	Forewing (mm)	0.78	0.77	0.77	0.77 (mm)

Fig. 5.2 Different nests of three species of honey bees from genus *Apis*

variability among *Apis* species. Other morphometric features also showed distinct variations among other phenotypic characters, viz. forelimbs and hindlimbs, and wing patterns, which support the criteria of using the food resources availability efficiently.

4 Nest Variation and Preferences

The nest architecture and location preference shown by two native and one exotic honey bee species were designed in a natural fashion and protected from predation. The hive architecture in the present study was suitable for each species for carrying out colony activities like food collection, storage, brood rearing and others. By studying honey bee species, a total number of 134 hives of *A. dorsata*, 13 of *A. florae* and 1 of *A. mellifera* respectively were found (Fig. 5.2). The nest variation was recorded, and the species of honey bees were observed to prefer to make more hives on trees than buildings (Figs. 5.3, 5.4, 5.5, and 5.6). Variation in number of hives throughout the year was recorded (Figs. 5.2, 5.3, and 5.4). The highest number

Fig. 5.3 Different nest shapes and location of *Apis dorsata*

Fig. 5.4 Different nest shapes and location of *Apis florea*

Fig. 5.5 Different nest shapes and location of *Apis mellifera*

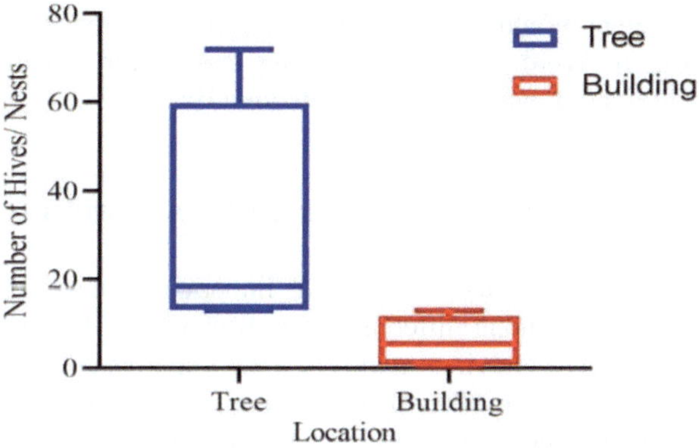

Fig. 5.6 The number and location of nests

of nests was found at Harenalli site followed by the least number at Siddaganga Mutt (Fig. 5.5). This would be because of people's visit to the mat on a daily basis. The *A. florea* and *A. mellifera* preferred the bushy and ground areas. After all, seven selected sites showed great variation in the number of nests where Harenalli showed the highest number of hives compared to other sites. *A. dorsata* was observed to prefer wild conditions compared to other species, while *A. florea* and *A. mellifera* were mild in nature. *A. dorsata* colonies preferred thick and the highest branches of trees and rocks of buildings under warm conditions.

The highest number of nests was recorded for the *A. dorsata*, followed by *A. florea,* and least represented by *A. mellifera*, respectively. The highest number of nests with respect to all species though other months showed a variation pattern. *A. dorsata* constructed both U- and V-shaped nests, but Ü-shaped nests were more. These nests were asymmetrical (A), roadside (RS), interior side (IS), open, covered and north east to south west direction (NE-SW), north to south (NS) in nature. While as nest architecture of *A. florea* was little bit different in characteristics. They constructed the nests of both U and V shapes, symmetrical, roadside (RS), interior side (IS), covered, north east to south west direction (NE-SW), north to south (NS) in nature. *A. mellifera* showed different nest architecture from both *A. dorsata* and *A. florea,* respectively. Nest of *A. mellifera* was asymmetrical in nature, found on the interior side, covered or hidden and north east to south west direction (NE-SW). Moreover, the number of nests was found to be more located on trees rather than buildings in general. *A. dorsata* constructed nests both on big branches of trees (preferably banyan trees and other big trees) in multiple number and buildings than *A. florea* and *A. mellifera* who preferred small branches and bushes. Further, *A. dorsata* honey bee species was found at the central regions of tree branches, while *A. florea* and *A. mellifera* preferred bushy areas and ground levels and built variously shaped combs among all species. During colony construction, it usually attaches the comb crest to medium-sized (4–6 inches in diameter) branches. The nesting trees of honey bee species vary from small to large trees and low to high elevated trees with respect to selected sites. Uneven distribution of colonies during different months was also observed. This difference in nest variation and preference was to avoid competition among species for foraging and predation. The same observation was recorded by Reddy (1980).

5 Foraging Sources and Pollination

Foraging rate and sources are important factors that determine the pollination efficiency among different honey bee species. Foraging of honey bee species viz. *A. mellifera* and *A. cerana* is of great importance to the pollination of plants as they comprise about 80% of total insect pollination (Robinson and Morse 1989). The foraging behaviour was directly related to the accessibility of flowers in the study area.

All three honey bee species preferred different and available foraging sources, i.e. plants or trees (Table 5.7, Fig. 5.6), and among the diversity of nectar and pollen plants, only a few plants were more utilized than others and preferred as foraging sources. This revealed that honey bees were more specialized in foraging than general type. The study showed that all three honey bee species visited one species of flower at once compared to multiple foraging sites and individually used to pick-up more pollen grains. *A. dorsata* preferred the high degree of mobility and foraged out to various places in search of food, shelter protection and favourable microclimate conditions. *A. florea* did not prefer the high degree of mobility, while *A. mellifera* preferred the ground levels for building nests.

Table 5.7 Diversity of nesting and foraging plants/trees for honey bees

Site/Family/Plant species	No. of trees	No. of nests	Types of characters of branches
1. Dabbaspette Family: Meliaceae *Melia azadhirach*	1	5	Wood, hard to peel
2. Hirehalli Family: Moraceae *Ficus bengalensis*	2	18	Wood, hard to peel
3. Tumkur university a. Family: Moraceae *Ficus bengalensis*	1	6	Woody, hard to peel

(continued)

Table 5.7 (continued)

Site/Family/Plant species	No. of trees	No. of nests	Types of characters of branches
b. Family: Fabaceae *Saraca asoca*	1	1	Woody
c. Family: Fabaceae *Leuceaena leucocephala*	1	1	Woody, hard to peel
d. Family: Fabaceae *Delonix regia*	1	1	Woody, hard to peel
4. Siddaganga Mutt **a**. Family: Oleaceae *Jasminium dichotomum*	1	1	Hard to Peel
b. Family: Moraceae *Ficus bengalensis*	1	1	Woody, hard to peel
5. SSCW **a**. Family: Fabaceae *Albbizia samana*	1	1	Woody
6. Hoskere **a**. Family: Aracaceae *Cocus nucifera*	9	9	Woody
b. Family: Moraceae *Ficus bengalensis*	2	12	Woody
c. Family: Fabaceae *Tamarindus indica*	1	1	Woody
d. Family: Oleaceae *Jasminum multiflorum*	2	3	Peel of
e. Family: Meliaceae *Aadhirachta indica*	1	1	Woody
f. Family: Rutaceaes *Citrus lemon*	1	1	Peel of
g. Family: Verbenaceae *Lantana camara*	3	3	Peel of
h. Family: Rutaceae *Citrus medica*	1	1	Peel of
7. Harenalli **a**. Family: Moreceae *Ficus bengalensis*	1	52	Woody and hard to peel
b. Family: Aracaceae *Cocus nucifera*	1	1	Woody and hard to peel

Some bees were seen foraging on different flowers of the plants, while some were seen to forage on one type of flower specifically. Some of the bees were observed to gather water from sources viz. bore wells, water taps, small pools, ditches, etc. (Fig. 5.7). Honey bees were noticed to prefer continuous water sources like water pools. Thus, honey bees can be grouped into categories based on foraging activity into water, nectar, pollen or resin foraging. Sometimes, bees can also collect wax from insects (scale insects) as reported by Dimou and Thrasyvoulou (2007) and

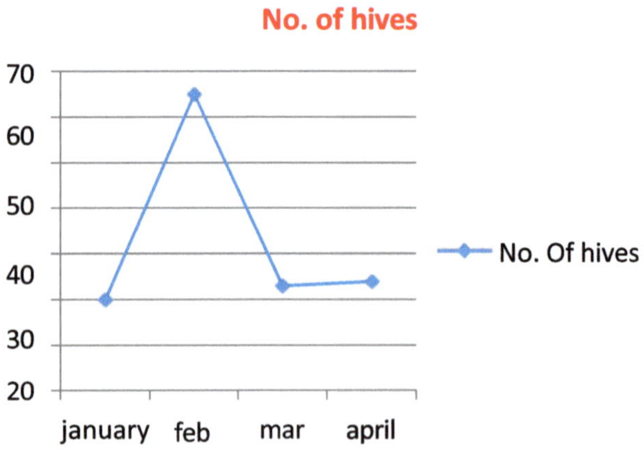

Fig. 5.7 Month-wise comb or nest variation of *Apis dorsata*

Abou-Shaara (2012). This study revealed unusual water collection by honey bees (e.g. roadside) as witnessed by Butler (1946). There is an individual and collective effort made by honey bees to maintain the colony. So, the foraging activity is linked with a genetic trait and depends on the genotype of the bees (Pankiw et al. 2002). Observations revealed that 19 plant species were used as forage food crops as a general preference, and among 4 plant species as special preference. The preferred plants for nectar and pollen collection included *Ipomea nil*, *Citrus lemon*, *Vernolia eleaeagnifolia* and *Muntingia calabura.* The nectar and pollen collection foragers sometimes preferred one food source over another and one flower over other. This foraging activity and preference are reported by other workers also (Sushil et al. 2013; Mattu and Bhagat 2015; Mattu et al. 2012).

Plant pollination is another important activity. Honey bees harvest pollen and nectar from flowers, and this helps in plant pollination (Dalio 2015; Kumar and Sharma 2016). So, different plant species act as pollinated plants, though some are primary pollinators while others as secondary sources. Thus, the foraging activity may be utilised like a biological indicator for different environmental studies, including pesticide effects (Balayianni and Balayiannis 2008). There has been increased demand for commercial crop pollination, and it has become possible because of beekeeping (Goulson et al. 2015; Painkra et al. 2021). Beekeeping facilitates information on pest infestation (Chamberlian et al. 2012); optimal foraging theories (Thuijsman et al. 1995), swarm intelligence (Kumar and Govindaraj 2013), honey bee foraging (Baig and Rashid 2007) and the amount and quality of their crops, respectively (Figs. 5.8 and 5.9).

6 Environmental Parameters and Foraging Behaviour

Bee foraging behaviour depends on a variety of environmental factors (Iwama 1977; Gouw and Gimenes 2013). There are different environmental factors that govern the foraging activity (Juan et al. 2022). These may be endogenous and exotic in nature.

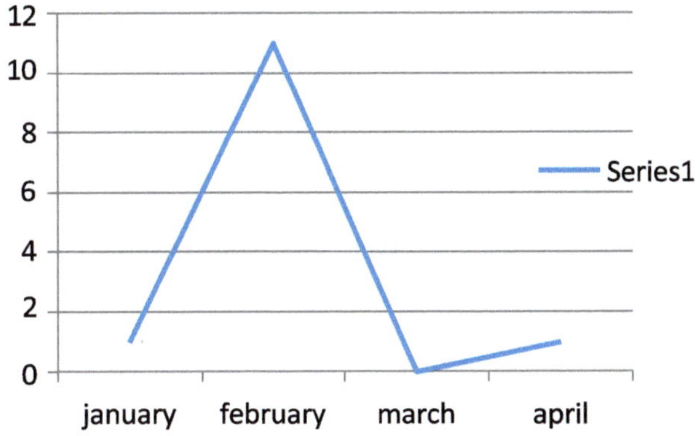

Fig. 5.8 Monthly-wise comb or nest variation of *Apis florea*

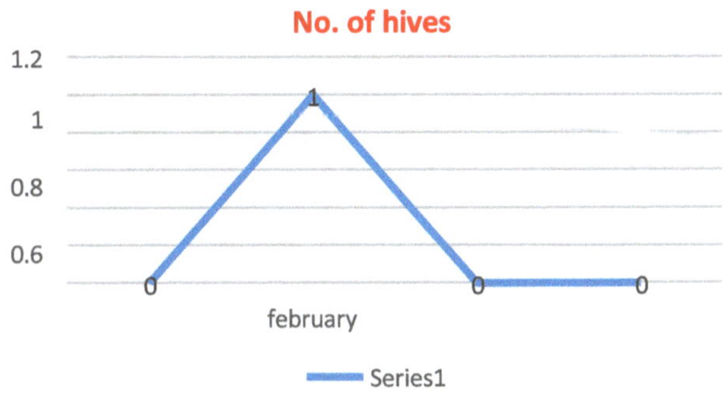

Fig. 5.9 Monthly-wise comb or nest variation of *Apis mellifera*

The endogenous includes those prevailing in the honey bee colony. The examples of endogenous parameters are queen presence and sexual behaviour, whether it is virgin or mated. Even foraging activity depends on colony strength and brood activity (Amdam et al. 2009; Abou-Shaara et al. 2012). Other endogenous factors include infection with diseases and parasites (especially *Nosema* species) (Abou-Shaara et al. 2012).

While exogenous factors include the availability of food plants and other environmental factors viz. temperature, humidity, sunlight, rainfall, elevation, etc. So, there is a close proximity among these factors, which affect foraging activity and preference. *A. dorsata* was found to prefer more sunlight compared to *A. florea* and *A. mellifera*. All honey bee species showed the adaptation to different environmental factors to utilise different ecological niches. Generally, the normal colonies were big in size during January and February compared to other months. However, all the selected sites revealed wide variations with regard to temperature, and the highest (38 °C) was recorded at Hosakere Station, while the lowest (29 °C) was recorded at Siddaganga Mutt, as shown in Table 5.8 and Fig. 5.6. Parameters, especially temperature, solar radiation and humidity, can have an effect on foraging activity of honey bees (Iwama 1977; Reddy et al. 2015; Clarke and Robert 2018; Soares et al. 2019). This leads to variation of commencement, distance, cessation time and

Table 5.8 Temperature range at seven selected study sites

Sl. no.	Selected sites	Temperature (average value)
1.	Dabaspeet	36 °C
2.	Hirehalli	30 °C
3.	Siddaganga Mutt (SM)	29 °C
4.	Tumkur University Campus (TUC)	35 °C
5.	SSCW	31 °C
6.	Hosakere	38 °C
7.	Harenalli	36 °C

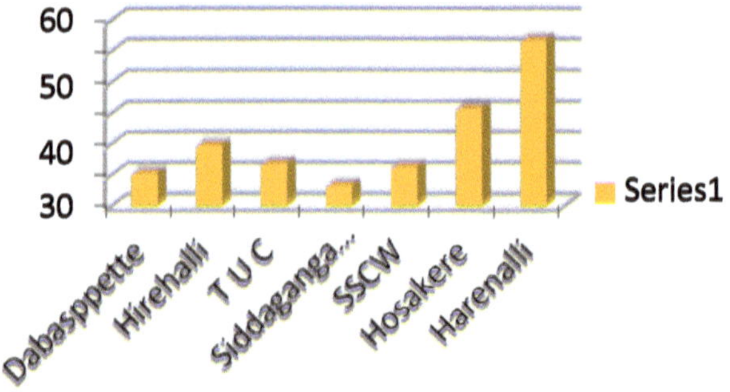

Fig. 5.10 Variation in number of hives at study sites

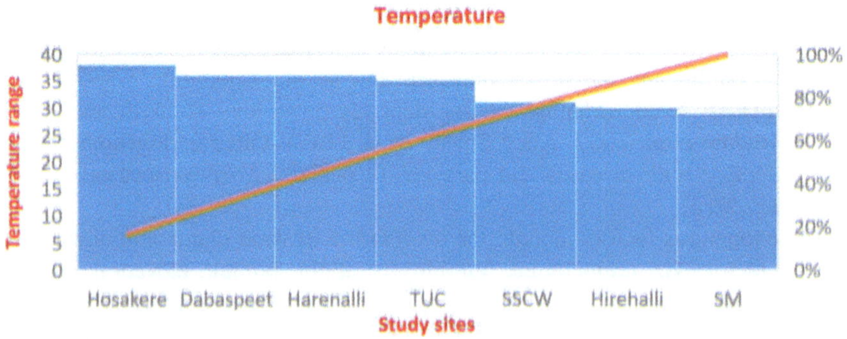

Fig. 5.11 Temperature range at study sites

foraging time, important for colony survival. So, the environmental factors don't act separately, and together they can form a suitable situation for honey bees and modify foraging behaviour (Willmer and Stone 2004; Polatto et al. 2014) (Figs. 5.10, 5.11, 5.12, and 5.13).

Fig. 5.12 Nectar and pollen grain collecting plants of honey bees

Fig. 5.13 Water sources for honey bees during foraging activity

7 Conclusion

Honey bees showed a wide variation with regard to nest, foraging preference and plant host range. Such a variation depends on a variety of environmental factors, both endogenous and exogenous. We must formulate a continuous monitoring system to sustain honey bees for better management of the terrestrial ecosystem, particularly for food and crop management purposes.

Acknowledgement The authors sincerely thank the Chairperson, DOSR in Zoology, Tumkur University, Karnataka, for providing laboratory facilities to carry out the research on honey bees.

References

Abou-Shaara HF (2012) Notes on water collection by honeybees. Bee World 89:50–51

Amdam GV, Rueppello O, Fondrk MK, Page RE, Nelson CM (2009) The nurses load: early-life exposure to brood-rearing affects behaviour and lifespan in honey bees (*Apis mellifera*). Exp Gerontol 44:447–452

Arias MC, Sheppered WS (2005) Phylogenetic relationships of Honeybees (Hymenoptera: Apinae: Apini) inferred from nuclear and mitochondrial DNA sequence data. Mol Phylogenet Evol 37(1):25–35

Baig AR, Rashid M (2007) Honey bee foraging algorithm for multimodal and dynamic optimisation problems. Proceedings of the 9th Annual Conference on Genetic and Evolutionary Computation, July 7-11, United Kingdom

Balayianni G, Balayiannis P (2008) Bee honey as an environmental bio indicator of pesticides' occurrence in six agricultural areas of Greece. Arch Environ Contam Toxicol 55(3):462–470

Butler CG (1946) Bee department, Report for 1939-45, pp 194–205. https://doi.org/10.23637/ERADOC-1-87

Cane JH, Sipes S (2006) Characterizing floral specialization by bees: analytical methods and a revised lexicon for oligolecty. In: Waser NM, Ollerton J (eds) Plant-pollinator interactions from specialization to generalization. University of Chicago Press, Chicago, IL, pp 99–122

Carvalho AT, Schlindwein C (2011) Obligate association of an oligolectic bee and seasonal aquatic herb in semi-arid north eastern, Brazil. Biol J Linn Soc 95:719–733

Chamberlian K, Briens M, Jacobs JH, Clark SJ, Pickett JA (2012) Use of honey bees (Apis mellifera L.) to detect the presence of Mediterranean fruit fly (Cera-titis capitate Wiedemann) larvae in Valencia oranges. J Sci Food Agric 92:2050–2054. https://doi.org/10.1002/jsfa.5742

Clarke D, Robert D (2018) Predictive modelling of honey bee foraging activity using local weather conditions. Apidologie 49:386–396

Claus R, Michael SE, Nicolas JV (2020) A primer of host-plant specialization in bees. In: Emerging tropics in life science. Portland Press. https://doi.org/10.1042/ETLS20190118

Dalio JS (2015) Foraging behaviour of *Apis mellifera* on *Trianthema portulacastrum*. J Entomol Zool Stud 3(2):105–108

Dimou M, Thrasyvoulou A (2007) Collection of wax scale (Ceroplastes sp.) by the honey bee *Apis mellifera*. J Apic Res 46:129

Engel MS (1999) The taxonomy of recent and fossil honeybees (Hymenoptera: Apidae; *Apis*). J Hymenopt Res 8(2):165–196

Engel MS (2002) The honey bees of India, Hymenoptera: Apidae. J Bombay Nat Hist 99:5–7

Ferrari RR (2015) A new species Geodiscelis Michener and Rozen (Hymenoptera: Colletidae: Xeromelissinae) from Peru. Zootaxa 4033:293–300

Gallai N, Salles JM, Settele J, Vaissierea BE (2009) Economic valuation of the vulnerability of world confronted with pollinator decline. Ecol Econ 68:810–821

Garibaldi LA, Carvalheiro LG, Vaissiere BE, Gemmill-Herren B, Hipolito J, Freitas BM (2016) Mutually beneficial pollinator diversity and crop yield outcomes in small and large farms. Science 351:388–391

Goulson D, Nicholis E, Botias C, Rotheray EL (2015) Bee declines driven by combined stress from parasites, pesticides and lack of flowers. Science 347(6229):1255957

Gouw MS, Gimenes M (2013) Differences of the daily flight activity rhythm in two Neotropical stingless bees (Hymenoptera, Apidae). Sociobiology 60:183–189

Greenleaf SS, Kremen C (2006) Wild bees enhance honey bee's pollination of hybrid sunflower. Proc Natl Acad Sci USA 103:13890–13895

Gupta RK, Reybroeck W, Van Veen JW, Gupta A (2014) Beekeeping for poverty alleviation and livelihood security, technological aspects of beekeeping, vol 1, pp 63–101. https://doi.org/10.1007/978-94-017-9199-1_2

Hadisoesilo S, Otis GW, Meixner M (1995) Two distinct populations of cavity-nesting honeybees (Hymenoptera: Apidae) in South Sulawesi, Indonesia. J Kans Entomol Soc 68:399–407

Iwama S (1977) A influencia dos fatores climaticos na atividade externa de *Tetragonisca angusta* (Apidae, Meliponinae). Bol Zool Univ S Paulo 2:189–201

Juan CT, Virginia MR, Yostin A, Anovel B (2022) Environmental conditions and bee foraging on watermelon crops in Panama. J Anim Behav Biometeorol. https://doi.org/10.31893/jab.22034

Klatt BK, Holzschuh A, Westphal C, Clough Y, Smit I, Pawelzik et al (2014) Bee pollination improves crop quality, shelf life and commercial value. Proc R Soc B Biol Sci 281:20132440

Klein AM, BE Vaissiere, JH Cane et al (2007) Importance of pollinators in changing landscapes for world crops. Proc Bio Sci 274(1608):303–313. https://doi.org/10.1098/rspb.2006.3721

Kremen C, Williams NM, Bugg RL, Fay JP, Thorp RW (2004) The area requirements of an ecosystem service, crop pollination. Ecol Lett 7:1109–1119

Kumar SS, Govindaraj M (2013) A detailed study about foraging behaviour of artificial bee colony (ABC) and its extensions. Int J Eng Technol 5:992–997

Kumar D, Sharma V (2016) Evaluation of Acacia species as honey bees forage potential. Int J Sci Res 5(1):1726–1727

Kumar M, Singh R (2016) Initiation-cessation and period of foraging activity of honey bees on coriander (*Coriandrum sativum L.*) flowers. Advances in Life Sci 5(23):11119–11121

Linksvayer TA, Fewell JH, Gadau J, Laubichler (2012) Developmental evolution in social insects: regulatory networks from genes to societies. J Exp Zool (Mol Dev Evol) 318:159–169

Mattu VK, Bhagat T (2015) Pollinator diversity and relative abundance of insect visitors on apple crop in Kullu hills of Western Himalaya, India. Int J Sci Res 4(6):264–268

Mattu VK, Raj H, Thakur ML (2012) Foraging behaviour of honeybees on crop and its variations with altitude in Shimla Hills of Western Himalaya, India. Int J Sci Nat 3(1):296–301

Michener CD (2000) The bees of the world. John Hopkins University Press, New York

Michener CD (2007) The bees of the world, 2nd edn. John Hopkins University Press, New York, Baltomore, MD

Muller A, Kuhlmann M (2008) Pollen hosts of Western palaerctic bees of the genus Collets (Hymenoptera: Colletidae): the Asteraceae paradox. Biol J Linn Soc 95:719–733

Neff JL (2004) Hooked hairs and not so narrow tubes: two new species of Collectes Latreille from Texas (Hymenoptera: Apidae: Colletidae). J Hymenopt Res 13:250–261

Painkra GP, Sachin KJ, Vinod KP (2021) Foraging behaviour of various species of Honey bees on Maize among its correlation with weather parameters. J Entomol Zool Stud 9(1):613–617

Pankiw T, Tarpy DR, Page RE (2002) Genotype and rearing environment affect honeybee perception and foraging behaviour. Anim Behav 64:663–672

Polatto LP, Chaud-Netto J, Alves-Junior VV (2014) Influence of abiotic factors and floral resource availability on daily foraging activity of bees. J Insect Behav 27:593–612

Raffiudin R, Crozier RH (2007) Phylogenetic analysis of honeybee behavioural evolution. Mol Phylogenet Evol 43(2):543–552

Reddy CC (1980) Observations on the annual cycle of foraging and brood rearing by *Apis cerana indica* colonies. J Apic Res 19(1):17–20

Reddy PR, Rashmi T, Verghese A (2015) Foraging activity of Indian honey bee *Apis cerana*, in relation to ambient climate variables under tropical conditions. J Environ Biol 36:577–581

Robinson WE, Morse RA (1989) The value of honeybees as pollinators of US crops. Am Bee J 129(1):477–487

Simpson BB, Neff JL (1981) Floral rewards: alternatives to pollen and nectar. Ann Mo Bot Gard 68: 301–322

Sivaram V (2001) Honeybee flora and beekeeping in Karnataka state, India. Proceedings of 37th International Apiculture Congress, Durban, South Africa

Soares KO, Lima MV, Evangelista-Rodrigues A, Silva AAF, Silva FJDA, Da Lima AIB, Costa CRG (2019) Factors influencing the foraging behaviour of *Trigona* spinipes (Apidae, Meliponinae). Biol Rhythm Res 52:1–11

Sushil SN, Stanley J, Hedau NK, Bhatt JC (2013) Enhancing seed production of three Brassica vegetables by honey bee pollination in North-Western Himalayas of India. Univers J Agric Res 1:49–53

Tanaka H, Roubik DW, Kato M, Liew F, Gunsalam G (2001) Phylogenetic position of *Apis nuluensis* of northern Borneo and phylogeography of A. cerana as inferred from mitochondrial DNA sequence. Insect Soc 48(1):44–51

Thorp RW (2000) Pollen collection by bees. Plant Syst Evol 222:211–223

Thuijsman F, Peleg B, Amitai M, Shmida A (1995) Automata, matching and foraging behaviour of bees. J Theor Biol 175:305–316

Willis LG, Winston ML, Honda BM (1992) Phylogenetic relationships in Honeybees (Genus *Apis*) as determined by the sequence of the cytochrome oxidase II region of mitochondrial DNA. Mol Phylogenet Evol 1(3):169–178

Willmer PG, Stone GN (2004) Behavioural, ecological, and physiological determinants of the activity patterns of bees. Adv Study Behav 34:347–466

Abiotic Stress and Physiological Changes in Flowers Can Mediate the Pollinator Behavior

6

Pritha Ghosh and K. S. Jagadish

Abstract

Unfavorable weather during flowering puts pollinator-plant interactions at risk and drastically lowers agricultural output. Flower metabolism serves as the foundation for these traits. Two frequent stresses that impact plant growth and development are drought and rising temperatures. Although plants must use their inadequate resources to generate features of the flower and attract pollinators, the reproductive stage is mainly prone to stress. Research indicates that abiotic stressors alter the physiological and floral physiology of *Impatiens glandulifera*, cross-pollinated by bee species, during the flowering season. Higher temperatures shortened the life span of flowers, whereas stress from dry circumstances can diminish bloom production by up to 90%. Higher temperatures led to lower content of nectar and pollen protein, indicating that both stresses influenced floral characteristics. Flowering may be negatively impacted even if plants use physiological resistance techniques in response to drought pressure and elevated temperatures. Pollinator behavior is mediated by all these effects.

Keywords

Mutualistic interaction · Abiotic stress · Floral provisioning · Bee-pollinated species

P. Ghosh (✉)
Department of Entomology, School of Agriculture, Lovely Professional University, Phagwara, Punjab, India

K. S. Jagadish
Department of Entomology, GKVK, UAS, Bengaluru, Karnataka, India

A. K. Chakravarthy (ed.), *The Role of Pollinators in Tropical Landscapes*,
https://doi.org/10.1007/978-981-95-9139-8_6

1 Introduction

Since the Devonian period, far before the Cretaceous and the subsequent bursts of angiosperms and higher orders of insects, there has been an ecological and evolutionary preference for insect-mediated spore dissemination (Smith 1975). Due to the bristles and retrorse hooks acting as physical attachments to arthropod setae, animals have been able to disseminate mid-Devonian spores with extremely elaborate ornamentation. Pollination ecosystem has been a rapidly studied field since the 1960s. Flowers' primary metabolism supports physiological and ecological processes involved in plant reproduction and floral development. Carbohydrates and amino acids play as vital food sources, while also supplying energy and precursors for secondary metabolic events, including the production of floral colors and volatile compounds (VOCs).

Before the Cretaceous and the subsequent bursts of angiosperms and higher orders of insects, there was an ecological and evolutionary preference for insect-mediated spore dispersal dating back to the Devonian period (Smith 1975). Animals have spread mid-Devonian spores with extremely intricate ornamentation because the hooks have functioned as attachments to arthropod setae. Pollination ecology has grown rapidly since the 1960s. Primary metabolism in flowers supports the physiological and ecological processes essential for floral growth and reproduction of plant. Sugars and amino acids not only serve as vital food sources for pollinators but also supply energy and act as precursors for secondary metabolic pathways that produce pigments and scent compounds in flowers. In animal-pollinated plants, mutualistic interactions between pollinators acting as signal receivers and flowers acting as signal senders lead to the evolution of floral characteristics. Pollinator-mediated selection is believed to create flower colors, which vary widely and affect pollinator attraction and foraging behavior. Pollinators and plants must regularly coordinate for cross-pollination to occur. Key characteristics of flowers, such as size, shape, color, orientation, markings, scent, and movement, developed in plants draw pollinators and influence pollen transport and fertilization (Kevan et al. 1975). According to Macior (1974), pollinators have a range of behavioral characteristics that enable them to forage on a variety of plants efficiently. The intrinsic circadian rhythms of several plant species regulate the production of floral characteristics and behavioral adaptations for pollinators (Moore 2001; Bloch et al. 2017). Animals can obtain and react to signals at the right period of day, for which circadian clocks play a crucial role in integrating external environmental inputs like light and temperature. For instance, the congregation and release of pigmented and fragrant VOCs, respectively, make flowers' hues and fragrances beautiful presentations of their metabolic resources (Khan and Giridhar 2015).

2 Flower Physiology

In seed plants, flower commencement marks a critical transition from vegetative to reproductive growth. This occurrence is especially significant in seed plants because of the unusual interplay between vegetative and reproductive growth, which is influenced by the flower's morphology. Flowers are basically modified shoots that emerge from shoot meristems, generating structures known as floral primordia. Flowering plants are responsible for 90% of the 350,000 known plant species. The most enigmatic stage of a plant's life is 0. It increases the variety of genetic recombination by providing plants with a mechanism for genetic outcrossing. Flowers are specialized structures that are very different from the vegetative plant body in terms of morphology and cell types. A shoot apex experiences physiological and biochemical changes as it gets ready to develop into a floral bud. The success of the plant's reproduction depends on the precise timing of flowering. To generate fruit and seeds, plants need to be able to recognize when to produce blossoms.

3 Flower Size and Timing of Anthesis

Various physiological impacts of temperature on floral features have been found in research thus far; the processes governing these variations are probably influenced by whether and when plants are exposed to stress of heat (Wahid et al. 2007). The size of individual flowers, for example, is affected by temperature. When artificially warmed to 23 °C, the flower diameter of pumpkin plants (*Cucurbita maxima*) was reduced (Hoover et al. 2012). The corolla of flower tubes of morning glory (*Lpomoea trichocarpa*) grown in colder climates was exhibiting shorter compared to those generated in warmer climates.

The timing of anthesis is also influenced by temperature; on warm mornings, flowers open 2–3 h earlier. Given their physical ability to access floral provisioning or rewards, pollinators are known to be influenced by floral dimensions, as demonstrated by the morphological correlations between proboscides and nectar spur length (Nilsson 1988). Variations in floral dimensions disturb pollinator foraging effectiveness even if rewards are still available because flower size affects the energy cost of obtaining that reward (Harder 1983).

Similarly, pollinators become active earlier and gain access to those rewards if anthesis happens earlier due to raised temperatures in the morning; however, this affects the availability of rewards to the pollinator's activity in the rest of the day (Murcia 1990). It's similarly significant to remember that pollinator attraction is influenced by flower size (Totland 2001).

4 Production of Pollen, Scent, and Nectar

Additionally, temperature affects the generation of pollen, nectar, and flower fragrance. According to some research, endogenous floral smell synthesis appears to diminish with increasing temperatures (Sagae et al. 2008), yet warmer temperatures enhance the emissions of organic compounds released by flowers (Yuan et al. 2009). Although temperature affects nectar content, composition, and production (Pacini et al. 2003), there has been little study relating these impacts to climate change. Rising temperatures may modify the emission or volatilization of floral smells, decreasing the detectability of blossoms for insects such as moths, which count on long-distance signals to locate floral rewards (Kevan and Baker 1983; Yuan et al. 2009).

Undoubtedly, changes in the composition of nectar and production affect pollinator energetics and activity as well as long-term impacts on pollinator fitness (Burkle et al. 2013), perhaps particularly for insects like wasps and lepidopterans that depend on nectar for both sugars and amino acids (Kevan and Baker 1983). Reduced pollen supply can have a severe impact on bee reproductive success, as many species rely on pollen from several plant sources to properly rear their offspring.

The preference of bumblebees for potato (*Solanum tuberosum*) blossoms with pollen grains over those with nonviable or shriveled pollen grains suggests that the previous one are more nutrient-dense, while there is a lack of clear data on whether pollinators find less viable pollen less appetizing (Batra 1993). A Mediterranean plant's nectar quality concentration rose with temperature up to 38 °C, but it decreased with temperature (23 °C vs. 19 °C). There is evidence of anthomophilous insects from various orders acting as pollinators (Scaven and Rafferty 2013).

5 Coleopteran Pollinators

According to Proctor and Yeo (1973), the most archaic pollinators are coleopterans. Angiosperms remained well-distinguished by the time they gained prominence till early Cretaceous period. According to Leppik (1975, 1977) and Gottsberger (1974), their evolutionary relevance resulted from their function as "mess-and-soil" pollinators linked to elaborately constructed, open bowl-shaped blooms, frequently exhibiting a characteristic aroma of fermented fruit or amino acids.

Beetles consume floral fluids, pollen, and floral components or unique food bodies in flowers. Many flowers are visited by specialist beetles that do not exhibit the cantharophily syndrome, and some adult beetles appear to be confined to floral diets. The Adephagan suborder of predatory beetles does not visit flowers, but the Elateridae, Scarabeidae, and Cleridae are among the Polyphaga that do.

6 Dipteran Pollinators

Diptera are another group of early-evolved pollinators distinguished by suctorial or lapping mouthparts. Most Nematocera have short but changeable proboscis, limiting them to bloom with straightforwardly reachable nectaries. These flowers, like those in Senecio, Polygonum, and other Cruciferae genera, have short or slightly recessed corolla tubes. The bulk of these dipteran pollinators, including those from the families Mycetophilidae, Cecidomyiidae, Simuliidae, Chironomidae, and Ceratopogonidae, are relatively small.

Flowers with deeper tubular corollas are visited by Nematocera with longer proboscides, such as the families viz., Culicidae and Bibionidae. These insects naturally search for nectar (Hocking 1953), while some families, such as Bibio, Scatopse, Sciara, and Atrichopogon, consume pollen (Downes 1971; Willis and Burkill 1895).

The Muscidae are a big family with some well-known anthophilous species. Scatophaga, for example, visits flowers and searches for nectar as well as their prey. However, not all Muscidae exhibit flower-visiting behavior; several have short proboscides and only visit the blooms' exterior sections. Despite their ubiquity and frequent flower visits, their importance in pollination is sometimes overlooked. Members of the Anthomyiidae family, on the other hand, are well-recognized for being frequent floral visitors, with some species having relatively long probosces. These flies eat both nectar and pollen, with the latter potentially providing a crucial source of protein for adult nutrition (Kevan et al. 1975).

7 Lepidopteran Pollinators

Most lepidopterans rely significantly on flower nectar, but the moths also consume a range of other liquids. Some micropterigidae are direct pollen feeders (Faegri and van der Pijl 1978). There are several Lepidoptera with unusually long proboscides of about 25–30 cm spur, named as *Xanthopan morgani f. praedicta*, thought to pollinate *Angraecum sesquipedale*, an orchid. Stimuli perceived with vision are the most important in flower seeking resources for the European hawk moth (*Macroglossa stellatarum*) (Knoll 1922). Other sphingids, such as Polyptychus, have short proboscides and are known to pollinate a wide variety of flowers. In general, Lepidoptera can only consume less gelatinous nectars, although some, such as *Plusia gamma* (Proctor and Yeo 1973), release saliva to dilute syrupy nectars, allowing for imbibition.

8 Hymenopteran Pollinators

Hymenoptera is the most significant anthophylous order. *Symphyta, sawiies*, have minute mouthparts and no adaptation to anthophily, but they commonly visit flowers with effortlessly reachable nectar, particularly Ranunculus and Compositae.

Many lay eggs in the blooms of the plant on which their larva prefer as host, and which include Salix, Symphoricarpos, and Scabiosa. They ingest flowers that get extensively dusted with pollen. Although most sawflies (hymenopterans) are generalist types of visitors to the flowers, some have explicit pollination links with orchids (Coleman 1932). The Formicidae are regular visitors of floras, although their involvement in pollination is little understood. Many of the flowers seen have closely related characteristics to reproduce. Apoidea are the most significant and adaptable insects for anthophily. They are physically, developmentally, and taxonomically varied and are highly improved for nectar absorption, and their bodies for pollen gathering. They learn floral intricacies fast and are interactively expert at deploying resources from flowers. Bees, including adults and larvae, are almost totally vegetarian, and they load cells with pollen and nectar for larvae or during their dearth period for the purpose of survival.

The processes of carbohydrate hydrolysis and processes related to successful fertilization by the pollen and nectar composition have been better validated by recent discoveries. Lastly, a new area of research is developing that investigates how pollinators' floral preferences are impacted by the chemical changes of sugars and proteins by invading bacteria.

Amino acid production pathways in flowers have been discovered, and the proteins that regulate germination of pollen tubes have been studied. Pollination is one of the most fascinating natural phenomena. Flowering plants reproduce by pollination. This procedure involves the transfer of pollen from male to female portions of the same or different plants. For some plants, pollen transfer requires assistance from another external agent.

Plants and pollinators both mutually benefit from each other. The pollinator benefits from the plant-pollinator connection by feasting on the flowers filled with nectar and pollen. The plant being sessile benefits as the pollinator travels from flower to bloom, sharing pollen while it forages for food. This pollen migration allows the plant to reproduce and transmit its genetic information to other plants. Most flowering plants rely on pollinators to reproduce.

Unfortunately, pollinator populations are declining worldwide, resulting in a negative impact on the flowering plants that rely on them. It also causes problems for people who rely on pollinators for a variety of reasons, including the food we consume and the air we breathe. This fall in pollinator populations is due in part to human activities that have contributed to the loss of wild and flower-rich habitats.

Anthropogenic greenhouse gas emissions have continually increased, resulting in the current high concentration of CO_2 in the atmosphere (CO_2) and ozone depletion, which are the causes of climatic variation (IPCC; https://www.ipcc.ch/). Global temperatures are anticipated to increase by 1.5 °C in the upcoming years (NOAA; https://www.noaa.gov/).

All these changes have a significant impact on the aptness of ecosystem health and plant pollinator interactions. Flower orientation, temperature, and phenology all influence phenology. False springs are becoming more common as heat waves coincide with late winters. Worldwide farmers follow to harvest early to avoid crop damage due to environmental challenges. Wild pollinator species are known to move through life cycles more quickly to avoid harsh weather (Balasubramanian et al. 2006; Crimmins et al. 2010), and many have dispersed to arctic and upland regions due to their host's altered phenology resulting from climatic shifts (Anderson et al. 2012).

Plants' nonreproductive tissues adapt quickly to variations in their surroundings; for example, foliage and cotyledons often get extended as a result of temperature rise (Quint et al. 2016), and formation of lateral roots has been observed in response to drought (Babe et al. 2012). Flowers have specialized metabolism, which regulates the emission of volatiles and pigment production, the signs used by flowers to attract pollinators, and responds to environmental changes. Abiotic stressors have an impact on basic and specialized metabolism, as well as physiological and ecological activities of flowers, such as pollination and fruit and seed set.

Climate change influences cross-species relationships. This suggests that insect-mediated pollination necessitates sophisticated signaling with respect to the circadian clock and environmental indicators to possess timing within and between organisms, ensuring plant reproduction. Floral temperature is an additional critical factor in managing and coordinating interactions, either directly or indirectly increasing floral reward indications in flowers to attract pollinators (Heinrich 1972; Harrap et al. 2017). Reports revealed that bumble bees use less energy to reach the optimal flying temperature while foraging from warmer blooms (Sapir et al. 2005; Rands and Whitney 2008).

An investigation found that pollinators transform their feeding habits depending on flower temperature, preferring cooler plants during hot spells and warmer plants during cool spells (Norgate et al. 2010). Floral temperature affects germ cells and seed viability, as well as pollinator visits. As a result, several plant species have advanced to regulate the microclimate of flowers to preserve normal physiology (Corbet 1990).

Flowering plants depend on a variety of pollinators for reproduction and employ several techniques to attract them. Pollinators are paid for their visits in plant-pollinator mutualisms, usually with carbohydrate-rich nectar, but nutritional incentives such as pollen or wax can also be provided. Plants not only offer nutrients to ectothermic pollinators, but they also warm them. Bumble bees manage their temperature, but if given the chance, they may graze at warmer blossoms, showing that there is some value to doing so (Whitney and Glover 2007).

9 Effects of Pollinator Physiology on Flowering Plant Fitness

Pollinators are also at risk from a variety of climate-related changes. There has been little investigation into how rising temperatures affect the physiology of pollinators. In a research study about insect pollinators' temperature ecology and how it may affect the long-term pollination success of blooming plants, it has been found that, when temperatures rise, thermoregulatory constraints determine which pollinators are active and when (Willmer and Stone 2004). Body size is proportionate to the capability to physiologically (rather than behaviorally) adjust to temperature changes. Pollinating insects larger in size are better able to regulate temperatures than smaller insects (Bishop and Armbruster 1999).

Bigger insects store more heat and release it slowly, thus increasing the risk of overheating. Indeed, reports also indicated that the passive convective loss of heat via the thorax of Asian honeybees was inversely proportional to the size of the body (Dyer and Seeley 1987).

10 Implications for Plant-Pollinator Networks

Heating up has a cumulative effect on flowering plants and insect pollinators, shaping pairwise connections and overall interaction networks (Fig. 6.1). Plant-pollinator interactions are somewhat understood (Olesen et al. 2008), and

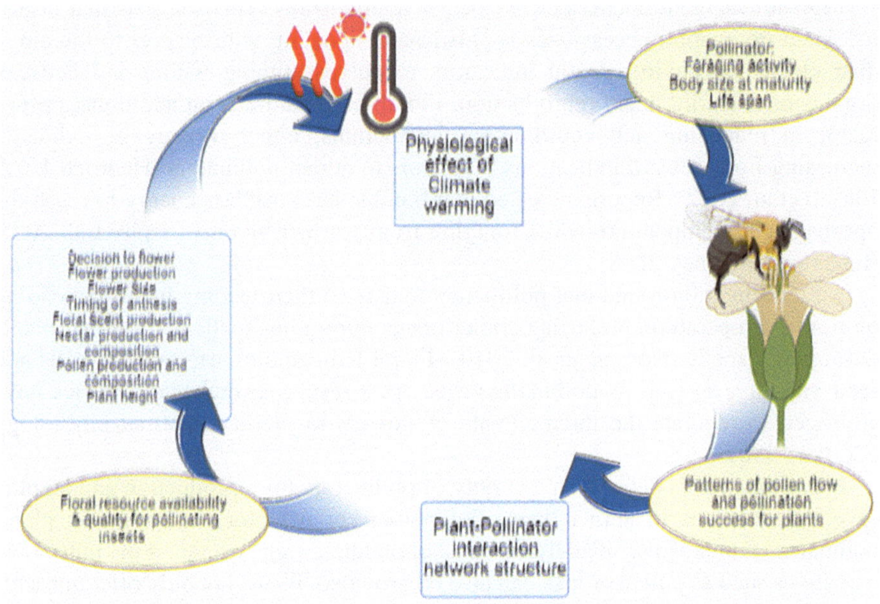

Fig. 6.1 Framework showing the impact of temperature rise and flower-pollinator interactions

researchers explored how these systems respond to disturbances induced by phenology due to altered climate (Memmott et al. 2007) and temperature shifts (Devoto et al. 2007).

Although generalization and the development of new interactions help to buffer networks, species loss and phenological mismatches degrade them over time (Burkle et al. 2013). Physiological responses to heating may be able to modify plant-pollinator links even in the absence of changes in species composition or seasonal overlap. Positive feedback loops between pollinators and plant populations appear to be likely, as direct plant physiological responses lower pollinator reproductive success, which leads to decreased pollination success.

If plant and pollinator responses are complementary, resulting in decreased flower and body sizes, or if anthesis and feeding occur earlier in the day, interactions may have minimal net impact. Although species responses may be less directed and more unpredictable, they give rise to additional interactions that reinforce the stability of the network.

11 Plant-Pollinator Interactions: Physiological Responses to Warming

Mature pollen is resistant to temperature changes, a feature that vanishes when stresses arise throughout pollen growth and development (Franchi et al. 1996). Pollen development is an energy-intensive process that is powered by sucrose from photosynthetic tissues and transported to the anthers via the filament. In the anthers, sucrose is hydrolyzed to glucose and fructose, which are then internalized by pollen grains and transformed and stored as starch and/or lipids (Taurino et al. 2018). In drought-stressed rice (25% soil water content), starch accumulated outside the anther wall rather than within pollen grains. Transcriptome data showed reduced expression of INVs, HXK, and glycoside hydrolase genes, suggesting that pollen abortion arises from disrupted sugar metabolism in the anthers rather than limited sugar transport (Jin et al. 2013). Heat-treated flowers (32 °C: 26 °C, day: night) have a reduced ability to consume carbohydrate because pollen grains contain less starch granules, although the concentration of soluble sugars in the locular fluid increases due to stress in tomato (Pressman et al. 2002). Abiotic stress affects the floral physiology of bee-pollinated species, including *Impatiens glandulifera*, during the flowering phase.

12 Orientation of the Flower/Capitulum Aligns Pollen Emergence with the Temporal Patterns of Pollinator Visits

According to a study, flower orientation influences the floral microclimate, which in turn impacts pollinator and plant interactions as well as reproductive fitness. The east-facing sunflower capitula exhibited earlier style elongation, pollen presentation, and pollinator visitation when the capitula were modified to face west. Compared to capitula facing west, east-facing capitula produced more progeny and, occasionally, heavier, more filled seeds. The timing of style elongation, pollen emergence, and foraging pollinators was found to be significantly influenced by changes in the capitulum ambient temperature.

The effectiveness of style extension and pollen emergence, pollinator visitation, and plant fitness are all directly impacted by these findings, which imply that eastward capitulum orientation helps regulate daily floral temperature oscillations (Creux et al. 2021).

For reproduction, flowering plants need a wide variety of pollinators, and they use a wide variety of tactics to draw them in. Pollinators are usually rewarded for their visit in plant-pollinator mutualisms by being given nectar, which is high in carbohydrates, although other food rewards like pollen or wax may also be provided. During the subsequent time window, insect visits to east-facing but not west-facing capitula increased dramatically; over the 20 min starting at ZT 1.0, east-facing capitula received >200 visits, whereas west-facing capitula saw only slight increases. Similarly, the study found that insect visits to wild sunflower populations' easterly-oriented capitula at ZT had increased. Following ZT 1.5, there were no statistically significant variations in pollinator visits to capitula oriented in opposite directions. In the morning, pollinators favored capitula facing east rather than west.

13 Physiological Responses of Floras Due to Exposure of Abiotic Stress (Table 6.1)

The pathways of global ecological alteration drivers (climate, nitrogen, and CO_2 interactions) may impact the pollination process. Dashed arrows show pathways (4, 5) found in other studies (see Fig. 6.2), and solid arrows indicate pathways

Table 6.1 Physiological changes in flowers due to drought stress

Plant	Stress during floral development	Floral response (physiological)	Floral response (metabolic)
A. thaliana	Drought	Arrested floral development. Sterility of male and female reproductive parts	Amplification of disaccharide catabolic genes Attenuation of fatty acid metabolism genes.
A. thaliana	Drought	Early arrest of flower development Sterility of anthers and ovules	Transcriptional changes of ABA-responsive genes, pectin esterase, amino acid transporter genes, and hexose transporter
Oryza sativa	Drought	Abnormal growth in pollen and anther	Peculiar carbohydrate metabolism Downregulation of metabolic genes
	Drought	Reduction in yield, healthy grain	Variations in the enzymatic action and genes responsible for ABA-expression in spikelets
Triticum aestivum	Drought	Sterility of Pollen	Alteration in metabolic activity
Ipomopsis aggregata	Drought	Reduction in pollen and nectar production	Changes in nectar quality

determined in this study. (1) Combined effects of factors on flower size, abundance, nectar volume, and sex ratio. (2) Flowering phenology was impacted by the above three drivers. (3) The drivers impacted nectar composition above. Bees more frequently forage (6) and feed on nectar (7) from the elevated N treatment, and impact interacted with CO_2 and temperature. (8) Nectar from the elevated N and elevated CO_2 treatments significantly decreased workers' longevity.

14 Disruption to Foraging Behavior

Like most arthropods, pollinators use olfactory signals through volatile organic compounds (VOCs), in a particular quantity, which is recognized by its host plants from a distance (Bruce et al. 2005; Beyaert and Hilker 2014). Based on the principle of the most effective pollinator, the floral characteristics should alter throughout time to fascinate and match the most competent and abundant pollinators (Stebbins 1970). From this, it is well evident and anticipated that pollinators will be flexible and adaptive when communications between plants and pollinators are chiefly generalist. Because of their low reaction plasticity, specialized pollinators are less able to acclimatize to climate variation (Forrest 2015; Kevan et al. 1975). The process of recognition collapses when the VOCs' composition or perception fluctuations happen. The signal quality is changed by variations in the ambient temperature, which also impacts the volatile release levels from florets (Farré-Armengol et al. 2014).

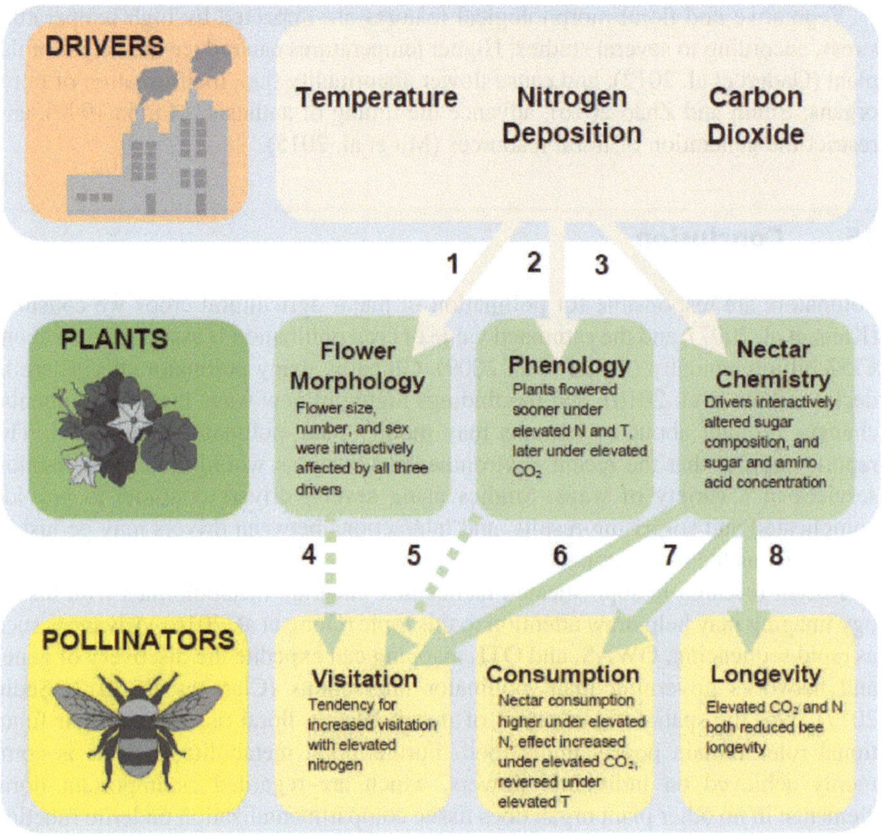

Fig. 6.2 Environmental change pathways

Warming alters the ratios of molecules that make up floral aromas in addition to elevating the rate and richness of floral volatile organic compounds (VOCs) emissions. Variable species and chemicals within species have variable responses to temperature in floral emissions (Scaven and Rafferty 2013; Farré-Armengol et al. 2014). Plant-pollinator interactions may be impaired by species-specific deviations to the volatile mix released because pollinators will no longer be drawn to host-plant floral cues if they are confused with a more fragrant environment.

Additionally, host selection is significantly influenced by plant visual stimuli (Döttrel and Vereecken 2010). Another indicator used by pollinators, especially bees, to recognize flowers is their warmth. Bees use thermal detectors in their tarsi and antennae to identify variations in blossom temperature (Hammer et al. 2009). When the temperature is low, bees choose to forage warm nectar from blooms (Dyer et al. 2006). As evidenced by species of stingless bees (Trigona carbonaria), which favored warmer nectars at lesser air temperatures (between 23 and 30 °C) but switched to ambient temperature nectar when the air temperature reached 34 °C, this fondness for warmer flowers may, however, differ on this background (Norgate et al. 2010).

Vegetative and floral morphological features are impacted by high temperature stress, according to several studies. Higher temperatures can reduce the height of the plant (Qaderi et al. 2012), and cause flower abnormality (i.e., malformation of floral organs; Smith and Zhao 2016), advance the timing of anthesis (Murcia 1990), and restrict the generation of floral resources (Mu et al. 2015).

15 Conclusion

Pollinators are responsible for pollination of major agricultural crops we consume (Klein et al. 2007), and the estimated value of crop pollination is assessed to be about €153 billion annually (Gallai et al. 2009). Globally, many pollinator species are in decline (Potts et al. 2010), and the findings highlight new ways that environmental changes brought about by humans may modify plant-pollinator mutualisms. The reports suggest that the recent environmental variations would disturb pollination services in a variety of ways. Studies using several drivers continue to provide complicated and surprising results, and interactions between drivers may be just as substantial as the major impact.

Recent research in sophisticated tactics that integrate metabolomics with histology imaging may help draw attention to this topic (Dong et al. 2016). Advances such as rapid sequencing, GWAS, and QTL mapping can expedite the discovery of genes and networks governing plant–pollinator interactions (Clare et al. 2013; Sedio 2017). Yet, the spatial organization of metabolites in floral tissues and their functional roles remain poorly understood. Furthermore, metabolite analysis is commonly achieved on individual flowers, which are regarded as important floral elements; in no other plant organ does tissue compartmentalization underlie function as much as in flowers. The last era has seen significant progress in considering global warming and its consequences on complex plant-pollinator networks. The field responses have been also documented significantly to a lesser extent with laboratory reports. These progresses should help to lessen the effects of climate change on pollination.

References

Anderson JT, Panetta AM, Mitchell-Olds T (2012) Evolutionary and ecological responses to anthropogenic climate change: update on anthropogenic climate change. Plant Physiol 160: 1728–1740

Babe A, Lavigne T, Severin JP, Nagel KA, Walter A, Chaumont F, Batoko H, Beeckman T, Draye X (2012) Repression of early lateral root initiation events by transient water deficit in barley and maize. Philos Trans R Soc Lond B Biol Sci 367:1534–1541

Balasubramanian S, Sureshkumar S, Lempe J, Weigel D (2006) Potent induction of Arabidopsis thaliana flowering by elevated growth temperature. PLoS Genet 2:e106

Batra SWT (1993) Male-fertile potato flowers are selectively buzz-pollinated only by Bombus terricola Kirby in upstate New York. J Kansas Entomol Soc 66:252–254

Beyaert I, Hilker M (2014) Plant odour plumes as mediators of plant-insect interactions. Biol Rev Camb Philos Soc 89:68–81. https://doi.org/10.1111/brv.12043

Bishop JA, Armbruster WS (1999) Thermoregulatory abilities of Alaskan bees: effects of size, phylogeny and ecology. Funct Ecol 13:711–724
Bloch G, Bar-Shai N, Cytter Y, Green R (2017) Time is honey: circadian clocks of bees and flowers and how their interactions may influence ecological communities. Philos Trans R Soc Lond B Biol Sci 372:20160256
Bruce TJ, Wadhams LJ, Woodcock CM (2005) Insect host location: a volatile situation. Trends Plant Sci 10:269–274. https://doi.org/10.1016/j.tplants.2005.04.003
Burkle LA, Marlin JC, Knight TM (2013) Plant-pollinator in- teractions over 120 years: loss of species, co-occurrences, and function. Science 339(6127):1611–1615
Clare EL, Schiestl FP, Leitch AR, Chittka L (2013) The promise of genomics in the study of plant-pollinator interactions. Genome Biol 14:207
Coleman E (1932) Pollination of Diuris pedunculata R. Br. Victorian Nat 49:179–186
Corbet SA (1990) Pollination and the weather. Isr J Bot 36:13–30
Creux NM, Brown EA, Garner AG, Saeed S, Scher CL, Holalu SV, Yang D, Maloof JN, Blackman BK, Harmer SL (2021) Flower orientation influences floral temperature, pollinator visits and plant fitness. New Phytol 232(2):868–879
Crimmins TM, Crimmins MA, Bertelsen CD (2010) Complex responses to climate drivers in onset of spring flowering across a semi-arid elevation gradient. J Ecol 98:1042–1051
Descamps C, Boubnan N, Jackquemart AL, Quinet Muriel (2021) Growing and flowering in a changing climate: effects of higher temperatures and drought stress on the bee-pollinated species Impatiens glandulifera Royle. Plants 10(5):988
Devoto M, Zimmermann M, Medan D (2007) Robustness of plant-flower visitor webs to simulated climate change. Ecol Austral 17:37–50
Dong Y, Li B, Aharoni A (2016) More than pictures: when MS imaging meets histology. Trends Plant Sci 21:686–698
Döttrel S, Vereecken N (2010) The chemical ecology and evolution of bee-flower interactions: a review and perspectives. Can J Zool 88:668–697. https://doi.org/10.1139/Z10-031
Downes AJ (1971) The ecology of blood-sucking D iptera: an evolutionary perspective. In: Fallis AM (ed) Ecology and physiology of parasites. Univ. Toronto Press, Toronto, pp 232–258
Dyer FC, Seeley TD (1987) Interspecific comparisons of endo- thermy in honey-bees (Apis): deviations from the expected size-related patterns. J. Exp Biol 127:1–26. Econ., 68, 810–821
Dyer AG, Whitney HM, Arnold SEJ, Glover BJ, Chittka L (2006) Bees associate warmth with floral colour. Nature 442:525. https://doi.org/10.1038/442525a
Faegri K, van der Pijl L (1978) The principles of pollination ecology, 3rd edn. Pergamon, Oxford, p 244
Farré-Armengol G, Filella I, Llusià J, Niinemets Ü, Peñuelas J (2014) Changes in floral bouquets from compound-specific responses to increasing temperatures. Glob Chang Biol 20:3660–3669. https://doi.org/10.1111/gcb.12628
Forrest JRK (2015) Plant–pollinator interactions and phenological change: what can we learn about climate impacts from experiments and observations? Oikos 124:4–13. https://doi.org/10.1111/oik.01386
Franchi GG, Bellani L, Nepi M, Pacini E (1996) Types of carbohydrate reserves in pollen: localization, systematic distribution and ecophysiological significance. Flora 191:143–159
Gallai N, Salles J-M, Settele J, Vaissière BE (2009) Economic valuation of the vulnerability of world agriculture confronted with pollinator decline. Ecol Econ 68(3):810–821
Gottsberger G (1974) The structure and function of the primitive angiosperm flower. Acta Bot Neerl 23:461–471
Hammer TJ, Hata C, Nieh JC (2009) Thermal learning in the honeybee, Apis mellifera. J Exp Biol 212:3928–3934. https://doi.org/10.1242/jeb.034140
Harder LD (1983) Flower handling efficiency of bumblebees: morphological aspects of probing time. Oecologia 57:274–280
Harrap MJM, Rands SA, Hempel de Ibarra N, Whitney HM (2017) The diversity of floral temperature patterns, and their use by pollinators. E Life 6:e31262

Heinrich B (1972) Temperature regulation in the bumblebee Bombus vagans: a field study. Science 175:185–187

Hocking B (1953) The intrinsic range and speed of flight of insects. Trans R Entomol Soc London 104:223–345

Hoover SER, Ladley JJ, Shchepetkina AA, Tisch M, Gieseg SP (2012) Warming, CO2, and nitrogen deposition interactively affect a plant-pollinator mutualism. Ecol Lett 15:227–234. [PubMed: 22221802]

Jin Y, Yang H, Wei Z, Ma H, Ge X (2013) Rice male development under drought stress: phenotypic changes and stage-dependent transcriptomic reprogramming. Mol Plant 6:1630–1645

Kevan PG, Baker HG (1983) Insects as flower visitors and pollinators. Annu Rev Entomol 28:407–453

Kevan PG, Chaloner W, Savile D (1975) Interrelationships of early terrestrial arthropods and plants. Palaeontology 18(2):391–417

Khan MI, Giridhar P (2015) Plant betalains: chemistry and biochemistry. Phytochemistry 117:267–295

Klein AM, Vaissiere BE, Cane JH, Steffan-Dewenter I, Cunningham SA, Kremen C et al (2007) Importance of pollinators in changing landscapes for world crops. Proc Biol Sci 274:303–313

Knoll F (1922) III. Lichsinn und Blumenbesuchd er Falters von Macroglossa stellatarum. See Ref 177. Abhandlungen der Kaiserlich-Königlichen Zoologisch-botanischen Gesallschaft in Wien 12(2):120–377

Leppik EE (1975) Morphogenic stagnation in the evolution of Magnolia flowers. Phytomorphology 25:451–464

Leppik EE (1977) Floral evolution in relation to pollination ecology. Today and Tomorrow's Print. Publ., New Delhi, India, p 164

Macior LW (1974) Pollination ecology of the front Range of the Colorado Rocky Mountains. Melanderia 15:1–59

Memmott J, Craze PG, Waser NM, Price MV (2007) Global warming and the disruption of plant-pollinator interactions. Ecol Lett 10:710–717

Moore D (2001) Honey bee circadian clocks: behavioral control from individual workers to whole-colony rhythms. J Insect Physiol 47:843–857

Mu J, Peng Y, Xi X, Wu X, Li G, Niklas KJ et al (2015) Artificial asymmetric warming reduces nectar yield in a Tibetan alpine species of Asteraceae. Ann Bot 116:899–906. https://doi.org/10.1093/aob/mcv042

Murcia C (1990) Effect of floral morphology and temperature on pollen receipt and removal in Ipomoea trichocarpa. Ecology 71:1098–1109. https://doi.org/10.2307/1937378

Nilsson LA (1988) The evolution of flowers with deep corolla tubes. Nature 334:147–149

Norgate M, Boyd-Gerny S, Simonov V, Rosa MGP, Heard TA, Dyer AG (2010) Ambient temperature influences Australian native stingless bee (Trigona carbonaria) preference for warm nectar. PLoS One 5:e12000. https://doi.org/10.1371/journal.pone.0012000

Olesen JM, Bascompte J, Elberling H, Jordana P (2008) Temporal dynamics in a pollination network. Ecology 89:1573–1582

Pacini E, Nepi M, Vesprini JL (2003) Nectar biodiversity: a short review. Plant Syst Evol 238:7–21

Potts SG, Biesmeijer JC, Kremen C, Neumann P, Schweiger O, Kunin WE (2010) Global pollinator declines: trends, impacts and drivers. Trends Ecol Evol 25:345–353

Pressman E, Peet MM, Pharr DM (2002) The effect of heat stress on tomato pollen characteristics is associated with changes in carbohydrate concentration in the developing anthers. Ann Bot 90:631–636

Proctor M, Yeo P (1973) The pollination of flower. Collins, London, p 418

Qaderi MM, Kurepin LV, Reid DM (2012) Effects of temperature and watering regime on growth, gas exchange and abscisic acid content of canola (Brassica napus) seedlings. Environ Exp Bot 75:107–113. https://doi.org/10.1016/j.envexpbot.2011.09.003

Quint M, Delker C, Franklin KA, Wigge PA, Halliday KJ, van Zanten M (2016) Molecular and genetic control of plant thermomorphogenesis. Nat Plants 2:15190

Rands SA, Whitney HM (2008) Floral temperature and optimal foraging: is heat a feasible floral reward for pollinators? PLoS One 3:e2007
Sagae M, Oyama-Okubo N, Ando T, Marchesi E, Nakayama M (2008) Effect of temperature on the scent emission and en- dogenous volatile profile of Petunia axillaris. Biosci Biotechnol Biochem 72:110–115
Sapir Y, Shmida A, Ne'eman G (2005) Morning floral heat as a reward to the pollinators of the Oncocyclus irises. Oecologia 147:53–59
Scaven VL, Rafferty NE (2013) Physiological effects of climate warming on flowering plants and insect pollinators and potential consequences for their interactions. Curr Zool 59(3):418–426
Sedio BE (2017) Recent breakthroughs in metabolomics promise to reveal the cryptic chemical traits that mediate plant community composition, character evolution and lineage diversification. New Phytol 214:952–958
Smith AP (1975) Insect pollination and heliotropismi n Oritrophiumli mno- philum (Composita¢) of the Andean paramo. Biotropica 7:284–286
Smith AR, Zhao D (2016) Sterility caused by floral organ degeneration and abiotic stresses in Arabidopsis and cereal grains. Front Plant Sci 7:1503. https://doi.org/10.3389/fpls.2016.01503
Stebbins GL (1970) Adaptive radiation of reproductive characteristics in angiosperms I: pollination mechanisms. Annu Rev Ecol Syst 1:307–326. https://doi.org/10.1146/annurev.es.01.110170.001515
Taurino M, Costantini S, De Domenico S, Stefanelli F, Ruano G, Delgadillo MO, Sanchez-Serrano JJ, Sanmartin M, Santino A, Rojo E (2018) SEIPIN proteins mediate lipid droplet biogenesis to promote pollen transmission and reduce seed dormancy. Plant Physiol 176:1531–1546
Totland O (2001) Environment-dependent pollen limitation and selection on floral traits in an alpine species. Ecology 82:2233–2244
Wahid A, Gelani S, Ashraf M, Foolad MR (2007) Heat tolerance in plants: an overview. Environ Exp Bot 61(3):199–223
Whitney HM, Glover BJ (2007) Morphology and development of floral features recognised by pollinators. Arthropod Plant Interact 1:147–158. https://doi.org/10.1007/s11829-007-9014-3
Willis JC, Burkill IH (1895) Flowers and insects in Great Britain. L-IV. Ann Bot 9:227–273; 17: 313-349, 539-70; 22: 603-4
Willmer PG, Stone GN (2004) Behavioral, ecological, and physio- logical determinants of the activity patterns of bees. Adv Stud Behav 34:347–466
Yuan JS, Himanen SJ, Holopainen JK, Chen F, Stewart CN (2009) Smelling global climate change: mitigation of function for plant volatile organic compounds. Trends Ecol Evol 24:323–331

Part II

Pollinators in the Wild and Managed Ecosystems

Impact of Climate Change on Insect Pollinators of Crops in Cultivated Ecosystems in Tropics and Subtropics

7

Pavithra Prakash Nayak and Bijitha Ramesh

Abstract

Many of the cultivated ecosystems depend on pollinators for essential ecosystem services. Only a few compilations, though, cover the state of the science and anticipated developments regarding how cultivation practices and global warming affect pollinators. This chapter summarizes the effects of climate change via species distribution, insect–crop interactions, and physiological and behavioral effects on pollinators. It also covers the effects of conventional and organic agricultural operations and agrochemical exposure, particularly in the tropics. Conservation strategies for mitigating adverse effects include addressing the sustainability of food systems and ecosystems on planet Earth.

Keywords

Climate change · Bee pollinators · Crops · Cultivated ecosystem · Conservation strategies

1 Insect Pollinators: Importance in Crop Pollination

Agricultural production is an important economic sector influenced by bee pollinators. Food crop cultivation has expanded with increasing reliance on insect pollinators (Osterman et al. 2021). Pollinators have a key part in a variety of cultivated ecosystems (Costanza et al. 2014; Bhattacharyya et al. 2017). Native species pollinate many crops effectively. Aside from bees, wasps, beetles, flies, butterflies, moths, birds, bats, and even nonflying mammals are on the list of

P. P. Nayak (✉) · B. Ramesh
Department of Climate Change, Environmental Management & Policy Research Institute, Doresanipalya Forest Campus, Bangalore, India

A. K. Chakravarthy (ed.), *The Role of Pollinators in Tropical Landscapes*,
https://doi.org/10.1007/978-981-95-9139-8_7

pollinators. These pollinators facilitate fruit setting and improve yield (Reddy et al. 2012). However, bees are used as pollinators, with a 70–80% success rate in the production of agricultural and horticultural crops (Reddy et al. 2012). Bee pollination makes up one-third of the human diet. The importance of bees as a core pollinator and their dependence on 39 crop species is described in detail (Thakur 2012; Klein et al. 2007). The bees include specialist and generalist flower visitors. Among the 80,000 reported species of insect pollinators, 17,533 species are identified as bee pollinators worldwide (Waykar and Baviskar 2016). In nature, these could be either solitary or parasocial in form (Michener 2007). It's known that crop pollination would decline more steeply without native wild pollinators. The importance of native pollinators in the production of crops like coffee, beans, chilies, and cardamom has been investigated in a landscape dominated by smallholder farmers (Klein et al. 2007). Bumble and mason bees for sweet cherries, Nomia and Osmia bees for orchard fruits, Megachile for alfalfa, and stingless bees for coffee crops are examples of non-Apis species studied to improve pollination services (Eeraerts et al. 2020; Tanda 2021; Bosch and Kemp 2002).

Das et al.'s (2018) recent compilation lists the vegetable and fruit crops that are pollinator-dependent and emphasizes the need to preserve nesting sites in agricultural landscapes. Therefore, a synergistic effect of different wild and domesticated bees is essential to supplement agricultural crop pollination, especially for pollinator-dependent species (Brittain et al. 2013; Abrol and Abrol 2012). Thus, exploiting the basic biology, nesting habitats, foraging preferences, and abundance in connection to climate change will be useful in addressing how Apis and non-Apis pollinators can better fit into agricultural plans.

2 Climate Change Impact on Bee Pollinators

Climate change is defined as an average change in the distribution of weather patterns over an extended period, ranging from decades to millions of years. This change interacts with changes in land use and land cover, pollution, disruption in biological systems, and resource over-extraction to hasten the effects on ecosystems (Mooney et al. 2009). Local economies, food security, biodiversity, and ecosystem services are all impacted by climate change. The situation will quickly get worse. These may have a variety of effects on pollinators. As a result of extreme weather, temperature changes, and heat stress, pollinator populations may go extinct. Hence, understanding the response of bees to a climate change scenario is essential. Rader et al. (2013) investigated the impact of temperature on the diurnal activity patterns of native bee pollinators on watermelon crop. Their findings showed that there will be a 14.5% decline in managed honey bee pollination services. In the absence of native wild pollinators, services would decline even more drastically, resulting in an estimated overall change in pollination services of +4.5% by 2099. The consequences of climate change characterized by elevated temperatures affect their distribution and abundance, physiology, behavior, and pollination activity (Kjøhl et al. 2011; Rinvoll 2016). In a tropical climate, there are distinct seasons and dry

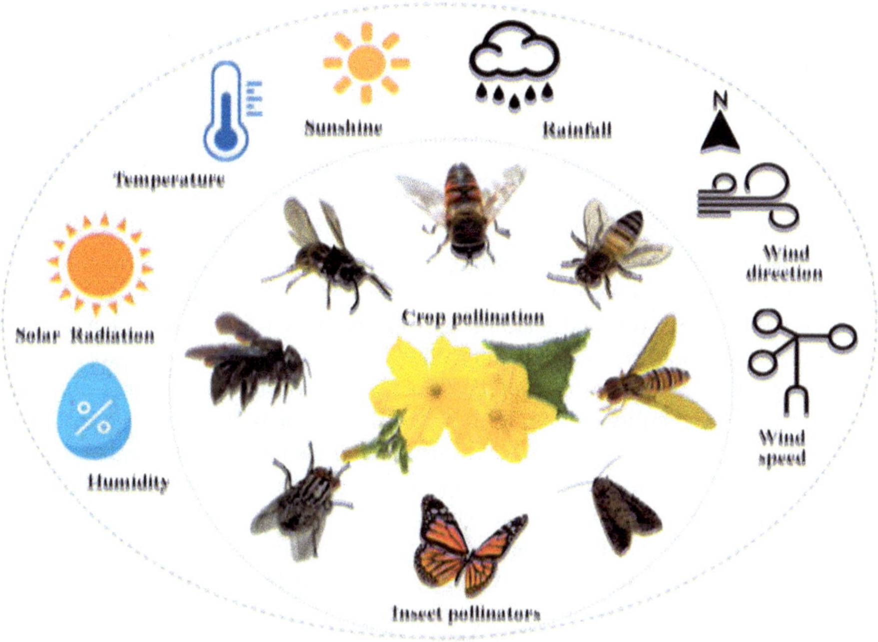

Fig. 7.1 An illustration depicting different climatic parameters to display the influence of pollinator activity and crop pollination

periods. So, Asian honey bees need to store enough food to endure periods of scarcity (Le Conte and Navajas 2008). A long-term trend in the honey bee populations and their by-products, beeswax and honey, was assessed (Phiri et al. 2022). Figure 7.1 provides an example illustrating different forms chosen to reflect the varieties of pollinators interacting with the cucumber flower.

2.1 Plant-Pollinator Interaction

Pollination depends on interactions between pollinators and flowering plants, and because of the mismatch, pollination has suffered (Hegland et al. 2009). Honey bees are resilient to climate change. Climate change has an effect on plant-pollinator interactions due to factors like rising temperatures, higher carbon dioxide levels and ozone levels, and extreme climatic situations like droughts and floods. Apart from changing temperature, alterations in greenhouse gases also play a role in how well plants perform and how they interact. According to Otieno et al. (2022), high levels of ozone caused bee visitation rates in a field bean crop to decrease, while high levels of carbon dioxide caused species of *Bombus terrestris* and *Osmia cornuta* to visit more flowers. Increasing the induced CO_2 level in an experimental setup resulted in less protein and nutritional value in pollen obtained from *Solidago canadensis* plant sources (Ziska et al. 2016). The existence of native bees in North America depends on these floral resources. Other food crops might experience a similar fate with rising atmospheric carbon dioxide levels, and bees consequently suffer from widespread nutritional deficiencies. On a similar measure, Descamps et al. (2021) quantified the floral nectar and pollen modifications due to temperature rise by 3–6 °C and water stress in a highly attractive bee-pollinated species, *Borago officinalis* (Boraginaceae). Due to increased stress, they observed changes in the ratios of amino acids in pollen and nectar. Bees use multimodal clues such as floral scents, color, patterns, and temperature for foraging and navigation (Buehlmann et al. 2020). Plants emit volatile organic compounds, referred to as VOCs, as a form of chemical signaling to draw pollinators. Démares et al. (2022) studied the effects of ozone levels on memory and learning as well as the sensitivity of honey bees to VOCs. Their findings suggested a connection between issues with olfactory coding in the honey bee brain and oxidative stress brought on by ozone. This obliquely implies that the reliability of long-term pollination services may be in jeopardy.

Being ectothermic, the temperatures surrounding the bees determine their day-to-day activities. This temperature has a profound effect on plant-pollinator activity (Hegland et al. 2009). The FAO (2011) has mentioned the impacts of climate change on plant pollination, i.e., the temperature sensitivity of entomophilous crops (Settele et al. 2016). Crops maintain a particular temperature range throughout their life cycle. Extremely low or high temperatures hurt crop development, but they have a larger impact during the anthesis period (Luo 2011). In 2013, Scaven and Rafferty discussed the physiological responses of plants and insect pollinators to climate warming. Aldridge et al. (2011) revealed that global warming has complex effects on phenology at spatial scales based on patch-level habitat differences. Klatt et al. (2014) demonstrated experimentally that the wind and self-pollination mediated by bees had more improvement in strawberry quality, quantity, and market value. Ramírez and Kallarackal (2018) described the state of tropical tree pollination interactions in the tropics and also suggested analyzing ecomorphology patterns to foresee pollination changes in response to climate change. Crop modeling studies have highlighted several climate-driven risks caused by interactions between plants and pollinators. Nevertheless, there are still many unknown areas. Research that aims to provide refuges for pollinators fleeing changing climates to move across the landscape is essential. The establishment of resilient food systems which balance crop demands and pollinator supplies would therefore be ensured by this standard (Settele et al. 2016). Being aware of alternative pollinator agents to be used in the tropics under different climate change scenarios is crucial. These could be obtained through a network analysis involving various species. These should include pollinators' visitation rate, diversity, and abundance data pertaining to crops. This helps to understand how variable precipitation rates and high temperatures affect crops. Also, it's essential for an efficient crop-pollinator dependency rate, as these are partially available.

2.2 Species Geographic Distribution

Climate change has shaped the bees' geographic range over the past century. This impact will be severe on the endemic species and endangered bees. Rader et al. (2013) illustrated the significance of native biodiversity in mitigating the effects of climate change. Substantial fluctuations in weather and climate are likely to affect bee pollinators in the species distribution pattern. The recent advances in crop insect modeling for distribution and risk mapping are explicitly accessible (Tonnang et al. 2017). Giannini et al. (2017a) assessed the relative impact of projected climate change on the geographic distribution of 95 pollinator species in 13 Brazilian crops. In turn, they estimated relative effects on crop production. The mapping was at the municipality level using species distribution modeling to reveal the reduction in the probability of pollinator occurrence by almost 0.13 by 2050 and 90% of the municipalities to suffer species loss. Giannini et al. (2017b) analyzed the effects of climate change on a native stingless bee (*Melipona subnitida* Ducke). The output depicted a shift in future areas, threatening species dispersal and gene flow. Attempts were made to understand how climatically suitable ranges are projected to change under different future climate scenarios and with different dispersal abilities in bumblebees (Sirois-Delisle and Kerr 2018). In Iran's north-western regions, Rahimi et al. (2021) investigated the distribution pattern of five wild bee species under various global warming scenarios. Gomez et al. (2021) studied the distribution of nine stingless bee species and predicted that seven of the nine species were experiencing a significant decline that would impact agriculture and rural life. Furthermore, not all areas within the potential distributional range have habitats suitable for wild bees. This could limit wild bees to a particular niche. This in turn creates an imbalance between farms and ecosystems fit for wild bees. Eventually, farmers will experience a paucity of wild bee pollination. Therefore, farmers should consider using managed bees and be aware of how climate change is affecting agricultural output. These studies aid in planning management activities to benefit different species across different climate scenarios for crop pollination.

2.3 Vegetation Structure: Food Resources

Another problem bees face in agricultural areas is a lack of year-round food sources (Bohart 1972). Eusocial bees that pollinate tropical crops are susceptible to pollination failure under conditions of habitat loss and reduced floral resource availability. It is possible to predict how changes in land use will impact pollinator communities and crop productivity by quantifying the interactions between these two. A study by Ricketts et al. (2008) estimated the relation between pollination services and geographic isolation from natural habitats and included 16 crop varieties from five continents. They observed a sharp decline in the number of native visitors and pollinator diversity in natural habitats. These dropping rates were higher among social bees than solitary bees, and they were higher in tropical compared to temperate areas. For comprehending the associations between non-*Apis* bees and seasonal crops, Abrol et al. (2012) conducted their investigations in two landscapes, namely the plains and hilly regions of Jammu. Smith et al. (2021) used the Normalized Difference Vegetation Index (NDVI) model to account for seasonal and annual variations in floral production in a study to understand how changing land cover affects the honey yield in the Northern Great Plains. They analyzed the relationship between the NDVI and foraging resources. It was found that geographic landscapes with more forage resources had higher honey yields, with an equivalent 0.01% increase in honey yield for every unit increase in forage resources.

The significance of pollen nutrition for bees must be emphasized to maintain a healthy bee colony and encourage stress resistance (Huang 2012). Polidori et al. (2010) investigated the floral resource utilization and nesting needs of the Mediterranean-regional bee species *Lasioglossum malachurum.* This helps to adopt the correct conservation plan and maintain bee pollination activity. Quantitative information on the habitat requirements and its preferences for cavity-nesting and ground-nesting bee pollinators are unclear, say Antoine and Forrest (2021). In a tropical climate, there are distinct seasons and dry periods. So, Asian honey bees store enough food to endure periods of scarcity (Le Conte and Navajas 2008). However, the mentioned studies depict how changes in land use, deforestation, and loss of food resources are reducing the amount and quality of natural habitats for bees, leading to declines in bee populations. Moreover, studies focusing on the native bee species and its relevant crops are critical for addressing regional species issues caused by climate change.

2.4 Thermoregulation Capacity and Physiological Response

Insect physiology is extremely sensitive to temperature changes, which affect each species' physiological limits and metabolic rate differently (Parmesan and Yohe 2003; Dukes et al. 2009). The inability of honey bees to adjust their body temperature in response to changes in their environment is well known. Their body temperature is significantly impacted by environmental factors like solar radiation and the temperature of the surrounding air. Body metabolism and the upper lethal thermal limits are crucial physiological traits adapted to the environment and climate conditions in the evolutionary process (Kjøhl et al. 2011). Stabentheiner et al. (2010) directed a study on the social thermoregulation of honey bee colonies under changing environmental temperatures. Individual bees are unable to regulate their body temperature within the hive. For normal hives, the brood nest temperature is kept constant at 32 °C. However, honey bees are exposed to various ambient temperatures during their nectar and pollen foraging, which affects their thermoregulatory ability (Kovac and Stabentheiner 2011). Käfer et al. (2011) carried out studies on the upper thermal limits of the honeybee (*A. mellifera*). Kovac et al. (2014) investigated two neighboring races of western honeybees, namely, *A.m. carnica* and *A.m. ligustica*, from different climatic regions. Their results revealed that *A.m. ligustica* populations tolerated higher temperature exposures than *A. m. carnica* populations. *Lepidotrigona hoozana* and *A. cerana*, two species of bees, were used in a diurnal and seasonal variation to apprehend the effects of climate on bee flight activities by Sung et al. (2011).

2.5 Monitoring of Foraging Behavior

Foraging activity is one of the distinctive behaviors of social and eusocial insects, e.g., honey bees, stingless bees, etc. This behavior is related to the honey bee colony and the ambient environmental variables in terms of food timing and location. Climate conditions, in particular, have a significant impact on foraging activity. Any variation within or outside the colony affects the behavior of the bees. These are essential for the survival of honeybees. This behavior can improve crop yield while improving pollination (Routray et al. 2017). Any investigations into these changes under different cultivation practices will be noteworthy for consideration in the near future. Since rainfall, low temperatures, and high winds restrict the foraging activity of bees. Nascimento and Nascimento (2012) have reported the influence of different seasons on the foraging activities and internal colonial parameters of *Melipona asilvai* in response to environmental variables as a case study in Atlantic forest areas of northeast Brazil. The team used video cameras to monitor departures, returns, and the types of materials used to describe foraging activities by foragers. Predominantly, the authors reported that foraging activity decreased by 90% from the dry to the wet seasons. Then, temperature and humidity had less influence on the departure activity among stingless bees and the extreme cutbacks in activity during the rainy season. This strongly advocates the seasonal diapause in this species. Abou-Shaara (2014) described the factors influencing the foraging preferences and regulation of foraging tasks, subspecies variations, and monitoring methods for controlling this behavior. The effect of four abiotic factors, viz., temperature, relative humidity, rainfall, and wind speed, on the foraging activity of the Indian honey bee, *A. cerana* F., was the focus (Reddy et al. 2014). But Abou-Shaara et al. (2013) reported a significant negative correlation ($r = -0.09$) between foraging activity and temperature. The studies were conducted to understand the influence of weather on bee foraging activity in *A. mellifera* to explore the utility of predictive modeling for monitoring commercial honey bee health. According to Clarke and Robert (2018), for predicting bee egress during foraging, an ordinary-least-squares GLM algorithm model was applied to weather data such as temperature, solar radiation, atmospheric pressure, humidity, rainfall, wind direction, and speed, from two hives for two consecutive years. Temperature and solar radiation were observed as being responsible for 78% of the variations in bees' activity. Real-time monitoring of the European honey bee, *A. mellifera*, and its responses to simulated heat waves revealed a 70% increase in foraging activity at the colony and individual levels (Bordier et al. 2017). The relationship between foraging activity and foraging performance was recorded in *A. mellifera* using an automated behavior monitoring system by Klein et al. (2019). The effects of microclimatic variations on the foraging habits of native bees from various tropical regions are still largely unknown. The alterations in temperature control this behavior. Research is required to comprehend the effects of species-specific reactions to extreme temperatures on the foraging of neglected bee pollinators that are local to a given region (Fig. 7.2a).

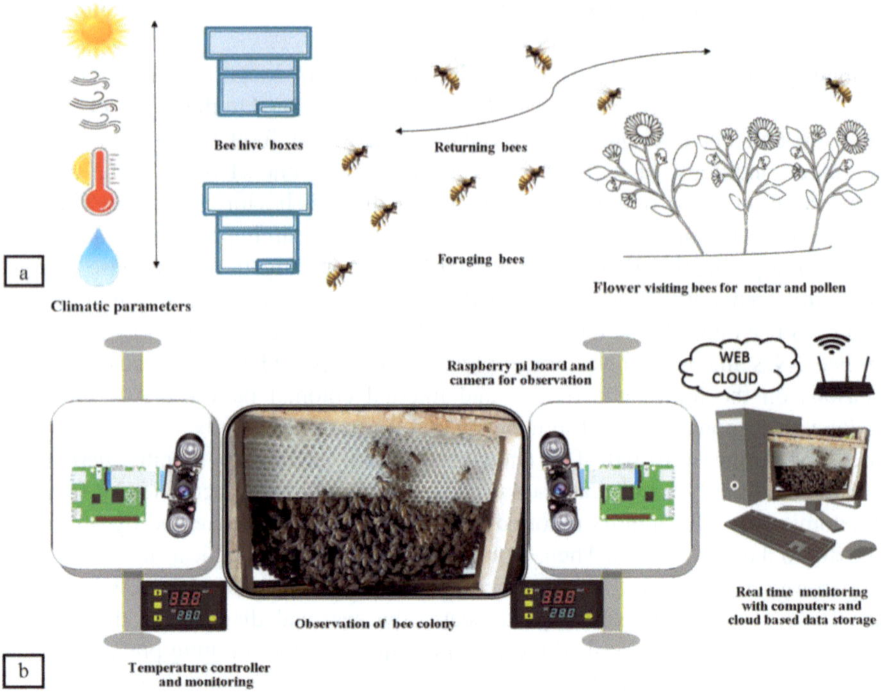

Fig. 7.2 An illustration depicting real-time behavioral changes with climatic parameters: (**a**) monitoring of foraging activity and (**b**) monitoring of egg-laying behavior under the influence of temperature in a bee colony using the Raspberry Pi board with a camera for visualization

2.6 Real-Time Monitoring of Behavioral Changes

When it rains, pollination stays inside, and in the extreme heat, they gather water to cool the colony (Le Conte and Navajas 2008). Climate and weather changes impact the availability of forage and thermoregulatory capacity during winter and the initial brood-rearing time during spring (Insolia et al. 2022). Stabentheiner et al.'s (2007) research examined the relationships between aggressive behavior and thermoregulation in *A. mellifera*. Ectotherm insects' fitness gets impacted by acute heat stress brought by seasonal and inter-annual temperature variation in both tropical and temperate regions (Kingsolver et al. 2013). Switanek et al. (2017) investigated the impact of long-term weather and climatic conditions on honeybee colony survival and winter mortality in a temperate climate. A study by Henneken et al. (2012) examined the impact of weather parameters on the appearance of honeybee swarms. Their findings showed that the emergence of swarm formation is affected by temperature, precipitation, and rainfall. Perhaps the swarming frequency increased dramatically on successive warm days. It did, however, significantly reduce cold events. The honeybee will cope relatively well with climate change; that may not apply to all bee species in the same way, and less is known about them. There are a few partially documented reports on climate-induced changes in bee pollinators and their responses to extreme temperature changes and weather events. For example, how egg laying of the queen bee and brood development could be impacted at different climatic conditions needs an elaborate assessment in a microclimatic condition (as shown in Fig. 7.2b).

2.7 Spread of Disease and Pest

Pests and diseases spread outside of their known active range as a result of climate change. The vulnerability of honeybees to pathogens (protozoa, bacteria, and viruses), ectoparasitic mites, and predators leads to poor colony health and colony loss. In tropical Asia, mites of the genus *Tropilaelaps* parasitize honey bee species, particularly *Apis dorsata* and *Apis mellifera*. Of the four *Tropilaelaps* species reported from Asia, only *Tropilaelaps clareae* has been documented to parasitize *A. mellifera*. Only *T. clareae* parasitizes *A. mellifera* among the four *Tropilaelaps* species reported in Asia. *Varroa destructor*, an external parasitic mite, affects both *A. mellifera* and *A. cerana* and can weaken or kill entire colonies. Microsporidian protozoan bacteria, namely, *Nosema apis* and *N. cerana*, infect *A. mellifera* and *A. cerana*, respectively. These affect the adult bees' digestive systems and cause diarrhea, a reduced lifespan, and weakened immune systems. American and European foulbrood are bacterial diseases that affect the honeybee brood (larvae and pupae), leading to the death of entire colonies. The deformed wing virus causes developmental abnormalities in bees, including deformed wings and a reduced lifespan. These pathogens frequently have various haplotypes with varying levels of virulence. The likelihood of these haplotypes reproducing more quickly has increased due to warmer temperatures and longer growing seasons, resulting in higher bee colony infestation rates (Le Conte and Navajas 2008). Giliba et al. (2020) used climate change modeling to assess the likelihood of a *V. destructor* infestation. In South Korea, Truong et al. (2023) conducted a five-year surveillance study to assess the prevalence and infection rate of pathogens and parasites. Infection rates for the Israeli acute paralysis virus, the chronic bee paralysis virus, the *Ascosphaera apis*, the *A. woodi*, the *Melissococcus plutonius*, and the *Paenibacillus* larvae were higher and could reach 16% in 2021, according to their report. Extreme temperature and precipitation events spread *Varroa destructor* mites, claims Insolia et al. (2022). They used information about the climate, stressors, and changes in land use patterns between 2015 and 2021. An association between meteorological factors and the frequency of risk among honeybees was found to depict that the risk of varroosis and sacbrood disease increases as the temperature rises (Rowland et al. 2021). The prevalence of Chalkbrood, Sacbrood, Varroa, European and American foulbrood, and chronic bee paralysis was reported in the same article. These sets of monitoring data help in identifying and predicting the outbreaks of honeybee diseases. Disease and weather parameters are partially understood about how they interact (Fig. 7.3).

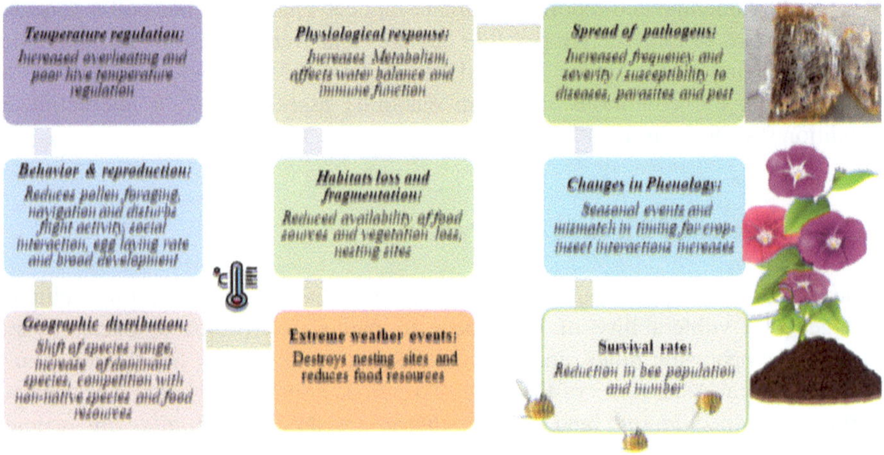

Fig. 7.3 Outline the various ways that climate changes affect the bee pollinators

3 Impact of Cultivation Practices on Pollinators

Agricultural farming practices have historically been unsuitable for pollinators through different means and mechanisms. In the process, pollinator populations have fluctuated and shown a signed decline over the past 30 years (Reynolds 2019; Sánchez-Bayoa and Wyckhuys 2019). The agricultural sector is suffering due to this decline (Basu et al. 2011). Pathogens, parasites, pesticides, habitat loss, insufficient food sources, and climate change are all posing additional risks (Brown and Paxton 2009; Tracy and Jessica 2015; FiBL 2016). According to Brittain et al. (2010), this could change the variety and abundance of wild bees in a specific area. In Bartomeus et al. (2014) evaluated the pollinator contribution to crop yield quality and quantity. They found that pollinator visitation rates decreased with increasing agricultural intensification. Langlois et al. (2020) evaluated the effect of extensive agricultural practices on floral resources and pollinators in a contrasting agricultural landscape. According to FAO (2015), crops reliant on pollinators have better growth and yield than those that don't have access. The main contributing factors are modifications in the cropping patterns, farming methods, and urbanization.

3.1 Conventional Farming Practices

The continued growth of farm sizes for crop or animal production is a trend that affects pollinators (Belfrage et al. 2015). Habitat loss and fragmentation are caused by intensified land use practices, aerial pesticide spraying, promoted mono-cropping (Öckinger and Smith 2007), destroyed hedgerows and nesting areas (New 2005), the use of genetically modified seeds, and implemented tillage practices (Steffan-

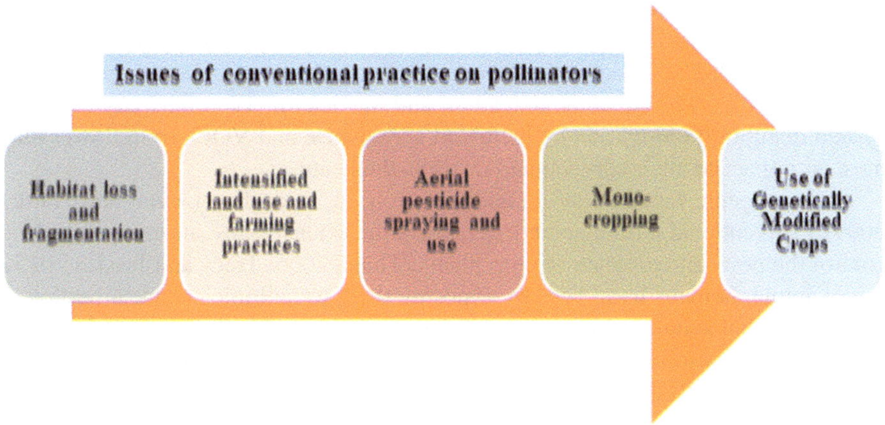

Fig. 7.4 Illustration of how conventional agricultural practices have a negative impact on bee pollinators

Dewenter et al. 2002; Biesmeijer et al. 2006; Steinhauer et al. 2018). Moreover, these affect the soil health, reducing the capacity of soil to sequester carbon and mitigate climate change. The poor-quality soil disturbs the ground-nesting bees (Fig. 7.4).

Cover crops and mulching affect bee larval survival and nesting habitats. Other agricultural practices that affect pollinator communities include crop rotation, weed control, and the use of genetically modified crops (Nicholls and Altieri 2012). These factors have a harmful effect on the diversity of pollinators. Agricultural land makes a good nesting ground for bees. Hence, the nesting preferences of ground-dwelling bees were assessed to compare their strengths. It indicated that there were fewer nesting bees on conventional farms than in natural habitats, indicating that agricultural intensification is harming bee populations (Kim et al. 2006). Conventional farming involves tillage practices, which modify bees' nesting requirements because of altered soil conditions. Bees are becoming sensitive to the negative externalities of the agricultural sector because of these factors (Bicknell 2018).

3.2 Agro-Chemical Exposure

Agrochemical usage has an indirect effect on climate change in the cultivated ecosystem. These chemicals release greenhouse gases such as carbon dioxide, methane, and nitrous oxide into the field, affecting bees' ability to navigate to and from their hives as elevated levels of CO_2 disrupt ability to detect and respond to floral scents and lead to reduced foraging. Hence, the relationship between ozone levels and the quantity of pollinator visits was studied by Rollin (2022) and team. They demonstrated the effects of temperature and ozone on pollinator reactions to pesticides. Additionally, they claimed that pesticides have an effect on pollinators that is more than two times greater when ozone levels are at their highest. Thus, management plans and policies must be created to reduce agrochemical-based air pollution, safeguard biodiversity, and implement environmentally friendly food production methods. Olfaction is important because bees use olfactory receptors to sense smell for foraging, communication, and navigation. Therefore, any olfactory system disruption brought on by increased greenhouse and VOCs from flower may have an impact on their survival and disturbs the health.

Anthropogenic factors play a significant role in its decline (Aizen and Harder 2009). Conventional farming practices involve the extensive use of agrochemicals to control the pests and diseases of crop plants (Thapa 2006; Tracy and Jessica 2015). For example, humans harm bees by exposing them to multiple chemical pesticides with both lethal and nonlethal effects. The extensive use of herbicides and fertilizers can indirectly harm pollinators by reducing the availability of floral resources for pollinators (Nicholls and Altieri 2012). Insecticides directly cause mortality through poisoning (Alston et al. 2007). According to Kearns et al.'s (1998) study, native bee populations consistently showed signs of recovery after the nonuse of the organophosphate insecticide fenitrothion, which was used to control the budworm but is extremely toxic to insects that pollinate bees, was discontinued. Specifically, studies have linked bee population decline to exposure to chemical pesticides (Tracy and Jessica 2015). In addition to increasing agricultural production, insecticides have damaging effects on nontarget species like honeybees and other pollinators. The insecticide affects the life cycle and reproduction, olfactory memory and learning, motor activity, and communicative ability (Medrzycki et al. 2003; Colin et al. 2004; Decourtye et al. 2005; Decourtye and Devillers 2010; Tosi et al. 2016; Azpiazu et al. 2019). A study by Sanchez-Bayo and Goka (2014) revealed the incidence of 161 different pesticides present in the pollen, wax, and honey of bee colonies. Insecticides made up 52%, followed by fungicides (25%), herbicides (17%), and acaricides (6%), extremely dangerous to bee health. The widespread use of insecticides like neonicotinoids, coumaphos, chlorpyrifos, and fipronil is harming bee populations (Leska et al. 2021). According to Desjardins et al. (2021), the fungicide Pristine® reduces in *A. mellifera* capacity for olfactory associative learning.

Neonicotinoids are a group of insecticides extremely toxic to bees (Matsuda et al. 2001). These primarily target postsynaptic nicotinic acetylcholine receptors (nAChRs), according to Phua et al. (2009). These groups of insecticides adhere to the surface of the bee's body or legs while the bees are foraging. Although fungicides and herbicides are not designed to harm insects, their use could still be harmful (Mullin 2015). According to Stanley et al. (2010), these particles are present in hives and bee products at concentrations of ppm/ppb and kill bees and increase bee mortality. Pollinators experience acute toxicity from even the degraded neonicotinoids, such as their olefin and hydroxy metabolites (Dively and Kamel 2012). Exposure to pesticides is negatively impacting pollinators (Krupke and Long 2015). Wintermantel et al. (2019) added that intensive conventional farming has resulted in lower survival rates for bee colonies. Lethal individual assessment is given considerable weight in the majority of pollinator health studies. Moreover, alternative pest management and nonchemical methods replace neonicotinoid

Fig. 7.5 Conventional farming practices that harm bee pollinators: (**a**) agrochemicals sprayed on crops disturb bees and enter the soil; (**b**) soil tillage practices disturb ground-nesting bees and in turn affect the soil health

pesticides by up to 96%, say Jactel et al. (2019). Tosi et al. (2022) reviewed qualitative and quantitative data regarding the lethal, sublethal, and combined toxicity of a wide range of pesticides on bees. Considering how pesticide exposure affects the host's microbiota and whether probiotics can lessen the sub-lethal effects of pesticides were determined by Chmiel et al. (2020) as a means of counteracting the effects of pesticides that are less lethal than fatal. These studies' findings may contribute to the long-term survival and success of pollinators. Douglas et al. (2020) evaluated the quantity and potency of the insecticides used to characterize the spatiotemporal patterns in bee-toxic load from oral and contact sources. They recently created an integrated method of spatial mapping using data on land use and pesticides to track the use of pesticides and protect bees from harmful insecticides (Douglas et al. 2022). Long-term and real-time monitoring initiatives are the need of the hour across the tropics, where they are lacking (Fig. 7.5).

3.3 Organic Farming

Climate-smart farming methods are promoted to cut GHG emissions and boost carbon sequestration. Therefore, agro-ecological techniques that require little in the way of agrochemical inputs are encouraged and promote sustainable agriculture. Organic fertilizers and pesticides have lower levels of toxicity and persistence than synthetic chemicals, say Durán-Lara et al. (2020). The authors, Hossain et al. (2017), have discussed a range of pesticide substitutes and pest management techniques, plant protection strategies and how they are integrated, market statistics for biopesticides, and details on their widespread use. The importance of native bee communities was studied under both organic and conventional management methods (Kremen et al. 2002; Madhurima and Sattagi 2018). It was recommended that pollinator-friendly tips from organic farmers be adopted under conventional farming practices. This improves the health of honeybees. Bees can benefit from organic farming practices, as they prohibit the use of synthetic pesticides and fertilizers (Misiewicz et al. 2015). The anthropogenic stress from conventional farming could be reduced for pollinators by shifting from conventional farming to organic-based principles. So, it might be possible to protect the abundance of native bees by restoring habitats close to areas with significant anthropogenic influence. Preserving the pollinators necessary for agriculture could be accomplished via organic farming, a bee-friendly alternative, as a pesticide application is not allowed under organic farming. Secondly, organic farmers should run their operations in a way that fosters biodiversity and saves the environment (Vinod et al. 2016). Agri-environment-friendly programs and organic farming are implemented to reduce the effects of agricultural intensification (Kleijn et al. 2006). To mitigate the negative impact of conventional agricultural practices and encourage environmental stewardship, organic farming and certification are growing in popularity. Organic farming aids in contributing toward pollinator health by adopting crop rotation, cover crops, and an insectary hedgerow approach. This creates pollinator strength during all seasons by contributing to the loss of floral resources (Langlois et al. 2020). Agricultural landscape diversification and sustainable practices contribute to bee protection. So that bees have access to floral resources and maintain a healthy population that leads to their future growth. The preservation of floral resources is one of the most popular strategies for safeguarding native bee habitats (Lane et al. 2020). The stress that conventional farming practices place on bees may be lessened by converting conventional farms to organic ones. Therefore, this protects the abundance of native bees by restoring habitats close to areas with significant anthropogenic influence. Bees could benefit from organic farming methods. However, there are restrictions and limitations to these benefits. Moreover, organic farming did not appear to mitigate the current bee decline issue (Bicknell 2018). In Xavier et al. (2015) evaluated the toxicity and sublethal effects of botanicals on *A. mellifera* adults and larvae. Among the tested herbal ingredients were rotenone, neem oil, andiroba oil, citronella oil, eucalyptus oil, and garlic extract. Though biopesticides are marketed as having little impact on the environment, their results showed the need to assess the risks pollinators face.

4 Conservation and Mitigating Strategies

Conservation strategies should consider how climate change is affecting the most notable species, i.e., bee pollinators. In this regard, an assessment report on pollinator pollination and food production for understanding biodiversity and ecosystem services is accessible (IPBES 2016). Pollination services can be preserved by protecting the nesting sites and habitats of wild pollinators within agricultural landscapes (Ricketts et al. 2008). Likewise, the Convention on Biological Diversity (2018) reviews pollinators and pollinator contests from their conservation perspective. Concerns have grown over the past decade with the alarming decline in the population of native honeybees, which is harming crop pollination and the global security of the food supply (Vinod et al. 2016). As per Lane et al. (2020), one popular method for protecting native bee habitats is to preserve floral resources. The crop varieties could increase the recruitment of pollinators in fallow lands (IPBES 2016). According to this, even small habitat patches are critical for maintaining the diversity of bee pollinators in agricultural landscapes (Brosi et al. 2008). Despite being crucial for structuring diverse bee communities in urban green spaces, nesting resources receive little attention. According to Fortel et al. (2016), *Osmia bicornis* L. and *O. cornuta* L. were the two species common in the hotel area. Recently, urban gardens and agricultural farms have used artificial nests like bee hotels and nest boxes to attract pollinators (Rahimi et al. 2021). However, the majority of wild bees build their nesting sites in the ground and wood (Vaughan and Black 2007). These species are vulnerable to habitat loss. There aren't many suitable nesting locations in these agricultural landscapes. As a result, man-made nesting structures offer ground-nesting bees in degraded landscapes vital support. The sharing of guidelines for "Building and Maintaining Nesting Hills" will be helpful for the restoration of bee habitats, said Neumüller et al. (2022).

Recovering pollinators from climatic events like drought is helped by areas known as reservoirs because they have a lot of tree cover and a variety of bee species. Hannah et al. (2017) gave an instance of how fine-scale impact models can be used to plan landscape-scale pollination service adaptation by gauging the health of the habitat and estimating native bee diversity. Imbach et al. (2017) looked into the connection between bee diversity and coffee suitability and found that it impacts coffee production. This improved comprehension of coupled impacts can aid in prioritizing adaptation policies and focusing on adaptation strategies for a globally significant crop that sustains households in bio-diverse regions. Landscape conservation initiatives would strike a balance between the need for native species to provide services to mankind and their conservation needs, say Ricketts et al. (2008) (Fig. 7.6).

At this stage, more investigation is needed to ascertain how conventional agricultural practices can lessen the harmful effects on pollinators. Without any mitigation strategies, the global "pollinator crisis" prevails to threaten managed and wild bee populations. It is critical to stop the decline of pollinators brought on by agricultural pesticides (Kevan 1999). Pesticides are prevalent in modern agriculture. Hence, the impact of toxicity on pollinators needs attention by strictly implementing pollinator-friendly measures. These synthetic pesticides get mixed with flower nectar and pollen content, soil, and water during farmers' spraying. Thus, stricter regulations on pesticides can reduce bees' direct exposure. However, indirect exposure is still a problem that continues to harm the fitness and survival of bee colonies. It is critical to use the appropriate foliar spray method, timing, and formulation to reduce pesticide toxicity on pollinators (Crane and Walker 1983). So the safety of bees from poisonous agro-chemicals is guaranteed. This will happen when pesticide exposure rates and its risk assessment indicators are known (Damalas and Eleftherohorinos 2011). The FAO has created an international code of conduct for pesticide management. Moreover, these give a framework for best practices to help reduce pollinator exposure to pesticides (Bicknell 2018). Farmers lack the essential

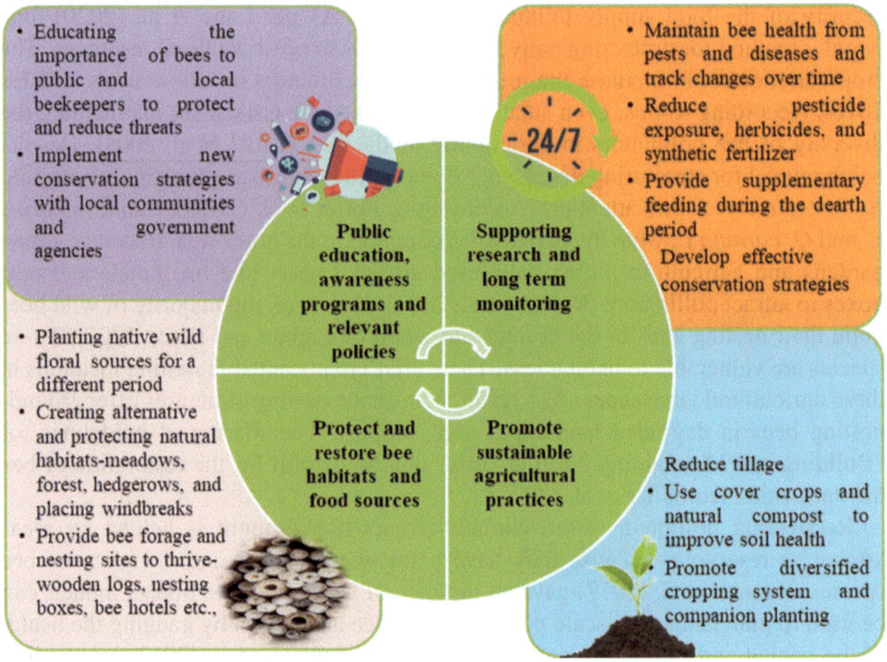

Fig. 7.6 A diagram illustrating the conservation strategies for reducing the damage that agricultural practices and climate change impose on bee pollinators

skills to operate equipment efficiently, and they do not apply the recommended chemical dosages as specified in warning labels. Since 1968, as per pesticide registration, a guideline for use in a county is prohibited under the Insecticide Act, unless it is confirmed that there won't be any residue after the crops are harvested. All that needs to be done by the government to manage pesticides is the adoption of new laws and policies, IPM training, funding the purchase of low-risk products, encouraging scientific research, launching public awareness campaigns, and giving farmers recurring training (Abhilash and Singh 2009).

According to Sethi et al. (2022), banning 14 dangerous pesticides has lessened the negative effects of pesticides on crop yield in India. Tembo et al. (2018) claim that the use of secondary plant metabolites as bio-pesticides will aid in developing crop production systems that are both ecologically sound and sustainable from an agronomic perspective. Inter-cropping, field border plantings, and riparian buffer strips enhance the floral and nesting resources for pollinators (Williams and Kremen 2007; Batary et al. 2011). It is crucial to concentrate on developing doable strategies to improve bee health and survival rates. According to a few studies, probiotics and inactivated yeast cultures of *S. cerevisiae* were suggested as dietary supplements for honeybees because they increase their metabolic activity (Chmiel et al. 2020; Sokolova et al. 2022). However, such studies need extended empirical investigation to understand whether superior performance prevails post-treatment of dietary supplements on honeybee species. Lastly, initiatives to reduce pollinator stress and protect the ecosystems by supporting research on the conservation of insect pollinators in the tropics and sub-tropics must concentrate on (i) assessing the efficiency of specific agro-based initiatives for limiting intensive agricultural expansion and protecting natural habitats; (ii) encouraging sustainable farming practices that will eliminate toxic chemicals, and (iii) creating long-term management and monitoring programs to track changes on pollinator through technological interventions (iv) restoring habitats such as floral resources, nesting areas, and wintering areas; (v) monitoring and educating on the disease management strategies; (vi) promoting agroforestry and crop diversification would be crucial to preserving and enhancing the insect pollinator population; and (viii) promoting knowledge in the formulation and implementation of policies for pollinator support, such as reducing the applications of pesticides and other chemicals that are harmful to their survival and health.

Acknowledgments The authors are thankful to EMPRI, Government of Karnataka, for providing the research grants to take up the research projects on the topic and review of literature.

References

Abhilash PC, Singh N (2009) Pesticide use and application: an Indian scenario. J Hazard Mater 165(1–3):1–12

Abou-Shaara HF (2014) The foraging behaviour of honey bees, *Apis mellifera*: a review. Vet Med 59(1):1–10

Abou-Shaara HF, Al-Ghamdi AA, Mohamed AA (2013) Honey bee colonies performance enhance by newly modified beehives. J Apic Sci 57(2):45–57

Abrol DP, Abrol DP (2012) Wild bees and crop pollination. In: Pollination Biology: Biodivers Conserv Agric Prod, pp 111–184

Abrol DP, Shankar U, Chatterjee D, Ramamurthy VV (2012) Exploratory studies on diversity of bees with special emphasis on non-*Apis* pollinators in some natural and agricultural plants of Jammu division, India. Curr Sci 103(7):780–783

Aizen MA, Harder LD (2009) The global stock of domesticated honey bees is growing slower than agricultural demand for pollination. Curr Biol 19(11):915–918

Aldridge G, Inouye DW, Forrest JR, Barr WA, Miller-Rushing AJ (2011) Emergence of a mid-season period of low floral resources in a montane meadow ecosystem associated with climate change. J Ecol 99(4):905–913

Alston DG, Tepedino VJ, Bradley BA, Toler TR, Griswold TL, Messinger SM (2007) Effects of the insecticide Phosmet on solitary bee foraging and nesting in orchards of Capitol Reef National Park, Utah. Environ Entomol 36:811–816

Antoine CM, Forrest JR (2021) Nesting habitat of ground-nesting bees: a review. Ecol Entomol 46(2):143–159

Azpiazu C, Bosch J, Vinuela E, Medrzycki P, Teper D, Sgolastra F (2019) Chronic oral exposure to field-realistic pesticide combinations via pollen and nectar: effects on feeding and thermal performance in a solitary bee. Sci Rep 9(1):1–11

Bartomeus I, Potts SG, Steffan-Dewenter I, Vaissiere BE, Woyciechowski M, Krewenka KM, Tscheulin T, Roberts SP, Szentgyörgyi H, Westphal C, Bommarco R (2014) Contribution of insect pollinators to crop yield and quality varies with agricultural intensification. PeerJ 2:e328

Basu P, Bhattacharya R, Ianetta P (2011) A decline in pollinator dependent vegetable crop productivity in India indicates pollination limitation and consequent agro-economic crises. Nat Precedings:1–1

Batary P, Baldi A, Kleijn D, Tscharntke T (2011) Landscape-moderated biodiversity effects of agri-environmental management: a meta-analysis. Proc Royal Soc B Biol Sci 278(1713):1894–1902

Belfrage K, Björklund J, Salomonsson L (2015) Effects of farm size and on-farm landscape heterogeneity on biodiversity: case study of twelve farms in a Swedish landscape. Agroecol Sustain Food Syst 39(2):170–188

Bhattacharyya M, Acharya SKR, Chakraborty SK (2017) Pollinators unknown: people's perception of native bees in an Agrarian District of West Bengal, India, and its implication in conservation. Trop Conserv Sci 10:1–14. https://doi.org/10.1177/1940082917725440

Bicknell J (2018) The effects of organic and conventional agriculture on bee populations in the United States. University Honors Theses. Paper 648. https://doi.org/10.15760/honors.663

Biesmeijer JC, Roberts SP, Reemer M, Ohlemuller R, Edwards M, Peeters T, Schaffers AP, Potts SG, Kleukers RJ, Thomas CD, Settele J (2006) Parallel declines in pollinators and insect-pollinated plants in Britain and The Netherlands. Science 313(5785):351–354

Bohart GE (1972) Management of wild bees for the pollination of crops. Annu Rev Entomol 17: 287–312

Bosch J, Kemp WP (2002) Developing and establishing bee species as crop pollinators: the example of *Osmia spp.* (Hymenoptera: Megachilidae) and fruit trees. Bull Entomol Res 92(1): 3–16

Brittain C, Williams N, Kremen C, Klein AM (2013) Synergistic effects of non-Apis bees and honey bees for pollination services. Proc Roy Soc B: Biol Sci 280(1754)

Brittain CA, Vighi M, Bommarco R, Settele J, Potts SG (2010) Impacts of a pesticide on pollinator species richness at different spatial scales. Basic Appl Ecol 11(2):106–115

Brosi BJ, Daily GC, Shih TM, Oviedo F, Durán G (2008) The effects of forest fragmentation on bee communities in tropical countryside. J Appl Ecol 45:773–783

Brown MJF, Paxton RJ (2009) The conservation of bees: a global perspective. Apidologie 40:410–416

Buehlmann C, Mangan M, Graham P (2020) Multimodal interactions in insect navigation. Anim Cogn 23:1129–1141

Chmiel JA, Daisley BA, Pitek AP, Thompson GJ, Reid G (2020) Understanding the effects of sublethal pesticide exposure on honey bees: a role for probiotics as mediators of environmental stress. Front Ecol Evol 8:22. https://doi.org/10.3389/fevo.2020.00022

Clarke D, Robert D (2018) Predictive modelling of honey bee foraging activity using local weather conditions. Apidologie 49:386–396

Colin ME, Bonmatin JM, Moineau I, Gaimon C, Brun S, Vermandere JP (2004) A method to quantify and analyse the foraging activity of honey bees: relevance to the sub-lethal effects induced by systemic insecticides. Arch Environ Contam Toxicol 47:387–395

Convention on Biological Diversity (2018) Review of pollinators and pollination relevant to the conservation and sustainable use of biodiversity in all ecosystems, beyond their role in agriculture and food productions. In: 22nd meeting, Montreal Canada, pp 1–70

Costanza R, De Groot R, Sutton P, Van der Ploeg S, Anderson SJ, Kubiszewski I, Farber S, Turner RK (2014) Changes in the global value of ecosystem services. Glob Environ Change 26:152–158

Crane E, Walker P (1983) The impact of pest management on bees and pollination. Trop Dev Res Inst, p 210

Damalas CA, Eleftherohorinos IG (2011) Pesticide exposure, safety issues, and risk assessment indicators. Int J Environ Res Public Health 8(5):1402–1419. https://doi.org/10.3390/ijerph8051402

Das A, Sau S, Pandit MK, Saha K (2018) A review on: importance of pollinators in fruit and vegetable production and their collateral jeopardy from agro-chemicals. J Entomol Zool Stud 6(4):1586–1591

Decourtye A, Devillers J (2010) Ecotoxicity of neonicotinoid insecticides to bees. Adv Exp Med Biol 683:85–95

Decourtye A, Devillers J, Genecque E, Menach KL, Budzinski H, Cluzeau S, Pham-Delegue MH (2005) Comparative sublethal toxicity of nine pesticides on olfactory learning performances of the honeybee *Apis mellifera*. Arch Environ Contam Toxicol 48:242–250

Démares F, Gibert L, Creusot P, Lapeyre B, Proffit M (2022) Acute ozone exposure impairs detection of floral odor, learning, and memory of honey bees, through olfactory generalization. Sci Total Environ 827:154342

Descamps C, Quinet M, Jacquemart A-L (2021) Climate change–induced stress reduce quantity and alter composition of nectar and pollen from a bee-pollinated species (*Borago officinalis*, Boraginaceae). Front Plant Sci 12:755843. https://doi.org/10.3389/fpls.2021.755843

DesJardins NS, Fisher A II, Ozturk C, Fewell JH, DeGrandi-Hoffman G, Harrison JF, Smith BH (2021) A common fungicide, Pristine®, impairs olfactory associative learning performance in honey bees (*Apis mellifera*). Environ Pollut 288:117720

Dively GP, Kamel A (2012) Insecticide residues in pollen and nectar of a cucurbit crop and their potential exposure to pollinators. J Agric Food Chem 60(18):4449–4456

Douglas MR, Baisley P, Soba S, Kammerer M, Lonsdorf EV, Grozinger CM (2022) Putting pesticides on the map for pollinator research and conservation. Sci Data 9(1):571. https://doi.org/10.1038/s41597-022-01584-z

Douglas MR, Sponsler DB, Lonsdorf EV, Grozinger CM (2020) County-level analysis reveals a rapidly shifting landscape of insecticide hazard to honey bees (*Apis mellifera*) on US farmland. Sci Rep 10:797. https://doi.org/10.1038/s41598-019-57225-w

Dukes JS, Pontius J, Orwig D, Garnas JR, Rodgers VL, Brazee N, Cooke B, Theoharides KA, Stange EE, Harrington R, Ehrenfeld J (2009) Responses of insect pests, pathogens, and invasive plant species to climate change in the forests of northeastern North America: what can we predict? Can J For Res 39(2):231–248

Durán-Lara EF, Valderrama A, Marican A (2020) Natural organic compounds for application in organic farming. Agriculture 10(2):41

Eeraerts M, Vanderhaegen R, Smagghe G, Meeus I (2020) Pollination efficiency and foraging behaviour of honey bees and non-Apis bees to sweet cherry. Agric Forest Entomol 22(1):75–82

FAO (2011) The state of the world's land and water resources for food and agriculture SOLAW—managing systems at risk. Food and Agriculture Organization of the United Nations, Rome, Italy

FiBL (Research Institute of Organic Agriculture) (2016) Bee health and threats to pollinators. Frick, Switzerland: FiBL

Food and Agriculture Organization (FAO) (2015) Climate change and food security: risks and responses

Fortel L, Henry M, Guilbaud L, Mouret H, Vaissière BE (2016) Use of human-made nesting structures by wild bees in an urban environment. J Insect Conserv 20(2):239–253. https://doi.org/10.1007/s10841-016-9857-y

Giannini TC, Costa WF, Cordeiro GD, Imperatriz-Fonseca VL, Saraiva AM, Biesmeijer J, Garibaldi LA (2017a) Projected climate change threatens pollinators and crop production in Brazil. PLoS One 12(8):e0182274. https://doi.org/10.1371/journal.pone.0182274

Giannini TC, Maia-Silva C, Acosta AL, Jaffe R, Carvalho AT, Martins CF, Zanella FC, Carvalho CA, Hrncir M, Saraiva AM (2017b) Protecting a managed bee pollinator against climate change: strategies for an area with extreme climatic conditions and socioeconomic vulnerability. Apidologie 48(6):784–794. https://doi.org/10.1007/s13592-017-0523-5

Giliba RA, Mpinga IH, Ndimuligo SA, Mpanda MM (2020) Changing climate patterns risk the spread of Varroa destructor infestation of African honey bees in Tanzania. Ecol Process 9(1):1

Gomez C, de Lima SCS, Siqueira JO (2021) Brazilian stingless bees are threatened by habitat conversion and climate change. Reg Environ Change 21(1):22

Gonzalez VH, Cobos ME, Jaramillo J, Ospina R (2021) Climate change will reduce the potential distribution ranges of Colombia's most valuable pollinators. Perspect Ecol Conserv 19(2): 195–206

Hannah L, Steele M, Fung E, Imbach P, Flint L, Flint A (2017) Climate change influences on pollinator, forest, and farm interactions across a climate gradient. Clim Chang 141:63–75. https://doi.org/10.1007/s10584-016-1868-x

Hegland SJ, Nielsen A, Lázaro A, Bjerknes AL, Totland Ø (2009) How does climate warming affect plant-pollinator interactions? Ecol Lett 12(2):184–195

Henneken R, Helm S, Menzel A (2012) Meteorological influences on swarm emergence in honey bees (Hymenoptera: Apidae) as detected by crowdsourcing. Environ Entomol 41(6):1462–1465

Hossain L, Rahman R, Khan MS (2017) Alternatives of pesticides. In: Pesticide residue in foods: sources, management, and control, pp 147–165. https://doi.org/10.1007/978-3-319-52683-6_9

Huang Z (2012) Pollen nutrition affects honey bee stress resistance. Terr Arthropod Rev 5(2): 175–189

Imbach P, Fung E, Hannah L, Navarro-Racines CE, Roubik DW, Ricketts TH, Harvey CA, Donatti CI, Läderach P, Locatelli B, Roehrdanz PR (2017) Coupling of pollination services and coffee suitability under climate change. Proc Natl Acad Sci 114(39):10438–10442

Insolia L, Molinari R, Rogers SR, Williams GR, Chiaromonte F, Calovi M (2022) Honey bee colony loss linked to parasites, pesticides and extreme weather across the United States. Sci Rep 12(1):20787. https://doi.org/10.1038/s41598-022-24946-4

IPBES (2016) In: Potts SG, Imperatriz-Fonseca VL, Ngo HT (eds) The assessment report of the intergovernmental science-policy platform on biodiversity and ecosystem services on pollinators, pollination and food production. Secretariat of the intergovernmental science-policy platform on biodiversity and ecosystem services, Bonn, Germany, p 552. https://doi.org/10.5281/zenodo.3402856

Jactel H, Verheggen F, Thiéry D, Escobar-Gutiérrez AJ, Gachet E, Desneux N (2019) Neonicotinoids Working Group. Alternatives to neonicotinoids. Environ Int 129:423–429

Käfer H, Kovac H, Stabentheiner A (2011) Upper thermal limits of honeybee (*Apis mellifera*) and yellowjacket (*Vespula vulgaris*) foragers, pp. 1–4

Kearns CA, Inouye DW, Waser NM (1998) Endangered mutualisms: the conservation of plant-pollinator interactions. Annu Rev Ecol Syst 29(1):83–112

Kevan PG (1999) Pollinators as bioindicators of the state of the environment: species, activity and diversity. Agric Ecosyst Environ 74:373–393

Kim J, Williams N, Kremen C (2006) Effects of cultivation and proximity to natural habitat on ground-nesting native bees in California sunflower fields. J Kansas Entomol Soc 79(4):309–320

Kingsolver JG, Diamond SE, Buckley LB (2013) Heat stress and the fitness consequences of climate change for terrestrial ectotherms. Funct Ecol 27(6):1415–1423

Kjøhl M, Nielsen A, Stenseth NC (2011) Pollination services for sustainable agriculture. In: Potential effects of climate change on crop pollination. Food And Agriculture Organization of the united nations, Rome, pp 1–36

Klatt BK, Holzschuh A, Westphal C, Clough Y, Smit I, Pawelzik E, Tscharntke T (2014) Bee pollination improves crop quality, shelf life and commercial value. Proc Royal Soc B Biol Sci 281(1775):20132440

Kleijn D, Baquero RA, Clough Y, Díaz M, De Esteban J, Fernandez F, Yela JL (2006) Mixed biodiversity benefits of Agri-environment schemes in five European countries. Ecol Lett 9(3): 243–254

Klein AM, Vaissière BE, Cane JH, Steffan-Dewenter I, Cunningham SA, Kremen C, Tscharntke T (2007) Importance of pollinators in changing landscapes for world crops. Proc R Soc B 274(1608):303–313. https://doi.org/10.1098/rspb.2006.3721

Klein S, Pasquaretta C, He XJ, Perry C, Søvik E, Devaud JM, Barron AB, Lihoreau M (2019) Honey bees increase their foraging performance and frequency of pollen trips through experience. Sci Rep 9(1):6778

Kovac H, Käfer H, Stabentheiner A, Costa C (2014) Metabolism and upper thermal limits of *Apis mellifera carnica* and *A. m. ligustica*. Apidologie 45:664–677

Kovac H, Stabentheiner A (2011) Thermoregulation of foraging honeybees on flowering plants: seasonal variability and influence of radiative heat gain. Ecol Entomol 36(6):686–699

Kremen C, Williams NM, Thorp RW (2002) Crop pollination from native bees at risk from agricultural intensification. Proc Natl Acad Sci 99(26):16812–16816

Krupke CH, Long EY (2015) Intersections between neonicotinoid seed treatments and honey bees. Curr Opin Insect Sci 10:8–13

Lane IG, Herron-Sweet CR, Portman ZM, Cariveau DP (2020) Floral resource diversity drives bee community diversity in prairie restorations along an agricultural landscape gradient. J Appl Ecol 57(10):2010–2018

Langlois A, Jacquemart AL, Piqueray J (2020) Contribution of extensive farming practices to the supply of floral resources for pollinators. Insects 11(11):818

Le Conte Y, Navajas M (2008) Climate change: impact on honey bee populations and diseases. Rev Sci Tech 27(2):485–497, 499–510

Leska A, Nowak A, Nowak I, Górczyńska A (2021) Effects of insecticides and microbiological contaminants on *Apis mellifera* health. Molecules 26:5080. https://doi.org/10.3390/molecules26165080

Luo Q (2011) Temperature thresholds and crop production: a review. Clim Chang 109:583–598. https://doi.org/10.1007/s10584-011-0028-6

Madhurima V, Sattagi HN (2018) Pollinator fauna and their relative abundance in guava under organic and conventional farming systems. J Exp Zool 21(2):1173–1179

Matsuda K, Buckingham SD, Kleier D, Rauh JJ, Grauso M, Sattelle DB (2001) Neonicotinoids: insecticides acting on insect nicotinic acetylcholine receptors. Trends Pharmacol Sci 22(11): 573–580

Medrzycki P, Montanari R, Bortolotti L, Sabatini AG, Maini S, Porrini C (2003) Effects of imidacloprid administered in sub-lethal doses on honey bee behaviour. Lab Tests Bull Insectol 56:59–62

Michener CD (2007) The bees of the world, 2nd edn. Johns Hopkins University Press, Baltimore, MD, p 953

Misiewicz T, Shade J (2015) The role of organic in supporting pollinator health. Org Cent. Retrieved January 2016, from http://www.organic-center.org

Mooney H, Larigauderie A, Cesario M, Elmquist T, Hoegh-Guldberg O, Lavorel S, Mace GM, Palmer M, Scholes R, Yahara T (2009) Biodiversity, climate change, and ecosystem services. Curr Opin Environ Sustain 1:46–54

Mullin CA (2015) Effects of 'inactive'ingredients on bees. Curr Opin Insect Sci 10:194–200

Nascimento DL, Nascimento FS (2012) extreme effects of season on the foraging activities and colony productivity of a stingless bee (*Melipona asilvai Moure*, 1971) in Northeast Brazil. Psyche 2012

Neumüller U, Burger H, Mayr AV, Hopfenmüller S, Krausch S, Herwig N, Burger R, Diestelhorst O, Emmerich K, Haider M, Kiefer M, Konicek J, Kornmilch JC, Moser M, Saure C, Schanowski A, Scheuchl E, Sing J, Wagner M, Witter J, Schwenninger HR, Ayasse M (2022) Artificial Nesting Hills promote wild bees in agricultural landscapes. Insects 14 13(8): 726

New TR (2005) Invertebrate conservation and agricultural ecosystems. Cambridge University Press

Nicholls CI, Altieri MA (2012) Plant biodiversity enhances bees and other insect pollinators in agroecosystems. Agron Suntain Dev 33:257–274

Öckinger E, Smith HG (2007) Semi-natural grasslands as population sources for pollinating insects in agricultural landscapes. J Appl Ecol 44(1):50–59

Ostenson H, Granatstein D, Benbrook C, Smith E, Daley C (2015) The role of organic in supporting pollinator health. Org Cent

Osterman J, Aizen MA, Biesmeijer JC, Bosch J, Howlett BG, Inouye DW, Paxton RJ (2021) Global trends in the number and diversity of managed pollinator species. Agric Ecosyst Environ 322: 107653

Otieno M, Peters MK, Duque L, Steffan-Dewenter I (2022) Interactive effects of ozone and carbon dioxide on plant-pollinator interactions and yields in a legume crop. Environ Adv 9:100285

Parmesan C, Yohe G (2003) A globally coherent fingerprint of climate change impacts across natural systems. Nature 421:37–42. https://doi.org/10.1038/nature01286

Phiri BJ, Fèvre D, Hidano A (2022) Uptrend in global managed honey bee colonies and production based on a six-decade viewpoint, 1961–2017. Sci Rep 12:21298. https://doi.org/10.1038/s41598-022-25290-3

Phua DH, Lin CC, Wu ML, Deng JF, Yang CC (2009) Neonicotinoid insecticides: an emerging cause of acute pesticide poisoning. Clin Toxicol 47(4):336–341

Polidori C, Rubichi A, Barbieri V, Trombino L, Donegana M (2010) Floral resources and nesting requirements of the ground-nesting social bee, *Lasioglossum malachurum* (Hymenoptera: Halictidae), in a Mediterranean semi-agricultural landscape. PSYCHAQ 2010:1–11

Rader R, Reilly J, Bartomeus I, Winfree R (2013) Native bees buffer the negative impact of climate warming on honey bee pollination of watermelon crops. Glob Change Biol 19(10):3103–3110. https://doi.org/10.1111/gcb.12264

Rahimi E, Barghjelveh S, Dong P (2021) How effective are artificial nests in attracting bees? A review. J Ecology Environ 45(16):1–11. https://doi.org/10.1186/s41610-021-00192-z

Ramírez F, Kallarackal J (2018) Climate change, tree pollination and conservation in the tropics: a research agenda beyond IPBES. Wiley Interdiscip Rev Clim Chang 9(1):e502

Reddy PV, Rashmi T, Verghese A (2014) Foraging activity of Indian honey bee, *Apis cerana* in relation to ambient climate variables under tropical conditions. J Environ Biol 36(3):577–581

Reddy PV, Verghese A, Rajan VV (2012) Potential impact of climate change on honeybees (*Apis* spp.) and their pollination services. Pest Manag Hortic Ecosyst 18(2):121–127

Reynolds S (2019) Pollinator ups and down. Bull Royal Entomol Soc 43(2):57–62

Ricketts TH, Regetz J, Steffan-Dewenter I, Cunningham SA, Kremen C, Bogdanski A, Gemmill-Herren B, Greenleaf SS, Klein AM, Mayfield MM, Morandin LA, Ochieng A, Potts SG, Viana BF (2008) Landscape effects on crop pollination services: are there general patterns? Ecol Lett 11(5):499–515. https://doi.org/10.1111/gcb.12264

Rinvoll AW (2016) Pollinator activity under climate change. M.Sc Dissertation. University of Oslo Norway, pp 1–46

Rollin O, Aguirre-Gutiérrez J, Yasrebi-de Kom IA, Garratt MP, de Groot GA, Kleijn D, Potts SG, Scheper J, Carvalheiro LG (2022) Effects of ozone air pollution on crop pollinators and pollination. Glob Environ Change 75:102529

Routray SK, Thakeran RM, Srikanth DN, Preeti P, Keerthi K (2017) Statistical analysis and modeling of honeybee foraging networks for apiculture. In: IEEE Technological Innovations in ICT for Agriculture and Rural Development (TIAR), Chennai, pp 164–167. https://doi.org/10.1109/TIAR.2017.8273708

Rowland BW, Rushton SP, Shirley MDF, Brown MA, Budge GE (2021) Identifying the climatic drivers of honey bee disease in England and Wales. Sci Rep 11(1):21953. https://doi.org/10.1038/s41598-021-01495-w

Sanchez-Bayo F, Goka K (2014) Pesticide residues and bees - a risk assessment. PLoS One 9(4): E94482

Sánchez-Bayoa F, Wyckhuys KAG (2019) Worldwide decline of the entomofauna: a review of its drivers. Biol Conserv 232:8–27

Sethi A, Lin CY, Madhavan I, Davis M, Alexander P, Eddleston M, Chang SS (2022) Impact of regional bans of highly hazardous pesticides on agricultural yields: the case of Kerala. Agric Food Secur 11(1):1–13

Settele J, Bishop J, Potts S (2016) Climate change impacts on pollination. Nat Plants 2:16092. https://doi.org/10.1038/nplants.2016.92

Sirois-Delisle C, Kerr JT (2018) Climate change-driven range losses among bumblebee species are poised to accelerate. Sci Rep 8(1):14464

Smith D, Davis A, Hitaj C, Hellerstein D, Preslicka A, Kogge E, Mushet D, Lonsdorf E (2021) The contribution of land cover change to the decline of honey yields in the northern Great Plains. Environ Res Lett 16(6):064050

Sokolova E, Mager S, Grizanova E, Kalmykova G, Akulova N, Dubovskiy I (2022) Stimulation effect of probiotic bacteria *Bacillus* spp. and inactivated yeast on the honey bees *Apis mellifera* physiology and honey productivity. Invert Surv J:85–90

Stabentheiner A, Kovac H, Brodschneider R (2010) Honeybee colony thermoregulation–regulatory mechanisms and contribution of individuals in dependence on age, location and thermal stress. PLoS One 5(1):e8967

Stabentheiner A, Kovac H, Schmaranzer S (2007) Thermal behaviour of honeybees during aggressive interactions. Ethology 113(10):995–1006

Stanley J, Chandrasekaran S, Preetha G, Kuttalam S (2010) Toxicity of diafenthiuron to honey bees in laboratory, semi-field and field conditions. Pest Manag Sci 66(5):505–510

Steffan-Dewenter I, Münzenberg U, Bürger C, Thies C, Tscharntke T (2002) Scale-dependent effects of landscape context on three pollinator guilds. Ecology 83(5):1421–1432

Steinhauer N, Kulhanek K, Antúnez K, Human H, Chantawannakul P, Chauzat MP (2018) Drivers of colony losses. Curr Opin Insect Sci 26:142–148

Sung IH, Yamane S, Lu SS, Ho KK (2011) Climatological influences on the flight activity of stingless bees (*Lepidotrigona hoozana*) and honeybees (*Apis cerana*) in Taiwan (Hymenoptera, Apidae). Sociobiology 58(3):835–850

Switanek M, Crailsheimb K, Truhetz H, Brodschneider R (2017) Modeling seasonal effects of temperature and precipitation on honey bee winter mortality in a temperate climate. Sci Total Environ 579:1581–1587

Tanda AS (2021) Non-Apis pollinators and their role in sustainable agriculture. Indian J Entomol: e21072. https://doi.org/10.5958/IJE.2021.92

Tembo Y, Mkindi AG, Mkenda PA, Mpumi N, Mwanauta R, Stevenson PC, Ndakidemi PA, Belmain SR (2018) Pesticidal plant extracts improve yield and reduce insect pests on legume crops without harming beneficial arthropods. Front Plant Sci 9:1425

Thakur M (2012) Bees as pollinators–biodiversity and conservation. Int Res J Agric Sci Soil Sci 2(1):1–7

Thapa RB (2006) Honeybees and other insect pollinators of cultivated plants: a review. Inst Agric Anim Sci 27:1–23

Tracy M, SJ (2015) The role of organic in supporting pollinator health. The Organic Center. Retrieved January 2016, from http://www.organic-center.org

Bordier C, Dechatre H, Suchail S, Peruzzi M, Soubeyrand S, Pioz M, Alaux C (2017) Colony adaptive response to simulated heat waves and consequences at the individual level in honeybees (*Apis mellifera*). Sci Rep 7(1):1–11

Tonnang HE, Hervé BD, Biber-Freudenberger L, Salifu D, Subramanian S, Ngowi VB, Guimapi RY, Anani B, Kakmeni FM, Affognon H, Niassy S (2017) Advances in crop insect modelling methods-towards a whole system approach. Ecol Model 354:88–103

Tosi S, Demares FJ, Nicolson SW, Medrzycki P, Pirk CW, Human H (2016) Effects of a neonicotinoid pesticide on thermoregulation of African honey bees (*Apis mellifera scutellata*). J Insect Physiol 93:56–63

Tosi S, Sfeir C, Carnesecchi E, Chauzat MP (2022) Lethal, sublethal, and combined effects of pesticides on bees: a meta-analysis and new risk assessment tools. Sci Total Environ 24:156857

Truong AT, Yoo MS, Seo SK, Hwang TJ, Yoon SS, Cho YS (2023) Prevalence of honey bee pathogens and parasites in South Korea: a five-year surveillance study from 2017 to 2021. Heliyon 9(1):e13494

Vaughan M, Black SH (2007) Enhancing nest sites for native bee crop pollinators. USDA National Agroforestry Center, pp 1–4p

Vinod M, Naik AK, Biradar R, Natikar PK, Balikai RA (2016) Bee pollination under organic and conventional farming systems: a review. J Exp Zool 19(2):643–652

Waykar B, Baviskar RK (2016) Diversity of pollinator bees from Paithan taluka of Aurangabad district (M.S.) India. J Entomol Zool Stud 5(1):697–700

Williams NM, Kremen C (2007) Resource distributions among habitats determine solitary bee offspring production in a mosaic landscape. Ecological App 17(3):910–921

Wintermantel D, Odoux JF, Chadoeuf J, Bretagnolle V (2019) Organic farming positively affects honeybee colonies in a flower-poor period in agricultural landscapes. J Appl Ecol 56(8): 1960–1969

Xavier VM, Picanço MC, Chediak M, Júnior PA, Ramos RS, Martins JC (2015) Acute toxicity and sublethal effects of botanical insecticides to honey bees. J Insect Sci 15(1):1–6

Ziska LH, Pettis JS, Edwards J, Hancock JE, Tomecek MB, Clark A, Dukes JS, Loladze I, Polley HW (2016) Rising atmospheric CO_2 is reducing the protein concentration of a floral pollen source essential for North American bees. Proc R Soc B Biol Sci 283(1828):20160414

Impact of Pollination of Bumblebees and the Productivity of Polyhouse Crops

8

Raj Kumar Thakur, Meena Thakur, Kiran Rana, Ruchi Sharma, Priyanka Thakur, Diksha Devi, Priyanka Rani, Mangla Ram Bajiya, and Monika

Abstract

Among the insect pollinators, honeybees and bumblebees are recognized as efficient pollinators. Crops that honeybees typically cannot pollinate are successfully pollinated by bumblebees. The population of insect pollinators has been falling recently, and several species of bumblebees are in danger of extinction, mainly prevalent in temperate regions. Bumblebees are more advantageous than honeybees as pollinating agents since pollinators frequent higher number of flowers in a minute, prefer blooms with lengthy corolla, and undertake buzz pollination. The majority of species prefer quiet, dark places to build nests, and the bumblebee queen often hibernates before establishing a colony in the spring. This chapter furnishes information on bumblebees' life cycle, biology, and pollination preferences and efficiency in agricultural crops grown under protected cultivation.

Keywords

Bumblebees · Honeybees · Pollination · Buzz pollination · Protected cultivation

R. K. Thakur (✉) · M. Thakur · M. R. Bajiya
Department of Entomology, Dr. YSP UHF Nauni, Solan, Himachal Pradesh, India

K. Rana · D. Devi
Department of Entomology, Dr. YS Parmar University of Horticulture and Forestry, Nauni, Solan, Himachal Pradesh, India

R. Sharma · P. Thakur · Monika
Senior Research Fellow, Department of Entomology, Dr. YSP UHF Nauni, Solan, Himachal Pradesh, India

P. Rani
PhD Research Scholar, Department of Entomology, Dr. YSP UHF Nauni, Solan, Himachal Pradesh, India

A. K. Chakravarthy (ed.), *The Role of Pollinators in Tropical Landscapes*,
https://doi.org/10.1007/978-981-95-9139-8_8

1 Introduction

Pollination is an important factor leading to high productivity and good quality fruit of any crop. Proper pollination is essential for good fruit set and yield. Insects, birds, bats, and small animals are the main biotic agents, although wind, water, and gravity are significant abiotic factors (Free 1993). Different insect species play an important role in the pollination of horticultural and agricultural crops. The most important groups of insect pollinators are Hymenoptera and Diptera. To provide food for their offspring, hymenopterans frequently gather pollen and nectar. This order includes some of the most significant pollinators, including honeybees, bumblebees, and solitary bees (McGregor 1976).

Due to their ability to function in inclement weather, low population size, and restricted flight range, bumblebees are considered the most effective and dependable pollinators for protected crops (Eijnde 1992). They are useful for buzz pollinating crops as well (Buchmann 1983). They are social insects establishing colonies with only one queen (Zaitoun et al. 2006). Bumblebees can be employed alone or in conjunction with honeybees to pollinate agricultural crops. Compared to honeybees, bumblebees make more consistent contact with the stigma; they are bigger, finer, and more robust, excellent foragers due to traits, viz., lengthy tongues, and ability to work at low temperatures and low light intensities (Heemert et al. 1990). Bumblebees are therefore promising wild pollinators (Goulson 2003; Winter et al. 2006).

They are successfully raised in nations including Japan, China, Israel, Turkey, and others for pollination. Many growers of fruits, vegetables, and seeds find that using bumblebees for pollination results in cost efficiency, more productivity, and higher-quality products (Velthius and van Doorn 2006). In this chapter, biology and the impacts of pollination by bumblebees on different crops have been discussed.

2 Life Cycle

2.1 Spring

Life commences in spring, when higher temperatures trigger a queen bumblebee hibernating in the soil. The queen spends the winter season in soil. When the queen first emerges, she feeds on pollens, sips nectar to get energy, and then searches for a nest site, which can be holes in the earth, bird boxes in garden, or unused niches in backyards. After choosing a nest site, the queen begins to gather pollen. The queen makes a mound of pollen and wax, lays a brood of eggs, and collects nectar stored in a bowl-like container. The queen maintains the eggs warm by sitting on a wax "nest" and facilitates the shaking of muscles in eggs for many days.

Fig. 8.1 Life cycle of bumblebee

2.2 Summer

The offspring of the first brood consist mostly of workers and females. They carry out work inside and outside the nest. Few workers will guard the nest, while others forage for nectar and pollen and clean the nest (Hemberger 2021). As the season progresses, nests begin producing offspring that are not workers. New queens (females) and drones (males) are produced to ensure the colony multiplies. The male bees leave the nest for good. They do not collect pollen and nectar (Hemberger 2021) (Fig. 8.1).

3 Importance of Bumblebees

According to Parker et al. (1987), honeybees are frequently given sole credit for the economic value of pollination, which is frequently used to support government subsidies for beekeeping, though there are other pollinators of which the farmers are unaware (Richards 1993; Batra 1995). Henslow (1867) discovered that honeybees could not forage on the flowers, yet until the 1980s, honeybees were encouraged for pollinating alfalfa (Olmstead and Wooten 1987; Robinson et al. 1989; Batra 1995). Bee species, particularly those from the Megachilidae, also provide efficient pollination (Brand and Westgate 1909), but through a combination of inertia and poor advice to farmers, it was not until the 1970s that the use of Megachilidae for alfalfa pollination became widespread.

As pollinators, honeybees have a variety of benefits, including the ability to pollinate large regions of crops, a large body of knowledge regarding colony management, and the production of honey. However, they also have disadvantages:

(a) fair weather foragers (Willmer et al. 1994), (b) unable to forage under cold conditions as well as rainy days, (c) unable to pollinate some crops (having deep corolla, e.g., *Trifolium pratense*) due to shorter tongue length, and (d) the pollen is delivered in poricidal anthers in some plants, including those in the Solanaceae family (which includes tomatoes and potatoes). In essence, they resemble an upside-down salt cellar; to collect the pollen, an insect must shake the anthers (known as buzz pollination). Due to all these disadvantages honeybees cannot efficiently pollinate these crops (Rick 1950).

Depending solely on one species to pollinate crops is a dangerous tactic by nature. The recent *Varroa destructor* mite pandemic, which virtually eliminated wild honeybees across most habitats and resulted in significant loss of managed hives, has brought this to light all too clearly. Recently, a relatively unknown malady called "Colony Collapse Disorder" destroyed commercial hives in North America (Belsky and Joshi 2019) and is threatening the harvests of crops relying on bee pollination.

Bumblebees are hardy and forage in extremely cold weather, even while it is raining (Corbet et al. 1993). There have been 164 species of bumblebees queens in North America, when the temperature was below freezing while, in the Scandinavian summer they forage 24 h each day. Bumblebees start feeding earlier in the day than honeybees due to higher tolerance for cold. For instance, on cucurbit crops, *Bombus impatiens* started foraging 15–40 min before honeybees (Stanghellini et al. 2002). They forage more quickly than honeybees under the same circumstances, pollinating more flowers per bee (Poulsen 1973; Free 1993; Stanghellini et al. 2002; Fuchs and Muller 2004). As a result, they offer dependable pollination services despite the weather's inconsistencies. Because different kinds of bumblebees have different tongue lengths, they can pollinate a variety of crops. As an example, short-tongued bumblebees like *B. terrestris* are crucial pollinators of oilseed rape, especially during inclement weather when honeybees are dormant (Delbrassinne and Rasmont 1988). To pollinate crops with deep flowers, such as field beans and red clover, species with medium or long tongues (*B. pascuorum* or *B. hortorum*) are required (Fussell and Corbet 1991). Because of their ability for buzz pollination, bumblebees are good pollinators of Solanaceae plants like tomatoes (Eijnde et al. 1991). The anthers of these flowers release pollens when they are vibrated, which bumblebees do by positioning thorax near the anthers and tightening wing muscles at a frequency of roughly 400 Hz (King 1993). Bumblebees pollinate Ericaceae plants like kiwifruit (*Actinidia deliciosa*), cranberries, and blueberries more efficiently than honeybees because they benefit from buzz pollination (Kevan and Gadawski 1984; Buchmann 1985; Mohr and Kevan 1987; Cane and Payne 1988; MacKenzie 1994) (Fig. 8.2).

In general, a close match between the pollinator's size and form and that of the flower is essential for pollination. Honeybees are poor at transferring pollen to some plants (Westerkamp 1991; Wilson and Thomson 1991). For example, honeybees obtain nectar while making little to no contact with the reproductive structures on some plants like cranberries (*Vaccinium* spp.), alfalfa (*M. sativa*), and delicious apples (*Pyrus malus*), making them poor pollinators (Gray 1925; Roberts and Struckmeyer 1942; Farrar and Bain 1946; McGregor 1976; Robinson 1979). Similar findings show that bumblebees pollinate apples, cucumbers, and watermelon

Fig. 8.2 Honeybee versus bumblebee

(*Citrullus lanatus*) more effectively than honeybees (Stanghellini et al. 1997, 1998; Thomson and Goodell 2001). Bumblebees' bodies are more hairy compared to the honeybees, which increases their pollen transfer efficacy; for example, when visiting raspberry flowers, bumblebees deposited significantly more pollen on the stigmas than did honeybees. Bumblebees are among the most significant wild pollinators of crops in Europe and North America (Corbet 1987; Plowright and Laverty 1987; Corbet et al. 1991). They visit and pollinate at least 25 important crops farmed in the EC, including field beans, red clover, alfalfa, oilseed rape, and a variety of hard and soft fruits (Corbet et al. 1991; Erickson and Buchmann 1983)

4 Bumblebee Pollination Under Protected Conditions

Compared to honeybees, bumblebees are more adaptable and climate resilient. Rearing of bumblebees is not manually intensive. Under protected cultivation, bumblebees outperform the honeybees. Bumblebees can forage under relatively lower temperatures and dim-light conditions compared to domesticated honeybee species. More techniques and studies are required to utilize them for various crops under protected cultivation conditions. Bumblebees are very efficient pollinators of agricultural crops and flowering plants. Bumblebees pollinate vegetable, fruit, nut, and field crops (Hobbs et al. 1961; Holm 1966; Free and Butler 1959; Thorp 2003). Approximately 95% of commercially reared colonies were used in the production of greenhouse tomatoes (Velthuis and van Doorn 2006) and sweet peppers (Ercan and Onus 2003). The need for bumblebees is required for pollination, producing quality seed, and increasing crop yields. Bumblebee pollination increased total yield and resulted in higher fruit quality (Velthuis and van Doorn 2006).

Bumblebees of the genus *Bombus* spp. are unique due to their foraging capabilities and adaptability to a wide range of conditions. For instance, buzz pollination is a mechanism that effectively releases pollen from flowers. In many countries bumblebees are now part of sustainable agricultural systems.

Bumblebees have been reared commercially for pollination since the 1980s in several nations. Worldwide, many bumblebee species, e.g., *Bombus terrestris* Linnaeus, *B. impatiens* Cresson, *B. occidentalis* Greene, *B. lucorum* Linnaeus, and

B. ignites Smith, are commercially used (Velthuis and van Doorn 2006), and there are more than 30 producers of bumblebee colonies.

Crop plants belonging to families, *Actinidiaceae*, *Boraginaceae*, *Aricaceae*, and *Solanaceae*, are termed as buzz pollinating crops as these plants were pollinated only through wing vibration frequency of bumblebees which causes pollen released in large amounts and therefore released pollen move onto the surface of the bee under the influence of an electrostatic field emanating from it (Corbet et al. 1982; Ericson and Buchmann 1983). The bumblebee pollination can increase the fruit quality in kiwi by visiting in large numbers (Pomeroy and Fisher 2002). The use of bumblebees as pollinators was found to be cheaper (9100 pounds) than mechanical pollination (10,000 pounds) in the Netherlands and Belgium (Roulston and Julier 2009). Several researchers have conducted different experiments to investigate the effect of bumblebee pollination on bell pepper and recorded an increase in fruit set and fruit weight (Ruijter et al. 1991; Yousef 2009). Pollination studies showed that bumblebees are more effective pollinators of solanaceous crops than honeybees (Neiswinder 1956; Asada and Ono 1996). Eijnde et al. (1991) reported that bumblebee pollination has increased the productivity of tomatoes up to 70% under protected conditions. Even in India, such experiments have given positive results, especially in tomatoes and bell peppers.

Introduction of bumblebees in hot pepper grown under greenhouses increased the fruit weight and seed numbers. Higher fruit size was also reported in bumblebee-pollinated hot pepper compared to the control (Kwon and Saeed 2003). Dasgan and Ozdogan (2004) compared the effect of bumblebees with plant growth regulators, hand vibration, and control on greenhouse tomatoes. They concluded that bumblebee pollinated flowers raise crop yields (90%) in comparison to other treatments such as growth regulators (41%) and vibration (9%). Yousef (2009) studied pollination efficiency of bumblebees (*B. terrestris*) and honeybees (*A. mellifera*) on the quality and productivity of eggplant and chili pepper. It was recorded that in eggplant and chili pepper cultivated in poly houses, enhanced fruits were produced in case of two honeybee nuclei (66%; 83%) compared to one honeybee nuclei (59%; 75%), bumblebee (52%; 61%), and control (30%; 35%).

Ahmad et al. (2015) conducted pollination studies in a glasshouse with two cultivars of tomato, i.e., 'Grandella' and 'Cherry' tomato, and studied different parameters. In 'Grandella' cultivar, the highest fruit gain (104 g/fruit) was recorded in bumblebee pollinated plants compared to manual pollination (74.4 g/fruit) and self-pollination (20.6 g/fruit).

5 Utilization of Bumblebees in Pollination

In India, studies have been conducted on *B. haemorrhoidalis* colonies on the pollination of crops cultivated in protected cultivation in a polyhouse. Laboratory-reared colony pollinated the cucumber crop effectively from 0600 to 1900 h, resulting in increased yield and better-quality parameters (Kashyap 2007). Chauhan (2011) used the colonies of *B. haemorrhoidalis* for the pollination of cucumber crop under protected conditions. According to these studies, the cost-to-benefit ratio was found to be 1:2.02 with an increase of 22% in the total returns over the control. The pollination effect of *B. haemorrhoidalis* was recorded by Thakur (2018) in Nauni, Himachal Pradesh, and noticed that bumblebee pollination positively affected the fruit quality characteristics with increase in fruits numbers, fruit size (g), fruit length (cm), fruit breadth (cm), healthy fruits (No.), seed number (No.), 1000 seed weight (g), and fruit yield (kg/m^2) by 3.77%, 24.60%, 13.51%, 21.52%, 23.84%, 113.65%, 5.44%, and 25%, respectively. Also, a reduction in misshapen fruits by 38.30% was observed. Yankit et al. (2018) determined the impact of bumblebee pollination on tomato yields (*Solanum lycopersicum* Mill.) in protected cultivation and observed that bumblebee pollination increased the nutritional value and yield of tomato. They reported the percentage enhancement in fruit numbers per plant, healthy fruits (Table 8.1).

B. haemorrhoidalis was used successfully for pollination of bell pepper grown under caged conditions, as bumblebee rearing proved to be superior to open pollination. The bumblebee pollination resulted in better pollination, which showed a positive effect on fruit quality parameters with significant enhancement in fruit quantity per plant (22.12%), fruit weight (g) (24.71%), and fruit production (kg/m^2) (82.35%), respectively, compared to open pollination (Devi 2019). Further, in case of strawberry cv. Sweet Charlie, an increase of 10.38%, 25.64%, 16.22%, 35.29% and 5.80% in fruit production (Kaushal 2020) (Fig. 8.3) was recorded.

6 Bumblebee Pollinated Crops

6.1 Bell Pepper

The laboratory-reared bumblebee (*B. haemorrhoidalis* Smith) queens used for pollination led to a rise in fruit quantity per plant (3.77%), fruit weight (24.60%), fruit length (13.51%), fruit breadth (21.52%), healthy fruits (23.84%), and seed number (113.64%) (Thakur et al. 2021). Devi (2019) also reported a positive effect on different fruit quality parameters with a significant rise in fruit quantity per plant (22.12%) and fruit weight (g) (24.71%).

6.2 Brinjal

Brinjal (Solanaceae) flowers are self-fertile. Under natural conditions, wind pollination is the only way of pollination for this crop. But it is not sufficient under polyhouse conditions, so pollinators like bumblebees are required to release the pollen from the anther efficiently. Brinjal flowers do not produce nectar; hence, a combination of bumblebees and honeybees can be used. The bumblebee pollinated eggplants gave a higher yield (25%), fruit size (14% in weight and 7% in length), and four times the quantity of seeds per fruit than vibration under unheated plastic houses (Abak et al. 2000; Al-Abbadi 2009).

Table 8.1 Effect of pollination by bumblebees on crops in India

S. No.	Crop	Type of structure used for trial	Temp (°C)	Month of trial initiation	Foraging activity		% increase in fruit set over open pollination	% increase in production over open pollination
					Max	Min		
1.	Tomato	Caged conditions (Insect nylon proof net)	21.5–33.4	July	8.33 (1800–1900)	4.67 (1200–1300)	–	64.79%
2.	Cucumber	Greenhouse	19.1–31.8	May	8.24 (1800 h)	4.49 (1200 h)	4.08%	22%
3.	Kiwifruit	Caged conditions	15.8–30.5	May	5.07 (1600–1800 h)	3.22 (1200–1400)	46.25%	46.93%
4.	Capsicum	Polyhouse	26.4–30.2	May	3.00 (1400–1600 h)	1.26 (1200–1400 h)		89.42%
5.	Capsicum	Caged conditions	24–26.5	May	1.42 (1800–1900 h)	0.70 (1200–1300 h)	8.23%	82.35%
6.	Strawberry	Caged conditions	1.9–29.7	October–May	5.33 (1600–1800 h)	2.50 (1200–1400 h)	7.87%	3.50%

Fig. 8.3 Utilization of *B. haemorrhoidalis* in different crops for pollination purposes

6.3 Chili

Bumblebees (*Bombus terrestris* L.) are known as effective pollinators in the cultivation of greenhouse hot pepper (*Capsicum annuum* L.). The colony traffic and foraging activity of bumblebees were highest at 25.7 °C in greenhouse, whereas at 32.7 °C, the foraging activity and colony traffic decreased by 69.7% and 40.0%, respectively. The bumblebee pollination enhanced the fruit mass and quantity of seeds by 27.2% and 47.8%, respectively, in comparison to the control (Kwon and Saeed 2003). Shipp et al. (1994) observed that the effect of bumblebees as pollinators was significant ($P < 0.05$) for parameters of fruit and seed 'Plutona' cultivar of sweet pepper. Other commercially pollinated crops by bumblebees are listed in Table 8.2.

6.4 Cucumber

Honeybees pollinate crops such as squash, cucumbers, and melons due to the large number of bees per hive. Honeybees tend to go down the rows, whereas bumblebees tend to go across the rows or are more erratic in foraging (Mader et al. 2010). Thakur (2010) found that increased fruit set (4.08 kg/plant) in bumblebee pollinated cucumber in comparison to control (no insect pollination) and found that bumblebee pollinated cucumbers produced more healthy fruits (83.62%), fruit weight (413.62 g), number of seeds/fruit (422), and 1000-seed weight (31.50 g) and less crooked fruits (16.18%) over control (Chauhan 2011).

Table 8.2 Crops pollinated by bumblebees

Crop	Scientific name	References
Tomato	*Solanum lycopersicum*	Yankit et al. (2018)
Pepper	*Capsicum annuum*	Shipp et al. (1994), Poporato et al. (1995), Abak et al. (1997), Meisels and Chaisson (1997), Dag and Kammer (2001), Kwon and Saeed (2003), Ercan and Onus (2003)
Egg plant	*Solanum melongena*	Abak et al. (1995)
Melon	*Cucumis melo*	Fisher and Pomeroy (1989)
Watermelon	*Citrullus lanatus*	van Ravestijin and van der Sande (1991), Stanghellini et al. (1997, 1998, 2002)
Cucumber	*Cucumis sativa*	Stanghellini et al. (1997, 1998, 2002)
Zucchini	*Cucurbita pepo*	–
Strawberry	*Fragaria* x *ananasa*	Paydas et al. (2000)
Raspberry	*Rubus idaeus*	Willmer et al. (1994)
Blackberry	*Rubus fruticosus*	–
Currant (red, black)	*Ribes sativum, R. nigrum*	–
Craneberry	*Vaccinium macrocarpon*	MacFarlane et al. (1994), MacKenzie (1994)
Blueberry	*Vaccinium corymbosum*	Stubbs and Drummond (2001), Sampson and Spiers (2002), Javorek et al. (2002)
Apple	*Malus domestica*	Goodell and Thomson (1997)
Pear	*Pyrus communis*	–
Cherry	*Prunus cerasus, P. avium*	–
Kiwifruit	*Actinidia deliciosa*	Pomeroy and Fisher (2002)
Peach	*Prunus persica*	–
Apricot	*Prunus armenica*	–
Plum	*Prunus domestica*	Calzoni and Speranza (1996)
Cardamom	*Amomum subulatum*	–

6.5 Strawberry

Paydas et al. (2000) found that 30% increase in yield per plant through bumblebee pollination as compared to the control. Zaitoun et al. (2006) conducted a study on the effects of bumblebee in strawberry pollination. The results showed that bumblebees increased 60% fruit formation over the control. They recorded that the proportions of malformed berries did not exceed 13.3% on plants pollinated by bumblebees. Bumblebees in the plastic houses increased the yield weight up to 29.6% compared to the control. The size and length of the berries were greater in plants pollinated by bumblebees than in other treatments. The bumblebees are beneficial for pollinating certain varieties of strawberry grown under greenhouses (Zaitoun et al. 2006; Dimou et al. 2008).

6.6 Tomato

Yankit (2016) conducted studies on the impact of bumblebee pollination in tomato and showed that bumblebee pollination increased tomato yield cultivated under protected conditions. Rise in quantity of fruits (75.80 fruits/plant), fruit weight (93.87 g), fruit yield (12.7 kg/m^2), healthy fruit (90.33%), number of seeds (102.85/fruit), 1000-seed weight (6.32 g), and number of locules (4.06/fruit) over control were observed. Thakur (2018) conducted a study on pollination efficiency of bumblebees in bell pepper and found that bumblebee foraging led to higher quantity of fruits per plant (3.77%) and fruit weight (g) (24.60%). Bumblebees are being successfully utilized for pollination in polyhouses for commercial crops in India and other tropical countries. This has proved beneficial from economic and environmental perspectives (Hatami et al. 2013; Dogterom et al. 1998; van Ravestijn and van der Sande 1991; Vergara and Buendia 2012; Tung and Kong, 1997).

7 Conclusion

Under protected conditions, bumblebees are very effective and efficient pollinators compared to honeybees due to longer tongue length, thick and branched hairs, as well as their ability to work under extreme conditions. This accrues multiple benefits to humans and is ecologically and economically sound.

Acknowledgments Authors sincerely thank the concerned scientists and publishers for their immense help.

References

Abak K, Sari N, Paksoy M (1995) Efficiency of bumble bees on the yield and quality of egg plant and tomato grown in unheated glasshouses. Acta Hortic 412:268–274

Abak K, Dasgan HY, Ikiz O, Uygun N, Sayalan M, Kaftanoglu O, Yeninar H (1997) Pollen production and quality of pepper grown in unheated greenhouses during winter and the effects of bumblebees (*Bombus terrestris*) pollination on fruit yield and quality. Acta Hortic 437:303–307

Abak K, Ozdogan AO, Dasgan HY, Derin K, Kaftanoglu O (2000) Effectiveness of bumble bees as pollinators for eggplants grown in unheated greenhouses. Acta Hortic 514:197–204

Ahmad M, Bodlah L, Mehmood K, Sheikh UAA, Aziz MA (2015) Pollination and foraging potential of European bumble bee, *Bombus terrestris* (Hymenoptera: Apidae) on tomato crop under greenhouse system. Pak J Zool 47:1279–1285

Al-Abbadi SYA (2009) Efficiency of different pollination treatments on Solanaceae yields grown in plastic house. J Biol Sci 9:464–469

Asada S, Ono M (1996) Crop pollination by Japanese bumblebee (Bombus spp): tomato foraging behavior and pollination efficiency. Appl Entomol Zool 31:581–586

Batra SWT (1995) Bees and pollination in our changing environment. Apidologie 26:361–370

Belsky J, Joshi NK (2019) Impact of biotic and abiotic stressors on managed and feral bees. Insects 10:233. https://doi.org/10.3390/insects10080233

Brand CJ, Westgate JM (1909) Alfalfa in cultivated rows for seed production in semiarid regions. United States Department of Agriculture, Bureau of Plant Industry, Circular No 24, pp 1–23

Buchmann SL (1983) Buzz pollination in angiosperms. In: Jones CE, Little RJ (eds) Hand book of experimental pollination biology. Van Nostrand-Rhein-hold, New York, pp 73–113

Buchmann SL (1985) Bees use vibration to aid pollen collection from non-poricidal flowers. J Kansas Entomol Soc 58:517–525

Calzoni GL, Speranza A (1996) Pear and plum pollination: honey bees, bumble bees or both? Acta Hortic 423:83–90

Cane JH, Payne JA (1988) Foraging ecology of the bee *Habropoda laboriosa* (Hymenoptera: Anthophoridae), an oligolege of blueberries (Ericaceae: Vaccinium) in the southeastern United States. Ann Entomol Soc Am 81:419–427

Chauhan A (2011) Refinement of bumble bee rearing technology and its use in cucumber pollination. M Sc Thesis. Department of Entomology, Dr Yashwant Singh Parmar. University of Horticulture and Forestry, Nauni, Solan. 152p

Corbet SA (1987) More bees make better crops. New Scientist 115:40–43

Corbet SA, Beament J, Eisikowitch D (1982) Are electrostatic forces involved in pollen transfer? Plant Cell Environ 5:125–129

Corbet SA, Williams IH, Osborne JL (1991) Bees and the pollination of crops and wild flowers in the European Community. Bee World 72:47–59

Corbet SA, Fussell M, Ake R, Fraser A, Gunson C, Savage A, Smith K (1993) Temperature and the pollinating activity of social bees. Ecol Entomol 18:17–30

Dag A, Kammer Y (2001) Comparison between the effectiveness of honey bee (*Apis mellifera*) and bumble bee (*Bombus terrestris*) as pollinators of greenhouse sweet pepper (*Capsicum annuum).* Am Bee J 141:447–448

Dasgan HY, Ozdogan AO (2004) Effectiveness of bumble bee pollination in anti-frost heated tomato greenhouses in the Mediterranean basin. Turkey J Agric Res 28:73–82

Delbrassinne S, Rasmont P (1988) Contribution à l'étude de la pollinisation du colza, *Brassica napus* L. var. *oleifera* (Moench) Delile, en Belgique. Bull Rech Agron Gembloux 23:123–152

Devi D (2019) Studies on nesting material and carbon dioxide narcosis on domiciliation of bumble bee (*Bombus haemorrhoidalis* Smith). M.Sc. Thesis. Department of Entomology, Dr Yashwant Singh Parmar University of Horticulture and Forestry, Solan. 67p

Dimou M, Taraza S, Thrasyvoulou A, Vasilakakis M (2008) Effect of bumble bee pollination on greenhouse strawberry production. J Apic Res 47:99–101

Dogterom MH, Matteoni JA, Plowright RC (1998) Pollination of greenhouse tomatoes by the North American *Bombus vosnesenskii* (Hymenoptera: Apidae). J Economic Entomol 91:71–75

Eijnde J (1992) Pollination of greenhouse strawberry by bumblebees and honeybees. Apidology:343–344

Eijnde J, Ruijter A, Steen J (1991) Method for rearing *Bombus terrestris* continuously and the production of bumblebee colonies for pollination purposes. Acta Hortic 88:154–158

Ercan N, Onus AN (2003) The effects of bumble bees (*Bombus terrestris* L.) on fruit quality and yield of pepper (*Capsicum annuum* L.) grown in an unheated green house. Israel J Plant Sci 51: 275–283

Erickson EH, Buchmann SL (1983) Electrostatics and pollination. In: Jones CE, Little RJ (eds) Handbook of experimental pollination biology I. Scientific and Academic Edition, New York, pp 173–183

Farrar CL, Bain HF (1946) Honey bees as pollinators of cranberry. 86:503–504

Fisher RM, Pomeroy N (1989) Pollination by greenhouse muskmelons by bumble bees (Hymenoptera: Apidae). J Econ Entomol Soc 62:581–589

Free J (1993) Insect pollination of crops. Acad, Press, London-New York, pp 172–180

Free JB, Butler CG (1959) Bumblebees. Collins, London

Fuchs R, Muller M (2004) Pollination problems in Styrian oil pumpkin plants: can bumblebees be an alternative to honeybees? Phyton-Annales Rei Botanicae 44:155–165

Fussell M, Corbet SA (1991) Forage for bumble bees and honey bees in farmland: a case study. J Apic Res 30:87–97

Goodell K, Thomson JD (1997) Comparisons of pollen removal and deposition by honey bees and bumble bees visiting apple. Acta Hortic 437:103–107

Goulson D (2003) Bumble bees, their behavior and ecology. Oxford Uniiversity Press, Oxford

Gray HE (1925) Observations of tripping of alfalfa blossoms. Can Entomol 57:235–237

Hatami M, Monfared A, Haghani M, Fahliani RA (2013) Effect of *Bombus terrestris* L. (Hymenoptera: Apidae) pollinating on flowers and fruiting trends of greenhouse tomato (*Lycopersicon esculentum*). Linzer Biol Beitr 45:1907–1119

Heemert CV, Ruijter AD, Eijnde JVD, Steen JVD (1990) Year-round production of bumblebee colonies for crop pollination. Bee World 71:54–56

Hemberger J (2021) A bumble bee's journey. Wild Life Preservation. https://wildlifepreservation.ca/blog/a-bumble-bees-journey

Henslow G (1867) Note on the structure of Medicago sativa, as apparently affording facilities for intercrossing of distinct flowers. J Linn Soc Bot 9(38):327–329

Hobbs GA, Nummi WO, Virostek JF (1961) Food- gathering behaviour of honey, bumble, and leaf-cutting bees (Hymenoptera: Apoidea) in Alberta. Can Entomol 93:409–419

Holm SN (1966) The utilization and management of bumble bees for red clover and alfalfa seed production. Annu Rev Entomol 11:155–182

Javorek SK, Macjenzie KE, Vander Kloet SP (2002) Comparative pollination effectiveness among bees (Hymenoptera: Apoidea) on lowbush blueberry (Ericaceae: *Vaccinium angustifolium*). Ann Entomol Soc Am 95:345–351

Kashyap L (2007) Domicilation of bumble bee*s (Bombus* sp.) and to study resource partitioning with honey bees. M Sc thesis. Department of Entomology, Dr Yashwant Singh Parmar University of Horticulture and Forestry, Nauni, Solan. 107p

Kaushal B (2020) Studies on *Bombus haemorrhoidalis* Smith pollination in Strawberry. MSc Thesis, Dr YS Parmar University of Horticulture and Forestry, Nauni, Solan, Himachal Pradesh, India-173230

Kevan PG, Gadawski SD (1984) Pollination of cranberries, Vaccinium macrocarpon, on cultivated marshes in Ontario. Proc Entomol Soc Ont 114:45–50

King MJ (1993) Buzz foraging mechanism of bumble bees. J Apic Res 32:41–49

Kwon YJ, Saeed S (2003) Effect of temperature on the foraging activity of *Bombus terrestris* L. (Hymenoptera: Apidae) on greenhouse hot pepper (*Capsicum annuum* L.). Appl Entomol Zool 38:275–280

Macfarlane RP, Patten KD, Royee LA, Wyatt BKW, Mayer DF (1994) Management potential of sixteen North American bumble bee species. Melanderia 50:1–12

MacKenzie KE (1994) The foraging behaviour of honey bees (*Apis mellifera* L) and bumble bees (*Bombus* spp) on cranberry (*Vaccinium macrocarpon* Ait). Apidologie 25:375–383

Mader TL, Johnson LJ, Gaughan JB (2010) A comprehensive index for assessing environmental stress in animals. J Anim Sci 89:2955

McGregor SE (1976) Insect pollination of cultivated crop plants Vol. 496. Agricultural Research Service, US Department of Agriculture, Washington, DC

Meisels S, Chiasson H (1997) Effcetiveness of *Bombus impatiens* CR. As pollinators of greenhouse sweet peppers (*Capsicum annuum* L.). Acta Hortic 437:425–429

Mohr NA, Kevan PG (1987) Pollinators and pollination requirements of lowbush blueberry (*Vaccinium angustifolium* Ait. and *V. myrtilloides* Michx.) and cranberry (*V. macrocarpon* Ait.) in Ontario with notes on highbush blueberry *(V. corymbosum* L.) and ligonberry (*V. vitis-ideae* L.). Proc Entomol Soc Ont 118:149–154

Neiswinder RB (1956) Pollination of glass house tomatoes by honeybees. Bee Craft 72:228–231

Olmstead AL, Wooten DB (1987) Bee pollination and productivity growth: the case of alfalfa. Am J Agric Econ 69:56–63

Parker FD, Batra SWT, Tepedino VJ (1987) New pollinators for our crops. Agric Zoolo Rev 2:279–304

Paydas S, Eti S, Kaftanoglu O, Yasa E, Derin K (2000) Effect of pollination of strawberries grown in plastic greenhouse by bumble bees on the yield and quality of the fruits. Acta Hortic 513:443–451

Plowright RC, Laverty TM (1987) Bumble bees and crop pollination in Ontario. Proc Entomol Soc Ont 118:155–160

Pomeroy N, Fisher RM (2002) Pollination of kiwi fruit (*Actinidia deliciosa*) by bumble bees (*Bombus terrestris*): effect of bee density and patterns of flower visitation. N Z J Entomol 25: 41–49

Poporato M, Pinna M, Manino A, Marletto F (1995) *Bombus terrestris* L. e *Apis mellifera* L. nell'impollinazione del peperone in coltura protetta. Apicott Mod 86:99–112

Poulsen MH (1973) The frequency and foraging behaviour of honeybees and bumble bees on field beans in Denmark. J Apic Res 12:75–80

Richards KW (1993) Non-Apis bees as crop pollinators. Rev Suisse Zool 100:807–822

Rick CM (1950) Pollination relations of Lycopersicon esculentum in native and foreign regions. Evolution 4:110–122

Roberts RH, Struckmeyer BE (1942) Growth and fruiting of the cranberry. Proc Am Soc Hortic Sci 40:373–379

Robinson WS (1979) Influence of 'Delicious' apple blossom morphology on the behavior of nectar-gathering honey bees. In: Proceedings of the IVth International Symposium on Pollination, Maryland Agricultural Experimental Station Special Miscellaneous Publications, vol 1, pp 393–399

Robinson WS, Nowodgrodzki R, Morse RA (1989) The value of bees as pollinators of U.S. crops. Am Bee J 129:411–423

Roulston TH, Julier HE (2009) Wild bee abundance and pollination service in cultivated pumpkins: farm management, nesting behaviour and landscape effects. J Economic Entomol 102:563–573

Ruijter A, Eijnde D, Van den J, Steen J (1991) Pollination of sweet pepper (*Capsicum annuum* L.) in greenhouses by honeybees. Acta Hortic 288:270–274

Sampson B, Spiers JM (2002) Evaluating bumblebees as pollinators of 'Misty' southern highbush blueberry growing inside plastic tunnels. Acta Hortic 574:53–61

Shipp JL, Whitfield GH, Papadopoulos AP (1994) Effectiveness of the bumble bee, *Bombus impatiens* Cr. (Hymenoptera: Apidae), as a pollinator of greenhouse sweet pepper. Sci Hortic 57:29–30

Stanghellini MS, Ambrose JT, Schultheis JR (1997) The effects of *A. mellifera* and *B. impatiens* pollination on fruit set and abortion of cucumber and watermelon. Am Bee J 137:386–391

Stanghellini MS, Ambrose J, Schultheis JR (1998) Seed production in watermelon: a comparison between two commercially available pollinators. HortScience 33:28–30

Stanghellini MS, Ambrose JT, Schultheis JR (2002) Diurnal activity, floral visitation and pollen deposition by honey bees and bumble bees on fi eld-grown cucumber and watermelon. J Apic Res 41:27–34

Stubbs CS, Drummond FA (2001) *Bombus impatiens* (Hymenoptera: Apidae): an alternative to *Apis mellifera* (Hymenoptera: Apidae) for lowlush blueberry pollination. J Econ Entomol 94: 609–616

Thakur RK (2010) Bumble bee rearing technology in India: present scenario and future strategies. In: National Symposium on perspectives and challenges of integrated Pest Management for Sustainable Agriculture, November 19-21 at Dr Yashwant Singh Parmar University of Horticulture and Forestry, Nauni, Solan

Thakur S (2018) Studies on bumble bee (*Bombus haemorrhoidalis* Smith) pollination in *Capsicum annuum* under protected cultivation. Department of Entomology, Dr Yashwant Singh Parmar University of Horticulture and Forestry, Nauni, Solan, M.Sc. Thesis, 54p

Thakur S, Sharma HK, Rana K, Thakur M, Sharma MK, Nayak RK (2021) Bumble bee (*Bombus haemorrhoidalis* Smith) -a potential pollinator in bell pepper under protected cultivation. Ind J Hortic 78:84–87

Thomson JD, Goodell K (2001) Pollen removal and deposition by honeybee and bumblebee visitors to apple and almond flowers. J Appl Ecol 38:1032–1044

Thorp RW (2003) Bumble bees (Hymenoptera: Apidae): commercial use and environmental concerns. In: Strickler K, Cane JH (eds) For non-native crops, whence pollinators of the future? Proceedings of Thomas Say Publications in Entomology. Entomological Society of America, Lanham, MD, pp 21–40

Tung CC, Kong HF (1997) Evaluation of the pollination efficiency of bumble bee (*Bombusterrestris*) on greenhouse tomatoes. Chinese J Entomol 16:167–175

van Ravestijin W, van der Sande J (1991) Use of bumble bee for the pollination of glass house tomatoes. Acta Hortic 288:204–212

Velthuis HHW, van Doorn A (2006) A century of advances in bumble bee domestication and the economic and environmental aspects of its commercialization for pollination. Apidologie 37: 421–451

Vergara CH, Buendia PF (2012) Pollination of green house tomatoes by the Maxican bumblebee *Bombus Ephippiatus* (Hymnoptera:Apidae). J Pollination Ecol 7:27–30

Westerkamp C (1991) Honeybees are poor pollinators—why? Plant System Evol 177:71–75

Willmer PG, Bataw AAM, Highes JP (1994) The superiority of bumblebees to honeybees as pollinators: insect visits to raspberry flowers. Ecol Entomol 19:271–284

Wilson P, Thomson JD (1991) Heterogeneity among floral visitors leads to discordance between removal and deposition of pollen. Ecology 72:1503–1507

Winter K, Adams L, Thorp W, Inouye DW, Day L, Ascher J, Buchmann S (2006) Importation of non-native bumble bees into North American potential consequences of using *Bombus terrestris* and other non-native bumble bees for greenhouse crop pollination in Canada, Mexico, and the United States. 33p

Yankit P (2016) Studies on bumblebee pollination in tomato (Solanum lycopercicum Mill.) under protected condition. M.Sc. Thesis. Department of Entomology, Dr Yashwant Singh Parmar university of Horticulture and Forestry, Solan. 35p

Yankit P, Rana K, Sharma HK, Thakur M, Thakur RK (2018) Effect of bumble bee pollination on quality and yield of tomato (*Solanum lycopersicum* ill.) grown under protected conditions. Int J Curr Microbiol Appl Sci 7:257–263

Yousef SAA (2009) Efficiency of different pollination treatments on solanaceae yields grown in plastic house. J Biol Sci 9:464–469

Zaitoun ST, Al-Ghzawi AA, Shannag HK, Al-Tawaha ARM (2006) Comparative study on the pollination of strawberry by bumblebees and honeybees under plastic house conditions in Jordanvalley. J Food Agric Environ 4:237

Pollination Entomology in West Coast Mangroves of South India

9

O. K. Remadevi, A. K. Chakravarthy, Debajyoti Chatterjee, and Shivakumara

Abstract

Mangrove forests stand as sentinels in the seashores of equatorial and tropical coasts, contributing to the ecological and economic security of mankind. Insects play diverse roles and assist in ecosystem processes, among which their role as pollinators is the most noteworthy. Baseline data on insects associated with pollination of select dominant mangrove species of the coastal tracts of Karnataka were generated. Thirty-six insect species on bloom foraging species were categorized in 18 families belonging to four insect orders and were recorded. Flower visiting insects were collected from three species, *Avicennia officinalis* L., *Rhizophora mucronata* L., and *Sonneratia caseolaris* (L.) Engler. Different species of flower frequenting pollinators and foraging sorties were recorded under field conditions. Based on the characteristics of insects and blooms, an effort was undertaken to compare the dominant pollinators from the just-flower-frequenting pollinator guild of mangroves. The flowering period of *R. mucronata* occurred from February to July in coastal Karnataka. *S. caseolaries* start flowering in October and the flowering season lasts up to May. Research showed that *S. caseolaries* and *A. officinalis* offer both nectar and pollen to the pollinators. In *R. mucronata* flowers, no nectar was found. A profuse amount of pollen was produced as reward in this species. Reproductive biological experiments revealed

O. K. Remadevi
Environmental Management Policy Research Institute, Bangalore, Karnataka, India

A. K. Chakravarthy (✉)
Society for Science and Technology Applications (SSTA), Bangalore, Karnataka, India

D. Chatterjee
Division of Entomology, ICAR - Indian Agricultural Research Institute, New Delhi, India

Shivakumara
Department of Entomology, College of Agriculture, V. C. Farm, Mandya, Karnataka, India

A. K. Chakravarthy (ed.), *The Role of Pollinators in Tropical Landscapes*,
https://doi.org/10.1007/978-981-95-9139-8_9

that *A. officinalis* reproduced through cross and selfing, with preference for cross-pollination. Field pollination biology experiments in *R. mucronata* showed self-pollination mode. Insect exclusion resulted in less fruit set (ca 5%) in *S. caseolaries.* This study highlighted interesting interactions and relationships of insects with plants in the mangrove ecosystems of the west coast of India.

Keywords

Pollination · Foraging · Nectar · *Avicennia officinalis* L. · *Rhizophora mucoronata* L. · *Sonneratia caseolaris* (L.) Engler · West coast, India

1 Introduction

Pollination refers to the actual mechanism of transferring pollen from one flower to another (Tomlinson 1986). It is a basic drive for gene recombination in flowering plants and has a key role in plant breeding programs (Bhattacharya and Mandal 2004). Pollination that occurs in the same flower is known as self-pollination/selfing. When the process of pollen transfer occurs between two individual flowers of the same species. It is referred to as cross-pollination or outcross. In outcross, an external agent carries the pollen from one flower to another conspecific flower, resulting in pollination. In India, mangrove pollination studies were mostly confined to the east coast. Scanty information is available regarding the reproductive biology and mode of pollination of select species from *Coringa* mangrove, Andhra Pradesh, and Bhitarkanika mangrove, Odisha. In most cases, the focus was to elucidate the reproductive biology of important mangrove species. Flower visitors, including insects, were documented as an important aspect of the system. Although observations on the activity of the visitors were reported plant species/groups-wise, no specific research was planned to elucidate the role of individual pollinators in pollination of a particular species or a group of mangrove plants. In Karnataka coast, Anand Rao and Suresh (2001) provided limited information on the pollination mechanisms of important mangrove species. Pollinating agents were not included in the study. In spite of attempts made for documentation of entomofauna of mangroves in India and studying their relationship with mangrove plants, research has been done to elucidate the services and activities of insects in the successful pollination of mangrove plants. This chapter brings out the results of studies on pollinators of mangroves of coastal Karnataka, South India. The mangroves extend over 1147 sq.km, amounting to 20% of the total mangrove spread in Karnataka in the West Coast of India. It is a coastal zone in the southern peninsula of the West Coast of India. Along the Arabian Sea, the Karnataka coastline stretches over 320 km from North to South (12°43°N, -15°N NS longitude 74°00F-75°00F) with a width of 8 to 25 km toward the inland (Ananda Rao and Suresh 2001). The average rainfall received by the coastal areas of Karnataka is estimated at 2741.7 mm. The climate of this area is characterized by high humidity around the year. Temperature varies from 35 °C to 26 °C in the summer to 25 °C to 14 °C in winter. Soil is mostly clayey. Four mangrove sites, viz., Karwar, Honnavar, Kundapura, and Mangrove, were selected for the study. Three species of mangrove, namely, *Avicennia officinalis* L., *Rhizophora mucronata* L., and *Sonnerata caseolories* (L.) Engler, were selected. These three species are dominant and abundant along the coastal tracts of Karnataka (Anand Rao and Suresh 2001; Chatterjee et al. 2008; Remadevi et al. 2008a). Another factor behind the selection of these three species was that they are widely exploited by local inhabitants for non-wood forest products. (NWFP) like fuel wood, charcoal, tannin, fodder, and food, and are hence subjected to severe anthropogenic stress. These plant species were selected by the Karnataka Forest Department for further studies (Remadevi and Raji 2003, 2006, 2007).

2 *Avicennia officinalis*

Avicennia officinalis starts the flowering process with the appearance of tiny floral buds in mid to late April in Karnataka coast. Matured buds are available at the end of April to early May. At the onset of monsoon rain, the blooming process triggers. With the recession of monsoon by August, blooming was found to come down to 11%. Flowering was found to cease completely in September. June and July were found to be the maximum flowering period for the species (Raji and Remadevi 2004; Raji and Remadevi 2005a; Raji et al. 2003; Raji 2003).

2.1 Pollination Biology

Pollination biology experiments performed in 2007 showed upto 20% fruit set in bagged conditions (when no flower visitors, including insects, were allowed to visit flowers). In 2008, an insect exclusion experiment showed 21.66% fruit set. To check the extent of cross-pollination, a substantial quantity of buds was emasculated prior to anther dehiscence. 71.6% fruit set was recorded in 2007 from emasculated buds. In 2008, the rate of fruit production decreased from 66.60% in May-June to 63.30% in July-August, 2008. Fruit set via natural process (control) was found to be low in both years. In 2007, the rate and extent of fruit set via pollination in selected mature bagged floral buds was 23.33%. The rate was found to be higher in 2008, showing 35% fruit production via the pollination process (Table 9.1).

Table 9.1 Reproductive biology (% fruit set)

Experiments	2007 (May–June)	2007 (July–August)	2008 (May–June)	2008 (July–August)
Insect exclusion by bagging	20%	15%	18.3%	21.66%
Emasculated before anther dehiscence	71.6%	68.33%	66.6%	63.3%
Natural pollination	18.3%	23.33%	16.6%	35%

2.2 Insect Pollination in *Avicennia marina*

The season of flowering, floral orientation, production of rewards, and guild of insect visitors were markedly different in all three mangal plant species investigated. In the southern part of the West Coast of India. *A. marina* was found to start the flowering process from May and continued until mid-August. Along the Karnataka coast, flower buds were first encountered in the northern part, gradually spreading south. Almost the same pattern of flowering in this species was reported on the east coast of India from the late summer up to the rainy season (Raju 1990). With small, disc-shaped, bright yellow flowers. A good amount of nectar and sticky pollen and 13–14 floral buds in one inflorescence, this species was found attractive to four groups of insects: Hymenoptera (Chief visitors), Diptera, Lepidoptera, and Coleoptera (occasional visitors). Flies belonging to the families Calliphoridae and Sarcophagidae comprised the major visitor groups, followed by individuals belonging to Calliphoridae, Cicindellidae, Chrysomelidae, Asilidae, Helomyzidae, Muscidae, Platystomatidae, Syrphidae, and Drosophilidae. Hymenopteran visitors were represented by six families: Apidae, Hakuetidae, Scollidae, Vespidae, Sphecidae, and Formicidae. Insects of two families of Lepidoptera, Papillionidae and Nymphalidae, were found to be major visitors of *A. marina* flowers, with Syntomidae as occasional visitor. A similarity of the insect-visitors group composition was found with that reported from the east coast of India (Subba et al. 1995; Subba and Raju 1997; Coupland et al. 2006; Dafni 1992, 2005).

Dipteran and Hymenopteran visitors were found to collect pollen and nectar, whereas Lepidopterans were found to restrict their foraging to collecting nectar only. Only one Coccinellidae (Coleoptera) was found to forage on *A. officinalis* flowers, taking nectar and pollen. The foraging activity of insects was the highest in the forenoon between 8.00 and 12.00 hrs. Calliphoridae and Sarcophagidae of the Dipterans were the most frequent visitors, followed by Vespidae (Hymenoptera) and Papillionidae (Lepidoptera). An insect-exclusion experiment, by bagging the flower buds before anthesis, resulted in a 20% fruit set, whereas 18.3% of fruit setting was found under natural conditions. Emasculated flowers after anthesis and before anther dehiscence were found to set 71% fruit. Subba et al. (1995) suggested that *A. marina* is compatible with both allogamy by vectors and autogamy by itself (Kathiresan and Bingham 2001; Raji and Remadevi 2005b).

3 Insect Pollination in *Avicennia alba*

Pollination biology experiment performed on *A. alba* in 2007 showed up to 26% fruit set in bagged conditions (when no flower visitors, including insect were allowed to visit flowers). To check the extent of cross-pollination, a number of buds were emasculated prior to anther dehiscence. Seventy percent fruit set was recorded from emasculated buds. The rate of fruit production from emasculated mature floral buds was 65%. The high fruit set showed the preference of the plant for cross-pollination. Production of fruits via natural process (control) was found to be low in both years of experiment. In 2007, the rate and extent of fruit set via pollination in selected matured tagged floral buds was 23.3%, whereas the rate was found to be higher in 2008, showing 35% fruit production via the pollination process.

Visitors of *A. alba* chiefly comprised Dipterans with a few Hymenopterans and Lepidopterans. Pale yellow color of the corolla, pungent smell, and trace amounts of nectar on the corolla tube attracted the attention of various visitors throughout the day. Dynamics of visitors were monitored, and peak visitation time was ascertained based on frequency of visitation at a particular time of the day during the maximum blooming months, June and July. For this study, insect visits were monitored from dawn to dusk. Among Dipterans, a few families, viz., Calliphoridae, Syrphidae, Muscidae, and Sarcophagidae, were found to be the dominant visitors.

Calliphorid flies were found to visit the flowers throughout the day during peak flowering seasons. Flower visitation started during morning hours around 7:30 am. The Peak foraging bouts were observed from 8:30 am to 11:30 am. Subsequently, the rate of visitation was found to increase from around 2:30 pm to 3:30 pm. The same trend of visitation dynamics was found in both the months of peak flowering season. There was no significant variation in visitation rate between the months of the peak flowering period.

Sacrophagid flies continued visitation activity in *A. alba* flowers throughout the day. But peak activity was recorded only in the forenoon. They started flower visitation around 8:30 am, and peak visitation was recorded between 10:30 am and 21:30 pm. Visitation activity gradually slowed down toward the afternoon and ceased around 4:30 pm. More flower visitation was recorded in July compared to June for this insect group. No significant variation was found in visitation rate between the months of the peak flowering period (Dafni and Maues 1998; Ge et al. 2003; Hodkinson 1992).

Flowering visiting activities of Muscidae were found to be higher in the forenoon of the day. Activity of Muscid flies in the flowers of *A. alba* was found to start at 7:30 am, and foraging was observed between 9:30 am and 11:30 am. Flower visiting activity was comparatively less in the afternoon and continued up to 5:30 pm. The trend of flower visitation was almost the same in both the months of peak flowering, with no significant difference in visitation rates between the months. Among Dipterans, comparatively, low visitation frequency was recorded in each case of Syrphidae. Peak foraging was recorded from 9:30 to 11:30 am. Very less visitation was recorded in the afternoon, and flowering visitation forays were absent subsequent to 3:30 pm. No significant variation in visitation was recorded between peak flowering months.

4 *Rhizophora mucronata*

The flowering period of *R. mucronata* stretched from February to July. The flower required about 2 months to attain maturity from tender bud condition. Opened flowers were encountered from February onwards. Twelve percent opened flowers

Table 9.2 Reproductive biology (% fruit set)

Experiments	March 2007	April 2007	March 2008	April 2008
Insect exclusion by bagging	18.1%	23.3%	21.6%	13%
Emasculated before anther dehiscence	3%	6.6%	8.3%	6.6%
Natural pollination	16.6%	25%	25%	18.1%

were recorded in February. In March and April, flowering reached peak, showing 41% and 59% of opened flowers, respectively. Less blooming was encountered from May onwards (17%). Almost no flowering was encountered from August to January. March to April was peak flowering for this species (Remadevi et al. 1997, 1999; 2008). Major pests of this mangrove ecosystem have been recorded by Kathiresan (1993) and Kathiresan and Bingham (2001). The viviparous nature of this mangrove has been studied by Mei et al (1998).

4.1 Pollination Biology

Insect exclusion experiment revealed that the fruit set was 23.3% in 2007 and 21.6% in 2008. Matured floral buds were emasculated before anther dehiscence, which showed less fruit set in both years. In 2007, the extent of fruit set from the emasculated buds was 6.6%, and in 2008, the same was 8.3%. This indicated that this plant species depended less on external agents for pollen transfer. Fruit production by pollination process was up to 23% in 2007. It showed a lower value in 2008, recording 21.6% fruit set. The results obtained clearly showed self-pollination mode in *R. mucronata* (Table 9.2).

5 *Sonneratia caseolories*

S. caseolories starts flowering in October and the flowering season lasts up to May. Less flowering was encountered in October, November, and December (flowering percentage 3% and 11%, respectively) for this species in the Karnataka coasts. Comparatively high flowering was encountered in January, February, and March (49%, 63%, and 29%, respectively). For *S. caseolories*, peak flowering was observed from January to March (Veena Kumari et al. 1997).

5.1 Pollination Biology

Insect exclusion resulted in very few fruits, as the maximum was only 5%. In 2008, no fruit production was found in bagging method. Emasculation of matured floral buds prior to anther dehiscence showed up to 52.5% fruit production in 2008. In

Table 9.3 Reproductive biology (% fruit set)

Experiments	March 2007	April 2007	March 2008	April 2008
Insect exclusion by bagging	18.1%	23.3%	21.6%	13%
Emasculated before anther dehiscence	3%	6.6%	8.3%	6.6%
Natural pollination	16.6%	25%	25%	18.1%

2009, fruit production was still higher, showing 53% fruit set. Natural fruit production was high in this species. Up to 70% of tagged mature buds have shown fruit set in 2008 and 72.5% in the experiment done in 2009 (Table 9.3). This result indicated predominance of cross-pollination in this species.

5.2 Insect Pollination in *Sonneratia alba*

Sonneratia alba, with a white, solitary flower, huge nectar supply, and a good quantity of pollen, was found blooming from January to July. It has been reported to bloom at night, with pollination by nocturnal pollinators (Pandit and Choudhury 2001; Tomlinson 1986). In the current study, the flowers were found blooming in the late night/early morning and attracted diurnal insect visitors. Mainly, Hymenopteran insects were found to build the major portion of the entomo-flower visitor guild for this plant species. *Apis dorsasta*, Trigon sp., *Ropalidia cynthiformis*, and *Rhynchium nitidulum* were the major Hymenopteran visitors documented. Only one Syrphidae collecting pollen represented Dipterans, and one Scarabidae (Coleoptera) collected nectar from *S. alba* during the study period. Though hawk moths were reported as major visitors to this flower (Tomlinson 1986), no fruit set (5%) was found in the insect exclusion experiment compared with 52.5% for flowers emasculated prior to anther dehiscence and 70% fruit production with pollination conditions. The results clearly depicted the trend of outcrossing with *S. alba* on the Karnataka coast, which was similar to the findings of Pandit and Choudhury (2001). The herbivory nature of insects has been investigated by Murphy (1990); Raju and Henry (2008); Raju et al (2006).

6 Research Data

Mangroves in nature have self and cross-pollinating mechanisms that depend on the species. For instance, *Aegiceras corniculatum* and Lumnitzera racemose are self-pollinated, *Avicennia officinalis* is self-fertile, but may be cross-fertilized too (Aluri and Subba Reddy 1995). In *Avicennia marina*, protandry makes self-pollination of an individual flower unlikely. However, fruits are set even when flowers are experimentally bagged to prevent cross-pollination of an individual flower unlikely. However, fruits are set even when flowers are experimentally bagged to prevent cross-pollination (between 4 and 41% of cross-pollinated flowers set fruit). Fruit abortion is significantly higher in self-fertilized flowers, indicating inbreeding depression (Clarke and Myerscough 1991a). Mangrove plants are visited by a group of animals, including bats, birds, and insects. Pollen is deposited on the animals as they deeply probe the flowers for nectar; they subsequently transfer the pollen grains to the stigma of another flower. The identity of the pollinator differs from species to species. *Lumnitzera littorea*, for instance, is pollinated basically by birds, while *Lumnitzera racemose* and small-flowered *Bruguiera* species are fertilized primarily by insects (Tomlinson 1986). For instance, sunbirds frequent and pollinate *Acanthus ilicifolius* (Aluri and Subba Reddy 1995) and big-flowered *Bruguiera hainesii* (Noske 1996). Birds indeed are key pollinators, particularly in the dry season. Bats are the major pollinators for *Sonneratia*, which opens flowers to expose the powdery stamens in the late night/early morning hours. If there are no bats, hawk moths become the primary night-time pollinators. Two lycaenid butterflies are key for the pollination of mangroves in Brisbane, Australia, where their abundance is correlated with the abundance of mangrove flowers (Hutchings and Recher 1982). Different types of pollinators are adapted to pollinating various species of mangrove plants. For instance, bees frequent and pollinate plants of *Avicennia*, *Acanthus*, *Excoecaria*, *Rhizophora*, *Scyphiphora*, and *Xylocarpus*. Wasps and flies frequent mangroves for building nests and are key pollinators of mangrove plants like *Ceriops decandra*, *Kandelia candel*, and *Lumnizera racemose* (Tomlinson, 1979, 1986). *Rhizophora* species produce large amounts of pollen and are chiefly wind-pollinated, although the stigma is not specifically modified to attract the windborne pollen (Tomlinson 1986). Emergence of leaves, flower buds, and flowers was documented in Thailand by Christensen and Wium-Andresen (1977) by marking procedure in *Rhizophora apiculata*. Time taken for sprouts of flower bud, primordia to mature propagules required almost 3 years. Several propagules dropped prematurely between June and July. Among flower buds, 7% remained as flowers and only 1–3% turned into fruits. The flowers of mangrove plants and insect pollinators continually interact to adapt for the pollination process to occur in a mutualistic way. This happens usually in an evolutionary period.

7 Decline of Pollinators

Currently, there is a growing global awareness of the extent to which agroecosystems and natural vegetation essentially depend on the pollination process (Klein et al. 2007; Huffaker et al. 1984). Several pollinating animals, including insects, bats, squirrels, and birds, are declining world over (Nabhan 1997; Şekercioğlu et al. 2004). Declines of pollinator populations adversely effect on plant reproductive process and fruit set in nature and cultivated systems (Cunningham 2000). Studies globally revealed that declines, particularly for insects, are generally related to habitat loss and use of pesticides (Kearns and Inouye 1993; Kremen et al. 2002; Goulson et al. 2005), and to a range of introduced pests and diseases (Semeniuk et al. 1978). Managed and feral populations of the honeybee,

Apis mellifera, are declining markedly in many countries, a process called Colony Collapse Disorder (CCD). This is a reason for serious concern, as a majority of agricultural and horticultural crops around the world rely on this species for pollination (Nabhan and Buchmann 1997; Klein et al. 2007). The most pressing factor, however, is climate change. Roles of pollinators are closely linked to the functioning and sustenance of ecosystems, whether managed or natural. It is also essential that we end poverty of people who matter in conserving habitats at large, world over.

References

Aluri RJS, Subba Reddy C (1995) Explosive pollen release and pollination in flowering plants. Proceedings In National Science Academy B61(4):323–332

Anand Rao T, Suresh PV (2001) Coastal ecosystems of Karnataka state, India: Mangroves. Karnataka Association for the Advancement of Science, Bangalore, p 308

Bhattacharya A, Mandal S (2004) Pollination, Pollen germination and stigma receptivity in Moringa oleifera Lamk. Grana 43(1):48–56

Chatterjee D, Remadevi OK, Abdul Latheef C (2008) Insect mediated pollination in three manor mangrove species in Karnataka coast, India. Proceedings of the FORTROP II International Conference "Tropical Forestry Change in a changing World". 17–20th November 2008 organized by faculty of forestry, Kasetsart University, Bangkok, Thailand (Proceedings in press)

Christensen B and Wium-Andersen S (1977) Seasonal growth of mangrove trees in Southern Thailand. In The Phenology of Rhizophora apiculata. Bt Aquatic Botany 3:281–286

Clarke PJ, Myerscough PJ (1991a) Floral biology and reproductive phenology of Avicennia marina in southeastern Australia. Aus J Bot 39:283–293

Coupland GT, Paling EI, Mc Guinness KA (2006) Floral abortion and pollination in four species of tropical mangroves from northern Australia. Aqua Bot 84:151–157

Cunningham SA (2000) Depressed pollination in habitat fragments causes low fruit set. Proc R Soc Lond B Biol Sci 267(1448):1149–1152

Dafni A (1992) Pollination ecology: a practical approach. Oxford University Press, New York

Dafni A (2005) Practical pollination biology. Enviroquest, Ltd., Cambridge, ON. 590 pp

Dafni A, Maues MM (1998) A rapid and simple procedure to determine stigma receptivity. Sex Plant Reprod 11:177–180

Ge J, Baiyan C, Peng L (2003) Mating system and outcrossing rates of four Bruguiera Gymnorrhiza populations of mangrove, China. Nat Sci 1(1):42–48

Goulson D, Hanley ME, Darvill B, Ellis JS, Knight ME (2005) Causes of rarity in bumblebees. Biol Conserv 122:1–8

Hodkinson ID (1992) Telmapsylla, new genus, an unusual psyllid from black mangrove in Florida and Costa Rica (Insectia, Homoptera, Psylloidea). Zool Seripta 21(3):307–309

Huffaker CB, Dahlsten DL, Janzen DH, Kenndy GG (1984) Insect influences in the regulation of plant populations and communities. In: Huffaker CB, Rabb RL (eds) Ecological entomology. John Wiley and Sons, New York, pp 659–691

Hutchings PA, Recher HF (1982) The fauna of Australian mangroves. Proc Linn Soc N.S.W. 106: 83–121

Kathiresan K (1993) Dangerous pest on nursery seedling on Rhizosphora. Indian For 119:1026

Kathiresan K, Bingham BL (2001) Biolog of mangroves and mangrove ecosystems. Adv Mar Biol 40:81–251

Kearns CA, Inouye DW (1993) Techniques for pollination biologists. University Press of Colorado

Klein AM, Vaissiere BE, Cane JH, Steffan-Dewenter I, Cunningham SA, Kremen C (2007) Importance of pollinators in changing landscapes for world crops. Proc R Soc Lond 274:303–313

Kremen C, Williams NM, Thorp RW (2002) Crop pollination from native bees at risk from agricultural intensification. Proc Nat Aca Sci 99(26):16812–16816. https://doi.org/10.1073/pnas.262413599

Mei S, Wong KC, Lee JSY (1998) Reproductive biology and population genetic structure of Kandelia candel (Rhizophoraceae), a viviparous mangrove species. Am J Bot 85(11): 1631–1637

Murphy DH (1990) The natural history of insect herbivory on mangrove trees in and near Singapore. Raffles Bull Zool 38:119–203

Nabhan and Buchmann (1997) The Forgotten Pollinators Book

Noske R (1996) Abundance, zonation and foraging ecology of birds in mangroves of darwin harbour, Northern Territory. Wildlife Research 23:443–474. https://doi.org/10.1071/WR9960443

Pandit S, Choudhury BC (2001) Factors affecting pollinator visitation and reproductive success in Sonneatia caseolaris and Aegiceras Corniculatum in a mangrove forest in India. J Trop Ecol 17(3):431–447

Raji B (2003) Insect diversity and insect plant relationship in mangroves of the West Coast of South India, Ph.D. thesis, Forest Research Institute Deemed University, India

Raji B, Remadevi OK (2004) Insect diversity in the mangrove forests along the West Coast of India. Contrib Biosci Felicit:77–87

Raji B, Remadevi OK (2005a) Entomofaunal diversity in the mangrove forests of West Coast (South India). Ann For 13(2):323–331

Raji B, Remadevi OK (2005b) Diversity indices of insects in the mangroves of Karnataka along the West Coast of India. Indian J Trop Biodivers 13(2):92–26

Raji B, Remadevi OK, Rao KS (2003) First report on the incidence of white flies from eumangroves. My Forest 39(4):371–374

Raju JSA (1990) Observations on the floral biology of certain mangroves. Proc Indian Natl Sci Acad B56(4):367–374

Raju JSA, Henry JK (2008) Reproductive ecology of mangrove trees *Ceriops decandra* (Griff.) Ding Hou and *Ceriops tagal* (Perr.) C.B. Robinson (Rhizophoraceae). Acta Bot Croat 67(2): 201–208

Raju JSA, Jonathan HK, Vijaylakshmi A (2006) Pollination biology of *Ceriops decandra* (Griff.) Ding Hou (Rhizophoraceae), an important true viviparous mangrove tree species. Curr Sci 91(9):1235–1238

Remadevi OK, Raji B (2003) Psychids as major pests of nursery plants of Rhizophora mucronata, an important mangrove species along the West Cost, Disease and Insects in Forests Nurseries. In: Lilja A, Sutherland JR, Potari M, Mohana C (eds) Proceedings of the 3th meeting of IUFRO working party. S7.03.04, May 6-8, 2003 at Peechi, Kerala, India, working papers of the Finnish Forest Research Institute. 11 93p-ISBN 951-40-1965-2. 37–40pp

Remadevi OK, Raji B (2006) Trophic composition of insect diversity in the Mangrive of West Coast of India. In: Ignachimuthu S, Jayaraj S (eds) Biodiversity and insect pest management. Narosa Publishing House, New Delhi. 399pp (190–194)

Remadevi OK, Raji B (2007) Inset herbivory as a major factor in the conservation and management of mangroves in India. In: Coastal Resource Management. 227–236pp

Remadevi OK, Raji B (2009) Insect plant relationships in the west coast mangroves. In: Invertebrate diversity and conservation in the Western Ghats. (In press)

Remadevi OK, Santhakumaran LN, Raja M (1997) Bio-ecology of *Pteroma plagiophleps* on *Rhizophora mucronata*. My Forest 33(1):375–379

Remadevi OK, Raja M, Raji B, Santhakumaran LN, Rao KS (1999) Insect pests of mangroves and coastal trees along east coast of India. Proceedings of Workshop on Forestry, Forest products and coastal population. 10–11 Febrauary, Institute of Wood Science and Technology, Bangalore, 93–96 pp

Remadevi OK, Abdul Latheef C, Chatterjee D, Raji B (2008a) Entomofauna of mangrove ecosystem of India-an annotated Cheak list. Biosystematica 2(2):33–80

Remadevi OK, Chatterjee D and Abdul Latheef C (2008b) Digital image analysis for the quantification of insect folivory in the mangroves of Karnataka, India. In: Proceedings of the FORTROP II International Conference "Tropical Forestry Change in a changing World". 17–20th November 2008 organized by faculty of forestry, Kasetsart University, Bangkok, Thailand (Proceedings in press)

Remadevi OK, Raji B, Chatterjee D, Abdul Latheef C (2008c) Entomopfauna of mangroves-an annotated bibliography. Zool Surv India, Kolkota. 1–80pp

Şekercioğlu CH, Daily GC, and Ehrlich PR (2004) Ecosystem consequences of bird declines. Proc Natl Acad Sci 101(52):18042–18047. https://doi.org/10.1073/pnas.0408049101

Semeniuk V, Kenneally KF, Wilson PG (1978) Mangrove of Western Australia. Handbook No.12. Western Australian Naturalists Club, Perth

Subba Reddi C, Raju JSA (1997) Reproductive biology of three mangrove plant species. Indian J For 29(2):153–157

Subba Reddi C, Aluri JSR, Reddy N (1995) Pollination ecology of Avicinnia officinalis L (Avicinniaceae). J Palynol 31:253–260

Tomlinson PB (1986) The botany of mangroves. Cambridge University Press, Cambridge. 419 pp

Tomlinson PB, Primack RB, Bunt JS (1979) Preliminary observation in floral biology in mangrove Rhizophoraceae. Biotropica 11:256–277

Veena kumari K, Mohanraj P, Bandyopadhyay AK (1997) Insect herbivores and their natural enemies in the mangals of the Andaman and Nicobar Islands. J Nat Hist 31(7):1105–1126

Diversity, Abundance, and Efficiency of Insect Pollinators of Avocado, Rambutan, and Litchi

10

A. T. Rani, G. S. Madhu, B. M. Muralidhara, S. Rajendiran, Nayan Deepak, B. R. Jayanthi Mala, P. V. R. Reddy, and V. Rakshith

Abstract

Avocado, rambutan, and litchi depend on insect pollinators for successful production. Unlike bee-pollinated flowers with hidden floral rewards, avocado flowers offer exposed nectar and pollen. This attracts a variety of insect taxa to flowers for pollination. Apart from the bees, insects of orders Diptera, Coleoptera, and Heteroptera also pollinate avocado flowers. Rambutan receives on cross-pollination facilitated by insects for successful fruit set. The primary insect pollinators for rambutan are from Diptera and Hymenoptera. Foraging activity of pollinators showed significant diurnal variability (10:00 AM–11:00 AM) in the morning hours. Litchi flowers typically bloom in late winter or early spring. Despite male and female flowers on the same tree, litchi flowers do not self-pollinate due to self-sterility. Apis bees are the dominant pollinators for better qualitative and quantitative yields; Pollinators activity in fruit orchards should be conserved.

Keywords

Pollinators · Fruits · Tropics · Conservation

A. T. Rani (✉) · G. S. Madhu · B. M. Muralidhara · S. Rajendiran · Nayan Deepak · V. Rakshith
ICAR-IIHR-Central Horticultural Experiment Station, Chettalli, Kodagu, Karnataka, India

B. R. Jayanthi Mala · P. V. R. Reddy
ICAR-Indian Institute of Horticultural Research, Hessaraghatta, Bengaluru, Karnataka, India

A. K. Chakravarthy (ed.), *The Role of Pollinators in Tropical Landscapes*,
https://doi.org/10.1007/978-981-95-9139-8_10

1 Introduction

Pollination is the process by which pollen is transferred from the male reproductive organs (anthers) of a flower to the female reproductive organs (stigma) of the same or different flower. This transfer occurs through agents such as wind, water, and insects, leading to fertilization, seed development, and the transmission of genetic information to the next generation. This fundamental process plays an important part in the evolution and conservation of diverse genetic traits, key to sustaining the health and balance of ecosystems.

A pollinator is an animal that transfers pollen from the male anther of a flower to the female stigma (Fig. 10.1), facilitating the fertilization of the ovules in the flower by the male gametes from the pollen grains. About 80% of flowering plant species rely on animal pollinators, which include vertebrates and mammals, although insects are the primary pollinators. Pollinators play a crucial role in the production of a diverse range of crops such as orchard fruits, agricultural crops, horticultural crops, and forage plants. Over 75% world's food crops depend to some extent on pollination by pollinators.

Insect pollinators play a part in food production by enhancing crop yield and quality. Their pollination activities exemplify a mutually beneficial relationship between plants and insects. Insects benefit from obtaining nectar and pollen from plants, while plants benefit from having their pollination needs met by these insects (Kuehn 2015). Entomophilous pollination, facilitated by insects, is key for global food production and is recognized as a key ecosystem service. Studies indicate that

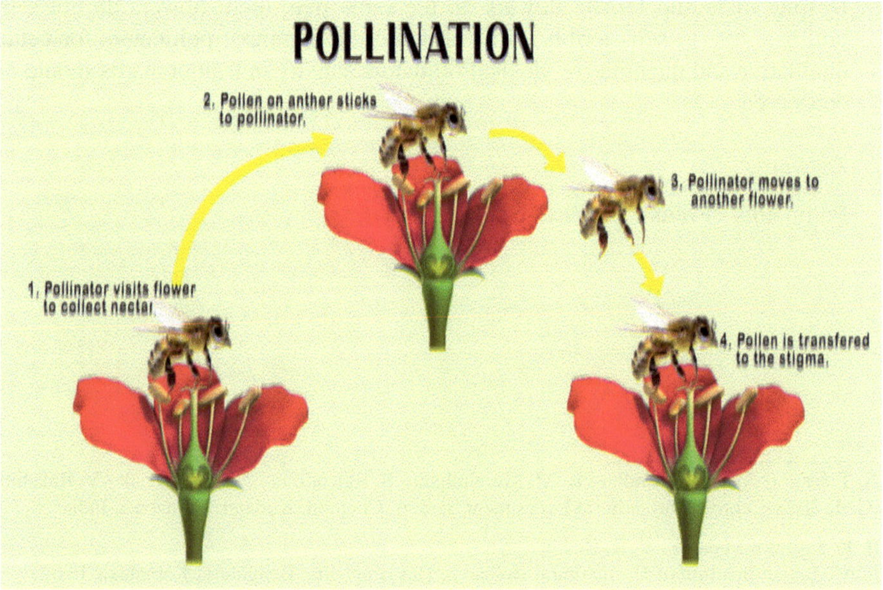

Fig. 10.1 The process of insect pollination

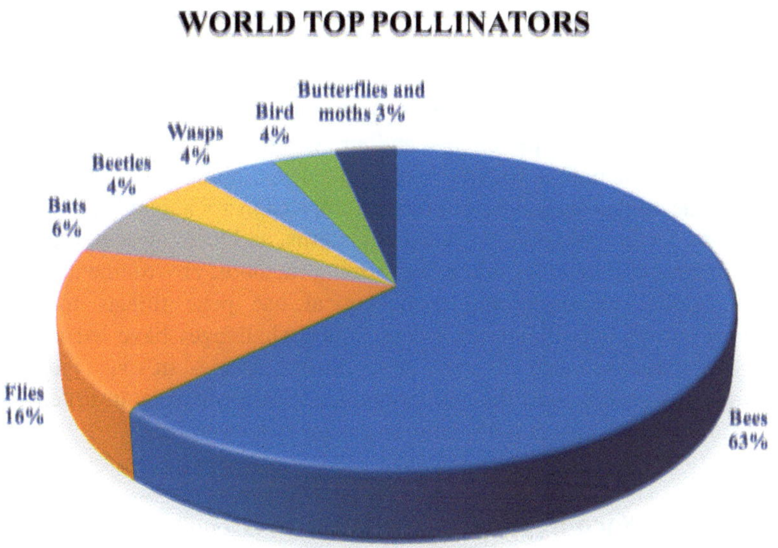

Fig. 10.2 World top pollinators (Modified, Gautam, Aastha, 2020) in crop production

up to 75% of food crop species rely on insect pollination (Klein et al. 2007). Over the past decade, extensive research on pollination has significantly advanced our understanding of the positive impact of animal pollination on crop yields worldwide (Garibaldi et al. 2011, 2013).

Entomophily, or pollination by insects, is the most prevalent method of pollen transfer and has been crucial in the evolutionary history of angiosperms. Globally, there are approximately 250,000 species of flowering plants that rely on around 200,000 species of animals for pollination. Cross-pollination occurs in about 95% of flowers, with over 85% of them depending on insects. Insect pollinators encompass a diverse range of species, including honeybees, bumblebees, pollen wasps, ants, flies such as bee flies, hoverflies, and mosquitoes, butterflies, moths, and flower beetles (Fig. 10.2). Around 50% of seed-propagated plant species and one-third of the food supply are either directly or indirectly reliant on these insect-pollinated plants. Moreover, over 65% of all flowering plants rely on insects for pollination. The sustainability of ecosystems, environmental quality, and biodiversity conservation are upheld through the propagation of wild species facilitated by insect pollination.

Pollinators play a crucial role in increasing crop yields and enhancing crop quality. They contribute to the stability of crop production by mitigating year-to-year and location-to-location variability caused by several factors. Many insect-pollinated crops are not only high-yielding but also rich in nutrients and have significant economic value. In developing countries, insect-pollinated crops such as coffee and cocoa are important sources of income. For instance, cocoa beans used in chocolate production are pollinated by midges. Likewise, mango flowers rely on a variety of insects for pollination, including wasps, ants, flies, butterflies, beetles, and bees, with flies (*Musa domestica*) being particularly important. Research conducted in Malaysia revealed that flies contributed 53% to the total fruit set of mangoes.

Different crops rely on various insect pollinators. Some crops have specific pollinators, such as beans, which heavily depend on bumblebees. On the other hand, crops like oilseed rape, apples, and strawberries are pollinated by a diverse range of insects, including solitary bees, honeybees, bumblebees, hoverflies, and other flies. The insects crucial for pollinating crops can vary annually and regionally, influenced by factors like climate, weather patterns, and local agricultural practices. Changes in land use, diseases, intensive farming methods, environmental pollution, invasives, global warming, and agrochemical use pose threats to pollinator populations and their pollination services. These challenges have led to a decline in both the abundance and diversity of pollinator species in managed and wild populations. As global food demand continues to rise, there is an urgent need to conserve and manage insect pollinators to ensure food security.

2 Importance of Insects' Pollination

In humid tropical regions, fruits such as avocado, litchi, rambutan, kokum, and garcinia are cultivated. Bees are the primary pollinators for more than 73% of cultivated crops, followed by other groups of insects like flies, wasps, butterflies, and thrips also playing a role (Zych and Jakubiec 2006). Fruit crops are largely cross-pollinated and require careful timing of crop protection measures to safeguard pollinators. Animal pollinators, like bees, birds, and bats, are crucial for pollinating over 90% of tropical flowering plants (Ollerton et al. 2011). They significantly impact global crop production, contributing to increased yields of 87 primary food crops worldwide (Das et al. 2018). Among pollinators, honeybees alone are estimated to contribute around 80% to insect pollination (Robinson et al. 1989), resulting in a 50% higher fruit production compared to wind pollination alone (Krishnan et al. 2012). For instance, in Arabica coffee, bee visits nearly double the rate of berry set compared to self-pollination (Belavadi and Ganeshaiah 2013). Furthermore, pollinators play a role in promoting biodiversity, and the diversity of plants and pollinators is interconnected. Monitoring the population of pollinators can serve as an indicator of ecosystem health (McGregor 1976).

The ecological and economic importance of insect-mediated natural pollination of crops, enhanced quality and yield, is increasingly recognized and offers new financial opportunities. In 2005, the global economic value of pollination services provided mainly by insect pollinators, especially bees, was estimated at €153 billion for key crops crucial for the world's food supply (Gallai et al. 2009). In India, insect pollination directly contributes an impressive ₹112,615.73 crores (USD 22.52 billion) annually to agriculture, accounting for 8.72% of its value. This contribution includes spill-over benefits such as improved quality traits, increased seed production, and enhanced breeding efficiency (Chaudhary and Chand 2017). The presence of native pollinators in areas near crops, such as apples, almonds, or coffee, can boost yields by around 20% (Dar et al. 2017). Native insect pollination in the United

States conserves about $3.1 billion annually in the agricultural economy by supporting natural crop production, as stated by the American Institute of Biological Sciences (Woteki 2013). Moreover, this pollination process contributes to the production of goods valued at approximately $40 billion annually in the United States alone (EFSA, Press Release 2014). Despite their significance, insect pollinators face threats like chemical insecticides, monoculture practices, and habitat destruction. Therefore, conserving these pollinators is crucial for maintaining life on planet Earth.

3 Avocado

Avocado (*Persea americana*) originated in Central America and southern Mexico, where the flowering phase has co-evolved with local insect species (Ish-Am 2005). It is one of over 105 crops that benefit from animal pollination (Rader et al. 2020), primarily relying on insect pollinators (Ish-Am and Lahav 2011). In its natural habitat, various insect taxa visit avocado flowers, including bees, wasps, and flies. Avocado's unique flowering behavior, known as protogynous diurnally synchronous dichogamy (PDSD) (Free 1993), involves female flowers becoming receptive

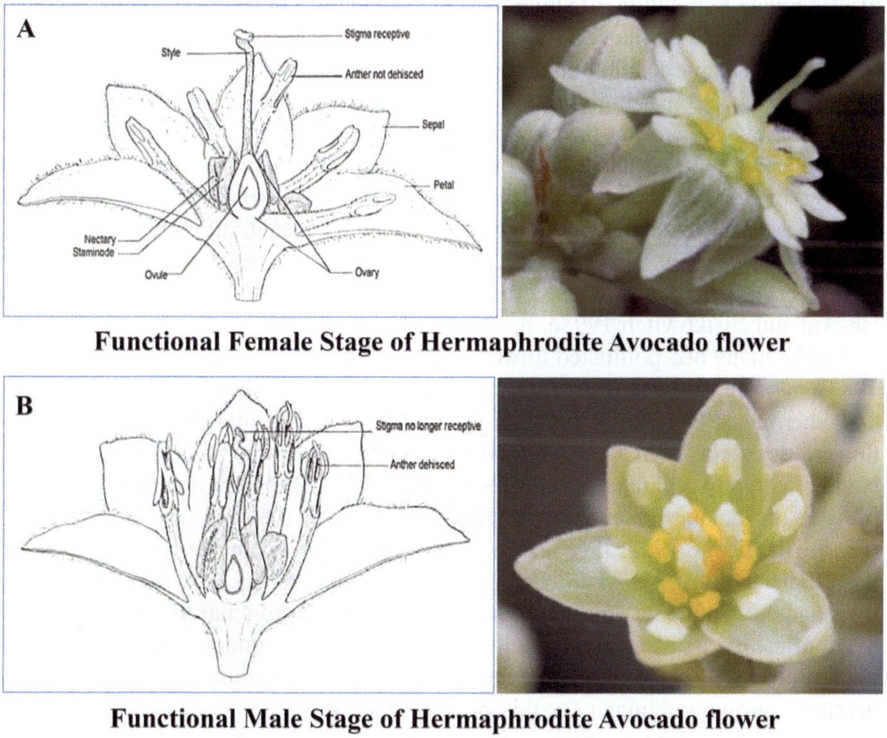

Fig. 10.3 Avocado A type and B type flowers

before male flowers release pollen. The bisexual flower opens twice: first as a female, then as a male the next day when pollen is released. This alternating pattern reduces self-fertilization and promotes cross-pollination by insects or other pollinators, underscoring the crucial role of pollinators in avocado fruit set.

Avocado cultivars are classified into two types (Fig. 10.3): type A and type B. Type A cultivars have flowers in the female or pistillate stage I during the morning of their initial opening and then reopen in the male or staminate stage II the following afternoon, while type B cultivars have flowers in female or pistillate stage I during the afternoon of their first opening and transition to male or staminate stage II the following morning the opposite timing (Lahav and Gazit 1991). In cool weather, the timing of male and female openings may shift, impacting pollination dynamics. Nectar secretion during both male and female flower openings attracts potential pollinators, such as nectar-collecting insects (Davenport 1986; Ish-Am and Eisikowitch 1992; Peterson 1955; Stout 1923, 1933). Despite a mature avocado tree producing around one million flowers in a typical season, only a small fraction (about 0.05%) result in fruit production due to mass flowering (Ish-Am 2005).

Research on avocado pollination and pollinators is primarily conducted in the USA and Israel, with limited studies from Mexico, the origin of avocados and the world's largest producer, and no studies from the second-largest producer, the Dominican Republic (FAO 2020; Dymond et al. 2021). There is a widespread issue of inadequate pollination affecting avocado production in Israel, the USA, Mexico, Central America, and likely in other regions as well. This issue comes from features such as unique composition of local pollinator communities, climate variations, and cultivar-specific traits, highlighting the need for region-specific research in these areas. Honeybees are the primary pollinators for avocados in many countries but tend to abandon avocado orchards when more attractive blooms are available elsewhere. Honeybees may not visit avocado flowers diligently in view of repellent properties of the nectar, containing high mineral concentrations and the alcohol perseitol (Afik et al. 2006, 2011). Additionally, honeybees' location constancy behavior may hinder efficient cross-pollination (Bekey 1986; Davenport 1986; Ish-Am and Eisikowitch 1998a, b, c).

Unlike many bee-pollinated flowers with hidden floral rewards (Kevan and Baker 1983), avocado flowers offer exposed nectar and easily collectible pollen (Vischer and Sherman 1998), attracting various insect taxa. The morphology of avocado flowers does not align with a specific pollinator, indicating that multiple insects could effectively pollinate them (Wysoki et al. 2002). Reported potential pollinators of avocado include honeybees, stingless bees, wasps, flies, beetles, and even bats (Fig. 10.4) (Crane 1992; Free and Williams 1976; Papademetriou 1976). However, the efficiency of these visitors as avocado pollinators has not been thoroughly investigated. Similar to many crops reliant on insect pollination, avocado yields may suffer due to widespread declines in pollinator populations (Biesmeijer et al. 2006; Potts et al. 2016). Enhancing avocado yields is becoming increasingly critical given the growing demand for this product, with global production reaching 32.6 million tons from 1999 to 2008 and 50.4 million tons from 2009 to 2018. Avocados have evolved from being a nutritious staple to becoming a significant export crop for

Fig. 10.4 Diversity of insect pollinators on avocado flowers

numerous countries, generating USD 6.84 billion in global exports in 2018 alone (FAO 2020).

Dymond et al. (2021) conducted a comprehensive review and meta-analysis of studies focusing on avocado pollination ecology. Their findings indicate that while research on avocado pollination has been observed across regions worldwide, there exists a notable scarcity of information from major avocado-producing countries like Mexico and the Dominican Republic. Across many studies, insects were identified as significant contributors to pollination, fruit set, and overall yield in avocado orchards. Honeybees (*Apis mellifera*) emerged as crucial pollinators in numerous regions due to their efficiency and high population numbers. However, many wild pollinators were also observed visiting avocado flowers and were the predominant visitors in more than 50% of the studies analyzed. Notably, stingless bees (Meliponini) and blow flies (Calliphoridae) were highlighted for their effectiveness as avocado pollinators. There remains a lack of comprehensive information on pollinators' efficiency of various flower visitors. To optimize avocado yields, growers are advised to ensure an adequate abundance of pollinators in their orchards. This can be realized by either increasing honeybee hive density or adopting sustainable practices that support wild pollinators by preserving or enhancing natural habitats.

3.1 Diversity of Avocado Pollinators

Several workers have reported a diverse range of insects pollinating the avocado, and Table 10.1 depicts the diversity of insect pollinators on avocado flowers. In South Africa, Eardley and Mansell (1996) observed 49 different insect species visiting avocado flowers, with honeybees being the most abundant and significant pollinators at a large avocado estate near Tzaneen called Westfalia. Research by

Table 10.1 Diversity of insect pollinators of avocado

Common name	Scientific name	Order	Family	References
Honeybee	*Apis mellifera* *Apis cerana* *Apis florea* *Apis dorsata* *Tetragonula* sp.	Hymenoptera	Apidae	Eardley and Mansell (1996), Ish-Am et al. (1999), Jayanthimalamala and Reddy (2020)
Social stingless bee	*Geotrigona acapulconis* *Trigona nigerrima* *Partamona bilineata* *Nannotrigona perilampoides* *Scaptotrigona pectoralis* *Trigona nigra* *Scaptotrigona mexicana* *Trigona fulviventris* *Plebeia frontalis*			
Social solitary bees	*Bombus* spp. *Exomalopsis* spp.		Apidae	
Small carpenter bee	*Allodape microstica* *Braunsapis* *Amegilla* sp.		Anthophpridae	
Hallictin bee	*Lasioglossum* sp. *Halictus (Seladonia)* sp.		Halictidae	
Leaf cutter bee	*Megachile* prob. *semiflava* Cockerell		Megachilidae	
Sphecid wasp	*Larra* sp. *Philanthus* sp.		Sphecidae	
Ants	*Crematogaster* sp. *castanea*-group *Camponotus cinctellus* (Gerstaecker)		Formicidae	
Paper wasp, Social wasp	*Belonogaster* sp. *Polistes* prob. *spilophorus* Schletterer *Brachygastra mellifica* *Polistes* spp.		Vespidae	
Dauber wasp	Unidentified		Eumenidae	
Cucko wasp	*Stilbum* sp.		Chrysididae	
	Pseudacraea lucretia tarquinia (Trimen)	Lepidoptera	Nymphalidae	
Brown-veined white butterfly	*Belenois aurota* (Fabricius)		Pieridae	

(continued)

Table 10.1 (continued)

Common name	Scientific name	Order	Family	References
Blow fly	*Chrysomya regalis* Robineau-Desvoid *Chrysomya megacephala* *Chrysomya* sp. *Rhyncomya forcipata* Villeneuve *Lucilia* sp. *Rhinia apicalis* (Wiedemann) *Stomorhina rugosa* (Bigot) *Stomorhina lunata* (Fabricius) *Stomorhina cribrata* (Bigot)	Diptera	Calliphoridae	
House fly	*Musca domestica* Linnaeus *Musca* sp.		Muscidae	
	Cylindromyia sp. Dejeania bombylans (Fabricius) Linnaemya sp.		Tachinidae	
	Physiphora clausa (Macquart)		Ulidiidae	
Hover fly	*Ischiodon aegyptius* (Wiedemann) *Phytomia incisa* (Wiedemann) *Asarkina* sp. *Allograpta calopus* (Loew) *Paragus* sp. *Allograpta fuscotibialis* (Macquart) *Syritta fasciata* (Wiedemann) *Eristalinus taeniops* (Wiedemann) *Eumerus axinecerus* Speiser *Allobaccha* sp.		Syrphidae	
Fruit fly	*Metasphenisca longulior* (Munro)		Tephritidae	
Bee fly	Undetermined		Bombyliidae	
Assasian bug	*Rhinocoris* sp.	Hemiptera	Reduviidae	
Stink bug	*Atelocera foveata* Dallas		Pentatomidae	

(continued)

Table 10.1 (continued)

Common name	Scientific name	Order	Family	References
Plant bug	*Stenomacra marginella*		Largidae	
Lady bird	*Cheilomenes propinqua* (Mulsant)	Coleoptera	Coccinellidae	
Leaf beetle	Undetermined		Chrysomelidae	
Net winged beetle	*Lycus* sp.		Lycidae	

Ish-Am et al. (1999) suggests that wild pollinators play a substantial role in avocado pollination in Central American countries due to the long-standing coevolution between avocados and their pollinators in this region. Studies conducted in Mexico have demonstrated that native species, such as stingless bees and flies, are equally effective as honeybees in pollinating avocados (Can-Alonzo et al. 2005; Ish-Am et al. 1999; Perez-Balam et al. 2012). Notably, Mexico and Cuba have a higher relative abundance of wild pollinators (averaging 63%) compared to other countries (averaging 47%) (Dymond et al. 2021).

Ish-Am et al. (1999) observed a significant difference in flower visitors between commercial avocado orchards treated with potent insecticides and unsprayed trees in small plots and backyards. In sprayed orchards, the predominant visitors were honeybees (*Apis mellifera*), while unsprayed areas attracted a diverse array of visitors from Hymenoptera, Diptera, Coleoptera, and Heteroptera. Notably, stingless bees, wasps, and flies were observed visiting avocado flowers in high numbers in unsprayed environments. Nine species of stingless bees (Apidae, Meliponinae) and the "Mexican honey wasp" (*Brachygastra mellifica*) were identified as the primary pollinators for all three avocado races: Mexican, Guatemalan, and West Indian. Although honeybees were active in avocado blooms across most sites, they sometimes abandoned avocado flowers to forage on competing nearby flowers for nectar and pollen. Many fly species were also observed visiting avocado flowers, occasionally in large numbers ranging from tens to hundreds per tree. Some species from the Calliphoridae, Muscidae, and Syrphidae families carried substantial amounts of avocado pollen and made effective contact with stamens and stigmas. However, their overall pollination efficiency was limited due to their lower flower visitation rates (Wysoki et al. 2002; Vischer and Sherman 1998). Similarly, in Australia, Vithanage (1990) documented a high abundance of Diptera visiting avocado flowers.

Flies are known regular visitors of over 100 cultivated plant species, including economically important crops (Heath 1982). While flies are generally considered generalist flower visitors, their presence across various ecosystems and sheer

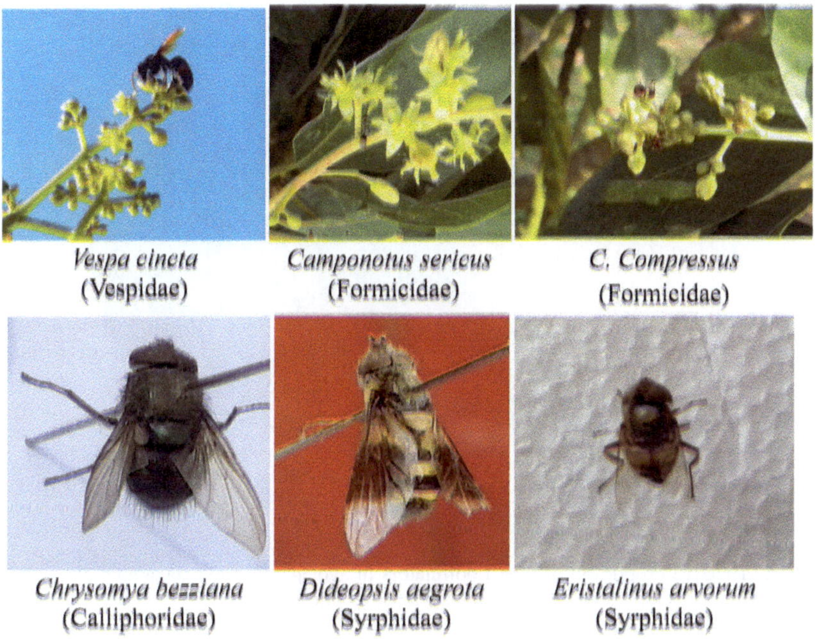

Fig. 10.5 Non-*Apis* pollinators recorded on avocado at the Central Horticultural Experiment Station, Chettalli, Kodagu, Karnataka, India. (Photo credits: Jayanthimala BR and Bhat PS)

numbers suggest they could have a considerable impact on pollen transfer (Ssymank et al. 2008). Pérez-Balam et al. (2012) documented two species of wasps and five species of flies from the families Calliphoridae (1 species), Syrphidae (2 species), and Tachinidae (2 species) visiting avocado flowers. This highlights the potential significance of flies in pollen transfer in view of their high abundance in orchards among avocado insect visitors. The role of Diptera, including flies, as pollinators cannot be overlooked. They rank as one of the most crucial groups of pollinating insects, second only to Hymenoptera (Kearns 2001; Ssymank et al. 2008).

The primary insect pollinators that visit avocado flowers belong to the orders Hymenoptera, Diptera, and Coleoptera at the Central Horticultural Experiment Station (CHES), Chettalli Kodagu (Fig. 10.5). Among these groups, Hymenopterans constitute the majority of pollinators at 52.66%, followed by Dipterans at 39.57%, and Coleopterans at 7.62% (Fig. 10.6). In the group of Hymenopterans, *A. cerena* was observed as the dominant pollinator with (17.56%) followed by *A. florea* (11.3%) and *Tetragonula iridipennis* (8.82%) whereas ants and wasps comprised 6.16% of total hymenopterans. In the Diptera order, three calliphorid species, two syrphid species, and one Musca species are observed, while in the Coleoptera order, three unidentified beetle species visit avocado blooms (Jayanthimala and Reddy 2020). Similarly, Read et al. (2017) noted honeybees as the predominant visitors in avocado orchards, representing 92.9% of all recorded insects in New Zealand. Other studies also mention various insect visitors to avocado flowers, including bees, flies,

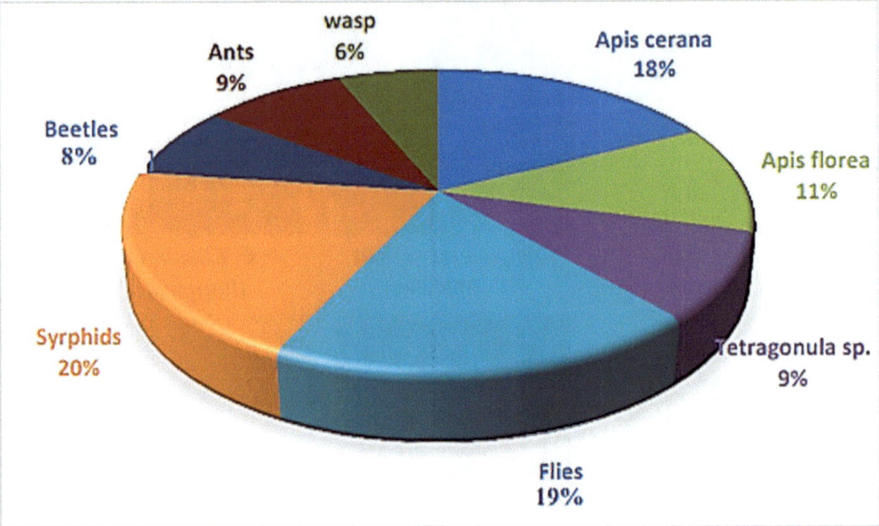

Fig. 10.6 Relative proportion of insect pollinators in avocado (Source: Jayanthimala and Reddy 2020)

wasps, beetles, and thrips (Vithanage 1990). In Colombia, common floral visitors to avocado include ants, bees such as *A. mellifera* and *Trigona* sp., flies, and wasps (Vasquez et al. 2011).

Dymond et al. (2021) mentioned that wild pollinators have demonstrated an important role in avocado pollination. Wild pollinators can be more abundant than managed honeybees and can be efficient pollinators for avocados. For instance, stingless bees have been observed carrying a substantial amount of pollen (around 500 grains), and their body size is well-suited to transfer pollen effectively in avocado flowers (Ish-Am et al. 1999). Blow flies have been identified as important pollinators due to their deposition of a high amount of pollen per visit, facilitated by the open structure of avocado flowers (Vithanage 1990). Managed buff-tailed bumblebees can contribute to avocado pollination as well, although their lower flower visitation rate compared to honeybees reduces their overall effectiveness (Ish-Am et al. 1998). Despite their potential, individual wild species contribute less to pollination due to their lower abundances relative to honeybees (e.g., 28% for stingless bees, 19% for blow flies, and 46% for honeybees). Therefore, active management strategies to increase the abundance of specific wild pollinators may enhance pollination efficiency. However, investigations are required to quantify the effectiveness of wild pollinators in different regions (Dymond et al. 2021).

In temperate regions, nocturnal pollinators from the diptera and lepidoptera orders also merit consideration. Lower night temperatures cause both male and female avocado flowers to open at night, highlighting the potential role of nocturnal pollinators (Dymond et al. 2021). Pattemore et al. (2018) demonstrated that pollinating insects such as wood gnats, crane flies, scarab beetles, capsid bugs, forest moths, and brown lacewings visit avocado flowers at night and carry avocado pollen. However, this aspect remains understudied, emphasizing the need for further investigations in this area. Table 10.1 illustrates the diverse range of insect pollinators observed on avocado flowers, as reported by several workers.

3.2 Abundance and Foraging Activity of Avocado Pollinators

A study focusing on the abundance and foraging patterns of pollinators on avocado flowers revealed specific insights into their behavior. *A. cerana* initiates foraging as early as 6:00 AM and continues until noon, while *A. florea* and *Tetragonula* sp. exhibit peak foraging activity between 9:00 AM and 12:00 PM (Fig. 10.7). Among Diptera species, syrphids were the most prominent (20.51%), followed closely by other fly species (19.06%). Regarding bee populations, *A. cerana* had the highest visitation rate (7.87 visits), followed by *A. florea* (5.06 visits) per panicle in a 30-minute timeframe. *Tetragonula iridipennis* and ant species exhibited 3.96 visits each per panicle in the same duration. Among Dipterans, syrphids showed an average of 9.19 visits per panicle in 30 minutes, while *Musca* sp. recorded 8.53 visits in the same timeframe, indicating their substantial foraging activity on avocado flowers compared to *Apis* species. Overall, the peak foraging hours of pollinators in avocado were observed throughout the day from 9.00 AM to 4.30 PM (Figs. 10.7 and 10.8) (Jayanthimala and Reddy 2020).

Another study conducted by Perez-Balam et al. (2012) in southern Mexico analyzed the contribution and performance of different insect taxa in avocado pollination. Flies were the most common visitors, after honeybees and wasps. Although bees and wasps frequented equal numbers of flowers, and even more than flies at the same timeframe, flies carried significantly more avocado pollen on their bodies. Interestingly, despite the higher pollen load, the number of pollen grains deposited on a stigma after a single visit was similar for all three taxa. However, honeybees and flies were deemed equally crucial for avocado pollination, with their importance surpassing that of wasps.

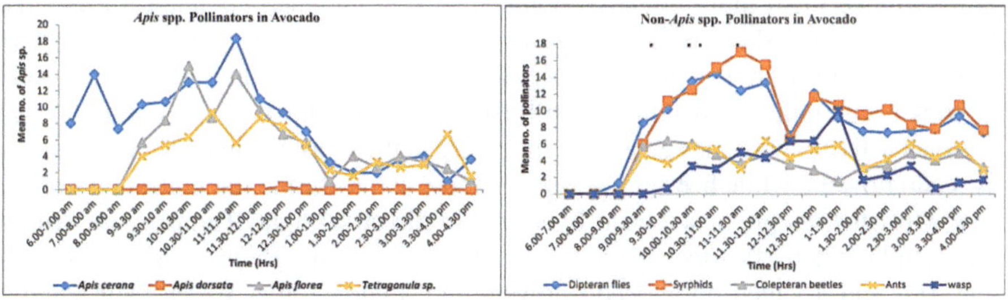

Fig. 10.7 Peak foraging time of *Apis* and non-*Apis* spp. pollinators in Avocado (Source: Jayanthimala and Reddy 2020)

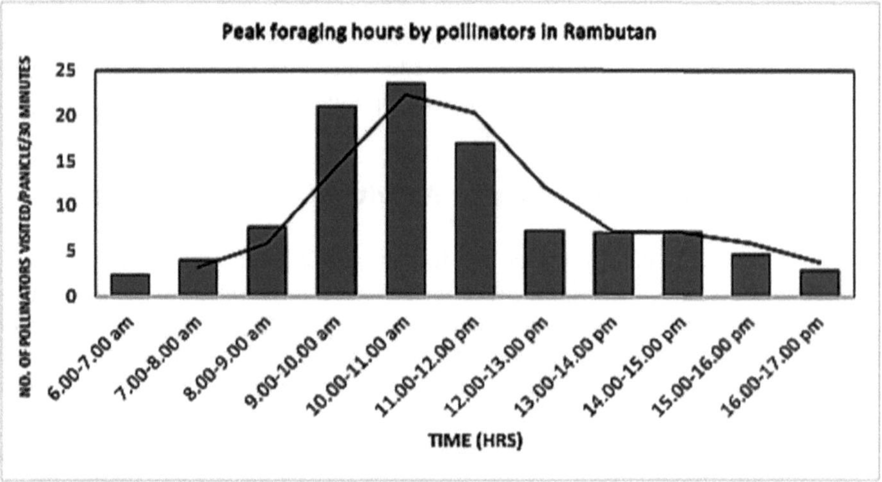

Fig. 10.8 Diurnal variations in the foraging activity of insect pollinators on avocado flowers (Source: Jayanthimala and Reddy 2020)

According to Ish-Am et al. (1999), honeybees displayed more activity on avocado blooms during the morning but gradually decreased their presence by noon, whereas stingless bees and wasps remained active, sometimes reaching up to a thousand individuals per tree. This shift in honeybee behavior could be interpreted in two ways: either as competitive exclusion due to other avocado visitors, especially stingless bees, or as a preference for alternative flower sources over avocados. The latter explanation is more plausible when honeybees ignore avocado blooms while other visitor numbers are low.

The honeybee is a common visitor to avocado flowers due to its abundance (Eardley and Mansell 1996). Honeybees are not efficient pollinators for avocados due to their grooming habits and large size, which make it challenging for pollen grains to stick to their bodies and transfer to small avocado flowers' stigmas. In contrast, small carpenter bees carry their pollen loosely, making all of it available for pollination. Their smaller size and behavior of dragging pollen-laden hind legs over flowers make them more effective pollinators. Certain small calliphorid flies, like *Stomorhinia*, also show effectiveness as pollinators. However, some insects, like certain hover flies, are less effective as pollinators because they hover around flowers to drink nectar rather than aiding in pollination. Temperature plays an important role in insect activity, with flies generally appearing in the morning and late afternoon on avocado flowers, while bees and wasps are more active during the hottest part of the day. Most insects prefer avocado blossoms in direct sunlight. Temperature effects are complicated by factors such as wind. Jayanthimala and Reddy (2020) observed that *Apis* spp. foraged on avocado flowers until midday, after which their activity declined. However, non-*Apis* pollinators such as dipterans remained active beyond noon. This peak activity of non-*Apis* pollinators in the afternoon is significant due to avocado flowers' protogynous diurnally synchronous dichogamy, aligning with the flower development stages and emphasizing the importance of pollinator attendance throughout the day.

Eardley and Mansell (1996) found that honeybees are not particularly attracted to avocado blossoms, as evidenced by the large number of honeybees observed on litchi flowers compared to avocado flowers nearby. This indicated that depending exclusively on honeybees for avocado pollination may not be optimal. Westerkamp (1991) suggests exploring more reliable alternative pollinators for crops like avocados, especially considering the limitations associated with honeybee pollination. Eardley and Mansell's study (1996) highlighted the presence of other pollinators for avocados, which could potentially be more efficient than honeybees if their population density is increased. Diversifying pollinators could reduce growers' dependence on a single pollinator.

3.3 Pollination Efficiency

Controlled pollination experiments conducted by Davenport et al. (1994) and Davenport (2019) showed no significant difference in pollination rates between open-pollinated treatments with high honeybee hive density and closed pollination treatments with no access to insect pollinators. This suggests that wind pollination may dominate in these systems. However, most studies indicate that the absence of insect pollinators leads to reduced pollination (Cabezas and Cuevas 2007), lower fruit set (Can-Alonzo et al. 2005), and decreased yields (Mulwa et al. 2019; Peterson 1955) compared to open-pollinated treatments.

Rose Nyakemiso Sagwe et al. (2021) in Kenya found that self- and wind-pollinated flowers had significantly lower fruit set rates (17.4% and 6.4%, respectively) compared to insect-pollinated flowers (89.5%). Honeybees have a pollen collection behavior where they primarily adhere pollen to their hind legs, with only accidentally attached pollen on other body parts available for pollination. This behavior reduces their effectiveness as pollinators, as noted by Westerkamp (1991). In another study by Perez-Balam et al. (2012), a small quantity of flowers set fruit in bagged inflorescences, indicating zero insect pollination. However, both open-pollination and cross-pollination treatments showed similar percentages of fruit set, which were higher than in permanently bagged flowers. This emphasizes the role of insect pollination for successful fruit production in avocados.

Jayanthimala and Reddy (2020) conducted a study that compared open pollination treatment with bagged avocado panicles, revealing a higher number of fruit sets in the open pollination treatment (Fig. 10.9). They found an 84.66% reduction in fruit set when flowers were excluded from insects, underscoring the crucial role of insect pollinators in avocado fruit set. Previous studies by Eisikowitch and Melamud (1982), Ish-Am (1994), Ish-Am and Eisikowitch (1998a, b, c), and Vithanage (1990) have also shown a positive correlation between honeybee visitation and avocado yield.

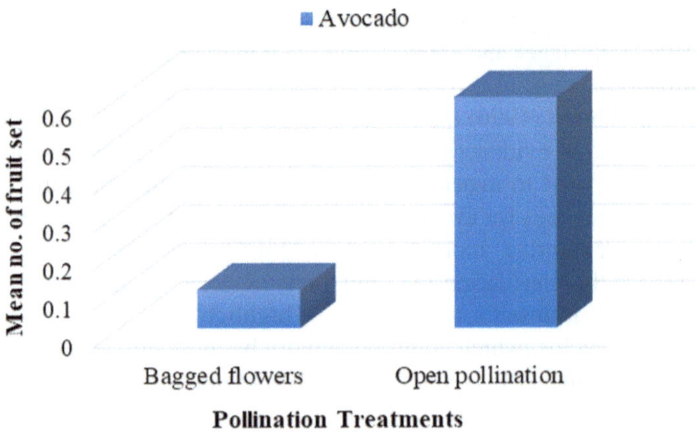

Fig. 10.9 Impact of pollinators on fruit set of avocado (Source: Jayanthimala and Reddy 2020)

In a meta-analysis by Dymond et al. (2021) regarding the role of insect pollinators in avocado production, they highlighted a higher percentage of fruit set and flowers pollinated in open-pollinated compared to closed-pollinated treatments. The disparities in yield between open-pollinated treatments are likely due to variations in pollination services and resource availability across different locations. Controlled pollination studies often do not report pollinator abundance or efficiency, which may change significantly based on factors such as managed beehive densities, proximity to natural areas and wild pollinators, and the availability of native avocado pollinators.

Managed honeybees are a significant contributor to avocado pollination across various regions, given their overall efficiency and high abundance, as indicated by Dymond et al. (2021). This observation aligns with previous reviews that have emphasized the pivotal role of honeybees in avocado pollination (Wysoki et al. 2002). However, honeybees can be influenced by adverse weather conditions such as wind, rain, and low temperatures, which may restrict their foraging activity and affect their contribution to avocado pollination (Can-Alonzo et al. 2005; Ish-Am and Eisikowitch 1998a). Additionally, honeybees might prioritize other nectar sources when available, potentially reducing their focus on avocado flowers. Given the variability in honeybee performance under different conditions, there is a growing recognition of the need for diverse pollinator communities to ensure consistent and effective avocado pollination (Woodcock et al. 2019). This is particularly crucial considering the rising global production of avocados (FAO 2020) and the importance of sustainable production practices that optimize resources like insect pollination (Garibaldi et al. 2011). In this context, Dymond et al. (2021) outlined three key strategies to enhance pollination in avocado orchards: increasing honeybee density, exploring the potential of other managed pollinators, and harnessing the capabilities of wild pollinators. These approaches collectively aim to improve pollination outcomes and support sustainable avocado production systems.

4 Rambutan

Rambutan (*Nephelium lappaceum* L.), a fruit crop belonging to Sapindaceae, relies on insect pollinators for pollination, although there is scarce information available on its specific pollinators. Native to Malaysia and Indonesia, rambutan is also known as the "hairy" lychee and is gaining commercial importance with its market expanding rapidly, especially in Southeast Asia, Australia, the Caribbean, India, Sri Lanka, Florida, and Central and South America (Zee et al. 1998; Sivakumar et al. 2002).

Rambutan typically flowers during March–April in India under humid tropical conditions. Its flowers exhibit three distinct types: staminate (M), hermaphrodite functional male (HFM), and hermaphrodite functional female (HFF) (Tindall 1994). Cross-pollination is crucial for fruit set in rambutan, as seen in the anatomy of its flowers. The HFF type has a bifid stigma for pollination, but its stamens do not release pollen; while the HFM type releases pollen, its stigma remains closed. This reliance on cross-pollination suggests a dependence on insects for effective pollination.

While rambutan plants are naturally dioecious, with roughly equal proportions of hermaphrodite and male plants (50:50), commercial orchards predominantly cultivate hermaphrodite plants, with most functioning as females (Tindall 1994). The rambutan inflorescence is erect and widely branched, bearing numerous flowers, mainly on shoot tips, although pseudo-terminal panicles can also be found. A notable feature is that female rambutan flowers produce nectar significantly more than male flowers, making them an important nectar source for bees (Fig. 10.10). Interestingly, hermaphrodite rambutan flowers do not show visible free pollen on their anthers, suggesting either the required for pollen transfer from male trees for fertilization or the potential for hermaphrodite trees to undergo apomictic fruit formation (Van Welzen et al 1988). In regions like Southeast Asia, hand cross-pollination significantly boosts fruit set compared to the low natural fruiting rate of 1%, whereas bagged flowers fail to bear fruit (Nakasone and Paull 1998), highlighting the importance of proper pollination practices for optimal fruit production in rambutan.

Rambutan relies on cross-pollination facilitated by insects for successful fruit set. The fragrant flowers of rambutan are highly attractive to a variety of insects,

Fig. 10.10 Rambutan inflorescence and individual hermaphrodite flowers

although specific information regarding the diversity of pollinator species and their foraging behaviors in Indian humid tropic conditions remains scarce. In rambutan orchards in Hawaii, issues such as poor fruit set and the development of fruits lacking arils have been recurrent, with studies linking these problems to insufficient pollination. Experiments that excluded bees during the flowering period confirmed the pivotal role of pollinators in Rambutan production (Nagao et al. 2002). Therefore, effective pollinator management is crucial for optimizing Rambutan yields.

4.1 Diversity of Pollinators

Table 10.2 illustrates the diverse range of insect pollinators recorded on rambutan flowers.

Table 10.2 Diversity of insect pollinators of rambutan

Common name	Scientific name	Order	Family	References
Honeybees	*Apis mellifera* *Apis cerana* *Apis florea* *Apis dorsata* *Tetragonula* sp.	Hymenoptera	Apidae	Shivaramu et al. (2012), Manuel Rincón-Rabanales et al. (2015), Jayanthimala and Reddy (2020)
Social stingless bee	*Scaptotrigona mexicana* *S. pectoralis* *T. (Tetragonisca) angustula* *Nannotrigona pelilampoides* *Trigona fulviventris* *T. nigerrima* *Oxitrigona mediorufa*			
Social Solitary bees	*Exomalopsis* sp. *Paratetrapedia* sp.			
Sweat bees	*Halictus hesperus* *Lasioglossum* sp. *Agapostemon aff. splendens*		Halictidae	
Wasps	Unidentified			Shivaramu et al. (2012), Jayanthimala and Reddy (2020)
Ants	Unidentified		Formicide	
Flies	Unidentified	Diptera	Calliphoridae	Shivaramu et al. (2012)
House fly	*Musca* sp.		Muscidae	Jayanthimala and Reddy (2020)
	Unidentified		Syrphidae	
Butterflies	Unidentified	Lepidoptera		

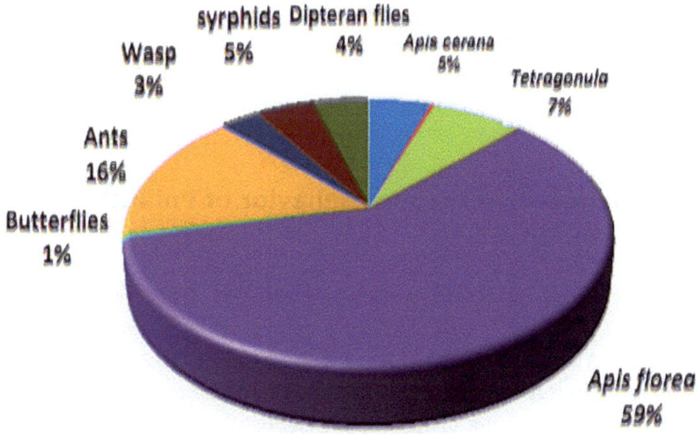

Fig. 10.11 Relative proportion of insect pollinators in rambutan (Source: Jayanthimala and Reddy 2020)

The primary insect pollinators on rambutan flowers were from the insect orders Hymenoptera and Diptera. Among them, hymenopterans are the predominant group, constituting 90% of the overall pollinators, followed by Diptera (9%) and Lepidoptera (1%) (Fig. 10.11). Within the hymenopterans, *Apis* spp. accounted for 71%, while non-*Apis* spp. constituted around 19% of the total hymenopterans. In Apis spp. group, *Apis cerana*, *A. florea*, *A. dorsata*, and *Tetragonula iridipennis*, whereas under non-*Apis* spp. some unidentified ants, wasps, and butterflies, in dipteran order, two species of syrphid flies and one Musca sp. and Calliphorid flies were recorded on avocado flowers. Among bee species, *A. florea* and *Tetragonula iridipennis* were the predominant insect foragers on rambutan (Jayanthimala and Reddy 2020).

Two bee families, Apidae (*Tetragonisca*, *Trigona*, *Oxytrigona*, *Nannotrigona*, *Apis*, *Exomalopsis*, *Paratetrapedia*, and *Scaptotrigona*) and Halictidae (*Halictus*), were identified as visitors to rambutan flowers. The main pollinator was *Scaptotrigona mexicana* (Apidae, Meliponini), followed by social halictid bees (*Halictus hesperus*). Feral Africanized honeybees did not show strong attraction to rambutan flowers in Mexico (Manuel Rincón-Rabanales et al. 2015). Similar studies conducted by Shivaramu et al. (2012) revealed that stingless bee, *Tetragonula irridipennis*, and Indian honeybee, *A. cerana*, were the most dominant foragers. Other species included *A. florea*, *A. dorsata*, an unidentified wasp, ants, and calliphorid flies also observed to visit rambutan flowers.

Small bees are often associated with pollination of small, generalized flowers (Heard 1994). Investigations recorded several social bees visiting rambutan (Nakasone and Paull 1998). However, Kiew (1995) proposed a contrasting view, suggesting that rambutan is not bee-pollinated but rather apomictic and parthenocarpic. This perspective arises from observations that hermaphrodite flowers do not release pollen, and male plants are not cultivated as they do not bear fruit. In contrast, Nakasone and Paull (1998) provided insights into the cultivation of rambutan in

Thailand, Indonesia, and Malaysia. They listed seven cultivars of rambutan and highlighted that even in the absence of male trees, mixed plantings with flowering synchrony resulted in acceptable yields. However, they also noted that plantings without such synchronized conditions yielded exceptionally low harvests.

4.2 Abundance and Foraging Behavior of Pollinators on Rambutan

Foraging activity of all the *Apis* sp. and non-*Apis* sp. foragers showed significant diurnal variability with maximum activity in the morning hours, with a peak activity at 10:00 to 11:00 AM (Fig. 10.12) (Shivaramu et al. 2012).

In a study conducted by Manuel Rincón-Rabanales et al. (2015) on rambutan pollination in Mexico, the peak visitation of insects was recorded between 10:00 AM and 11:00 AM. This observation is beneficial for determining the optimal timing for applying plant protection chemicals to mitigate any negative impact on pollinators of bee species, *A. florea* emerged as the dominant bee species among all visitors, with its peak foraging time recorded between 9:00 AM and 11:00 AM. *A. cerena* exhibited its highest foraging activity from 6:00 AM to 8:00 AM, while *Tetragonula* sp. showed peak foraging hours from 9:00 AM to 12:00 noon (Fig. 10.13) (Jayanthimala and Reddy 2020). This is a valuable information for timing plant protection measures to minimize their impact on pollinators and reduce losses.

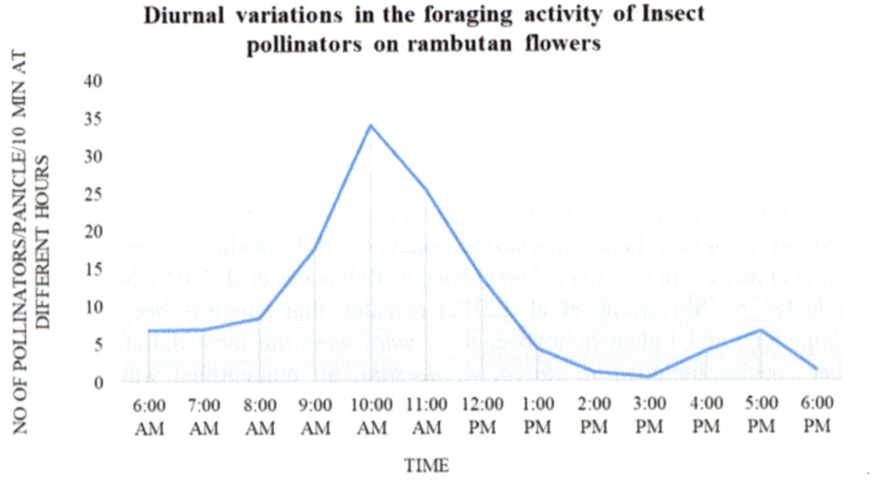

Fig. 10.12 Diurnal variations in the foraging activity of insect pollinators on rambutan flowers (Source: Shivaramu et al. 2012)

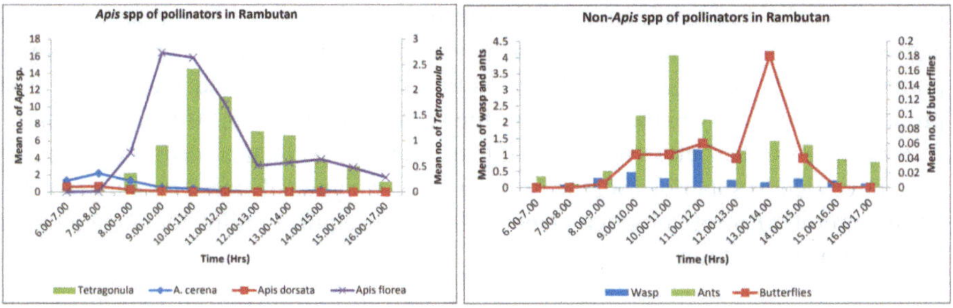

Fig. 10.13 Peak foraging time of *Apis* and non-*Apis* spp. pollinators in rambutan flowers (Source: Jayanthimala and Reddy 2020)

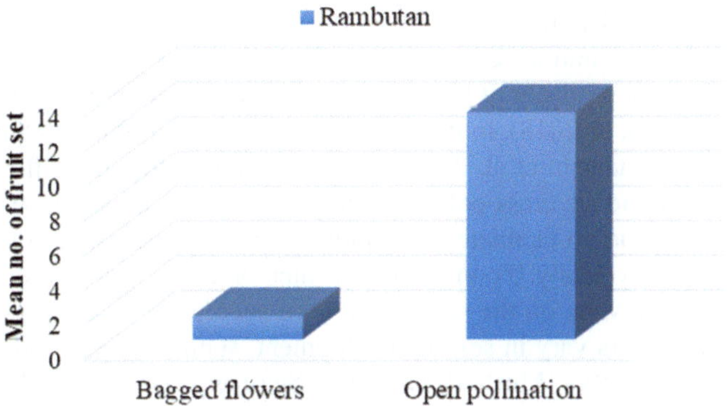

Fig. 10.14 Impact of pollinators on fruit set of rambutan (Source: Jayanthimala and Reddy 2020)

4.3 Pollinator Efficiency

Bagged flowers in rambutan showed significantly lower fruit set compared to open-pollinated flowers, highlighting the crucial role of insects in pollination (Fig. 10.14) (Jayanthimala and Reddy 2020). Protecting native pollinators is essential for better fruit production in rambutan. Shivaramu et al. (2012) reported that open-pollinated rambutan panicles had the highest fruit set at 29.35 fruits per panicle, while bagged flowers did not produce any fruits. This result was consistent with findings from Lim (1984) and Nagao et al. (2002) in Malaysia and Hawaii, highlighting the crucial role of pollinators in rambutan pollination.

Manuel Rincón-Rabanales et al. (2015) on rambutan pollination demonstrated that trees caged with pollinators (induced pollination) and those in open pollination treatments had 9.1 times more mature fruits compared to trees with bagged flowers (no pollinators). Fruits from induced pollination treatments (caged trees with pollinators) also had significantly higher fruit weight. Over a 10-year period, yields exceeding 7 tons per hectare were achieved. Both stingless bee species in open pollination treatments and caged trees showed nearly a tenfold increase in fruit production, especially in the hermaphrodite variety of rambutan. Although the initial fruit set was similar between outcrossing and selfing, pollinator-mediated selfing in caged trees led to superior fruit yields in terms of weight and size. This suggests that hermaphroditic flowers produce viable pollen that requires pollinator assistance for mature fruit production.

5 Litchi

Litchi, *Litchi chinensis* Sonn., (Sapindaceae) is often referred to as the "queen of fruits." Originating from Southeast Asia (Nakasone and Paull 1998; Priyadarshi et al. 2018), this tropical fruit is widely enjoyed both in its raw form and processed, often finding its way into ice cream recipes (Salunkhe and Desai 1984; Hui 2006). China is the biggest producer of litchi in the world, followed by India, Vietnam, and Thailand (Kumari et al. 2018). In India it is cultivated in 92,100 ha, yielding annually 583,400 tons (Priyadarshi et al. 2018). It's worth noting that litchi flowers exhibit high self-sterility and are cross-pollinated crops. Therefore, they rely on an adequate number of pollinators to facilitate pollination and ensure successful fruit set.

Litchi flowers typically bloom in late winter or early spring, producing three distinct types of flowers on the same panicle in successive order (Figs. 10.15 and 10.16). These flowers vary in sexual development, stamen length, and pistil development and functionality (Mustard et al. 1953; Stern and Gazit 1998). Type I flowers

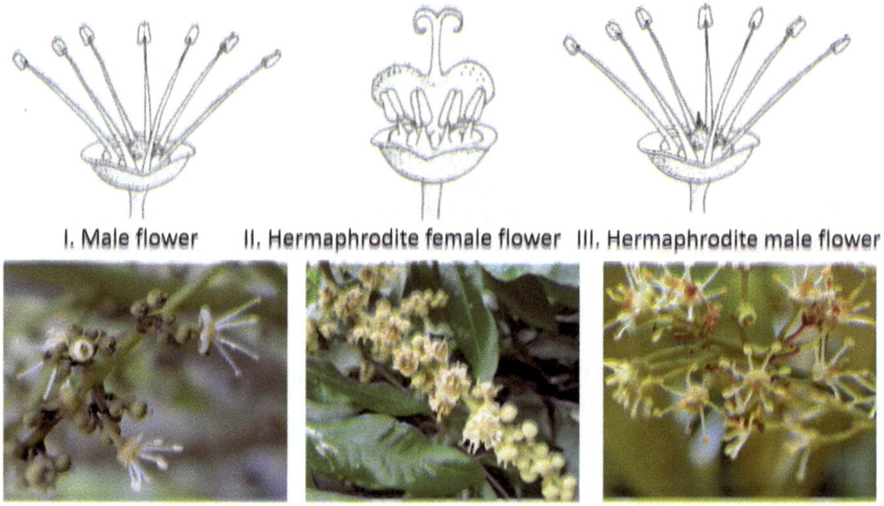

Fig. 10.15 Litchi flower types Source: https://growables.org/information/TropicalFruit/lychee.htm)

Fig. 10.16 Different types of litchi inflorescence

(M1) are functionally male without ovules having 6–8 stamens which produce abundant pollen. Type II flowers are hermaphrodite but functionally female (F), possessing well-developed pistils and stigmas, with 5–8 stamens that do not release pollen. Type III flowers are male (M2) with ample viable pollen production from 6 to 8 stamens and a rudimentary pistil lacking a style and stigma (Stern and Gazit 1998). Flowering cycles overlap, typically lasting 10 days for Type I, 7–10 days for Type II, and 7–10 days for Type III, although this duration varies with weather and cultivar (Mustard et al. 1953; Stern and Gazit 1998).

Despite having male and female flowers on the same tree, litchi does not self-pollinate due to the self-sterility of hermaphrodite flowers, necessitating insect-mediated pollination (Pandey and Yadava 1970; Phadke and Naim 1974). The abundant nectar in self-sterile flowers attracts insects like honeybees, flies, ants, and wasps, facilitating entomophilic cross-pollination (Figs. 10.15 and 10.16). Consequently, litchi, being an entomophilous plant, relies on sufficient insect pollinator visits for effective pollination. Insects, including Hymenoptera, Diptera, Lepidoptera, and Coleoptera, visit litchi flowers, with honeybees, flies, and wasps playing significant roles in pollination (Chaturvedi 1965). The low fruit set in self-sterile cultivars is attributed to insufficient pollinator presence, resulting in reduced production and fruiting failure (Free 1993; UJI 1989). Cross-pollination significantly improves fruit set, yield, and quality in litchi compared to unassisted pollination (Badiyala and Awasthi 1991; Mahanta and Rahman 1997; Rai and Srivastav 2012).

In India, beekeepers often relocate honeybee colonies to litchi plantations during March and April to benefit from litchi nectar, which yields high-quality honey. Despite the presence of wild honeybee colonies near litchi orchards, the precise contribution of bees to litchi pollination in India remains understudied. In India, *Apis mellifera* and *Apis cerana* are the predominant pollinators, with *A. cerana* more prevalent in South India and *A. mellifera* contributing over 80% of pollination in North India. In addition to honeybees, syrphids and non-*Apis* bees play crucial roles in litchi pollination. Syrphids, in particular, are among the most significant groups of pollinators. The most prevalent syrphids include *Episyrphus balteatus* DeGeer, *Melanostoma orientale* We., *Syrphus corollae* Fab., and *Erista listenax* L. Among non-*Apis* bees, species such as *Trigona* spp. (Melliponinae) and *Xylocopa aestuans* L. (Xylocopinae) are also recognized as highly efficient insect pollinators of litchi (Usha et al. 2009). However, comprehensive scientific studies on litchi pollination and its effects on yield are limited in India.

Pollination research is of paramount importance to validate the impact of pollinators on litchi cultivation, including percentage pollination, fruit set, productivity, and yield. Understanding the part of pollinators and effectiveness in pollinating litchi flowers significantly influences the success of litchi orchards. By studying pollination dynamics, researchers can gain insights into how different pollinators contribute to fruit production and optimize yield.

5.1 Diversity of Pollinators

Regular visitors to litchi flowers include honeybees (*Apis dorsata*, *Apis cerana indica*, *Apis florea*), *Melipona* spp., hoverflies, black ants, *Musca* spp., and *Vespa* spp. Among these, *Apis* and *Melipona* species are the prominent, constituting 98–99% of litchi flower visitors, with *A. mellifera* being the primary pollinator throughout the flowering period. Pandey and Yadava (1970) observed a significant increase in fruit productivity due to pollinators. Abrol (2006) documented that litchi flowers attract insects from 6 orders, 12 families, 22 genera, and 29 species. Among these, the four major *Apis* species were the dominant and efficient pollinators, making up over 65% of total pollinating insects. In a study by Shahid Ali et al. (2013) recorded from Haripur, Pakistan, 20 insect species from 16 genera in 8 families, Diptera and Hymenoptera, were identified. Dipteran families included Calliphoridae, Muscidae, Sarcophagidae, and Syrphidae, while Hymenopteran families included Andrenidae, Apidae, Halictidae, and Vespidae.

Rai et al. (2017) recorded 12 insect species visiting litchi flowers: Hymenoptera (6 species), Diptera (5 species), and Coleoptera (1 species). Among these, Hymenopteran insects were dominant pollinators, comprising 73.51% of the total. Notably, species such as *A. dorsata* Fab., *A. mellifera* L., and *A. cerana indica* Fab. represented 21.82%, 26.84%, and 7.60%, respectively, of the Hymenopteran pollinators. Non-*Apis* bees like *Trigona laeviceps* Smith, *Xylocopa aestuans* L., and *Halictus* sp. were also significant pollinators, constituting 17.25% of the total pollinators. However, Heard (1999) noted that non-*Apis* bees, especially *Trigona* species, while participating in litchi pollination, are not as efficient as *Apis* species due to their smaller size.

Dipterans are the second most abundant visitors to litchi flowers, constituting 26.47% of total pollinators. Major species of syrphid flies such as *Episyrphus balteatus* DeGeer, *Melanostoma orientale* L., *Syrphus corolla* Fab., and *Erista listenax* L. are among the most frequent visitors (Jarlan et al. 1997; Usha et al. 2009; Rai et al. 2017). Bhatia et al. (1995) reported 34 insect species from litchi flowers, with the highest numbers from Diptera, followed by Hymenoptera and Coleoptera. Among Coleopterans, a considerable population of *Coccinella septempunctata* was recorded on litchi blossoms. While these beetles are important predators of aphids and mites, they may also play a role in pollination (Rai et al. 2017). Similarly, Das et al. (2019) recorded 13 insect pollinators visiting litchi flowers in West Bengal, with Hymenoptera (7 species), Diptera (5 species), and Coleoptera (1 species).

Kumari et al. (2023) conducted a study documenting 75 insect species that visited litchi flowers during the flowering season, categorized under Hymenoptera (14 species), Diptera (19 species), Lepidoptera (33 species), and Coleoptera (9 species). Major pollinators among Hymenoptera included *Apis mellifera*, *A. cerana*, and *Mellipona* sp., compared to Lepidoptera and Diptera. Lepidoptera species included insects from the Lycaenidae, Nymphalidae, Papilionidae, Arctiidae, Hesperiidae,

Table 10.3 Diversity of insect pollinators of litchi

Common name	Scientific name	Order	Family	References
Giant honeybee	*Apis dorsata* Fab.	Hymenoptera	Apidae	Abrol (2006), Rai et al. (2017), Das et al. (2019), Kumari et al. (2023)
Indian honeybee	*Apis cerana indica* Fab.			
European honeybee	*Apis mellifera* L.			
Little bee	*Apis florea*			
Stinglees bee	*Trigona laeviceps* Smith *Bombus haemorrhoidalis* *Ceratina* sp. *Amegilla* sp. *Anthophora* sp.			
Bumble bee	*Xylocopa aestuans* L., *Xylocopa fenestrata*			
	Andrena sp.		Andrenidae	
Wasp	*Vespa tropica* *V. orientalis* *V. cincta* *V. mandarina* *Polistes hebraeus*		Vespidae	
Ant	*Camponotus compressus* *Formica* sp.		Formicidae	
Carder Bee	*Megachile lanata*		Megachilidae	
Sweat bee	*Halictus* sp., *Nomia* spp. *Lassioglossum* sp.		Halictidae	
Parasitoids	*Bracon* sp.		Braconidae	
	Pimpla sp.		Ichneumonidae	
Hover fly	*Episyrphus balteatus* *Syrphus corollae* *Eristalis tenax* *Eristalis* spp. *Eristalinus megacephalus* *Eristalinus taeniops* *Melanostoma orientale* *Sporophoria indiana* *Metasyrphus corollae* *Melanostoma univitatum* *Dasysyrphus* sp. *Brachypalpoides* sp.	Diptera	Syrphidae	Abrol (2006), Rai et al. (2017), Das et al. (2019), Kumari et al. (2023)

(continued)

Table 10.3 (continued)

Common name	Scientific name	Order	Family	References
	Sphaerophoria sp. *Melanostoma* sp. *Copestylum* sp.			
	Stomorhina sp. *Rhiniid* sp.		Rhiniidae	
	Exoprosopa sp.		Bombyliidae	
	Diopsis sp.		Diopsidae	
	Rhagoletis sp.		Tephritidae	
	Tachinid sp.		Tachinidae	
Flesh fly	*Sarcophaga* sp.		Sarcophagidae	
Blow fly	*Lucilia sericata* *Calliphora* sp.		Calliphoridae	
Hover fly	*Musca domestica* L., *Musca* spp.		Muscidae	
Lady bird beetle	*Coccinella septumpunctata*, *C. septumpunctata var divaricate*, *Menochilus sexmaculata Coleophora sanzeti*, *Leia* spp.	Coleoptera	Coccinellidae	Rai et al. (2017), Das et al. (2019), Kumari et al. (2023)
Longhorned beetle	*Chlorophorus* sp.		Cerambycidae	
	Lycus sanguineus *Dictyoptera simplicipes*		Lycidae	
	Epicauta sp.		Meloidae	
Lycanid butterfly	*Horaga* spp. *Horaga onyx* *Tarucus nara* *Udara dilecta* *Pseudozizeeria maha* *Euchrysops cnejus* *Zizeeria karsandra*	Lepidoptera	Lycaenidae	Abrol (2006), Kumari et al. (2023)
	Junonia almana *Kaniska canace* *Ypthima huebneri* *Aglais urticae* *Vanessa cardui* *Symbrenthia hippoclus* *Neptis hylas* *Junonia iphita* *Junonia lemonias* *Euploea core* *Junonia hierta* *Cyrestis thyodamas* *Danaus chrysippus*		Nymphalidae	Kumari et al. (2023)
	Papilio machaon *Papilio bianor* *Papilio polytes*		Papilionidae	

(continued)

Table 10.3 (continued)

Common name	Scientific name	Order	Family	References
	Lithosiina sp.		Arctiidae	
	Pieris brassicae *Pieris rapae* *Eurema* sp.		Pieridae	
	Amata bicincta *Asota* sp. *Syntomoides imaon* *Asota plaginota* *Lymantria marginata*		Erebidae	
	Hesperiid sp. *Sarangesa dasahara*		Hesperiidae	
Jewel bug	*Chrysocoris partricius*	Hemiptera	Scutelleridae	Abrol (2006)
Thrips	Thrips	Thysanoptera	Thripidae	

Noctuidae, and Pieridae families. Syrphid flies were prominent among Diptera. Coleoptera visitors included species from the Coccinellidae, Lycidae, Meloidae, and Cerambycidae. The total diversity index value was calculated at 1.15, indicating low diversity among insect pollinators visiting litchi flowers (Kumari et al. 2023). Table 10.3 illustrates the diverse range of insect pollinators observed on litchi flowers.

5.2 Abundance and Foraging Activity

Effective cross-pollination in litchi relies on the synchronization of pollen availability and stigma receptivity with the timing of pollinator visits to the flowers. Studies by, Das and Chaudhury (1958) have emphasized the importance of having sufficient pollen available when pistillate flowers are receptive. They noted that anther dehiscence, when a maximum number of flowers release pollen, typically occurs between 07:00 and 12:00 hours. Based upon these findings on floral biology, Pandey and Yadava (1970) observed maximum pollinator activity between 06:30 and 12: 00 hours at two locations. This observation suggests that pollinators are adapted to the floral biology of litchi, leading to effective cross-pollination during these critical hours.

Rai et al. (2017) observed that pollinator visits to litchi flowers commenced shortly after the flowers opened, starting in the first week of March. The peak activity of syrphids and *Apis* spp. was noted in mid-March. Das et al. (2019) reported that Hymenopterans were the most abundant pollinators, accounting for 77.71%, followed by Dipterans at 22.27%. Among Hymenopterans, *A. dorsata* was the predominant visitor, constituting 50.11%, followed by *A. cerana indica* (11.80%), *A. florea* (8.68%), and *A. mellifera* (7.12%) (Fig. 10.17a). The study also observed

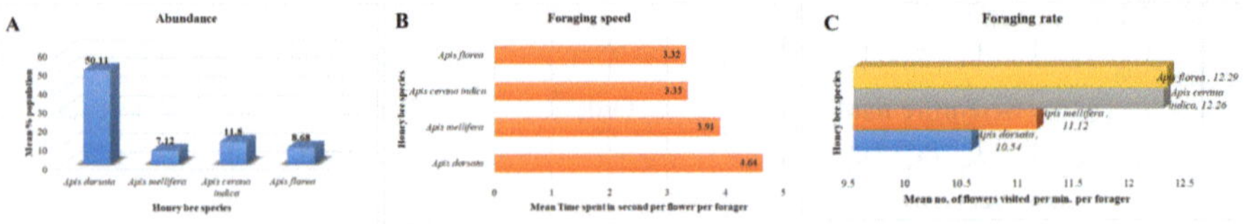

Fig. 10.17 Abundance and foraging activity of honeybees on litchi flowers (Source: Adopted and Modified from Das et al. 2019)

the foraging behavior of honeybees, noting that the maximum foraging speed (time spent in seconds per flower per forager) occurred between 9:00–11:00 AM, while the minimum was during 3:00–5:00 PM. *A. dorsata* had the highest mean foraging speed at 4.64 seconds, while *A. florea* exhibited the lowest at 3.32 seconds (Fig. 10.17b). Conversely, the highest foraging rate (number of flowers visited per minute per forager) was recorded during 3:00–5:00 PM for all honeybees, with *A. florea* having the highest mean foraging rate at 12.29 and *A. dorsata* at 10.54 (Fig. 10.17c).

Kumari et al. (2023) reported a higher abundance of insect visitors from Hymenoptera during morning and evening periods (50.25% and 44.89%, respectively) compared to Lepidoptera (21.83% and 26.67%) and Diptera (24.37% and 23.33%). Similarly, natural pollination facilitated by one *Apis mellifera* colony also showed a similar trend.

5.3 Weather Factors and Pollinator Activity

Studies by Kapil and Kumar (1974), Abrol (2006), and shed light on the ecological thresholds influencing the flight activity of honeybee species. Generally, temperatures between 15.5 and 18.5 °C, light intensities ranging from 600 to 1700 lx, and solar radiation levels of 9–20 mW/cm2 were identified as the minimum conditions for the commencement of flight activity in *Apis* species. Flight cessation in honeybee species was primarily driven by declines in light intensity and solar radiation, regardless of other factors. Throughout their flight activity, honeybee abundance and foraging behavior aligned with fluctuations in temperature, light intensity, solar radiation, nectar sugar concentration, and inversely with relative humidity. Path analysis conducted by Abrol (2006) revealed that each honeybee species responded differently to temperature, light intensity, and solar radiation, which influence foraging behavior.

Kumari et al. (2023) noted that during morning hours, Dipteran population was related positively to wind speed and ultraviolet (UV) conditions. Increases in temperature and rainfall were associated with decreased abundance of syrphid flies, *A. dorsata*, and *A. mellifera*. On the other hand, *A. cerana indica* indicated a positive correlation with maximum temperature. Non-*Apis* bee abundance increased with higher relative humidity percentages (Rai et al. 2017). Research conducted at the National Research Centre on Litchi (NRCL), Bihar, highlighted the foraging behavior of bees and non-bee pollinators. *A. mellifera* was the dominant forager on litchi, followed by *A. cerana*. Peak activity for *A. mellifera* occurred at 8:00 AM and 10:00 AM, while non-bee pollinators were active at 12:00 noon and 2:00 PM (Kumar and Kumar 2014).

5.4 Pollination Efficiency

Lal et al. (2021) conducted a study on the effect of open cross-pollination and self-pollination on fruit set and fruit characteristics in litchi. Results indicated that open cross-pollination facilitated by pollinators significantly increased fruit set, improved fruit characteristics, and boosted yield compared to self-pollination. Kumari et al. (2023) evaluated treatments, and it was observed that fruit characteristics such as length, breadth, weight, and total soluble solids (TSS) were notably larger in the natural pollination treatment provided with an *A. mellifera* colony compared to a caged tree with *A. mellifera* or natural pollination alone. Moreover, the proportion of fruit set and fruit weight was higher in the natural pollination treatment with an *A. mellifera* colony and other pollination modes compared to the pollinator exclusion treatment (bagged panicles) (Fig. 10.18).

Rai and Srivastav (2012) conducted a study on the effects of bee pollination on fruit set, yield, and quality characteristics of litchi. Their findings revealed that the highest fruit setting occurred in trees enclosed with cages containing *Apis mellifera*

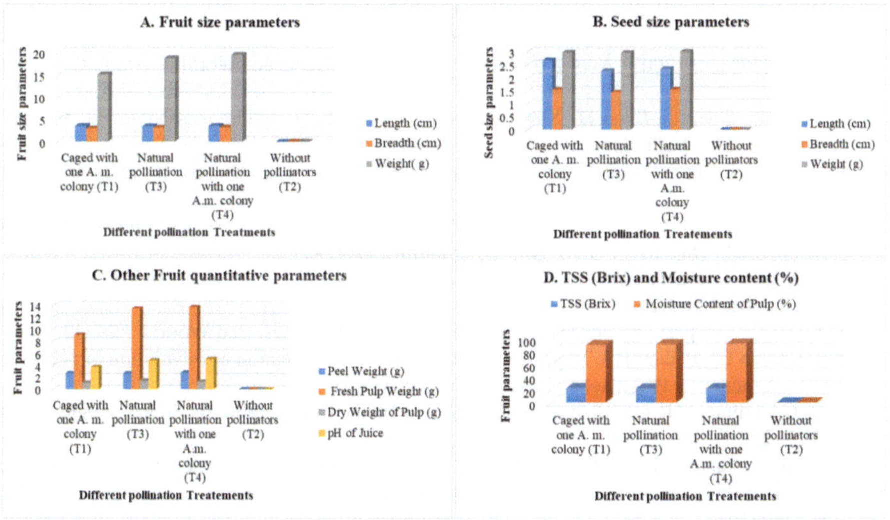

Fig. 10.18 Impact of insect-pollination on the quantitative parameters of litchi fruits (Source: Adopted and modified from Kumari et al. 2023)

boxes, both during the initial stage (8.4 fruits/panicle/tree) and the maturity stage (4.53 fruits/panicle/tree). In comparison, the fully open tree had 5.8 fruits/panicle/tree at the initial stage and 3.73 fruits/panicle/tree at the maturity stage. Minimal fruit setting was observed in the caged tree without *A. mellifera* boxes. Furthermore, besides fruit size, the weight and TSS content were also found to be higher in the caged tree with *A. mellifera* boxes compared to other conditions.

References

Abrol DP (2006) Diversity of pollinating insects visiting litchi flowers (*Litchi chinensis* Sonn.) and path analysis of environmental factors influencing foraging behaviour of four honeybee species. J Apic Res 45(4):180–187

Afik O, Dag A, Kerem Z, Shafir S (2006) Analyses of avocado (Persea americana) nectar properties and their perception by honey bees (*Apis mellifera*). J Chem Ecol 32:1949–1963

Afik O, Delaplane KS, Moo-Valle H, Quezada-Euán JJG (2011) Nectar minerals as regulators of flower visitation in stingless bees and nectar hoarding wasps. In: Memorias VII Seminario Mesoamericano sobre abejas nativas, Cuetzalan México, p 237

Badiyala SD, Awasthi RP (1991) Physico-chemical characteristics of some litchi (Litchi chinensis Sonn.) cultivars grown in Kangra valley. Indian Food Pack 45:42–45

Bekey R (1986) Pollination of avocado - some new insights with special reference to the 'Hass' variety. Calif Avocado Soc Yearbook 70:91–97

Belavadi VV, Ganeshaiah KN (2013) NICRA project on insect pollination manual. Bangalore University of Agriculture Sciences

Bhatia R, Gupta D, Chandel JS, Sharma NK (1995) Relative abundance of insect visitors on flowers of major subtropical fruits in Himachal Pradesh and their effect on fruit set. Indian J Agric Sci 65(12)

Biesmeijer JC, Roberts SPM, Reemer M, Ohlemüller R, Edwards M, Peeters T, Schaffers AP, Potts SG, Kleukers R, Thomas CD (2006) Parallel declines in pollinators and insect-pollinated plants in Britain and The Netherlands. Science 313(5785):351–354. https://doi.org/10.1126/science.1127863

Cabezas C, Cuevas J (2007) Avocado pollinators in Southeast Spain. World Avocado Congress, Chile

Can-Alonzo C, Quezada-Euan JJ, Xiu-Ancona P, Moo-Valle H, Valdovinos-Nunez G, Medina-Peralta S (2005) Pollination of 'criollo' avocados (Persea americana) and the behaviour of associated bees in subtropical Mexico. J Apic Res 44:3–8. https://doi.org/10.1080/00218839.2005.11101138

Chaturvedi RB (1965) Preliminary studies in the sex distribution, pollination and fruit development in litchi (Litchi chinensis Sonn.). Allahabad Farmer 39:49–51

Chaudhary OP, Chand R (2017) Economic benefits of animal pollination to Indian agriculture. Indian J Agric Sci 87(9):1117–1138

Crane E (1992) The past and present beekeeping with stingless bees. Bee World 73:29–42

Dar SA, Hassan GI, Padder BA, Wani AR, Parey SH (2017) Pollination and evolution of plant and insect interaction. J Pharmacogn Phytochem 6(3):304–311

Das CB, Chaudhury KR (1958) Blossom biology of litchi (Litchi chinensis Sonn). Indian Hortic 6(1):17–22

Das A, Sau S, Pandit MK, Saha K (2018) A review on: importance of pollinators in fruit and vegetable production and their collateral jeopardy from agro-chemicals. J Entomol Zool Stud 6(4):1586–1591

Das R, Jha S, Halder A (2019) Insect pollinators of litchi with special reference to foraging behaviour of honey bees. J Pharmacogn Phytochem 8(4):396–401

Davenport TL (1986) Avocado flowering. Hortic Rev 8:257–289

Davenport TL (2019) Cross- vs. self-pollination in 'Hass' avocados growing in coastal and inland orchards of Southern California. Sci Hortic 246:307–316. https://doi.org/10.1016/j.scienta.2018.10.051

Davenport TL, Parnitzki P, Fricke S, Hughes MS (1994) Evidence and significance of self-pollination of avocados in Florida. J Am Soc Hortic Sci 119(6):1200–1207. https://doi.org/10.21273/JASHS.119.6.1200

Dymond K, Celis-Diez JL, Potts SG, Howlett BG, Willcox BK, Garratt MPD (2021) The role of insect pollinators in avocado production: A global review. J Appl Entomol 145:369–383. https://doi.org/10.1111/jen.12869

Eardley CD, Mansell MW (1996) The natural occurrence of insect pollinators in an avocado orchard. South African Avocado Growers' Assoc Yearbook 19:36–38

EFSA (2014) Press Release 2014: EFSA identifies risks to bees from neonicotinoids. Efsa.europa.eu. Retrieved.

Eisikowitch D, Melamud H (1982) Preliminary experiments for increasing avocado pollination rates (In Hebrew). Alon Hanotea 37:19–29

FAO (2020) FAOSTAT statistical database. FAO

Free JB (1993) Insect pollination of crops, 2nd edn. Academic Press, London

Free JB, Williams IH (1976) Insect pollination of *Anacardium occidentale* L., *Mangifera indica* L., *Blighia sapida* Koenig and *Persea americana* Mill. Trop Agric 53:125–136

Gallai N, Salles JM, Settele J, Vaissière BE (2009) Economic valuation of the vulnerability of world agriculture confronted with pollinator decline. Ecol Econ 68(3):810–821

Garibaldi LA, Steffan-Dewenter I, Kremen C, Morales JM, Bommarco R, Cunningham SA, Carvalheiro LG, Chacoff NP, Dudenhöffer JH, Greenleaf SS, Holzschuh A, Isaacs R, Krewenka K, Mandelik Y, Mayfield M, Morandin LA, Potts SG, Ricketts TH, Szentgyörgyi H, Viana BF, Westphal C, Winfree R, Klein AM (2011) Stability of pollination services decreases with isolation from natural areas despite honey bee visits. Ecol Lett 14(10): 1062–1072. https://doi.org/10.1111/j.1461-0248.2011.01669

Garibaldi LA, Steffan-Dewenter I, Winfree R, Aizen MA, Bommarco R, Cunningham SA, Kremen C, Carvalheiro LG, Harder LD, Afik O, Bartomeus I, Benjamin F, Boreux V, Cariveau D, Chacoff NP, Dudenhöffer JH, Freitas BM, Ghazoul J, Greenleaf S, Hipólito J, Holzschuh A, Howlett B, Isaacs R, Javorek SK, Kennedy CM, Krewenka KM, Krishnan S, Mandelik Y, Mayfield MM, Motzke I, Munyuli T, Nault BA, Otieno M, Petersen J, Pisanty G, Potts SG, Rader R, Ricketts TH, Rundlöf M, Seymour CL, Schüepp C, Szentgyörgyi H, Taki H, Tscharntke T, Vergara CH, Viana BF, Wanger TC, Westphal C, Williams N, Klein AM (2013) Wild pollinators enhance fruit set of crops regardless of honey bee abundance. Science. 339(6127):1608–1611. https://doi.org/10.1126/science.1230200

Heard AT (1994) Behaviour and pollinator efficiency of stingless bees and honey bees on macadamia flowers. J Apicul Res 33:191–198

Heard TA (1999) The role of stingless bees in crop pollination. Annu Rev Entomol 44:183–206

Heath ACG (1982) Beneficial aspects of blowflies (Diptera: Calliphoridae). N Z Entomol 7:343–348

Hui YH (2006) Handbook of fruits and fruit processing, Wiley-India ed. John Wiley and Sons, Hoboken

Ish-Am G (2005) Avocado pollination: a review. In: New Zealand and Australia avocado grower's conference

Ish-Am G, Eisikowitch D (1992) New insight into avocado flowering in relation to its pollination. Calif Avocado Soc Yearbook 75:125–137

Ish-Am G (1994) Interrelationship between avocado flowering and honey bees and its implication on the avocado fruitfulness in Israel. (In Hebrew, English Abst.). Ph.D. Thesis Tel Aviv Univ Tel Aviv 157

Ish-Am G, Eisikowitch D (1998a) Low attractiveness of avocado (*Persea americana* Mill.) flowers to honeybees (Apis mellifera L.) limits fruit set in Israel. J Hortic Sci Biotechnol 73:195–204

Ish-Am G, Eisikowitch D (1998b) Mobility of honey bees (Apidae, *Apis mellifera*) during foraging and agricultural pollination. Apidologie 29:209–219

Ish-Am G, Eisikowitch D (1998c) Quantitative approach to avocado pollination. In: Proceedings of the Third World Avocado Congress, vol 1995, pp 46–51

Ish-Am G, Regev Y, Peterman Y, Lahav E, Degani C, Elbatzri R, Gazit S (1998) Improving avocado pollination with bumblebees: 3 seasons summary. Calif Avocado Soc 1998 Yearb 82: 119–135

Ish-Am G, Barrientos Priego AF, Castaneda-Vildozola A, Gazit S (1999) Avocado (Persea americana Mill.) pollinators in its region of origin. Rev Chapingo Ser Hortic 5:137–143

Ish-Am G, Lahav E (2011) Evidence for a major role of honeybees (Apis mellifera) rather than wind during avocado (Persea americana Mill.) pollination. J Hortic Sci Biotechnol 86(6):589–594. https://doi.org/10.1080/14620316.2011.11512808

Jarlan A, Olivers D, Gingras J, De Olivers D (1997) Effects of *Erista listenax* (Diptera: Syrphidae) pollination on characterstics of greenhouse sweet pepper fruits. J Economic Entomol 90(6): 1650–1654

Jayanthimalamala BR, Reddy PV (2020) Diversity and abundance of insect pollinators on avocado (Persea americana Mill.) and rambutan (Nephelium lappaceum L.) and their contribution to fruit set. Pest Manag Hortic Ecosyst 26(1):76–82

Kapil RP, Kumar S (1974) Foraging activity of *Apis dorsata* F. on *Brassica juncea* Czern and Coss. J Bombay Nat Hist Soc 71(2):327–331

Kearns CA (2001) North American dipteran pollinators: assessing their value and conservation. Conserv Ecol 5(1):5. http://www.consecol.org/vol5/iss1/art5

Kevan PG, Baker HG (1983) Insects as flower visitors and pollinators. Annu Rev Entomol 28:407–453

Kiew R (1995) Bee botany in tropical Asia: with special reference to peninsular Malaysia. In: Kevan PG (ed) The Asiatic hive bee: apiculture, biology and role in sustainable development in tropical and subtropical Asia. Enviroquest, Cambridge, Ontario, pp 117–124

Klein AM, Vaissiere BE, Cane JH, Steffan-Dewenter I, Cunningham SA, Kremen C, Tscharntke TD (2007) Importance of pollinators in changing landscapes for world crops. Proc Royal Soc B: Biol Sci 274:303–313. https://doi.org/10.1098/rspb.2006.3721

Krishnan S, Kushalappa CG, Shaanker RU, Ghazoul J (2012) Status of pollinators and their efficiency in coffee fruit set in a fragmented landscape mosaic in South India. Basic Appl Ecol 13(3):277–285

Kuehn FC (2015) Farming for native bees. World Wide Web electronic publication. Retrieved

Kumar R, Kumar V (2014) Impact of pollination by European honey bee, Apis mellifera L. on the yield and quality of litchi (Litchi chinensis Sonn.) fruits in India. Pest Manag Hortic Ecosyst 20(2):127–132

Kumari S, Rani R, Chandola JC, Mir H, Ahmad MF, Bharti A (2018) Effect of pollination method on fruit set in commercial cultivars of Litchi. Int J Curr Microbiol 7:603–606

Kumari P, Rana S, Bhargava B, Reddy SE (2023) Diversity, abundance, and impact of insect visitors in Litchi chinensis production. Agronomy 13(2):298

Lahav E, Gazit S (1991) World listing of avocado cultivars according to flowering type. Fruits 49: 299–313

Lal N, Sahu N, Marboh ES, Gupta AK, Kumar A, Nath V (2021) Effect of pollen grain and pollination period on fruit set in litchi (Litchi chinensis Sonn.). Natl Acad Sci Lett 1–4. https://doi.org/10.1007/s40009-020-01027-6

Lim AL (1984) The reproductive biology of rambutan, *Nephelium lappaceum* L. (Sapindaceae). Garden Bull Singapore 37(2):181–192

Mahanta M, Rahman A (1997) Studies on pollination of litchi (Litchi chinensis Sonn.) by honeybees (Apis cerana F.). Indian Bee J 59:74–75

Manuel Rincón-Rabanales M, Roubik DW, Guzmán MA, Salvador-Figueroa M, Adriano-Anaya L, Ovando I (2015) High yields and bee pollination of hermaphroditic rambutan (Nephelium lappaceumL.) in Chiapas, Mexico. Fruits 70(1):23–27

McGregor SE (1976) Insect pollination of cultivated crop plants, vol 496. Agricultural Research Service, US Department of Agriculture, Washington (DC)

Mulwa J, Kahuthia-Gathu R, Kasina M (2019) Avocado (*Persea americana*) yield as influenced by pollinators in Murang'a County. Kenya Agric Res Adv 01:34–41

Mustard MJ, Liu S, Nelson RQ (1953) Observation of floral biology and fruit setting in lychee varieties. Proc Fla State Hort Soc 66:212– 220

Nagao MA, Leite HM, Kawabata AM, Terada AY (2002) Abstract: update of tropical fruit research in Hawai¥i. In: Proceedings of the 12th Annual International Tropical Fruit Conference. Keauhou Kona, Hawaii, p 8

Nakasone HY, Paull RE (1998) Tropical fruits, vol 16. CAB Int., Wallington, UK, p 445

Ollerton J, Winfree R, Tarrant S (2011) How many flowering plants are pollinated by animals? OIKOS 120:321–326. https://doi.org/10.1111/j.1600-0706.2010.18644.x

Pandey RS, Yadava RPS (1970) Pollination of litchi (Litchi chinensis) by insects with special reference to honeybees. J Apic Res 9(2):103–105

Papademetriou MK (1976) Some aspects of the flower behavior, pollination and fruit set of Avocado (*Persea americana* Mill.) in Trinidad. Calif Avocado Soc Yearbook 60:106–153

Pattemore DE, Buxton MN, Cutting BT, McBrydie HM, Goodwin RM, Dag A (2018) Low overnight temperatures delay 'Hass' avocado (Persea americana) female flower opening, leading to nocturnal flowering. J Pollination Ecol 23(14):127–135

Perez-Balam J, Quezada-Euan JJ, Alfaro-Bates R, Medina S, McKendrick L, Soro A, Paxton RJ (2012) The contribution of honey bees, flies and wasps to avocado (Persea americana) pollination in southern Mexico. J Pollinat Ecol 8:42–47

Peterson PA (1955) Avocado flower pollination and fruit set. Calif Avocado Soc Yearbook 39:163–169

Phadke KG, Naim M (1974) Observations on the honeybee visitation to the litchi blossoms at Pusa, Bihar, India. Indian Bee J 36:9–12

Potts SG, Imperatriz-Fonseca V, Ngo HT, Aizen MA, Biesmeijer JC, Breeze TD, Dicks LV, Garibaldi LA, Hill R, Settele J, Vanbergen AJ (2016) Safeguarding pollinators and their values to human well-being. Nature 540(7632):220–229. https://doi.org/10.1038/nature20588

Priyadarshi V, Hota D, Karna AK (2018) Effect of growth regulators and micronutrients spray on chemical parameter of litchi (Litchi chinensis Sonn.) cv. Calcuttia. Int J Econ Plants 5:99–103

Rader R, Cunningham SA, Howlett BG, Inouye DW (2020) Non-bee insects as visitors and pollinators of crops: biology, ecology, and management. Annu Rev Entomol 65:391–407. https://doi.org/10.1146/annurev-ento-011019-025055

Rai VL, Srivastav P (2012) Studies on the impact of bee pollination on yield and quality of litchi (Litchi chinensis Sonn.). Progress Hortic 44(2):262–264

Rai VL, Srivastava P, Bisht K, Mishra VK (2017) Diversity and relative abundance of pollinating insects visiting litchi (Litchi chinensis Sonn.) inflorescence under tarai agro-climatic condition. J Exp Zool India 20(1):221–227

Read SFJ, Howlett BG, Jesson LK, Pattemore DE (2017) Insect visitors to avocado flowers in Bay of Plenty, New Zealand. In: Forest pest and beneficial insects, pp 38–44

Robinson WS, Nowogrodzki R, Morse RA (1989) The value of honey bees as pollinators of U.S. crops. Part II. Am Bee J 129(7):477–487

Sagwe RN, Peters MK, Dubois T, Steffan-Dewenter I, Michael H, Lattorff G (2021) Pollinator supplementation mitigates pollination deficits in smallholder avocado (*Persea americana* Mill.) production systems in Kenya. Basic Appl Ecol. https://doi.org/10.1016/j.baae.2021.08.013

Salunkhe DK, Desai BB (1984) Postharvest biotechnology of fruits. CRC Press, Boca Raton, pp 1–2

Shahid Ali, Shehzad A, Rafi MA, Zia A (2013) Insect Pollinators of Litchi (*Litchi chinensis*) from district Haripur, Pakistan. Pakistan J Agric Res 26(3):220–229

Shivaramu K, Sakthivel T, Reddy PVR (2012) Diversity and foraging dynamics of insect pollinators on rambutan (*Nephelium lappacum* L.). Pest Manag Hortic Ecosyst 18(2):158–160

Sivakumar D, Wijeratnam W, Wijesundera R, Abeyesekere M (2002) Control of postharvest diseases of rambutan using cinnamaldehyde. Crop Prot 21:847–852

Ssymank A, Kearns CA, Pape T, Thompson FC (2008) Pollinating flies (Diptera): a major contribution to plant diversity and agricultural production. Biodiversity 9:86–89

Stern RA, Gazit S (1998) Pollen viability in Lychee. J Amer Soc Hort Sci 123(1):41–46

Stout AB (1923) A study in cross-pollination of avocado in southern California. Calif Avocado Assoc Annu Rep 7:29–45

Stout AB (1933) The pollination of avocados. Florida Agric Exp Station Bull 257:1–44

Tindall H (1994) Rambutan cultivation, FAO, Plant production and protection paper

UJI, T. (1989) Pollination in *Nephelium lappaceium var lappaceum*. Ber Biol 1989(3):31–34

Usha, Srivastava P, Khan MS (2009) Biodiversity of insect pollinators on litchi at Pantnagar. J Eco-Friendly Agric 4(2):196–197

Van Welzen PC, Lamb A, Wong WWW (1988) Edible sapindaceae in Sabah. Nat Malaysiana 13: 10–25

Vasquez R, Ballesteros H, Tello J, Castañeda S, Calvo N, Ortega N, Riveros L (2011) Polinización dirigida con abejas Apis mellifera: Tecnología para el mejoramiento de la producción de cultivos con potencial exportador. Corporacion Colombiana de Investigacion Agropecuaria, Bogota

Vischer PK, Sherman G (1998) Insect visitors to avocado flowers. Subtrop Fruit News 6:7–10

Vithanage V (1990) The role of the European honeybee (*Apis mellifera* L.) in avocado pollination. J Hortic Sci 65:81–86

Westerkamp C (1991) Honeybees are poor pollinators - why? Plant Syst Evol 177:71–75

Woodcock BA, Garratt MPD, Powney GD, Shaw RF, Osborne JL, Soroka J, Lindström SAM, Stanley D, Ouvrard P, Edwards ME (2019) Meta-analysis reveals that pollinator functional diversity and abundance enhance crop pollination and yield. Nat Commun 10(1):1–10. https://doi.org/10.1038/s41467-019-09393-6

Woteki C (2013) The road to pollinator health. Science 341(6147):695. https://doi.org/10.1126/science.1244271

Wysoki M, van den Berg MA, Ish-Am G, Gazit S, Peña JE, Waite GK (2002) Pests and pollinators of avocado. In: Peña JE, Sharp JL, Wysoki M (eds) Tropical fruit pests and pollinators: biology, economic importance, natural enemies and control. Cabi Publishing International, New York, pp 223–293

Zee FT, Chan HT, Yen CR (1998) Lychee, longan, rambutan, and pulasan. In: Shaw PE, Chan HT, Nagy S (eds) Tropical and subtropical fruits. AgScience Inc., Florida, pp 290–335

Zych M, Jakubiec A (2006) How much is a bee worth? Economic aspects of pollination of selected crops in Poland. Acta Agrobot 59(1):289–299

Enhancement of Crop Productivity Through Bee Pollination in Pomegranate (*Punica granatum* L.)

11

K. T. Vijayakumar, T. N. Rakshitha, D. Rajanna, and H. L. Nithin Kumar

Abstract

Pomegranate, *Punica granatum*, is valued for its juice-containing arils and is consumed as fresh fruit. The quality and quantity parameters of fruit in cross-pollinated crops depend on honeybees due to its body modification for pollen collection and flower fidelity behavior. Pomegranate is self-pollinated; however, it is cross-pollinated by insects, mainly honeybees. The crop yields due to various pollinators' activity in pomegranate field would increase 38.38%. This chapter will discuss the aspects of pomegranate production, including flower biology, pollination, pollinator diversity, and yield enhancement with bee pollination.

Keywords

Pomegranate · Productivity · Bee pollination · fruit yield

1 Introduction

Pomegranate (*Punica granatum* L., family *Punicaceae*) is a fruit crop of major economic value, widely grown for its nutrient-rich fruit and juice. The term "pomegranate" comes from the Latin *Malum granatum*, which translates to "grainy apple." Originally native to Central Asia, pomegranate trees have adapted well to diverse soils and climatic conditions, supporting successful cultivation in places like the Mediterranean, parts of Asia, California, and other regions. In India, as per advance

K. T. Vijayakumar (✉) · T. N. Rakshitha · H. L. Nithin Kumar
AICRP on Honey Bees and Pollinators, UAS, GKVK, Bangalore, Karnataka, India

D. Rajanna
AINP on Soil Arthropod Pests, UAS, GKVK, Bangalore, Karnataka, India

A. K. Chakravarthy (ed.), *The Role of Pollinators in Tropical Landscapes*,
https://doi.org/10.1007/978-981-95-9139-8_11

estimates from the National Horticultural Board (NHB) for 2019–2020, pomegranate is cultivated over an area of about 2.83 lakh hectares, producing roughly 31.46 lakh metric tons (MT) of fruit. From this, around 67.89 thousand MT were exported, bringing in nearly ₹6885 million in foreign exchange earnings. This crop provides an important source of livelihood for more than 2.5 lakh families, especially in semi-arid and arid zones. Maharashtra is the primary producer, covering close to two-thirds of the country's pomegranate area. Other contributing states include Himachal Pradesh, Rajasthan, Karnataka, Telangana, and Gujarat. Over time, several improved and high-yielding cultivars have been developed and adopted across the country. Some popular varieties are Bhagwa, Alandi, and others. Flowering in pomegranate can occur as single flowers on branch spurs, in pairs, or as clusters at shoot tips (Varun et al. 2020). Previous studies, like those by Wetzstein et al. (2013), have highlighted that flower size and ovule development can differ depending on flower type and position. Beyond being valued for fruit production, pomegranate is used in traditional medicine. Studies established that different plant parts—including fruit peel, bark, flowers, and leaves—contain bioactive compounds known for antimicrobial effects, blood pressure reduction, and possible roles in managing or preventing diseases such as diabetes and cancer. These scientific insights have helped increase public interest in pomegranate as a functional food with notable health-promoting properties.

Although pomegranate is largely self-pollinated, insect-mediated cross-pollination, especially by bees, significantly improves fruit and seed set. Because pomegranate pollen grains are not easily carried by wind, insects become the main agents transferring pollen between flowers. Among these, honeybees stand out as particularly effective pollinators thanks to their hairy bodies, which trap and carry substantial amounts of pollen. Their rapid colony growth and active foraging behavior further enhance pollination success. It's estimated that insects facilitate roughly 80% of pollination in flowering plants, with bees alone responsible for the majority of this work. Remarkably, a honeybee hive individuals can visit and potentially pollinate hundreds of millions of flowers daily. Beyond their pollination services, honeybees also produce valuable hive products like honey, which provide farmers with an additional source of income.

In fruit crops such as pomegranate, the degree of cross-compatibility among different genotypes has important part in determining both fruit set and quality. While pomegranate is often described as an often cross-pollinated species, most genotypes exhibit self-compatibility, though the extent of this self-compatibility can differ among cultivars. Fruits that develop through self-pollination typically have fewer arils and are smaller in size, whereas cross-pollination between compatible genotypes can enhance both aril number and overall fruit size (Karimi and Mirdehghan 2015). Pollination by bees leads to a marked increase in fruit set percentage and average fruit weight compared to purely self-pollinated fruits (Derin and Eti 2001; Tao et al. 2010). Furthermore, even self-fertile varieties tend to produce larger seeds and higher yields when pollinated by honeybees and other insect pollinators (Verma and Partap 1994). Despite uniform management practices within orchards, fruit size often shows substantial variation, and factors like flower type and position might also influence fruit development, although these aspects remain insufficiently studied. Given this, a deeper understanding of floral traits and their variability, particularly regarding flower size and spatial position on the plant, becomes essential. The current chapter, therefore, provides an overview of insect visitor diversity, the foraging behavior of key pollinators, floral biology, and the influence of bee pollination on fruit set and yield in pomegranate.

2 Area, Production, and Productivity of Pomegranate

Pomegranate (*Punica granatum* L.) is cultivated extensively across tropical and subtropical regions of India. According to data from 2016 to 2017, the total area under pomegranate cultivation in the country is about 2.16 lakh hectares, with an overall production of approximately 25.21 lakh metric tons. Maharashtra stands as the leading producer, accounting for around 1.75 lakh hectares of cultivated area and producing about 18.50 lakh metric tons of fruits. Other pomegranate-producing states are given in Fig. 11.1. In India, pomegranate is commercially cultivated in Nasik, Solapur, Ahmednagar, Pune, Aurangabad, Sangli, Satara, Osmanabad, Dhule, and Latur districts of Maharashtra; Belgaum, Bijapur, and Bagalkot districts of Karnataka; Anantapur in Andhra Pradesh; Bhavnagar, Banaskantha, and Ahmedabad in Gujarat; and Jodhpur, Jalore, Barmer, Sirohi, and Sikar districts of Rajasthan.

There is substantial increase in area of pomegranate cultivation in India from 1.1 lakh ha in 2011–12 to 2.16 in 2016–17, and production enhanced from 7.7 lakh tons in 2011–12 to 25.21 lakh tons in 2016–17 (NHB, 2016–17) (Fig. 11.3). The state of Maharashtra alone contributes 66.21% of total pomegranate production followed by Karnataka (20%), Gujarat (9%), and Andhra Pradesh and Telangana (5%). Harvesting season of pomegranate is depicted in Fig. 11.2. By adopting proper *bahar* treatments, pomegranate in Maharashtra and Gujarat states is available round the year (Table 11.1 and Fig. 11.3).

3 Status of Cultivar Bhagwa and Other Varieties

The cultivar Bhagwa was identified by the ICAR–All India Coordinated Research Project on Arid Zone Fruits center at Mahatma Phule Krishi Vidyapeeth, Rahuri (Maharashtra), in 2003. Among growers, it is also known by various local names such as *Jai Maharashtra*, *Shendri*, *Kesar*, *Red Diana*, *Asthagandh*, and *Mastani* in Maharashtra. Bhagwa is now extensively cultivated in Solapur, Nasik, Sangli, Satara, Ahmednagar, Pune, and Dhule. The Bhagwa plant is a spreading-type, evergreen bush characterized by dark green foliage and spiny branches. Leaves are opposite, ovate to lanceolate in shape. The flowers are dark saffron in color and can appear terminally or axillary, either singly or in clusters. Three sex forms of flowers are observed in this cultivar: hermaphrodite, male, and intermediate. Bhagwa flowers in all three flowering seasons (bahars): *Ambe*, *Mrig*, and *Hastha*.

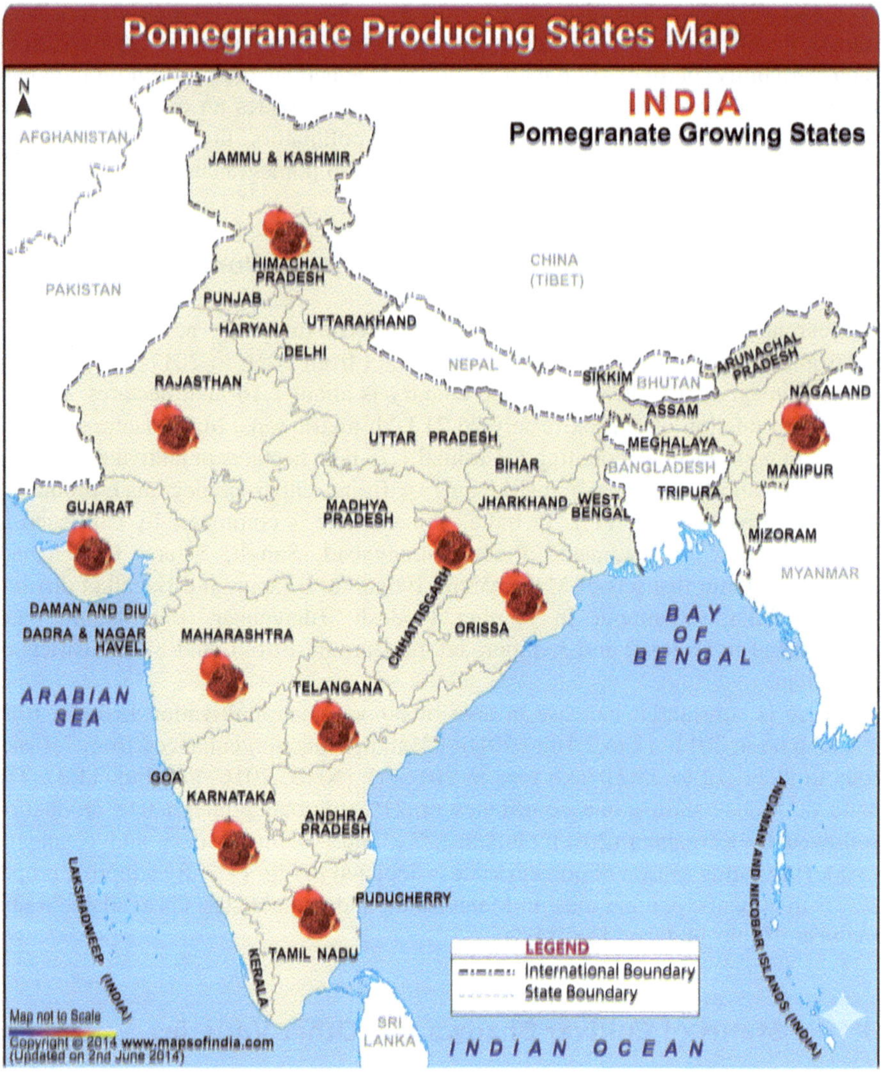

Fig. 11.1 Pomegranate growing states in India

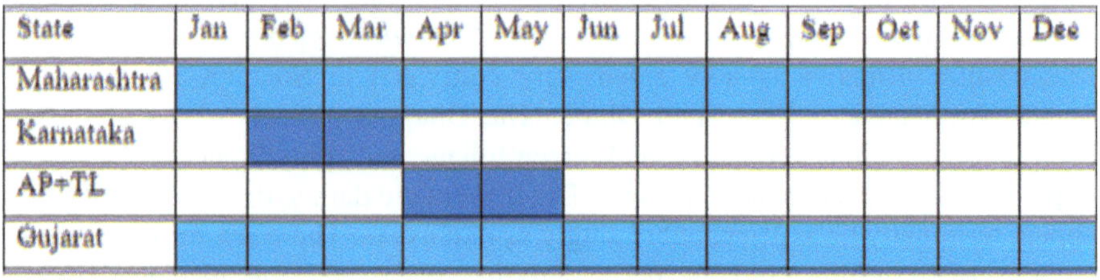

Fig. 11.2 Fruit harvesting season of pomegranate in different places

Table 11.1 Production of pomegranate in India

Sl no.	State	Production (tons)	Share (%)
1	Maharashtra	1763.99	54.85
2	Gujarat	684.32	21.28
3	Karnataka	305.72	9.51
4	Andhra Pradesh	283.54	8.82
5	Madhya Pradesh	75.16	2.34
6	Rajasthan	70.31	2.19
7	Telangana	16.96	0.53
8	Chhattisgarh	6.14	0.19
9	Tamil Nadu	3.35	0.10
10	Himachal Pradesh	2.83	0.09

Source: National Horticulture Board (2021–22)

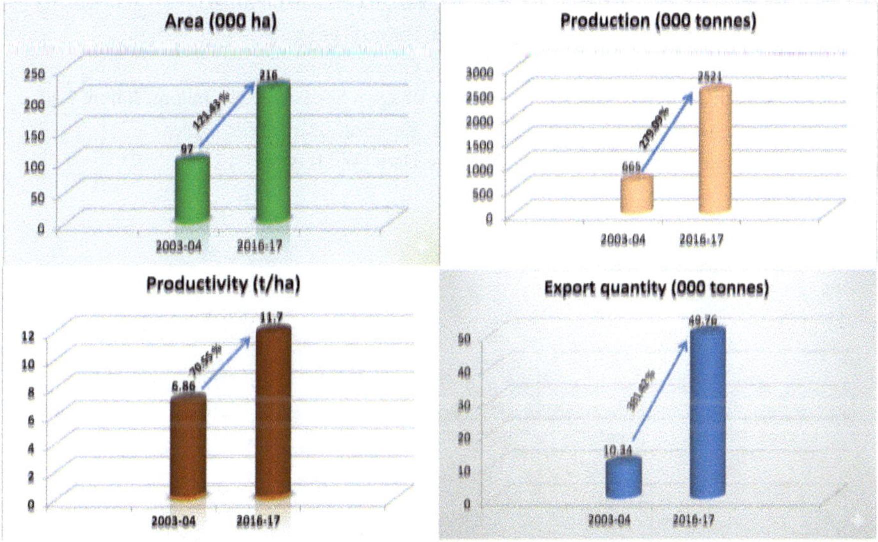

Fig. 11.3 Area, production, productivity, and export of pomegranate from India

The fruits of Bhagwa are particularly attractive, featuring a smooth, glossy, thick rind of dark saffron color and bold, dark red arils with soft seeds. This appealing appearance contributes to the cultivar's strong demand in both domestic and export markets, especially in European countries. Bhagwa fruits have excellent keeping quality (15–20 days under ambient conditions) and a thick rind, making them well-suited for long-distance transport. It is a late-maturing cultivar, requiring about 180–190 days for fruit maturity. Fruits are medium to large, with an average weight of around 405.97 g. Under drip irrigation, the average yield per plant ranges from 30 to 40 kg, while under high-density planting and improved horticultural practices, farmers have achieved yields of 25–30 tons per hectare. The arils are sweet, dark red in color, with a total soluble solids (T.S.S.) content of 15.38% and an acidity of 0.37%. Thanks to its attractive fruit and aril color and excellent fruit quality, Bhagwa fetches a market price that is often 2–3 times higher than that of the *Ganesh* cultivar. Moreover, Bhagwa is less susceptible to fruit cracking and fruit spot (PDI—4.68), is relatively free from aril blackening, and shows tolerance to mites and thrips (Saroj and Sharma 2017) (Table 11.2 and Fig. 11.4).

Table 11.2 Leading varieties of pomegranate in India

Sl. no.	State	Varieties
1.	Maharashtra	Muskat G-137, Ganesh, Phule Bhagwa Mridula, Bhagwa, Phule Arakta, Super
2.	Karnataka	Jyoti, Ruby, Bhagwa, Super, Madhugiri Phule Ganesh, Bhagwa
3.	Gujarat	Dholka, Bhagawa, Ganesh, Muskat
4.	Andhra Pradesh/Telangana	Bhagwa, Ruby, Ganesh, G-137
5.	Madhya Pradesh	Bhagwa, Ganesh, Jyoti
6.	Rajasthan	Jodhpur Bhagwa Seedless, Red, Jallore, Jallore Red
7.	Himachal Pradesh	Bhagwa, Kandhari, Daru (Wild) types
8.	Jammu & Kashmir	Bhagwa, Daru (Wild) types

Fig. 11.4 Pomegranate crop: variety, Bhagwa

4 Flower Biology and Plant Phenology of Pomegranate

Pomegranate is an andro-monoecious species, meaning that both staminate (male) and hermaphrodite (bisexual) flowers develop on the same plant (Shulman et al. 1984; Wetzstein et al. 2011). The hermaphrodite flowers have a characteristic vase shape, whereas staminate flowers exhibit a bell shape, making them easily distinguishable (Fig. 11.5). In hermaphrodite flowers, the stigma is positioned at the same height as the anthers or slightly above (Wetzstein et al. 2013). This floral structure facilitates both self-pollination and insect pollination. Although bees are recognized as the primary pollinators, wind can contribute to pollination. Experiments conducted revealed that fruit set increases when flowers are exposed to open and cross-pollination (Derin and Eti 2001). Research involving emasculation and pollination in different pomegranate cultivars conducted in India, Turkmenistan, and Tunisia has demonstrated that fruits are often naturally produced through self-pollination (Nalawadi et al. 1973; Levin 2006). However, to achieve fruit sizes that meet export market standards, successful pollination which affects seed number is crucial (Derin and Eti 2001). It is also noted that sometimes issues in fertilization and fruit set can arise either between different flower types of the same cultivar or between flower types of different cultivars (Gozlekci and Kaynak 1998).

Pomegranate produces large, showy flowers with delicate petals that range in color from vivid orange-red to deep red. The petals are typically orange-red or pink, though rarely white (Feng et al. 1998; Levin 2006). The number of petals usually equals the number of sepals. Flowers are generally short-peduncled or subsessile, and bisexual (hermaphrodite) flowers may appear solitary or in clusters. Typically, solitary flowers are borne on spurs with branches, while clusters are located terminally (Varun et al. 2020). As flowers mature, the sepal color turns to orange-red, with variation among different varieties. The hermaphrodite flowers have a well-developed ovary with a broad base, whereas male flowers possess only a rudimentary ovary. The multiple long stamens are arranged in a circle and inserted into the calyx walls, often numbering over 300 per flower. These stamens have orange-red filaments and yellow bilocular anthers that remain attached to the prominent calyx. Nectaries are situated between the stamens and the ovary base (Fahan 1976). In the

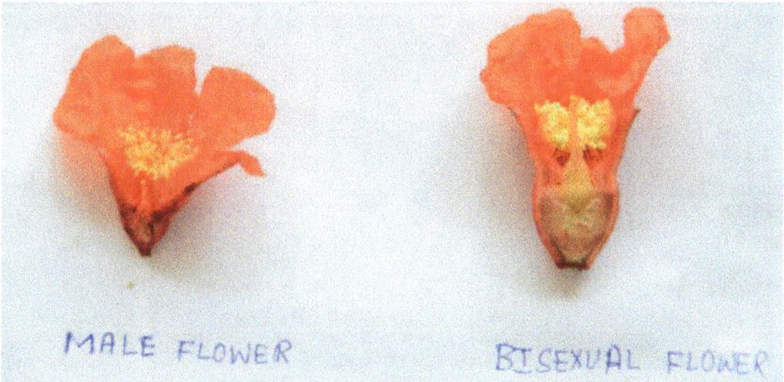

Fig. 11.5 Flower types in pomegranate

popular cultivar Bhagwa, flowers have six sepals and six petals, and the number of stamens per flower typically ranges from 190 to 221, with each stamen measuring about 0.72 cm in length (Vijayakumar et al. 2024). Hermaphrodite flowers are notably longer and wider than male flowers. For example, in the cultivar 'Beautiful,' hermaphrodite flowers measure about 3.6 cm in length, compared to 2.7 cm for male flowers (Wetzstein et al. 2011). Both male and hermaphrodite pomegranate flowers remain open and functional for approximately 3–4 days. The duration a flower remains functional can be a critical determinant of both male and female reproductive success (Steinacher and Wagner 2010).

Pomegranate flowers have an attractive red calyx and corolla, contributing to their ornamental appearance. The red calyx itself is thick, leathery, fused at the base, and ends in triangular lobes that persist after flowering, eventually forming the characteristic crown seen on mature fruits. In staminate (male) flowers, there are about 221.3 ± 75.31 anthers per flower, each composed of yellow anthers attached to long red filaments. Each anther contains an average of 62,500 ± 4949.74 pollen grains at the beginning of flowering, decreasing to around 53,000 ± 7071.06 pollen grains later. Notably, the male flowers contain maximum pollen grains (Deniz and Zeynel 2019). In contrast to the large number of stamens in male flowers, hermaphrodite (bisexual) flowers have a single elongate pistil measuring approximately 2.79 ± 0.14 cm in length, which ends in a stigma. The inferior ovary of hermaphrodite flowers contains about 1403 ± 149.34 ovules (Vijayakumar et al. 2024).

Pomegranate flowers can be borne either as single flowers or in clusters. Typically, these clusters consist of one terminal, more advanced flower, subtended by one or two pairs of oppositely arranged lateral flowers in a decussate orientation. Based on position and arrangement, three types of flowers are observed in pomegranate: single flowers, terminal flowers within a cluster, and lateral flowers within a cluster.

Single flowers and terminal flowers within a cluster are generally larger and contain more ovules compared to lateral flowers (Vijayakumar et al. 2024). In contrast, lateral flowers tend to exhibit a higher frequency of poor ovule development, which negatively affects their potential to set fruit (Fig. 11.6). Dissections

Fig. 11.6 Pomegranate flower types: (**a**) terminal flower, (**b**) lateral flower, and (**c**) single flower

have shown that almost half of the lateral flowers possess abortive ovules, making them incapable of fruit set (Wetzstein et al. 2013). The size and position of flowers significantly influence fruit size and overall flower quality (Table 11.3). Larger single and terminal flowers, which open earlier, often lead to better fruit set and larger fruits compared to the smaller lateral flowers. In pomegranate, the development of lateral flowers is typically delayed compared to terminal flowers. It remains unclear whether the earlier opening and pollination of terminal flowers might impact the differentiation and ovule development of subsequent lateral flowers, contributing to their reduced fertility (Wetzstein et al. 2013). Additionally, the presence of developing fruits can exert an inhibitory effect on the fruit set and growth of later flowers and fruits. This effect is thought to be mediated by changes in hormonal activity and synthesis (Leopold and Kriedemann 1975). Studies have also found that the development of functional hermaphroditic flowers correlates with adequate starch content in floral tissues (Reale et al. 2009). Finally, pollination studies have confirmed that both fruit set and final fruit size are significantly influenced by flower size at the time of pollination, underlining the importance of floral characteristics and position within the inflorescence in determining yield and fruit quality.

4.1 Flowering Habit

The flowering habit of pomegranate is strongly influenced by the prevailing climatic conditions of the region where it is cultivated (Pareek and Sharma 1993). In tropical climates, Pomegranate blooms throughout the year, while in sub-tropics it may be seasonal. In regions with colder winters, pomegranate trees tend to be deciduous, shedding their leaves completely. By contrast, under tropical conditions, the trees are generally evergreen or partially deciduous. For instance, in southern India, evergreen cultivars exhibit three distinct periods which usually coincide with June, October, and March (Nalawadi et al. 1973). In the northern hemisphere, flowering generally occurs in April–May. However, flowering can sometimes extend until the end of summer, especially in younger trees. While such late flowers are often fertile, the resulting fruits usually fail to mature properly because the trees enter dormancy during the cooler season typical of Mediterranean regions. Flower bud development takes place at varied times under tropical conditions. Based on the prevailing weather conditions, the pomegranate blooms in varied seasons in the tropics (Gur 1986). In sub-tropics, flowering usually occurs from March to May (Singh et al. 1978; Fouad et al. 1979). Several distinct flushes of flowering on the same tree occur quite frequently. The time required for completion of flower development in Indian cultivars is between 20 and 27 days (Nalawadi et al. 1973). The time taken for flower bud development was maximum (20.4 days) in 'Ganesh' compared to 'Bhagwa' (19.3 days), 'Mridula' (17.5 days), and 'Arakta' (16.2 days) in June (Babu et al. 2009b).

Table 11.3 Length measurements for different floral characteristics in three bisexual flower types found in pomegranate

	Flower type					
	Single		Terminal		Lateral	
Floral characters	Mean (mm)	Range (mm)	Mean (mm)	Range (mm)	Mean (mm)	Range (mm)
Ovary width	14.6^a	11.7–18.2	14.5^a	10.4–22.0	13.6^b	10.4–16.6
Base to sepal notch	30.0^a	22.3–36.9	30.9^a	19.7–40.5	25.1^b	16.5–33.1
Base to sepal tip	41.1^a	31.4–48.8	42.0^a	31.3–50.8	35.1^b	25.0–46.2
Total pistil length	31.2^a	23.2–37.1	30.6^a	21.4–40.0	26.1^b	15.8–37.1
Stigma + style + stylopodium	20.2^a	15.4–28.3	20.1^a	15.7–31.5	17.6^b	8.6–23.3
Stigma + style	12.7^a	9.0–16.9	12.4^a	7.3–16.5	10.9^b	5.1–14.5
Stigma diameter	1.1^b	0.3–1.6	1.0^c	0.2–1.6	1.3^a	0.8–1.8

4.2 Flowering Period

In North India, pomegranate typically exhibits two flowering seasons, whereas in western India, there are generally three distinct flowering seasons (Nalawadi et al. 1973). Specifically, in the subtropical regions of central and western India, the flowering seasons are known as:

- Ambe bahar (January–February)
- Mrig bahar (June–July)
- Hasth bahar (September–October)

Among these, Ambe bahar is the most commonly preferred by growers, as it usually results in higher yields due to more profuse flowering compared to the other seasons (Singh et al. 1967; Prasannakumar 1998). In Karnataka, flowering has been observed over a period of about 80–87 days during June–August (Nalawadi et al. 1973). In contrast, pomegranate flowers only once in Punjab (Josan et al. 1979). Under Delhi conditions, flowering may take place once or twice a year, depending on the cultivar (Nath and Randhawa 1959). Flowering is usually confined to the middle of April (Parmar and Kaushal 1982) in Himachal Pradesh. Meanwhile, in Bihar, two flowering seasons are observed: one during February–March and another during July–August.

4.3 Time of Anthesis and Dehiscence and Flower Longevity

Anthesis is the period during which a flower is fully open and functional. It typically starts in the morning, marked initially by the appearance of a split at the tip of the bud. Following this, the sepals gradually separate, slowly exposing the crumpled scarlet petals, which then begin to swell outward (Babu 2010). After reaching full bloom, the corolla takes about 3–4 h to completely open (Babu et al. 2009a). Importantly, flower buds do not open during the night. The maximum opening rates have been observed between 10:00 and 11:00 h (Game 1987). The exact timing of anthesis varies depending on environmental factors and the variety grown. For example, in the cultivar ‘Bhagwa,’ one study reported the peak period of anthesis to occur between 12:00 and 13:00 h (Vijayakumar et al. 2024). Interestingly, another study found the peak anthesis for the same cultivar to be between 10:00 and 12:00 h (Babu et al. 2009b). This variation highlights how the floral biology of the same cultivar can change in response to different environmental conditions (Fig. 11.7).

The optimum temperature for flower bud opening varies among pomegranate cultivars. The final function of the anther, which leads to the release of pollen grains, is driven by the endothecium tissue, which is responsible for anther splitting (Gur 1986). The timing of anther dehiscence (release of pollen) can occur a few hours before anthesis or much later, depending on the cultivar. In some cases, dehiscence begins between 7 and 15 h after anthesis, while in others, it may take up to 24 h (Gur 1986). Under Bangalore conditions, anther dehiscence was observed immediately

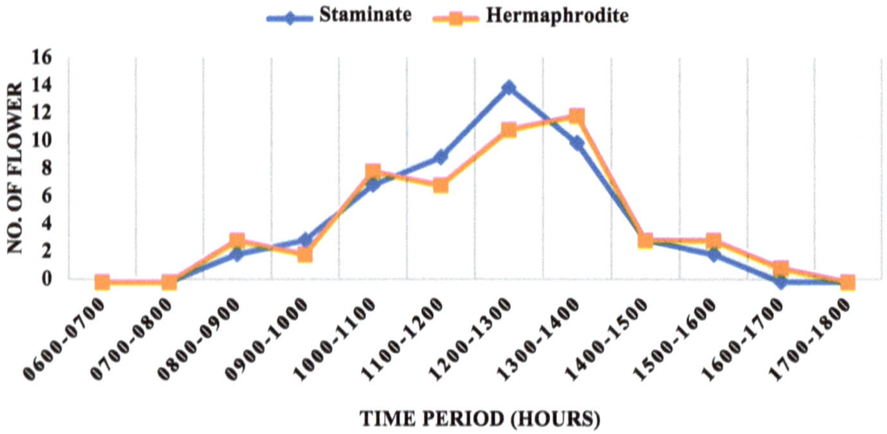

Fig. 11.7 Time and rate of anthesis in pomegranate flower

after anthesis. Environmental factors, such as humidity and higher temperatures, can accelerate the process of dehiscence (Josan et al. 1979). Observations show that anthesis for both staminate (male) and hermaphrodite (bisexual) flowers starts around 0800 h. In staminate flowers, anthesis typically ends by 16:00 h, whereas in hermaphrodite flowers, it can continue until 17:00 h (Fig. 11.4). Peak anthesis occurs between 1200 and 1300 h in staminate flowers ($n = 14$) and 13:00–14:00 h in hermaphrodite flowers ($n = 12$). The time window for anther dehiscence varies too: from 08:00 to 14:00 h in staminate flowers and 08:00 to 15:00 h in hermaphrodite flowers. Maximum anther dehiscence was recorded between 11:00 and 12:00 h in both flower types (Vijayakumar et al. 2024). Regarding the duration of floral functionality, the hermaphrodite flower remains open and functional for up to 4 days, whereas the staminate flower lasts for about 3 days.

4.4 Stigma Receptivity, Pollen Viability, and Pollen Tube Germination

Receptivity refers to the ability of the stigma to support the emergence of viable pollen grains. This process can begin as early as 16–18 h before anthesis and can extend up to 5–6 days after anthesis. Typically, the stigma gets receptive before anthesis and remains receptive until about the time when anthesis is over. The period of highest receptivity coincides closely with the time of anthesis itself. After this peak, receptivity gradually declines, as shown by pollination tests, and eventually ceases completely by the end of the third or fifth day (Gur 1986). Overall, the stigma is generally receptive from the day before anthesis up until the fourth day after anthesis.

The viability of pollen is a critical factor that directly affects fruit set in pomegranate. Viable pollen grains can be identified by their heavy staining with 1.0%

acetocarmine, while nonviable pollen grains remain unstained (Babu et al. 2009a). Studies have shown that in hermaphrodite flowers, pollen fertility tends to be higher compared to that of male (staminate) flowers (Nath and Randhawa 1959). According to Evreinoff (1957), stamen production occurs naturally in both hermaphrodite and male flowers, and both flower types produce fertile pollen. However, there is significant seasonal variability in pollen fertility. In observations across 3 days of flower life, pollen grains in both flower types generally remained viable throughout. Interestingly, in hermaphrodite flowers, pollen viability was initially absent during the morning hours on the first day of flower opening. However, viability developed later the same day and was maintained until the flower senesced (Sharma et al. 2022) (Fig. 11.8).

The highest pollen viability was observed in hermaphrodite flowers during the early stage of flowering. In staminate (male) flowers, maximum pollen viability reached 95.13% on the day of anther dehiscence. By contrast, in hermaphrodite flowers, pollen viability was initially lower at 64.81% on the first day but increased significantly to 95.94% on the second day. In both flower types, the highest pollen viability was typically recorded during the morning hours of the day (Fig. 11.9). Regarding pollen tube growth (Fig. 11.10), the average length of the pollen tube was

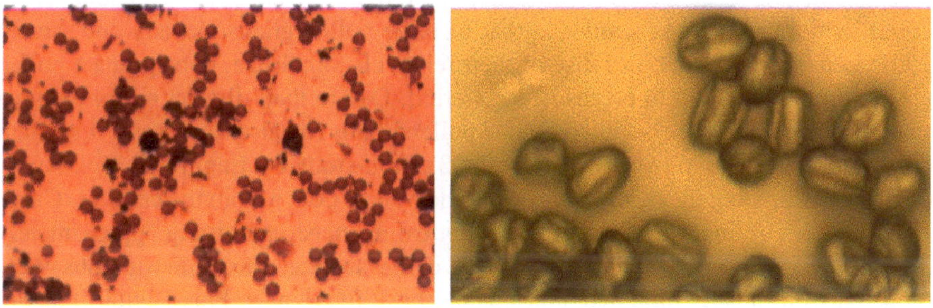

Fig. 11.8 Pollen viability test

Fig. 11.9 Pollen tube germination

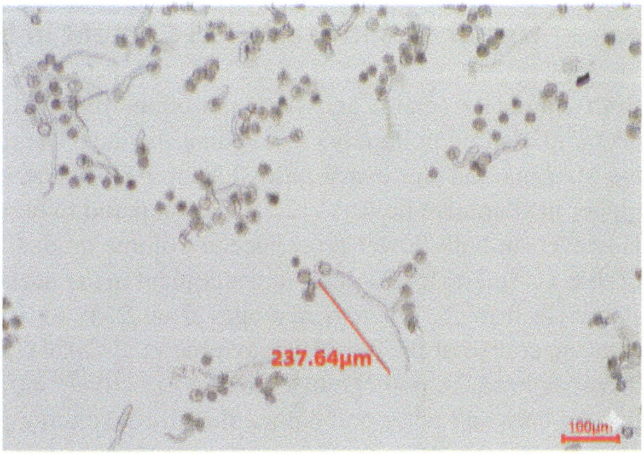

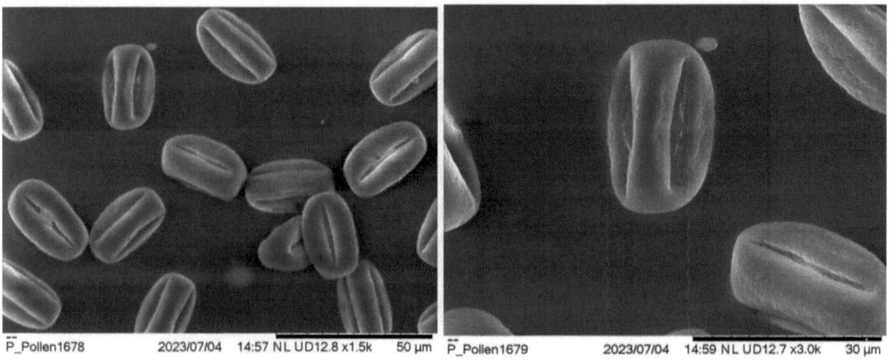

Fig. 11.10 Scanning electron microscopy images of pomegranate pollen. (**a**) Pollen grains (1000×). (**b**) Whole grain shape (3000×)

620.27 μm ($n = 20$) in pollen grains from staminate flowers, whereas it was slightly shorter in pollen grains from hermaphrodite flowers, averaging 577.64 μm ($n = 20$) (Vijayakumar et al. 2024).

Pomegranate pollen grains are smaller in size compared to those of pear (Westwood and Challice 1978) and apricot (Arzani et al. 2005; Dezhong et al. 1995), while their size is similar to that of olive pollen grains (Javady and Arzani 2001; Lanza et al. 1996). Because of their relatively small size, pomegranate pollen grains are light enough to be easily transferred. Due to their distinct morphological and varietal characteristics, pomegranate pollen grains can also serve as an important tool in modern taxonomy (Ghazaeian 2003). In a study by Engin and Gokbayrak (2019) investigating the morphology of pomegranate pollen from the cultivar 'Caner II,' the pollen grains were found to be elliptical in shape, measuring approximately 22 μm in size, and possessing a smooth exine surface. Photomicrographs were prepared from freshly mounted slides using the Biovis Image Plus Digital Imaging System (Fig. 11.10).

4.5 Nectar Quantity and TSS Content in Nectar

Both staminate (male) and hermaphrodite (bisexual) pomegranate flowers offer nectar to visiting pollinators throughout the day. The nectaries are situated between the stamens and the ovary base (Fahan 1976). The quantity of nectar is slightly higher in staminate flowers (12.50 μl) compared to hermaphrodite flowers (8.90 μl). However, in both flower types, nectar volume gradually declines by evening hours (Table 11.4). The total soluble sugar content in the nectar ranges from 27% to 31% in staminate flowers, whereas it ranges from 25% to 30% in hermaphrodite flowers over the course of the day (Vijayakumar et al. 2024). By providing both pollen and nectar as rewards, pomegranate flowers effectively attract and support higher visitation by bees and other pollinators, thereby enhancing pollination success.

Table 11.4 Nectar quantity and TSS content in nectar

	Nectar quantity (μl)		TSS (%)	
Time period	Staminate	Hermaphrodite	Staminate	Hermaphrodite
9.00–10.00	12.50	8.90	27.00	25.00
13.00–14.00	4.00	2.90	27.00	27.00
16.00–17.00	2.20	3.10	31.00	30.00
Total	**18.70**	**14.90**	–	–

5 Pollinators' Diversity

A pollinator is a biotic agent that transfers pollen from the anther to the stigma of a flower, enabling fertilization and thereby contributing to the production of fruits and seeds. Pollinators include a diverse group of animals such as bees, flies, bats, moths, and birds. Many fruits, vegetables, and field crops rely heavily on these pollinators for successful yield (Klein et al. 2007). Numerous studies have highlighted insect pollination as a vital ecosystem service essential for global agricultural food production (Gallai et al. 2009). In recent years, insect pollinators have gained increasing importance among farming communities as a strategy to boost crop productivity and help meet the demands of the world's growing population (Hajjar et al. 2008). While various insects contribute to pollination, honeybees are often regarded as the most significant pollinators, providing the bulk of agricultural pollination services worldwide (Breeze et al. 2011).

5.1 Species Richness, Diversity, and Relative Abundance of Flower Visitors

A total of 19 flower-visiting insect species were recorded on pomegranate flowers, with the highest contribution to pollination from honeybees and other hymenopterans (Derin and Eti 2001; Chavhan et al. 2023). Among the different bee species observed, *Apis dorsata* was the most dominant, showing the highest relative abundance (31%), followed by *A. florea* (28%), *A. cerana* (11%), and *Tetragonula iridipennis* (2%). These species were identified as the primary pollinators of pomegranate. Although lepidopterans (such as butterflies and moths) frequently visited the flowers, they were found to be poor pollinators (Fig. 11.5) (Vijayakumar et al. 2024). The pomegranate plant (*Punica granatum*) produces more pollen than nectar, and this pollen is especially important for bee colonies during colony build-up, when bees require large amounts of protein for brood rearing. This characteristic makes bees highly efficient pollinators. In addition to honeybees like *Apis cerana* and *A. mellifera*, pomegranate flowers are also visited by other insects such as butterflies, moths, and beetles. However, these insects are considered minor or occasional visitors and do not significantly contribute to pollination (Figs. 11.11, 11.12 and Table 11.5).

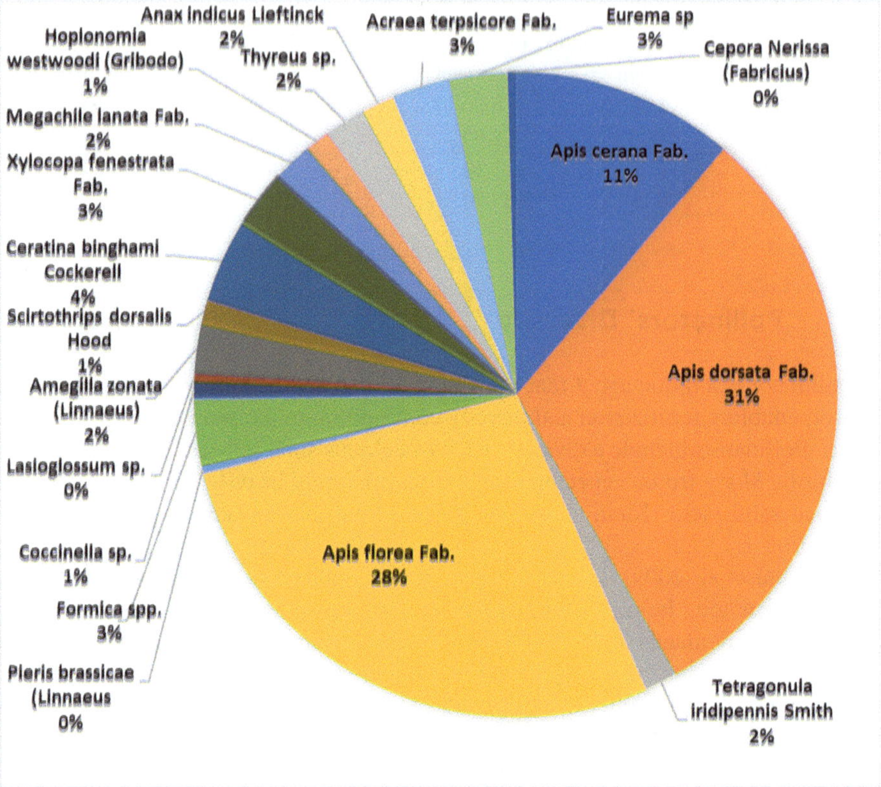

Fig. 11.11 Relative abundance of flower visitors in pomegranate

6 Foraging Behavior of Bees

The foraging activity in bees is closely linked to the nectar flow in pomegranate flowers. Besides, the temperature existing in the atmosphere should be around 20–28 °C.

During the hottest periods of the day, foragers tend to carry relatively little fluid, and pollen foraging activity declines at high ambient temperatures. The major bee pollinators recorded on pomegranate flowers are *Apis dorsata*, *A. cerana*, and *A. florea* (Fig. 11.13). The peak foraging activity of *A. dorsata* was observed between 10:00 and 11:00 h, with a secondary peak at 17:00 h. Meanwhile, *A. cerana* showed peak activity around 13:00 h (Fig. 11.14). For *A. florea*, foraging activity on the pomegranate variety 'Bhagwa' typically began after 10:00 h. Foragers were seen visiting both staminate and hermaphrodite flowers, collecting both pollen and nectar. The foraging rate and speed of bees were influenced by the time of day and the type of flower being visited.

Fig. 11.12 Diversity of flower visitors on pomegranate

Hoplonomia westwoodi (Gribodo)

Lasioglossum sp.

Formica spp.

Pieris brassicae (Linnaeus)

Eurema sp

Achraea terpiscore Fab.

Cepora Nerissa (Fabricius)

Fig. 11.12 (continued)

7 Bee Pollination on Fruit Yield

Honeybee pollination is not only known to improve fruit production and fruit quality, but it is also considered cost-effective (Kozin 1976). Proper pollination contributes to larger fruit size, higher yield, and more symmetrical fruit shape and can also hasten fruit maturity. Several studies have highlighted the crucial role of honeybees in enhancing fruit and seed yields across a variety of crops and cultivars: Assam lemon *Citrus limon* (L) Burm. (Gogoi et al. 2007), apple *Malus domestica* Borkh., pear *Pyrus communis* L., Japanese plum *Prunus salicina* Lundl. (Stern et al. 2007), rabbit eye blueberry *Vaccinium ashei* Reade (Dedej and Delaplane 2003). A

Table 11.5 List of flower visitors in pomegranate

Sl. no	Species names	Family	Order
1	*Apis cerana* Fab.	Apidae	Hymenoptera
2	*Apis dorsata* Fab.		
3	*Apis florea* Fab.		
4	*Tetragonula iridipennis* Smith		
5	*Amegilla zonata* (Linnaeus)		
6	*Ceratina binghami* Cockerell		
7	*Xylocopa fenestrata* Fab.		
8	*Thyreus* sp.		
9	*Megachile lanata* Fab.	Megachilidae	
10	*Hoplonomia westwoodi* (Gribodo)	Halictidae	
11	*Lasioglossum* sp.		
12	*Formica* spp.	Formicidae	
13	*Scirtothrips dorsalis* Hood	Thripidae	Thysanoptera
14	*Pieris brassicae* (Linnaeus)	Pieridae	Lepidoptera
15	*Eurema* sp		
16	*Cepora Nerissa* (Fabricius)		
17	*Acraea terpsicore* Fab.	Nymphalidae	
18	*Coccinella* sp.	Coccinellidae	Coleoptera
19	*Anax indicus* Lieftinck	Libellulidae	Odonata

a. *Apis dorsata* Fab. **B. *Apis cerana* Fab.**

Fig. 11.13 Bees foraging. (**a**) *Apis dorsata* Fab. (**b**) *Apis cerana* Fab

well-pollinated flower typically develops more seeds with greater germination potential, resulting in larger and better-shaped fruits. Improved pollination also tends to reduce the time between flowering and fruit set. In India, generally, four species of honeybees of commercial importance are found, i.e., *Apis florea*, *Apis dorsata*, *Apis cerana indica*, and *Apis mellifera*. Among these, first three are indigenous to India (Fig. 11.15).

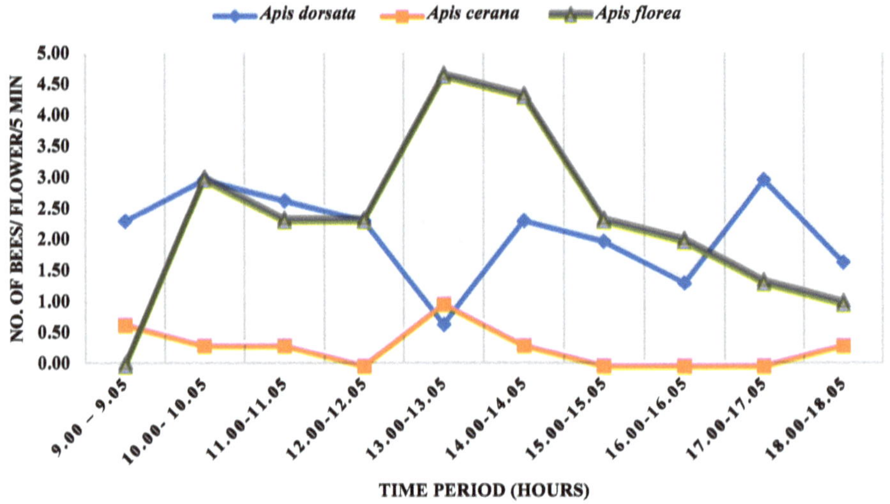

Fig. 11.14 Line diagram depicting the pattern of foraging

Fig. 11.15 Pomegranate crop with *A. cerana* colony

7.1 Fruit Set Under Various Modes of Pollination in Pomegranate

Proper pollination increases fruit size at yield, hastens maturity, and produces a more symmetrical fruit shape. A well-pollinated flower typically develops more seeds with higher germination potential, resulting in larger and better-shaped fruits. Improved pollination can also shorten the period between flowering and fruit set, which reduces the risk of fruit being exposed to pests, diseases, unfavorable weather, and agrochemicals and also helps in saving water. Additionally, quantitative traits, such as fruit weight, diameter, and length, as well as aril characteristics, were shown to improve under open and supplementary pollination compared to self-pollination. Bee and open pollinated fruits are with better quantitative and qualitative traits, i.e., average fruit weight (311.67 g, $P = 0.0004$), rind thickness (0.42 cm, $P = 0.025$), fruit diameter (8 cm, $P = 0.003$), test weight (31.46 g, $P = 0.008$), and arils per fruit (620, $P = 0.00000003$) were higher in open-pollinated conditions. Fruits obtained from bee (*A. cerana*) pollination were heavier (284.5 g, $P < 0.05$), larger (7.87 cm, $P < 0.05$), test weight (29.83 g, $P < 0.05$), and aril per fruit (485, $P < 0.05$) than those from self-pollinated plants (Vijayakumar et al. 2024). The total yield per plant was found to be highest under open-pollinated conditions (24.93 kg), which was statistically on par with bee-pollinated conditions (22.76 kg). In contrast, self-pollination resulted in a significantly lower yield per plant (15.36 kg) (Table 11.6). Pollinators, particularly honeybees, play a significant role in enhancing fruit set, fruit quality, and overall pomegranate yields (Derin and Eti 2001; Tao et al. 2010) (Fig. 11.16).

Total soluble solids (TSS) measure the concentration of soluble solids in fruit juice and are an important indicator of produce quality because they directly impact taste. In pomegranate, TSS was maximum in trees pollinated by *Apis cerana* (Table 11.7). Open pollination generally showed superior performance in both qualitative and quantitative traits (Sabbahi et al. 2005; Munawar et al. 2009; Jauker et al. 2012; Hudewenz et al. 2014). This advantage arises because, under open pollination, plants benefit not only from some degree of self-pollination but also from visits by diverse pollinators, including Apis bees, non-Apis bees, butterflies, and others (Chavhan et al. 2023). In many parameters, the effect of *A. cerana* pollination was comparable to that of open pollination treatments. Remarkably, 82% of the pollination was attributed solely to *A. cerana*, suggesting that introducing *A. cerana* colonies into the field can significantly increase pollinator density, thereby boosting pomegranate yield either directly or indirectly. Additionally, when pomegranate plants were caged with *A. cerana*, the resulting fruits were observed to have better shape and form, which enhances their marketability. Thus, incorporating *A. cerana* colonies is an effective practice to improve both the quality and quantity of pomegranate production.

Installing an *Apis cerana* bee colony can potentially increase crop yield and generate additional income for farmers. *A. cerana* is considered a highly efficient pollinator compared to other Apis and non-Apis species. Its compact colony size also makes it easier for beekeepers to manage, resulting in improved crop productivity. Additionally, *A. cerana* colonies are known for convenient management and low cost, a practical choice for integration into agricultural systems. In general, the presence of these bees has been shown to enhance pomegranate yields when compared to self-pollination and hand pollination. The maximum pollination efficiency index was recorded for *A. cerana* (Reddy et al. 2022; Kedswin et al. 2023). However, unfruitfulness and poor seed development in flowering plants include

Table 11.6 Effect of modes of pollination on yield parameters of pomegranate

Parameter	Fruit weight (g)	Moisture content (%)	Rind thickness (cm)	Diameter (cm)	Test weight (g)	Total number of arils/fruit	Yield per plant (kg)	Yield per hectare (tons)
Open pollination	311.67^{a}	78.77	0.42^{a}	8.00^{a}	31.46^{a}	620.00^{a}	24.93^{a}	20.77^{a}
Caged (control)	192.03^{b}	78.67	0.30^{b}	7.27^{b}	28.07^{b}	312.00^{c}	15.36^{b}	12.80^{b}
Caged with *Apis cerana* (bee pollination)	284.50^{ab}	79.00	0.37^{ab}	7.87^{a}	29.83^{ab}	485.33^{b}	22.76^{ab}	18.96^{ab}
P-value	0.0004*	0.084	0.025*	0.003*	0.0080*	0.00000003*	0.0004*	0.0004*
SEM±	10.19	0.33	0.02	0.09	0.49	7.60	0.81	0.68
CD (0.05)	35.25	1.14	0.075	0.32	1.70	26.31	2.82	2.35

Source: Vijayakumar et al. (2024)

*p value <0.05, samples are significantly different. Treatment means followed by the same alphabet are statistically on par with each other

Fig. 11.16 Pomegranate fruits pollinated by *A. cerana*

Table 11.7 Qualitative parameters of pomegranate (var. Bhagwa) as influenced by modes of pollination

Treatments	Total soluble sugars (%)	Reducing sugars (%)	Non-reducing sugars (%)	Refractive index of fruit juice	Ash content (%)	Total soluble solids
T1	44.12[b]	36.91[b]	7.20[a]	1.34[b]	0.59[b]	16.75[a]
T2	33.56[c]	34.05[b]	6.29[a]	1.35[b]	0.54[c]	14.75[b]
T3	31.23[c]	32.40[b]	4.44[b]	1.35[b]	0.67[a]	14.92[b]
T4	36.09[c]	34.90[b]	11.63[a]	1.34[b]	0.57[c]	15.15[a]
T5	55.29[a]	42.23[a]	13.06[a]	1.39[a]	0.58[b]	15.25[a]
Mean	40.05	36.10	5.89	1.35	0.60	15.36
CV	5.76	5.58	3.43	1.43	1.43	3.94
CD @ 5%	3.43	3.13	7.66	0.097	0.097	1.34

Source: Chavhan et al. (2023)

T_1: pollination by *A. cerana*; T_2: hand pollination with emasculation; T_3: self-pollination; T_4: hand + self-pollination; and T_5: open pollination

insufficient maternal resources (e.g., limited carbon allocation and nutrition), flower age effects on stigma receptivity, and inadequate pollen delivery due to a lack of pollinators, inviable pollen, or sexual incompatibility. Therefore, the deliberate use of *A. cerana* for pomegranate pollination is strongly recommended to achieve higher fruit set and yield. Beyond yield benefits, the establishment of bee colonies can also provide farmers with a supplementary source of income through the sale of honey and other hive products.

The most common flower visitors observed were *Apis dorsata*, *A. florea*, and *A. cerana*. These bees spend considerable time on each flower to collect floral rewards such as nectar and pollen, thereby contributing significantly to pollination. The visits of pollinators, particularly honeybees, have been shown to enhance both qualitative and quantitative traits of pomegranate fruit, although the degree of improvement may vary depending on the variety and local environmental conditions. Under open field conditions, the activity of various pollinators led to an increase in crop yield of 38.38% over the control (without pollinators).

Indian Beekeeping Under Threat of Dreaded Pest, Small Hive Beetle, *Aethina tumida* (Coleoptera: Nitidulidae)

The small hive beetle (SHB), *Aethina tumida* Murray (Coleoptera: Nitidulidae), is a significant invasive pest of honeybee colonies that has expanded its range from its native sub-Saharan Africa to multiple continents over the past three decades. The first Indian detection of *A. tumida* occurred in 2022 in *Apis mellifera* colonies. Subsequently, the pest spread rapidly across other Indian states, with confirmed infestations of the colonies, signaling a threat to beekeeping in the Indian subcontinent.

The grub causes the majority of damage to active hives by burrowing into combs, eating brood, honey, and pollen. Beetle larvae tunnel through combs of honey, feeding and defecating, causing discoloration and fermentation of the honey (Figs. 11.17, 11.18, and 11.19).

Fig. 11.17 Presence of SHB larva on the comb of *A. cerana* colony

(continued)

Fig. 11.18 Observations in *Apis cerana* colony at Hediyala, Mysore district

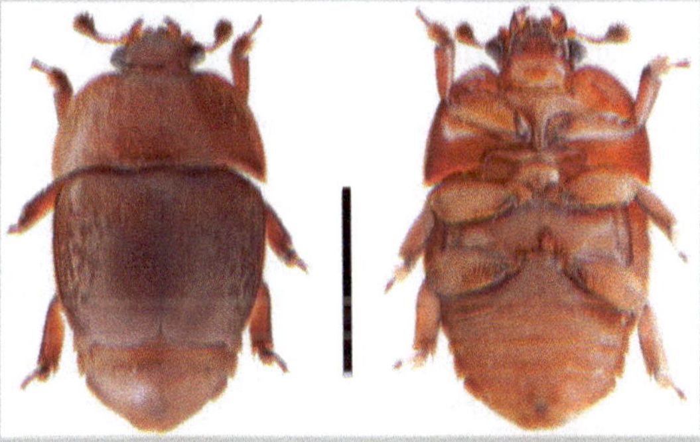

Fig. 11.19 Photograph of small hive beetle, *Aethina tumida* M. (dorsal and ventral views). (Source: Vijayakumar et al. 2025)

References

Arzani K, Nejatian MA, Karimzadeh G (2005) Apricot (*Prunus armeniaca*) pollen morphological characterization through scanning electron microscopy, using multivariate analysis. New Zealand J Crop Hort Sci 33:381–388. https://doi.org/10.1080/01140671.2005.9514374

Babu KD (2010) Flower biology of pomegranate (*Punica granatum* L.). Fruit Vegetable Cereal Sci Biotechnol 4(2):45–50

Babu KD, Chandra R, Jadhav VT, Sharma J (2009a) Blossom biology of pomegranate cv. 'Bhagawa' under semiarid tropics of western India. In: Abstracts of the 2nd international symposium on pomegranate and minor including mediterranean fruits, June 23–27, 2009, University of Agricultural Sciences, Dharwad, India, pp 88–89

Babu KD, Chandra R, Sharma J, Jadhav VT (2009b) Floral biology of pomegranate cultivar Ganesh under Solapur conditions of Maharashtra. In: Souvenir and Abstracts of 2nd International Symposium on Pomegranate and Minor Including Mediterranean Fruits, June 23–27, 2009, University of Agricultural Science, Dharwad, India, pp 86–87

Breeze TD, Bailey AP, Baelcombe KG, Potts G (2011) Pollination services in the UK: How important are honeybees? Agric Eco Env 142:137–143

Chavhan KY, Jagadish KS, Uthappa AR, Shishira D, Eswarappa G, Vijayakumar KT (2023) Study on insect diversity and their activity in *Punica granatum* var. Bhagwa. Pharma Innov J 12:1638–1642

Dedej S, Delaplane KS (2003) Honey bee (Hymenoptera: Apidae) pollination of rabbit eye blueberry *Vaccinium ashei* var. 'Climax' is pollinator density-dependent. Hort Entomol 96(4): 1215–1220

Deniz A, Zeynel D (2019) Determination of blooming, pollen and fruit set characteristics in *Punica granatum*. Not Bot Horti Agrobo 47(4):1258–1263

Derin K, Eti S (2001) Determination of pollen quality, quantity and effect of cross pollination on the fruit set and quality in the pomegranate. Turkish J Agri 25:169–173

Dezhong T, Boaming W, Gaixiu D, Xiaofung F (1995) Studies on the pollen morphology and ultrastructure of cultivated varieties of apricot (*Armenica vulgaris* Lam). Acta Hort 403:144–144

Engin H, Gokbayrak Z (2019) The biology of pomegranate pollen: all about formation, morphology, viability, germination and events relating to sperm nuclei. Alinteri J Agric Sci 34(2): 188–193

Evreinoff VA (1957) A contribution to the study of the pomegranate. J Agric Trop Appl Bot 4:124–138

Fahan (1976) Plant anatomy. Hakkibutz Hameuhad Publishers, Jerusalem, pp 321–394

Feng YZ, Chen DJ, Song MT, Zhao YL, Li ZH (1998) Assessment and utilization of pomegranate varieties resources. J Fruit Sci 15:370–373

Fouad MM, Barkat MR, El-yazal SA (1979) Bud burst activity, flowering and fruit set of 'Grenouillere' and 'Manfaloti' pomegranate cultivars under Giza conditions. Research Bulletin of Faculty of Agriculture and Animal Science, Shams University, pp 1089–1109

Gallai N, Salles JM, Settele J, Vaissiere BE (2009) Economic valuation of the vulnerability of world agriculture confronted with pollinator decline. Ecol Econ 68(3):810–821

Game RV (1987) Studies on the floral biology of some cultivars of pomegranate (*Punica granatum* L.) M.Sc. Agri. Thesis, MPAU, Rahuri

Ghazaeian M (2003) Study on pollen of some Asian pear genotypes by scanning electron microscopy (SEM) and suitable pollen germination medium. MSc. Thesis, Tarbiat Modares Univ., Tehran, Iran

Gogoi B, Rahman A, Rahman S, Deka MK (2007) Foraging behaviour and effect of *A. cerana* pollination on fruit set and yield of Assam lemon (*Citrus limon*). Indian J Agric Sci 77(2): 120–122

Gozlekci S, Kaynak L (1998) Investigations on the pollen production and quality in some standard pomegranate (*Punica granatum* L.) cultivars. Options Mediterraneennes 42:79–86

Gur A (1986) *Punica granatum*. In: Halevy AH (ed) CRC handbook of flowering, vol 4. CRC Press, Boca Raton, pp 147–150. https://doi.org/10.1201/9781351072564-18

Hajjar R, Jarvis DI, Gemmill-Herren B (2008) The utility of crop genetic diversity in maintaining ecosystem services. Agric Ecosys Env 123(4):261–270

Hudewenz A, Pufal G, Bogeholz AL, Klein AM (2014) Cross-pollination benefits differ among oilseed rape varieties. J Agric Sci 152(5):770–778

Jauker F, Bondarenko B, Becker HC, Steff, Ewenter I (2012) Pollination efficiency of wild bees and hoverflies provided to oilseed rape. Agric Forest Entomol 14:81–87

Javady T, Arzani K (2001) Pollen morphology of five Iranian olive (*Olea europaea* L.) cultivars. J Agric Sci Technol 3:37–42

Josan SS, Jawanda JS, Uppal D (1979) Studies on floral biology of pomegranate II. Anthesis, dehiscence, pollen studies and receptivity of stigma. Punjab Hort J 19:66–70

Karimi HR, Mirdehghan SH (2015) Effects of self, open, and supplementary pollination on growth pattern and characteristics of pomegranate fruit. Int J Fruit Sci 15:382–391. https://doi.org/10.1080/15538362.2015.1009974

Kedswin KS, Srinivasan MR, Saminathan VR, Preetha G, Thangamani C (2023) Pollination potential of (*Apis cerana indica*) in pumpkin (*Cucurbita moschata* Duchex Poir). Biol Forum An Int J 15(5a):181–186

Klein AM, Vaissiere BE, Cane JH, Steffan-Dewenter I, Cunningham SA, Kremen C, Tscharntke T (2007) Importance of pollinators in changing landscapes for world crops. Proc R Soc Ser B Biol Sci 274:303–313

Kozin RB (1976) Insect pollination of crops. In: Freitas BM, Pereira JOP (eds) Solitary bees: conservation, rearing and management for pollination: management and policy. Amsterdam Publishing Press, Amsterdam, pp 125–134

Lanza B, Marsilio V, Martinolli N (1996) Olive pollen ultrastructure: Characterization of exine pattern through image analysis-scanning electron microscope (IA-SEM). Sci Hort 65:283–294

Leopold AC, Kriedemann PE (1975) Plant growth and development. McGraw-Hill, New York

Levin GM (2006) Pomegranate roads: a soviet botanis's exile from Eden, 1st edn. Floreat Press, Forestville, pp 15–183

Munawar MS, Raja S, Siddique M, Niaz S, Amjad M (2009) The pollination by honeybee (*Apis mellifera* L.) increases yield of canola (*Brassica napus* L.). Pak Entomol 31:103–106

Nalawadi VG, Farooqui AA, Dasgupta N, Reddy MA, Gubbaiah Sulikeri GS, Nalini AS (1973) Studies on the floral biology of pomegranate (*Punica granatum* L.). Mysore J Agric Sci 7:213–225

Nath N, Randhawa GS (1959) Studies on floral biology of pomegranate. II. Anthesis, dehiscence, pollen studies and receptivity of stigma. Indian J Hort 16:121–135

Pareek OP, Sharma S (1993) Genetic resources of under exploited fruits. In: Chadha KL, Pareek OP (eds) Advances in horticulture, vol 1. Malhotra Publication, New Delhi, pp 189–225

Parmar C, Kaushal MK (1982) Wild fruits of the Sub-Himalayan region. Kalyani Publishers, Ludhiana, pp 74–79

Prasannakumar B (1998) Pomegranate. In: Chattopadhyay TK (ed) A text- book on pomology, vol 3. Kalyani Publishers, Ludhiana, p 232

Reale L, Sgromo C, Ederli L, Pasqualini S, Orlandi F, Fornaciari M, Ferranti F, Romano B (2009) Morphological and cytological development and starch accumulation in hermaphrodite and staminate flowers of olive (*Olea europa* L.). Sex Plant Reprod 22:109–119

Reddy P, Chauhan A, Singh HK (2022) Impact of bee pollination in Brinjal. Indian J Entomol:687–689

Sabbahi R, De Oliveira D, Marceau J (2005) Influence of honey bee (Hymenoptera: Apidae) density on the production of canola (Crucifera: Brassicacae). J Econ Entomol 98:367–372

Saroj PL, Sharma BD (2017) Cultivar Bhagwa: revolutionizing pomegranate scenario in India. All India Coordinated Research Project on Arid Zone Fruits, Bikaner, pp 1–19

Sharma R, Sharma VK, Sharma SD, Shireen K (2022) Assessment of pollen morphology and hybridization in pomegranate (*Punica granatum*). Indian J Agric Sci 92:1342–1347. https://doi.org/10.56093/ijas.v92i11.124062

Shulman Y, Fainberstein L, Lavee S (1984) Pomegranate fruit development and maturation. J Hort Sci 59:265–274

Singh SS, Krishnamurthi S, Katyal SL (1967) Fruit culture in India. Indian Council of Agricultural Research, New Delhi, pp 207–214

Singh RP, Kar PL, Dhuria HS (1978) Studies on the behaviour of flowering and sex expression in some pomegranate cultivars. Plant Sci 10:29–31

Steinacher G, Wagner J (2010) Flower longevity and duration of pistil receptivity in high mountain plants. Flora 205:376–387. https://doi.org/10.1016/j.flora.2009.12.012

Stern RA, Sapir G, Shafir S, Dag A, Goldway M (2007) The appropriate management of honey bee colonies for pollination of Rosaceae fruit trees in warm climates. Middle Eastern Russian J Plant Sci Biotechnol 1(1):13–19

Tao D-s, Dong X, Dong K, Zhang X, Yu Y-s (2010) Study on the effects of pollination by honeybees on pomegranate (*Punica granatum* L.). J Bee 3:10–11

Varun KN, Ankush G, Mirza A (2020) Characteristics of flowering and fruiting description of pomegranate (*Punica granatum* L.). Int J Curr Microbiol Appl Sci 9:401–412. https://doi.org/10.20546/ijcmas.2020.911.049

Verma LR, Partap U (1994) Foraging behavior of *Apis cerana* on cabbage and cauliflower and its impact on seed production. J Apicul Res 33:231–236

Vijayakumar KT, Nithin kumar HL, Rakshitha TN, Nayimabanu T (2024) Impact of flower position and pollinator diversity on yield parameters of pomegranate (*Punica granatum* L.). Sociobiol 71(2):e10368

Vijayakumar KT, Nayimabanu T, Rakshitha TN, Nithin Kumar HL (2025) Small hive beetle invades Karnataka: a growing threat to India's beekeeping industry. IE Blog No: 262. https://insectenvironment.com/f/invasive-small-hive-beetle-a-high-threat-to-indias-beekeeping

Westwood MN, Challice JS (1978) Morphology and surface topography of pollen and anthers of *Pyrus* species. J Am Soc Hort Sci 103:28–37

Wetzstein HY, Zhang Z, Ravid N, Wilkins E, Wetzstein ME (2011) Characterization of attributes related to fruit size in pomegranate. Hort Sci 46:908–912. https://doi.org/10.21273/HORTSCI.46.6.908

Wetzstein HY, Weiguang Y, Porter JA (2013) Flower position and size impact ovule number per flower, fruit set, and fruit size in pomegranate. J Am Soc Hort Sci 138:159–166. https://doi.org/10.21273/JASHS.138.3.159

Bee Pollinators of Medicinal Plants

12

S. Sardar, B. Majumdar, A. Rameshkumar, and S. I. Kazmi

Abstract

Plants have been continuously used in traditional medicine. Understanding the interactions between medicinal plants and pollinators is essential for conserving both. This study explores the therapeutic uses of 35 plant species belonging to 15 families—the bee pollinators from 5 families, namely, Andrenidae (02 species), Apidae (42 species), Colletidae (01 species), Halictidae (42 species), and Megachilidae (52 species), associated with these medicinal plants have also been reviewed.

Keywords

Medicinal plant · Pollinators · Bees · Floral association · Conservation

1 Introduction

The study of animal-plant pollination involves the complex interplay between animals and plants where the mutual relationship reaches maximum potential and benefits. Since their survival is interdependent, a detailed study of the types of pollinators of economically important plants is needed. Apart from pollination, plants are important sources of medicine (Girishkumar et al. 2016). Medicinal plants

S. Sardar
High Altitude Regional Centre, Zoological Survey of India, Saproon, Solan, Himachal Pradesh, India

B. Majumdar · A. Rameshkumar (✉)
Zoological Survey of India, Prani Vigyan Bhawan, New Alipore, Kolkata, West Bengal, India

S. I. Kazmi
Plant Protection Division, Aligarh Muslim University, Aligarh, Uttar Pradesh, India

A. K. Chakravarthy (ed.), *The Role of Pollinators in Tropical Landscapes*,
https://doi.org/10.1007/978-981-95-9139-8_12

have played a crucial part in medicine throughout history. In traditional and modern systems of medicine, medicinal plants are commonly utilized in health management worldwide. According to the WHO, about 80% of the population in Asian and African countries have been using traditional medicines based on plants for curing ailments (Oyebode et al. 2016). As time progressed, the development of synthetic antibiotics narrowed down the knowledge of the medicinal value of several plants among common people. Nevertheless, these synthetic drugs are expensive, have target-oriented effects, and their administration may cause side effects.

Overuse of medicines has led to resistant microbes (Zoomi et al. 2019) and a worldwide interest in bioactive compounds derived from plants and their antimicrobial properties (Miraj et al. 2020). Collection and curing of medicinal plants and plant products make a significant annual contribution to the Indian economy. Mazid et al. (2012) estimated that approximately 3000 plant species in India have medicinal properties. India, a megadiverse country, is home to a great diversity of medicinal plants across habitats. For instance, as the sediments brought by the Ganga were deposited in Uttar Pradesh, Bihar, and Bengal, the Ganga plains formed, and floral diversity has been found in the Gangetic plains since ancient times. As a result, this region has several medicinal plants growing in the wild due to favorable climatic and ecological conditions. This wide availability of plants with medicinal properties led to the progress of Ayurveda since the second century B.C. Since then, plants and plant-derived medicines have been an integral part of Indian culture.

In terrestrial ecosystems, pollinators are important for plant growth and development (Kevan 1999). Flowering plants are largely zoophilous (Ollerton et al. 2011), of which approximately 35% of crop production is pollinated by insects (Klein et al. 2007). Further, about 73% of the world's crops are pollinated by bees (Abrol 2009). Pollination contributes to food security, so plant species are the key sources of nectar in beekeeping since bees are dependent on nectar and pollen from flowers for survival. The plants from which bees forage, determine the quality and health benefits of bee products like honey, pollen, royal jelly, and propolis, and all other products derived from the beehives. Natural antioxidants found in bee products counteract the effects of oxidative stress associated with many diseases.

Intensive change in agricultural activity in the past half-century has been associated with factors like changes in land use systems, loss and fragmentation of pollinator habitats, and intensive pesticide use involved in the commercial mode of agriculture, has been causing a significant decline in bees' population (Deeksha et al. 2021). Moreover, the consideration of several medicinal plants as weeds, because of paucity of knowledge among the common public, has resulted in these plants being removed from and around agricultural fields, which negatively affects the vast diversity of pollinators (mainly solitary bee species) supported by them (Rameshkumar et al. 2022). Thus, understanding the importance of the medicinal "weeds" and their pollinators is essential for conservation. Thirty-five plant species belonging to 15 families have been reviewed for traditional uses as therapeutic agents. In addition, the variety of bee pollinators from five families of Apoidea supported by these plants has been explored.

2 Medicinal Plants and Pollinators

2.1 Family Acanthaceae

2.1.1 *Justicia adhatoda* Linn.

Common name: Vasaka/Malabar nut

Medicinal Uses Ayurveda claims that Vasaka is efficacious in the treatment of leprosy, hematological and cardiac ailments, thirst, fever, emesis, amnesia, leucoderma, jaundice, tumors, and oral infections. It is effective in the treatment of bronchitis, tuberculosis, asthma, and other pulmonary conditions. A decoction of the leaves is utilized to alleviate cough and other cold symptoms, while ointments may be employed for cleansing wounds due to its antibacterial and anti-inflammatory characteristics, as well as for easing rheumatic symptoms when applied to joints. Vasaka has been utilized to manage internal and external hemorrhaging resulting from colon illnesses such as peptic ulcers and hemorrhoids, as well as oral conditions (bleeding gums and mouth), and bleeding from the nostrils, genitals, or urinary tract. The juice of Vasaka can provide significant relief to ladies with menorrhagia. The plant has also been utilized to alleviate pain during labor and delivery (Kumar et al. 2010).

Pollinators *Apis cerana*, *Amegilla* sp. *Xylocopa verticalis*, *Xylocopa* sp. (Apidae) (Shivanna 2009); *Euaspis carbonaria*, *Megachile bicolor*, *Megachile cephalotes*, *Megachile gathela*, *Megachile lanata* (Megachilidae) (Gupta 1993).

2.2 Family Anacardiaceae

2.2.1 *Anacardium occidentale* Linn.

Common name: Cashew

Medicinal Uses Bark infusion and bark soaked in water are used to cure dysentery and stomach ache; bark macerations and decoctions are utilized to treat infectious processes, rheumatism, and ulcers. Leaves are used to cure skin infections and wounds, while fruits are helpful in curing asthma, wounds, and oral syphilis. Cashew gum is used as an analgesic, antidiabetic, and inflammatory agent (Salehi et al. 2020).

Pollinators *Apis cerana*, *Apis dorsata*, *Apis florea*, *Ceratina binghami*, *Ceratina smaragdula*, *Tetragonula iridipennis*, *Braunsapis* sp. (Apidae) (Sundararaju 2011); *Eoanthidium riparium* (Megahilidae) (Kumar 2015).

2.2.2 *Mangifera indica* Linn.

Common name: Mango

Medicinal Uses Various parts of the plant are utilized for the treatment of diarrhea, dysentery, anemia, asthma, bronchitis, cough, hypertension, insomnia, rheumatism, and toothaches. Additionally, they serve as a dentifrice, antiseptic, astringent, diaphoretic, stomachic, vermifuge, tonic, laxative, and diuretic. All components are employed in the management of abscesses, bites from rabid canines or jackals, tumors, snakebites, stings, datura poisoning, heat stroke, miscarriage, anthrax, blisters, oral injuries, tympanitis, glossitis, indigestion, bacillosis, liver conditions, excessive urination, and tetanus. Ripe mango juice serves as a treatment for heat stroke and functions as a restorative tonic. The seeds serve as an astringent and are utilized in the treatment of asthma. Inhalation of fumes from burned leaves is utilized as a remedy for throat infections and hiccups. The astringent bark is utilized in the treatment of rheumatism and diphtheria, and it is believed to exert a restorative effect on mucous membranes. Gum is employed in the treatment of scabies and as a dressing for cracked feet, and it is reported to possess anti-syphilitic properties (Shah et al. 2010).

Pollinators *Apis cerana*, *Apis dorsata*, *Apis florea*, *Tetragonula iridipennis* (Apidae) (Karunaratne et al. 2005); *Megachile disjuncta* (Megachilidae) (Kumar 2015).

2.3 Family Asteraceae

2.3.1 *Bidens pilosa* Linn.

Common name: Beggar tick/Spanish needle

Medicinal Uses Plant parts possess antibacterial, antidysenteric, anti-inflammatory, antimalarial, antipyretic, hepatoprotective, blood-lowering, hypoglycemic, and antidiabetic effects. It is an important medicine all over the world and is used to treat headaches, ear infections, and kidney problems. The leaf extract is used to cure malaria, stomach and mouth ulcers, diarrhea, and hangover, and as a poison antidote. *B. pilosa* is used to cure enteritis, bacterial dysentery, and pharyngitis. Young leaves and flowers are used to treat stomach disorders, hemorrhoids, and diabetes. In Japan, a traditional drug is made from dried *B. pilosa* powder and used as an ingredient in tea for livedo reticularis with summer ulceration (Xuan and Khanh 2016).

Pollinators *Apis cerana*, *Ceratina binghami*, *Ceratina hieroglyphica* (Apidae), *Halictus lucidipennis*, *Lasioglossum alphenum*, *Lasioglossum amblypygus*, *Lasioglossum bidentatum*, *Lasioglossum carinifrons*, *Lasioglossum* sp., *Leuconomia* sp., *Gnathonomia nasicana*, *Pachyhalictus* sp. (Halictidae) (Karunaratne et al. 2005).

2.3.2 *Guizotia abyssinica* (L.f.) Cass.

Common name: Niger

Medicinal Uses A paste of leaf and root mixed with water is utilized for the treatment of burns and rheumatoid arthritis. Seeds are utilized for birth control purposes. Seed oil is utilized for post-burn healing. A decoction of seeds or leaves is beneficial for syphilis. Garlic mixed with sprouted seeds is utilized for cough treatment. Seed pastes and oils are employed in the treatment of microbial infections (Dwivedi and Kohli 2012).

Pollinators *Apis cerana*, *Apis dorsata*, *Apis florea*, *Ceratina hieroglyphica*, *Ceratina lieftincki*, *Ceratina simillima*, *Ceratina smaragdula*, *Tetragonula laeviceps* (Apidae), *Halictus constrictus*, *Halictus fimbriatellus*, *Halictus funebris*, *Halictus gutturosus*, *Halictus lucidipennis*, *Halictus rugolatus*, *Homalictus* sp., *Lasioglossum serenum*, *Nomia crassipes*, *Pseudapis oxybeloides* (Halictidae), *Lithurgus atratus*, *Megachile hera*, *Megachile bicolor*, *Megachile stirostoma* (Megachilidae) (Patel and Pastagia 2021); *Coelioxys basalis*, *Coelioxys confusa*, *Megachile disjuncta*, *Megachile lerma* (Megachilidae) (Kumar 2015).

2.3.3 *Helianthus annus* Linn.

Common name: Sunflower

Medicinal Uses Sunflower seeds and sprouts possess significant antioxidant, antimicrobial, anti-inflammatory, and cardiovascular properties. This is utilized for the treatment of diseases. The seed provides a variety of nutritious components, including protein, unsaturated fats, fiber, vitamins (notably E), selenium, copper, zinc, folate, iron, and sulfur, among others (Guo et al. 2017).

Pollinators *Anthophora* sp., *Apis mellifera*, *Apis cerana*, *Apis dorsata*, *Apis florea*, *Tetragonula* sp., *Xylocopa* cf. *iridipennis* (Apidae), *Nomia* sp., *Halictus* sp. (Halictidae), *Megachile disjuncta* (Megachilidae) (Kumar and Srivastava 2021); *Amegilla* sp. (Apidae), *Lasioglossum* sp. (Halictidae) (Muhammad et al. 2021); *Coelioxys basalis*, *Euaspis carbonaria*, *Lithurgus* sp., *Megachile albifrons*, *Megachile carbonaria*, *Megachile cephalotes*, *Megachile lanata*, *Megachile stirostoma*, *Megachile umbripennis*, *Megachile vigilans* (Megachilidae) (Kumar 2015); *Megachile vera* (Megachilidae) (Gupta 1993); *Anthidium florentinum* (Megachilidae) (Sardar et al. 2021).

2.3.4 *Tridax procumbens* Linn.

Common name: Coatbutton/Tridax daisy

Medicinal Uses Various potential therapeutic activities have been identified, including antimicrobial and antifungal properties, antioxidant effects, antibiotic efficacy, hypotensive effects, anti-arthritic properties, wound healing capabilities, insecticidal activity, anti-inflammatory effects, and treatments for diarrhea and dysentery. Leaf juice is utilized for the treatment of fresh wounds, hemorrhage control, and as a hair tonic. In India, it is primarily utilized for wound healing, as an anticoagulant, antifungal agent, and insect repellent. Leaf extract is utilized in folk medicine for the treatment of infectious skin diseases. This medicine is recognized for its efficacy in treating liver disorders, as well as gastritis and heartburn (Chaudhari and Pati 2022).

Pollinators *Apis cerana*, *Apis dorsata*, *Apis mellifera*, *Ceratina propinqua*, *Thyreus* sp. (Apidae), *Pseudapis* sp., *Curvinomia iridescens* (Halictidae), *Pseudoanthidium* sp. (Megachilidae) (Deeksha et al. 2021); *Anthidium ardens*, *Coelioxys capitata*, *Coelioxys confusa*, *Coelioxys fuscipennis*, *Eoanthidium salemense*, *Eoanthidium semicarinatum*, *Megachile bicolor*, *Megachile hera*, *Megachile lerma*, *Pseudoanthidium orientale* (Megachilidae) (Kumar 2015); *Euaspis carbonaria*, *Icteranthidium sinapinum*, *Megachile coelioxysides*, *Megachile creusa*, *Megachile gathela*, *Megachile lanata*, *Megachile rugicauda*, *Megachile vigilans* (Megachilidae) (Gupta 1993); *Austronomia* sp., *Hoplonomia westwoodi*, *Lipotriches taprobanae*, *Lipotriches pulchriventris*, *Maynenomia* sp., *Nomia crassipes* (Halictidae) (Karunaratne et al. 2005).

2.4 Family Bignoniaceae

2.4.1 *Campsis radicans* Seem.

Common name: Trumpet vine

Medicinal Uses Plant parts are utilized for wound healing, treatment of fungal and bacterial infections, and as anti-itch agents. In China, the plant is recognized for its anticoagulant properties and has been traditionally employed to address abnormal menstruation and dysmenorrhea (Islam et al. 2019).

Pollinators *Apis cerana*, *Apis florea*, *Bombus* sp., *Xylocopa* sp. (Apidae), sweat bees (Halictidae) (Van Nest et al. 2021); *Megachile bicolor*, *Megachile creusa*, *Megachile lanata* (Megachilidae) (Gupta 1993).

2.4.2 *Tecoma stans* (Linn.) Juss. ex Kunth

Common name: Yellow elder/Yellow trumpet

Medicinal Uses In Latin America, it is traditionally utilized for the reduction of blood glucose levels. Bark exhibits smooth muscle relaxant, mild cardiotonic, and chloretic properties. Applications encompass the management of digestive disorders, regulation of yeast infections, investigational treatment of diabetes, and various other therapeutic uses. The plant's root is recognized as a potent diuretic, vermifuge, and tonic. External application of the root paste extracted after grinding when combined with lemon juice serves as a remedy for snake and rat bites (Singh et al. 2013).

Pollinators *Amegilla* sp., *Apis dorsata*, *Apis cerana*, *Ceratina simillima*, *Ceratina binghami*, *Tetragonula iridipennis*, *Xylocopa latipes*, *Xylocopa pubescens* (Apidae) (Jonathan et al. 2009); *Megachile carbonaria*, *Megachile cephalotes*, *Megachile conjuncta*, *Megachile gathela*, *Megachile lanata* (Megachilidae) (Gupta 1993).

2.5 Family Euphorbiaceae

2.5.1 *Euphorbia heterophylla* Linn.

Common name: Wild poinsettia

Medicinal Uses *Euphorbia heterophylla* leaf is used in traditional medical practices as laxative and anti-gonorrheal, migraine, and wart cure. The plant lattices are used as fish poison, insecticide, and oral poison. The leaves are effective against convulsions, seizures, and cough (James and Friday 2010).

Pollinators *Apis florea* (Apidae), *Megachile nana* (Megachilidae), *Austronomia krombeini*, *Halictus lucidipennis*, *Hoplonomia westwoodi*, *Lasioglossum* sp., *Leuconomia* sp., *Lipotriches comberi*, *Lipotriches* sp., *Nomia crassipes*, *Gnathonomia aurata*, *Pachynomia* sp., *Pseudapis oxybeloides*, *Sphecodes crassicornis* (Halictidae) (Karunaratne et al. 2005).

2.6 Family Fabaceae

2.6.1 *Cajanus cajan* (Linn.) Millsp.

Common name: Pigeon pea

Medicinal Uses Pigeon pea serves as a significant source of protein, vitamin B, and essential minerals. Leaf concentrates are beneficial for conditions such as jaundice, diarrhea, wounds, cough, bronchitis, bladder stones, and diabetes. The young leaves are utilized for the treatment of aphtha, while a decoction of the leaves has demonstrated efficacy in treating cough and diarrhoea. Pigeon pea leaves are utilized in the treatment of traumatism, burn infections, and bedsores. Pigeon pea leaves exhibit significant anti-inflammatory, antibiotic, and abirritation properties, as well as the ability to inhibit capillary permeability. This substance is recognized as an effective traditional Chinese medicine for the treatment of ischemic necrosis of the femoral head, as well as for its abilities to stop bleeding, alleviate pain, and eliminate parasites (Gupta 2021). The plant has the potential to prevent and treat human diseases (Syed and Wu 2018).

Pollinators *Apis cerana*, *Apis dorsata*, *Apis florea*, *Tetragonula iridipennis*, *Tetragonula laeviceps* (Apidae), *Halictus constrictus*, *Halictus fimbriatellus*, *Halictus funebris*, *Halictus gutturosus*, *Halictus lucidipennis*, *Halictus rugolatus*, *Homalictus* sp., *Lasioglossum serenum*, *Nomia crassipes*, *Pseudapis oxybeloides*

(Halictidae), *Megachile hera* (Megachilidae) (Patel and Pastagia 2021); *Ceratina binghami*, *Ceratina lieftincki*, *Ceratina simillima*, *Ceratina smaragdula* (Apidae), *Hoplonomia elliotii* (Halictidae), *Megachile laticeps* (Megachilidae (Rameshkumar et al. 2022); *Megachile albifrons*, *Megachile anthracina*, *Megachile bicolor*, *Megachile cephalotes*, *Megachile conjuncta*, *Megachile disjuncta*, *Megachile lanata*, *Megachile lerma*, *Megachile stirostoma*, *Megachile umbripennis*, *Megachile vigilans* (Megachilidae) (Kumar 2015); *Euaspis carbonaria*, *Megachile creusa* (Megachilidae) (Gupta 1993).

2.6.2 *Crotalaria juncea* Linn.

Common name: Sunn hemp

Medicinal Uses The plant extract is traditionally utilized as a blood purifier, contraceptive, astringent, demulcent, emetic, and purgative, as well as in the treatment of anemia, impetigo, menorrhagia, and psoriasis (Al-Snafi 2016).

Pollinators *Xylocopa latipes*, *Xylocopa pubescens* (Apidae) (Rakesh et al. 2019); *Xylocopa aestuans*, *Xylocopa amethystina*, *Xylocopa fenestrata*, *Xylocopa latipes* (Apidae), *Halictus* sp., *Nomia* sp. (Halictidae), *Coelioxys fuscipennis*, *Coelioxys sexmaculatus* (Megachilidae) (Patel and Pastagia 2021), *Megachile albifrons*, *Megachile anthracina*, *Megachile bicolor*, *Megachile cephalotes*, *Megachile disjuncta*, *Megachile lanata* (Megachilidae) (Kumar 2015); *Megachile conjuncta*, *Megachile lerma*, *Megachile umbripennis*, *Megachile vera* (Megachilidae) (Gupta 1993); *Megachile fraterna*, *Megachile laticeps* (Megachilidae) (Rameshkumar et al. 2022); *Apis florea*, *Braunsapis* sp., *Ceratina binghami*, *Ceratina hieroglyphica*, *Xylocopa tenuiscapa*, *Tetragonula iridipennis* (Apidae) (Karunaratne et al. 2005).

2.6.3 *Crotalaria pallida* Aiton

Common name: Rattle box

Medicinal Uses The leaf extract was found to have estrogenic effects and can be used in hormone replacement therapy (Boldrin et al. 2013); stem extracts displayed antioxidant as well as antibacterial activity (Islam et al. 2018); both leaf and stem extracts displayed anti-inflammatory activity (Govindappa et al. 2011). The plant is utilized for addressing urinary issues and fever; a poultice made from the roots is applied to joint swelling, while an extract from the leaves is consumed to eliminate intestinal worms (Kiruthiga et al. 2014).

Pollinators *Apis cerana*, *Ceratina binghami*, *Ceratina hieroglyphica*, *Xylocopa fenestrata* (Apidae) (Karunaratne et al. 2005); *Megachile albifrons*, *Megachile anthracina*, *Megachile bicolor*, *Megachile disjuncta*, *Megachile lanata* (Megachilidae) (Kumar 2015); *Megachile laticeps* (Megachilidae) (Rameshkumar et al. 2022).

2.6.4 *Guilandina bonduc* Linn.

Common name: Fever nut/Gray nicker

Medicinal Uses Powdered roasted seeds alleviate flatulence and post-partum stomach pain. Seed powder combined with milk is utilized to manage diarrhea. The astringent seed coat is advantageous as a treatment for diarrhea, dysentery, and colitis. Seeds have antihelminthic and antimalarial effects and can be utilized for treatment purposes. A tonic derived from seeds may also be utilized to alleviate cough and asthma. The leaves, when fried in ghee, alleviate gas and cure constipation, therefore being beneficial for piles. The seed's outer layer is highly advantageous for managing leucorrhea. The seeds exhibit contraceptive properties. The oil derived from the leaves is a good nervine tonic (Pandey et al. 2018). The seeds act as a uterine stimulant, enhance menstrual flow in oligomenorrhea, and alleviate lower abdominal pain, particularly in cases of PCOS (Kandasamy and Balasundaram 2021).

Pollinators *Apis cerana*, *Apis dorsata*, *Xylocopa* sp. (Apidae) (Singh and Chauhan 2010); *Megachile bicolor*, *Megachile lanata* (Megachilidae) (Kumar 2015).

2.6.5 *Medicago sativa* Linn.

Common name: Alfalfa

Medicinal Uses Traditionally, leaves and seeds were utilized in Ayurvedic and homeopathic treatment for problems of the central nervous system, renal discomfort, diabetes, inflammation, and obesity. Traditionally, alfalfa sprouts or leaves were utilized for the treatment of arthritis, fever, as an anticancer agent, antirheumatic, cardiotonic, depurative, galactagogue, emmenagogue, and antiscorbutic; for the management of boils (Al-Snafi et al. 2021), renal calculi, pyrexia, lithiasis, and dysuria; and to alleviate edema and fluid retention. In South America, the plant is utilized as a diuretic for kidney issues, muscular soreness, scurvy, and vascular edema (Bora and Sharma 2011).

Pollinators *Apis cerana*, *Apis dorsata*, *Apis florea*, *Apis mellifera*, *Bombus* sp., *Xylocopa* sp. (Apidae), *Nomia* sp. (Halictidae), *Megachile rotundata*, *Osmia* sp. (Megachilidae) (Haedo et al. 2022), *Ceratina binghami* (Apidae) (Rameshkumar et al. 2022); *Megachile cephalotes*, *Megachile hera*, *Megachile lanata*, *Megachile umbripennis* (Megachilidae) (Gupta 1993).

2.6.6 *Pongamia pinnata* (L.) Pierre

Common name: Karanja

Medicinal Uses *Pongamia pinnata* has been utilized as a crude medicinal agent for the therapy of tumors, hemorrhoids, dermatological conditions, and ulcers. The root is efficacious in treating gonorrhea, cleansing gums, teeth, and ulcers and is utilized for vaginal and dermatological conditions. The seed oil is efficacious in treating keratitis, urinary discharges, hemorrhoids, ulcers, chronic fever, rheumatism, leukoderma, lumbago, scabies, leprosy, bronchitis, whooping cough, chronic dermatological conditions, wound healing, hypertension, and hepatic discomfort. The leaves are utilized for rheumatism, gonorrhea, and dermatological conditions. The bark is utilized for diabetes, malaria, hemorrhoids, beriberi, and hemorrhages (Al-Muqarrabun et al. 2013).

Pollinators *Amegilla* sp., *Apis dorsata*, *Apis cerana*, *Apis florea*, *Ceratina binghami*, *Ceratina simillima*, *Tetragonula iridipennis*, *Xylocopa pubescens*, *Xylocopa latipes* (Apidae) (Raju and Rao 2006); *Apis mellifera*, *Ceratina smaragdula* (Apidae), *Nomia* sp. (Halictidae), *Coelioxys confusa*, *Megachile amputata*, *Megachile hera*, *Megachile rotundata*, *Megachile vigilans* (Megachilidae), (Shankar et al. 2017); *Ceratina binghami*, *Xylocopa aestuans*, *Xylocopa latipes* (Apidae), *Coelioxys* sp., *Megachile cephalotes*, *Megachile disjuncta* (Megachilidae) (Veereshkumar Kaushik et al. 2020); *Euaspis edentata*, *Megachile anthracina* (Megachilidae) (Kumar 2015); *Euaspis carbonaria*, *Megachile bicolor*, *Megachile conjuncta*, *Megachile creusa*, *Megachile lanata*, *Megachile lerma*, *Megachile nana*, *Megachile umbripennis*, *Megachile vera* (Megachilidae) (Gupta 1993).

2.6.7 *Tephrosia purpurea* (Linn.) Pers.

Common name: Fish poison/Wild indigo

Medicinal Uses It is traditionally employed to treat various external injuries and gastro-duodenal ailments. The medication is utilized for cough and chest constriction. A decoction of the root is beneficial for the enlargement and obstruction of the liver, spleen, and kidney, as well as for dyspepsia and chronic diarrhea. Gargling cleanses the oral cavity. The infusion of seeds is utilized as anthelmintic oil and for skin conditions such as scabies and leucoderma. Leaves are utilized in the treatment of dyspepsia, respiratory ailments, hemorrhoids, syphilis, and gonorrhea. An extract of pods is efficacious as an analgesic and anti-inflammatory agent (Dalwadi et al. 2014).

Pollinators *Amegilla* cf. *zonata*, *Amegilla niveocincta*, *Apis florea*, *Apis dorsata*, *Braunsapis mixta*, *Ceratina sexmaculata*, *Xylocopa amethystina*, *Xylocopa fenestrata*, *Xylocopa latipes*, *Thyreus ramosus*, *Tetragonula iridipennis* (Apidae), *Hoplonomia elliotii* (Halictidae), *Andrena* sp., *Icteranthidium saltatore*, *Megachile coelioxysides*, *Megachile gathela* (Megachilidae) (Kachhawa et al. 2021), *Anthidiellum latipes*, *Coelioxys basalis*, *Coelioxys fuscipennis*, *Euaspis edentata*, *Megachile albifrons*, *Megachile anthracina*, *Megachile carbonaria*, *Megachile cephalotes*, *Megachile disjuncta*, *Megachile lanata*, *Megachile lerma*, *Megachile vigilans*, *Pseudoanthidium flaviventre* (Megachilidae) (Kumar 2015); *Euaspis carbonaria*, *Icteranthidium sinapinum*, *Icteranthidium tshangiricum*, *Megachile bicolor*, *Megachile creusa*, *Megachile rugicauda*, *Megachile stirostoma*, *Megachile umbripennis*, *Pseudoanthidium orientale* (Megachilidae) (Gupta 1993).

2.7 Family Lamiaceae

2.7.1 *Hyptis suaveolens* (Linn.) Poit.

Common name: Pignut

Medicinal Uses Botanical components are helpful in the treatment of the respiratory system, gastrointestinal tract, gynecological system, and integumentary system. It is frequently utilized for fever and headache, as an insect repellent, for gastrointestinal disorders, dermatological diseases, injuries, fatigue, abnormal leucorrhea, renal ailments, diarrhea, and malaria (Li et al. 2020).

Pollinators *Amegilla* sp., *Apis cerana*, *Ceratina binghami*, *Ceratina* sp., *Thyreus histrio*, *Trigona* sp., *Xylocopa latipes* (Apidae), *Pseudapis oxybeloides* (Halictidae), *Megachile* sp. (Megachilidae) (Aluri and Reddy 1989); *Coelioxys basalis*, *Coelioxys confusa*, *Coelioxys fuscipennis*, *Euaspis basalis*, *Megachile anthracina*, *Megachile bicolor*, *Megachile disjuncta*, *Megachile lerma* (Megachilidae) (Kumar 2015); *Amegilla comberi*, *Amegilla fallax*, *Amegilla niveocincta*, *Amegilla violacea*, *Ceratina smaragdula*, *Thyreus ceylonicus*, *Thyreus takaonis*, *Thyreus ramosellus*, *Tetragonula iridipennis*, *Xylocopa amethystina*, *Xylocopa collaris*, *Xylocopa fenestrata* (Apidae), *Austronomia krombeini*, *Austronomia notiomorpha*, *Curvinomia formosa*, *Curvinomia iridescens*, *Halictus lucidipennis*, *Hoplonomia westwoodi*, *Gnathonomia aurata*, *Lasioglossum amblypygus*, *Leuconomia* sp., *Lipotriches comberi*, *Lipotriches exagens*, *Lipotriches taprobanae*, *Lipotriches fulvinerva*, *Lipotriches pulchriventris*, *Nomia crassipes*, *Pachynomia* sp., *Pseudapis oxybeloides*, *Sphecodes crassicornis* (Halictidae), *Anthidiellum* sp., *Coelioxys capitata*, *Euaspis carbonaria*, *Megachile hera*, *Megachile lanata*, *Megachile vigilans*, *Pseudoanthidium* sp. (Megachillidae) (Karunaratne et al. 2005).

2.7.2 *Lavandula latifolia* Medik.

Common name: Broadleaved lavender

Medicinal Uses Lavender oil possesses potent analgesic, carminative, sedative, anti-inflammatory, and antidepressant properties, as well as antibacterial benefits. It is efficacious against various microorganisms, including those exhibiting resistance to antimicrobial agents. Lavender oil is efficacious in treating fungal infections, facilitating wound healing, mitigating hair loss, alleviating anxiety, addressing canker sores, managing post-menstrual issues, and relieving post-surgical pain (Bagri and Savita 2019).

Pollinators *Anthophora* sp., *Apis mellifera*, *Ceratina* sp., *Xylocopa* sp., *Bombus* sp. (Apidae), *Halictus* sp. (Halictidae), *Anthidium cingulatum*, *Anthidium florentinum* (Megachilidae) (Herrera 1989); *Coelioxys capitata*, *Coelioxys confusa*, *Coelioxys fuscipennis*, *Euaspis carbonaria*, *Megachile albifrons*, *Megachile cephalotes*, *Megachile disjuncta*, *Megachile lerma* (Megachilidae) (Kumar 2015).

2.7.3 *Leucas aspera* (Willd.) Link

Common name: Thumbai

Medicinal Uses Headaches, asthma, bronchitis, inflammation, dyspepsia, jaundice, and anthelmintic, laxative, and diaphoretic effects are some of the many oral uses for hot water extract. The complete plant extract is administered orally to address scabies, psoriasis, snake bites, venomous bug bites, and scorpion stings. The herb is utilized externally as an insect repellent. A small quantity of flowers sautéed in ghee is administered orally for the treatment of coughs and colds. The blossoms are crushed, and their perfume is inhaled to alleviate migraine symptoms. The sap of leaves is utilized for otalgia and purulent discharge. A paste of crushed leaves mixed with chalk is applied to dental cavities to inhibit deterioration. The leaf decoction is administered nasally as an antivenin. Leaf paste combined with turmeric is utilized for the treatment of wounds and boils. The decoction of roots, stems, and inflorescences is administered orally for high fevers, influenza, and malaria (Das et al. 2012).

Pollinators *Amegilla* sp., *Apis cerana*, *Apis florea*, *Apis dorsata*, *Anthophora* sp., *Trigona* sp., *Ceratina* sp. (Apidae), *Megachile* sp. (Megachilidae) (Rao 2014); *Xylocopa pubescens*, *Xylocopa latipes* (Apidae) (Prasad and Sunojkumar 2014); *Pachyanthidium lachrymosum* (Smith) (Kumar 2015), *Thyreus* sp. (Apidae), *Curvinomia* sp., *Steganomus* sp., *Lasioglossum* sp. (Halictidae), *Heriades* sp. (Megachilidae) (Chakraborty et al. 2021).

2.7.4 *Ocimum americanum* Linn.

Common name: Wild basil/Hoary basil

Medicinal Uses The plant has multiple native uses in East Africa, including mint in tea, tobacco flavoring, and personal scent. Either leaves are pulverized between the palms and inhaled, or hot aqueous vapor is inhaled to relieve nasal congestion and treat bronchial catarrh. Decoctions derived from the plant are utilized for the treatment of coughs, ocular and aural disorders, hemorrhoids, TB, and abdominal pain. Plant extracts are utilized to address ulcers and anticathartic conditions. The Swahilis utilize the aerial components of the plant to reduce blood pressure and alleviate stomach pain. In India, leaf extract decoction is traditionally employed to address constipation, diabetes, dysentery, diarrhea, and hemorrhoids (Ali et al. 2021).

Pollinators *Amegilla* sp., *Apis florea*, *Apis cerana*, *Ceratina binghami*, *Ceratina* sp. (Apidae), *Pseudapis oxybeloides* (Halictidae) (Raju 1989); *Coelioxys confusa*, *Coelioxys fuscipennis*, *Euaspis carbonaria*, *Heriades* sp., *Megachile albifrons*, *Megachile bicolor*, *Megachile cephalotes*, *Megachile disjuncta*, *Megachile lerma* (Megachilidae) (Kumar 2015); *Thyreus* sp. (Apidae), *Curvinomia* sp., *Pseudapis* sp., *Steganomus* sp. (Halictidae) (Chakraborty et al. 2021).

2.7.5 *Ocimum basilicum* Linn.

Common name: Common basil

Medicinal Uses Asthma, bronchitis, cough, and gastrointestinal problems are among the common respiratory tract ailments treated with this plant. The plant extract and its derivatives are efficacious for dermatological issues such as acne, eczema, boils, psoriasis, and rashes. Floral and foliar components are ingested as infusions, syrups, or decoctions, serving as sudorific, stimulant, carminative, diuretic, and febrifuge agents, and are commonly recommended for bronchitis and coughs (Aminian et al. 2022). It has also been utilized in the treatment of conditions such as anxiety, fever, infections, insect stings, abdominal pain, migraines, and constipation; it can additionally regulate and lower blood glucose levels. Basil seeds possess diuretic, antipyretic, antispasmodic, and stomachic effects. Basil polysaccharides exhibit antitumor, antioxidant, antiaging properties, alongside antibacterial benefits, anti-atherosclerotic effects, immune stimulation, and efficacy against diabetic mellitus (Shahrajabian et al. 2020).

Pollinators *Amegilla* sp., *Apis florea*, *Apis cerana*, *Ceratina binghami*, *Ceratina* sp., *Thyreus histrio* (Apidae) (Raju 1989); *Andrena* sp. (Andrenidae), *Amegilla* cf. *zonata*, *Apis dorsata*, *Ceratina smaragdula*, *Xylocopa latipes*, *Xylocopa pubescens* (Apidae), *Halictus* sp., *Nomia* sp. (Halictidae), *Megachile bicolor*, *Megachile cephalotes*, *Megachile disjuncta*, *Megachile hera*, *Megachile lanata*, *Pseudoanthidium orientale* (Megachilidae) (Abrol et al. 2017); *Megachile laticeps* (Megachilidae) (Rameshkumar et al. 2022).

2.7.6 *Thymus vulgaris* Linn.

Common name: Garden thyme

Medicinal Uses Thyme is utilized for respiratory ailments such as whooping cough, bronchitis, and asthma, in the form of tea, ointment, tincture, syrup, or via steam inhalation. It is utilized to avert arteriosclerosis, alleviate toothache, treat urinary tract infections, and address dyspepsia. It also eliminates fungi, germs, and other parasites from the stomach and intestines and acts as an appetite stimulant. This plant is utilized to combat physical and mental debility, as well as to alleviate sleeplessness. Many inflammatory conditions, including rheumatism, edema, bug stings, and aches, can be alleviated with its use (Dauqan and Abdullah 2017).

Pollinators *Bombus* sp., *Apis mellifera* (Apidae), *Lasioglossum* sp. (Halictidae) (Arnan et al. 2014); *Coelioxys capitata*, *Coelioxys fuscipennis*, *Euaspis carbonaria*, *Megachile cephalotes*, *Megachile disjuncta*, *Megachile lerma*, *Megachile stirostoma*, *Megachile vigilans* (Megachilidae) (Kumar 2015).

2.8 Family Lythraceae

2.8.1 *Cuphea hyssopifolia* Kunth

Common name: Mexican false heather

Medicinal Uses It is cytotoxic, antiviral, antimicrobial, antiprotozoal, cardiovascular, antioxidant, diuretic, and hepatoprotective (Moke et al. 2020).

Pollinators *Amegilla* sp., *Apis cerana, Apis florea, Apis mellifera, Tetragonula* sp., *Thyreus* sp., *Trigona* sp. (Apidae), *Megachile* sp. (Megachilidae) (Peniwidiyanti et al. 2020); *Coelioxys capitata, Coelioxys confusa, Euaspis carbonaria, Lithurgus* sp., *Megachile vigilans* (Megachilidae) (Kumar 2015).

2.9 Family Malvaceae

2.9.1 *Abelmoschus esculentus* (Linn.) Moench

Common name: Okra/lady's finger

Medicinal Uses Dysentery, diarrhea, acute inflammation, dysuria, catarrhal kidney infection, ardor urinae, dysuria, and high fever are all treated with infusions of fruit mucilage. Plasma replacement and antipyretic qualities are further uses for this infusion. A decoction of unripe fruit, roots, and leaves serves as a demulcent and emollient poultice. Extract of leaves combined with egg albumin and applied to hair results in black, silky hair. The juice extracted from roots is utilized to remedy cuts, wounds, and boils. The infusion of roots is utilized for the treatment of diabetes, ulcers, jaundice, syphilis, and as a laxative and stomachic. Seeds possess antispasmodic, cordial, stimulant, antidiabetic, and sudorific effects and are utilized in the treatment of spermatorrhea. A decoction of leaves and flowers is utilized to treat bronchitis and pneumonia (Chanchal et al. 2018).

Pollinators *Tetragonula irridipennis* (Apidae) (Perera and Karunaratne 2019); *Amegilla* sp., *Apis mellifera, Xylocopa* sp. (Apidae), *Lasioglossum* sp., *Lipotriches* sp. *Seladonia* sp. (Halictidae) (Auguste et al. 2021); *Lithurgus atratus, Lithurgus* sp. (Megachilidae) (Kumar 2015); *Ceratina smaragdula* (Apidae) (Rameshkumar et al. 2022).

2.9.2 *Abelmoschus moschatus* Medik.

Common name: Musk mallow

Medicinal Uses The seeds are beneficial in treating ophthalmopathy, cardiac weakness, cough, asthma, bronchitis, excessive thirst, burning sensation, nausea, dyspepsia, flatulent colic, diarrhea, strangury, gonorrhea, spermatorrhea, calculi, halitosis, respiratory ailments, ptyalism, vomiting, various neural disorders, pruritus, leukoderma, and general weakness. Vapors are inhaled to alleviate throat dryness.

Alcohol extracts of powdered seeds are utilized for snake bites, whereas aqueous extracts are employed for asthma, colds, influenza, and helminthic infections. Seeds ground into a paste with milk are utilized to alleviate irritation and soften the skin. The seeds' decoction is utilized in the treatment of stomach cancer. Mucilage derived from roots and leaves is advised for gonorrhea and sexual disorders. The poultice composed of leaves and roots is utilized for cystitis, fever, headache, rheumatism, varicose veins, and boils (Pawar and Vyawahare 2017).

Pollinators *Lithurgus atratus* (Megachilidae) (Kumar 2015).

2.9.3 *Grewia asiatica* Linn.

Common name: Phalsa

Medicinal Uses The fruits are advantageous for cardiovascular, hematological, and hepatic problems. The root bark is utilized for rheumatism and urinary system disorders. The leaves are utilized on the skin for treating wounds and lacerations (Zia-Ul-Haq et al. 2013).

Pollinators *Apis cerana*, *Apis dorsata*, *Apis florea*, *Apis mellifera*, *Tetragonula iridipennis*, *Xylocopa fenestrata*, *Xylocopa pubescens* (Apidae), *Halictus* sp. (Halictidae), *Coelioxys apicata* (Megachilidae) (Udikeri and Chandra 2019); *Megachile lanata* (Fabricius), *Megachile phaola*, *Megachile vigilans* (Megachilidae) (Gupta 1993).

2.10 Family Pedaliaceae

2.10.1 *Sesamum indicum* Linn.

Common name: Sesame

Medicinal Uses Sesame oil possesses mild laxative, emollient, and demulcent properties. A poultice made from the seeds and young leaves has further uses. The oil possesses extensive medicinal and pharmacological applications. Sesamin in the oil safeguards the liver against oxidative damage. It possesses inherent antibacterial and antifungal properties against prevalent skin infections. It possesses antiviral, anti-inflammatory, and wound-healing effects. Oil is utilized for chronic ailments such as hepatitis, diabetes, and migraines (Anilakumar et al. 2010).

Pollinators *Lithurgus atratus* (Kumar 2015); *Megachile lanata*, *Megachile stirostoma*, *Xylocopa aestuans*, *Xylocopa amethystina*, *Xylocopa fenestrata*, *Xylocopa latipes* (Patel and Pastagia 2021); *Apis cerana*, *Amegilla comberi*, *Ceratina binghami* (Apidae), *Gnathonomia aurata*, *Lasioglossum serenum*, *Systropha* sp. (Halictidae) (Ekanayake and Karunaratne 2022).

2.11 Family Plantaginaceae

2.11.1 *Bacopa monnieri* (Linn.) Pennell

Common name: Water hyssop/Brahmi

Medicinal Uses This herb's efficacy as a cognitive enhancer surpasses that of other therapeutic plants, and it is utilized to alleviate weariness and sadness. Besides stimulating the central nervous system and supporting the circulatory system, it alleviates and reduces varicose veins and assists in diminishing scarring. Additionally, it aids in the regeneration of skin and connective tissues, and it diminishes the appearance of cellulite (Khansari et al. 2009).

Pollinators *Apis cerana*, sp., *Ceratina* sp. (Apidae), *Lasioglossum* sp., *Steganomus* sp. (Halictidae) (Chakraborty et al. 2021).

2.12 Family Rubiaceae

2.12.1 *Spermacoce ocymoides* Burm. f.

Common name: Purple-leaved button weed

Medicinal Uses The leaves of this plant possess antibacterial and antifungal properties and are also utilized in the treatment of dysentery and diarrhea (Jeyachandran et al. 2013). Leaf juice remedies ocular infections (Reang et al. 2016).

Pollinators *Apis cerana, Apis dorsata, Apis florea, Ceratina* sp. (Apidae), *Nomia* sp. (Halictidae) (Raju and Radhakrishnan 2017); *Coelioxys* sp., *Euaspis carbonaria, Megachile hera, Megachile lerma, Megachile vigilans* (Megachilidae) (Kumar 2015).

2.13 Family Talinaceae

2.13.1 *Talinum portulacifolium* (Forssk.) Asch. Ex Schweinf

Common name: Fame flower/pink baby's-breath

Medicinal Uses It is utilized as a green leafy vegetable. Plant components are utilized for antihistaminic, anticholinergic, spasmolytic, and antiulcerogenic purposes. Leaf powder is utilized for the treatment of diabetes, hepatitis, as an aphrodisiac, and for mouth ulcers. The leaves serve as a stomachic. The root is utilized for treating cough, pulmonary tuberculosis, and gastritis. It is additionally utilized for the treatment of dehydrating diarrhea. The ethanolic extract is beneficial in addressing issues related to diabetes mellitus and lipid metabolism (Hemalatha and Abirami 2018).

Pollinators *Apis cerana*, *Ceratina* sp., *Thyreus* sp. (Apidae), *Curvinomia* sp., *Lasioglossum* sp., *Steganomus* sp. (Halictidae), *Heriades* sp. (Megachilidae) (Chakraborty et al. 2021).

2.14 Family Verbenaceae

2.14.1 *Lantana camara* Linn.

Common name: Common lantana

Medicinal Uses The leaves serve as an antitumoral, antimicrobial, and antihypertensive tonic, and expectorant. The fruits are beneficial for fistula, pustules, tumors, and rheumatism, while the roots are employed to combat malaria, rheumatism, and skin rashes (Rana et al. 2019).

Pollinators *Amegilla* cf. *zonata*, *Thyreus* sp., *Xylocopa aestuans* (Apidae), *Hoplonomia elliotii* (Halictidae) (Deeksha et al. 2021); *Pseudoanthidium flaviventre* (Megachilidae) (Kumar 2015).

2.14.2 *Lippia alba* (Mill.) N.E.Br. ex. Britton & P.Wilson

Common name: Bushy lippia

Medicinal Uses The leaves of *L. alba* are extensively utilized in Indian traditional medicine for their antidepressant, somatic, and analgesic attributes. Tribal individuals utilize leaf paste as a treatment for earache and dizziness induced by heat waves (Rameshkumar et al. 2022). Leaves are utilized as an infusion to alleviate states of agitation, hypertension, digestive disorders, nausea, and colds, to promote local wound healing, and as a syrup for cough and bronchitis. An infusion of the roots is utilized for hypertension, flatulence, discomfort, cold, cough, and as a sedative. Plant extracts are advantageous for alleviating stomach pain and digestive issues (Mamun-Or-Rashid et al. 2013).

Pollinators *Amegilla* cf. *zonata*, *Ceratina binghami*, *Ceratina lieftincki*, *Ceratina simillima*, *Ceratina smaragdula*, *Thyreus takaonis* (Apidae), *Hylaeus* sp. (Colletidae), *Austronomia* sp., *Hoplonomia elliotii*, *Hoplonomia westwoodi*, *Lasioglossum* sp., *Nomia curvipes* (Halictidae), *Coelioxys angulatus*, *Megachile conjuncta*, *Megachile laticeps*, *Megachile bicolor*, *Megachile fraterna*, *Megachile* sp. (Megachilidae) (Rameshkumar et al. 2022).

2.14.3 *Vitex negundo* Linn.

Common name: Chinese chaste tree

Medicinal Uses Extracts have been utilized for their anti-inflammatory, astringent, demulcent, deobstruent, tranquilizing, anti-arthritic, anthelmintic, vermifuge, antifungal, antipyretic, retinoprotective, wound-healing, desiccant, cleansing, resolvent,

Fig. 12.1 *Vitex negundo* flowers attractive to pollinators

antispasmodic, antibacterial, analgesic, antiseptic, rejuvenating, and antirheumatic properties, and are beneficial as a demulcent in dysentery, cephalgia, otalgia, colic, uropathy, wounds, and ulcers. Bark is beneficial in treating odontalgia, veminosis, and ophthalmopathy. Fruit possesses nervine, cephalic, aphrodisiac, emmenagogue, and vermifuge properties. The seeds are advised for managing premature ejaculation, alleviating swellings, treating spermatorrhea, and serving as an aphrodisiac. The essential oil derived from the leaves is also efficacious in the treatment of venereal illnesses and certain syphilitic dermatological conditions. Leaf decoction is utilized in catarrhal fever accompanied by cephalalgia and diminished auditory perception. A tincture of the root bark alleviates bladder irritation and rheumatism. In China, fruit is utilized for the treatment of erythematous, uncomfortable, and edematous eyes, as well as headaches and arthritic joints (Bano et al. 2015) (Fig. 12.1).

Pollinators *Apis dorsata*, *Apis florea*, *Ceratina smaragdula*, *Ceratina* sp., *Tetragonula iridipennis* (Apidae) (Kumar et al. 2017); *Bombus trifasciatus* (Apidae), *Andrena flavipes* (Andrenidae) (Khan et al. 2021); *Anthidiellum latipes*, *Coelioxys confusa*, *Heriades* sp., *Megachile disjuncta*, *Megachile lanata* (Megachilidae) (Kumar 2015); *Megachile bicolor*, *Megachile conjuncta*, *Megachile hera*, *Megachile lerma*, *Megachile umbripennis* (Megachilidae) (Gupta 1993); *Coelioxys angulatus* (Megachilidae) (Rameshkumar et al. 2022).

2.15 Family Zygophyllaceae

2.15.1 *Tribulus terrestris* (Linn.)

Common name: Gokharu/puncture vine

Medicinal Uses This botanical extract has been utilized by males as a dietary supplement since the 1980s, on the assumption that it elevates testosterone levels for bodybuilding and sexual enhancement. A controlled study has not consistently demonstrated a change in testosterone levels; it is not safe, and prescription medications may interact negatively with it (Pokrywka et al. 2014).

Pollinators *Ceratina* sp., *Thyreus* sp. (Apidae), *Lasioglossum* sp., *Steganomus* sp. (Halictidae) (Chakraborty et al. 2021).

3 Discussion

Megachilid bees are major pollinators of Asteraceae, Fabaceae, Lamiaceae, and Verbenaceae, while halictids are major pollinators of Asteraceae and Euphorbiaceae. Apidae bees are mostly generalists and major pollinators of most families. Several plants, mainly within the families Fabaceae and Verbenaceae, bear flowers that need tripping by insects to release anthers. These plants solely rely on insects, especially bees, for pollination. Several other plants bear flowers with poricidal anthers. These anthers need vibration for the release of pollens. Buzz-pollinating bees can only pollinate such flowers as the blue-banded bees (*Amegilla* spp.), large carpenter bees (*Xylocopa* spp.), alkali bees (*Nomia* spp.), and leafcutter bees (*Megachile* spp.). The study of Apidae and Megachilidae floral associations has been studied extensively in India. The knowledge on associations of halictids, colletids, and andrenids is lacking (Fig. 12.2). Apidae genera like *Amegilla*, *Braunsapis*, *Thyreus*, and *Xylocopa* include species that are specialist foragers.

In contrast, the honey bee genus *Apis* and small carpenter bee genus *Ceratina* are generalist foragers. Within Megachilidae, most genera within the tribe Anthidiini, viz., *Anthidium*, *Anthidiellum*, *Euaspis*, etc., and some species of *Coelioxys*, *Lithurgus*, *and Megachile*, are specialists. Although many halictide genera, viz., *Lasioglossum*, *Hoplonomia*, and *Curvinomia*, are known to be generalist foragers on plants, their plant associations in India are understudied. Same goes for colletid and andrenid genera, *Colletes*, *Hyleaus*, and *Andrena.* Several genera like *Thyreus* (Apidae), *Coelioxys*, *Euaspis* (Megachilidae), *Sphecodes*, and *Systropha* (Halictidae) are cleptoparasites of other bees. Although they do not actively collect pollen, they cause accidental pollination via adherence of pollen to body hairs while

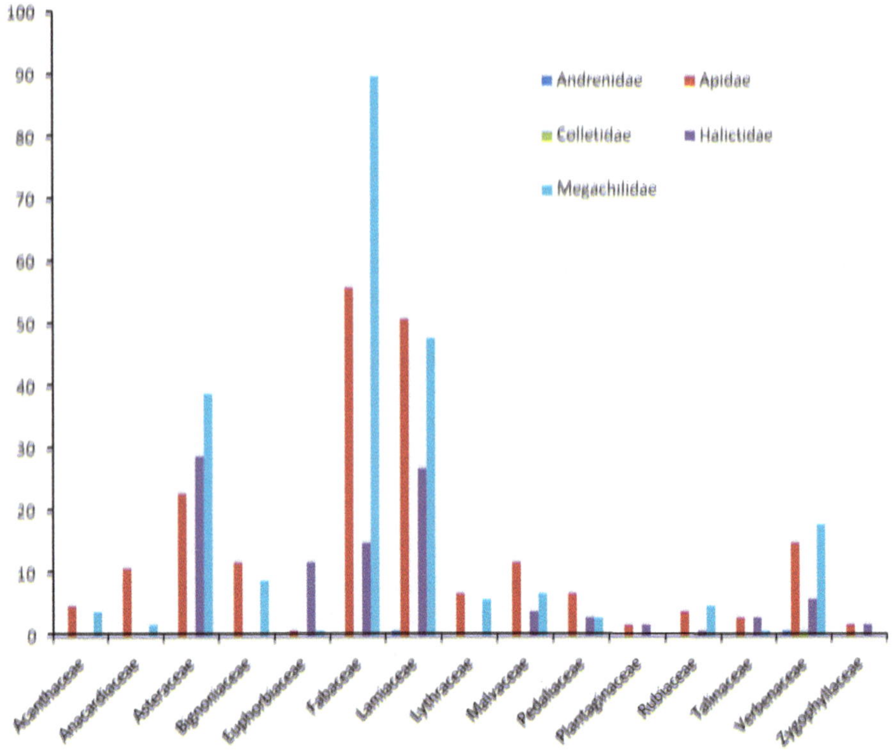

Fig. 12.2 Association of pollinators belonging to different families of bees and medicinal plants

foraging for nectar. Therefore, along with medicinal plants' phenology and floral biology, their associations with pollinators should also be studied.

Acknowledgments The authors are thankful to the Director, Zoological Survey of India, Kolkata, for providing facilities.

References

Abrol DP (2009) Plant–Pollinator interactions in the context of climate change– an endangered mutualism. J Palynol 45:1–25

Abrol DP, Shankar U, Chatterjee D (2017) Foraging rhythm of bees in relation to flowering of sweet basil, *Ocimum basilicum* L. Curr Sci 113(12):2359–2362

Al=Snafi AE, Khadem HS, Al-Saedy HA, Alqahtani AM, Batiha GE, Abolfazi J (2021) A review of *Medicago sativa*: a potential medicinal plant. Int J Biol Pharm Sci Arch 1(2):22–33

Ali H, Nguta J, Ole-Mapenay I, Musila F, Omambia V, Matara D (2021) Ethnopharmacological uses, biological activities, chemistry and toxicological aspects of *Ocimum americanum* var. *americanum* (Lamiaceae). J Phytopharmacol 10(1):56–60

Al-Muqarrabun LMR, Ahmat N, Ruzaina SAS, Ismai NH, Sahidin I (2013) Medicinal uses, phytochemistry and pharmacology of *Pongamia pinnata* (L.) Pierre: a review. J Ethnopharmacol 150(2):395–420

Al-Snafi AE (2016) The contents of pharmacology of *Crotalaria juncea*– a review. IOSR J Pharm 6(6):77–86

Aluri RJS, Reddy CS (1989) Pollination ecology of *Hyptis suaveolens* (Lamiaceae). Proc Indian Natl Sci Acad B55(5&6):411–416

Aminian AR, Mohebbati R, Boskabady MH (2022) The effect of *Ocimum basilicum* L. and its main ingredients on respiratory disorders: an experimental, preclinical, and clinical review. Front Pharmacol 12:805391

Anilakumar KR, Pal A, Khanum F, Bawa AS (2010) Nutritional, medicinal and industrial uses of Sesame (*Sesamum indicum* L.) seeds– an overview. Agric Conspec Sci 75(4):159–168

Arnan X, Escolà A, Rodrigo A, Bosch J (2014) Female reproductive success in gynodioecious *Thymus vulgaris*: pollen versus nutrient limitation and pollinator foraging behavior. Bot J Linn Soc 175(3):395–408

Auguste PM, Chantale D, Dounia Fohouo FNT (2021) Impact of Hymenoptera on *Abelmoschus Esculentus* (L.) Moench, 1794 (Malvaceae) seed yields at Bilone (Obala, Cameroon). Int J Res Stud Agric Sci 7(8):1–12

Bagri S, Savita (2019) Medicinal uses of Lavender Oil (*Lavandula latifolia*) and its effect on human health. Plant Arch 19(2):1163–1166

Bano U, Jabin A, Ahmed A, Siddiqui MA (2015) Therapeutic uses of *Vitex Negundo*. World J Pharm Res 4(12):589–606

Boldrin PK, Resende FA, Höhne APO, de Camargo MS, Espanha LG, Nogueira CH, Melo MSF, Vilegas W, Varanda EA (2013) Estrogenic and mutagenic activities of *Crotalaria pallida* measured by recombinant yeast assay and Ames test. BMC Complement Altern Med 13:216

Bora KS, Sharma A (2011) Phytochemical and pharmacological potential of *Medicago sativa*: a review. Pharm Biol 49(2):211–220

Chakraborty P, Mukherjee PA, Laha S, Gupta SK (2021) The influence of floral traits on insect foraging behaviour on medicinal plants in an urban garden of eastern India. J Trop Ecol 37:200–207

Chanchal DK, Alok S, Kumar M, Bijauliya RK, Rashi S, Gupta S (2018) A brief review on *Abelmoschus esculentus* Linn. Okra. Int J Pharm Sci Res 9(1):58–66

Chaudhari HC, Pati KP (2022) A review on medicinal importance of *Tridax Procumbens* Linn. Res Rev Pharm Pharm Sci 11(2):1–20

Dalwadi PP, Patel JL, Patani PV (2014) *Tephrosia purpurea* Linn (Sharpunkha, Wild Indigo): a review on phytochemistry and pharmacological studies. Indian J Pharm Biol Res 2(1):108–121

Das SN, Patro VJ, Dinda SC (2012) A review: ethnobotanical survey of genus *Leucas*. Pharmacogn Rev 6(12):100–106

Dauqan EMA, Abdullah A (2017) Medicinal and functional values of Thyme (*Thymus vulgaris* L.) herb. J Appl Biol Biotechnol 5(2):17–22

Deeksha MG, Guleria N, Khan MS (2021) Evaluating the association of pollinators' diversity with scrubland weed flora. J Entomol Zool Stud 9(2):663–669

Dwivedi S, Kohli S (2012) Folk–lore uses of *Guizotia abyssinica* (L.F) Cass. among tribal and rural people of Madhya Pradesh. Int J Pharm Teach Pract 3(4):434–437

Ekanayake EMD, Karunaratne WAIP (2022) Native Sri Lankan bees enhance pollination of sesame, Sesamum indicum. Ceylon J Sci 51(1):29–36

Girishkumar E, Sasikala K, Anwar NP (2016) An evaluation of pollinators of the medicinal plant– *Helicteres isora* (Sterculiaceae). J Glob Biosci 5(3):3823–3827

Govindappa M, Bharath N, Shruthi HB, Sadananda TS, Sharanappa P (2011) Antimicrobial, antioxidant and *in vitro* anti–inflammatory activity and phytochemical screening of *Crotalaria pallida* Aiton. Afr J Pharm Pharmacol 5(21):2359–2371

Guo S, Ge Y, Na Jom K (2017) A review of phytochemistry, metabolite changes, and medicinal uses of the common sunflower seed and sprouts (*Helianthus annuus* L.). Chem Cent J 11:95

Gupta RK (1993) Taxonomic studies on the Megachilidae of North–Western India. Scientific Publishers (India), Jodhpur, pp 1–294

Gupta RK (2021) Pharmacology and phytochemicals activities of *Cajanus cajan*: review. J Emerg Technol Innov Res 8(11):307–319

Haedo JP, Martínez LC, Graffigna S, Marrero HJ, Torretta JP (2022) Managed and wild bees contribute to alfalfa (*Medicago sativa*) pollination. Agric Ecosyst Environ 324:107711

Hemalatha K, Abirami P (2018) Pharmacognostic studies on *Talinum portulacifolium* (Forssk.) Asch. Ex Schweinf. Asian J Pharmacol Clin Res 11(10):470–473

Herrera CM (1989) Pollinator abundance, morphology, and flower visitation rate: analysis of the "quantity" component in a plant–pollinator system. Oecologia 80:241–248

Islam M, Hossain M, Hossen F, Mukharjee SK, Sultana N, Paul SC (2018) Evaluation of antioxidant and antibacterial activities of *Crotalaria pallida* stem extract. Clin Phytoscience 4:8

Islam M, Jannat T, Kuddus MR, Rashid MA, Haque MR (2019) In vitro and in vivo evaluation of pharmacological potentials of *Campsis radicans* L. Clin Phytoscience 5:42

James O, Friday ET (2010) Phytochemical composition, bioactivity and wound healing potential of *Euphorbia heterophylla* (Euphorbiaceae) leaf extract. Int J Pharm Biol Res 1(1):54–63

Jeyachandran M, Kumar TA, Gandhimathi S (2013) Phytochemical investigation of ethnomedicinal *Spermacoce ocymoides* roots. J Pharmacogn Phytochem 2(3):86–88

Jonathan KH, Raju AJS, Branham KS, Devi DS (2009) Interactions between insect pollinators and the ornamental tree, *Tecoma stans* (L.). J Threat Taxa 1(2):126–127

Kachhawa G, Charan S, Meena R (2021) Natural insecticide plants: *Tephrosoria Purpurea* (L.) Pers. and *Trichodesma amplexicaule* Roth in ecological relationship with hymenopteran pollinators in the Thar desert of Rajasthan, India. Int J Bot Stud 6(3):761–766

Kandasamy V, Balasundaram U (2021) *Caesalpinia bonduc* (L.) Roxb. as a promising source of pharmacological compounds to treat Poly Cystic Ovary Syndrome (PCOS): a review. J Ethnopharmacol 279:114375

Karunaratne WAIP, Edirisinghe JP, Gunatilleke CVS (2005) Floral relationships of bees in selected areas of Sri Lanka. Ceylon J Sci (Biol Sci) 34:27–45

Kevan PG (1999) Pollinators as bioindicators of the state of the environment: species, activity and diversity. Agric Ecosyst Environ 74:373–393

Khan S, Kumari P, Wani IA, Verma S (2021) Pollination biology and breeding system in *Vitex negundo* L. (Lamiaceae), an important medicinal plant. Int J Plant Reprod Biol 13(1):77–82

Khansari M, Shakiba Y, Mahmoudi M (2009) Chronic inflammation and oxidative stress as a major cause of age–related diseases and cancer. Recent Patents Inflamm Allergy Drug Discov 3(1):73–80

Kiruthiga R, Rakkimuth R, Aravinthan KM (2014) Antibacterial activity of *Crotalaria pallida* Aiton. (Fabaceae). Indian J Pharm Biol Res 2(1):82–85

Klein AM, Vaissiere B, Cane JH, Steffan–Dewenter I, Cunningham SA, Kremen C, Tscharntke T (2007) Importance of crop pollinators in changing landscapes for world crops. Proc R Soc London Ser B: Biol Sci 274:303–313

Kumar V (2015) Taxonomic studies on leafcutter bees (Hymenoptera: Megachilidae) of Karnataka (thesis). University of Agriculture, Bangalore, p 160

Kumar R, Srivastava P (2021) Abundance and diversity of pollinator fauna of sunflower (*Helianthus annuus* L.) at honey bee research and training center, Pantnagar, Uttarakhand. Pharma Innov J 10(10):845–848

Kumar KPS, Bhowmik D, Chiranjib, Tiwari P, Kharel R (2010) Indian traditional herbs *Adhatoda vasica* and its medicinal application. J Chem Pharm Res 2(1):240–245

Kumar BD, Deepika DS, Raju AJS (2017) Reproductive ecology of the semi–evergreen tree *Vitex negundo* (Lamiaceae). Phytol Balcan 23(1):39–53

Li R, Tang G, Liu X, Li J, Wang D, Ji S (2020) An ethnopharmacological review of *Hyptis suaveolens* (L.) Poit. Trop J Pharm Res 19(7):1541–1550

Mamun-Or-Rashid ANM, Sen MK, Jamal MAHM, Nasrin SA (2013) Comprehensive ethnopharmacological review on *Lippia Alba* M. Int J Biomed Mater Res 1(1):14–20

Mazid M, Khan TA, Mohammad F (2012) Medicinal plants of rural India: a review of use by Indian folks. Indo Global J Pharm 2(3):286–304

Miraj SS, Parveen N, Chaudhari SA (2020) In the back drop of overuse of synthetic drugs, can botanicals be one of the answers: a pilot study on the medicinal use of *Chlorophytum* and *Curcuma* by tribals of Central India. Curr Tradit Med 6:75–83

Moke EG, Mordi JC, Umukoro EK (2020) Effects of methanol leaf extract of *Cuphea hyssopifolia* Kunth on liver enzymes activity and antioxidant indices of paracetamol⍰induced hepatotoxicity in Wistar Rats. Afr J Biomed Res 23:123–126

Muhammad W, Saeed S, Ahmad A, Ishfaq MY, Anjum MM (2021) Pollinator community of Sunflower (*Helianthus annus* L.) and its role in crop reproductive success. Asian J Agric Biol 2: 202007398

Ollerton J, Winfree R, Tarrant S (2011) How many flowering plants are pollinated by animals? Oikos 120:321–326

Oyebode O, Kandala NB, Chilton PJ, Lilford RJ (2016) Use of traditional medicine in middle–income countries: a WHO–SAGE study. Health Policy Plan 31:984–991

Pandey DD, Jain AP, Kumar A (2018) *Caesalpinia bonducella*: A pharmacological important plant. J Pharm Innov 7(12):190–193

Patel JD, Pastagia JJ (2021) Insect pollinator's diversity in the Dangs district of Gujarat. Pharma Innov J SP–10(5):603–606

Pawar A, Vyawahare NS (2017) Phytopharmacology of *Abelmoschus moschatus* Medik.: a review. Int J Green Pharm 11(4):648–653

Peniwidiyanti WIF, Rinandio DS, Hutabarat PWK, Hariri MR, Setyanti D, Saripudi (2020) The selection of ornamental plant for landscape design of pollination garden at Bogor Botanic Gardens. Jurnal Biodjati 5(2):223–235

Perera RASN, Karunaratne WAIP (2019) Floral visits of the wild bee, *Lithurgus atratus*, impact yield and seed germinability of Okra, *Abelmoschus esculentus in* Sri Lanka. J Pollinat Ecol 25(1):1–6

Pokrywka A, Obmiński Z, Malczewska-Lenczowska J, Fijałek Z, Turek-Lepa E, Grucza R (2014) Insights into supplements with *Tribulus terrestris* used by athletes. J Hum Kinet 41:99–105

Prasad ER, Sunojkumar P (2014) Pollination biology of medicinally important plant *Leucas aspera*. J Entomol Zool Stud 2(4):341–344

Raju AJS (1989) Reproductive ecology of *Ocimum americanum* L. and *O. basilicum* L. (Lamiaceae) in India. Plant Species Biol 4:107–116

Raju AJS, Radhakrishnan J (2017) Contribution to the knowledge of three Indian *Spermacoce* L. (Rubiaceae) and some preliminary information about their pollination ecology. An Biol 39: 111–126

Raju AJS, Rao SP (2006) Explosive pollen release and pollination as a function of nectar–feeding activity of certain bees in the biodiesel plant, *Pongamia pinnata* (L.) Pierre (Fabaceae). Curr Sci 90(7):960–967

Rakesh SR, Ganesan NM, Srinivasan MR, Kumar M (2019) Studies on diversity, abundance and pollination efficiency of insect pollinators and pollination mechanism involved in sunnhemp (*Crotalaria juncea* L.). *Electr J Plant Breed* 10(2):838–851

Rameshkumar A, Sardar S, Majumder B, Dey S, Kazmi SI (2022) Observation of bee pollinators (Apoidea) on a medicinal plant *Lippia alba* (Mill.) (Verbenaceae). Curr Sci 123(5):703–707

Rana SK, Chakraborthy GS, Mazumder A, Das S, Sodhi G, Kumar K, Kumar P (2019) A review of *Lantana Camara*, a herbal medication with diverse clinical pharmacology. Int J Allied Med Sci Clin Res 7(2):435–441

Rao CB (2014) Reproductive ecology of *Leucas aspera* (Wild.) Link (Lamiaceae). Indian J Res Pharm Biotechnol 2(4):1332–1338

Reang I, Goswami S, Pala NA, Kumar M, Bussmann RW (2016) Ethnoveterinary applications of medicinal plants by traditional herbal healers in Reang Tribeo South District Tripura, India. Med Aromat Plants 5(2):100234

Salehi B, Gültekin-Özgüven M, Kirkin C, Özçelik B, Morais-Braga MFB, Carneiro JNP, Bezerra CF, da Silva TG, Coutinho HDM, Amina B, Armstrong L, Selamoglu Z, Sevindik M, Yousaf Z, Sharifi-Rad J, Muddathir AM, Devkota HP, Martorell M, Jugran AK, Cho WC, Martins N (2020) Antioxidant, antimicrobial, and anticancer effects of *Anacardium* plants: an ethnopharmacological perspective. Front Endocrinol Sec Cancer Endocrinol 11:295

Sardar S, Kasparek M, Rameshkumar A, Kazmi SI (2021) Four species of anthidiine bees (Apoidea: Megachilidae) new to India. Can Entomol 154:1–10

Shah KA, Patel MB, Patel RJ, Parmar PK (2010) *Mangifera indica* (Mango). Pharmacogn Rev 4(7): 42–48

Shahrajabian MH, Sun W, Cheng Q (2020) Chemical components and pharmacological benefits of Basil (*Ocimum basilicum*): a review. Int J Food Prop 23(1):1961–1970

Shankar U, Abrol DP, Singh A (2017) Plants for Bees *Pongamia Pinnata* (L.) Pierre. J Palynol 53: 133–137

Shivanna KR (2009) Pollination biology, breeding system and reproductive success of *Adhatoda vasica*, an important medicinal plant. Curr Sci 96(3):408–412

Singh R, Chauhan S (2010) Phenology and reproductive biology of *Caesalpinia bonducella* (L.) Flem. (Fabaceae). Int J Plant Reprod Biol 2:2

Singh A, Nagori BP, Mathur K (2013) *Tecoma stans*: an important medicinal plant. Int J Pharm Erudit 3(2):13–21

Sundararaju D (2011) Diversity of bee pollinators and flora in cashew. J Hortic Sci 6(1):52–55

Syed R, Wu Y (2018) A review article on health benefits of Pigeon pea (*Cajanus cajan* (L.) Millsp). Int J Food Nutr Res 2:15

Udikeri A, Chandra U (2019) Diversity and abundance of pollinators fauna in Phalsa *Grewia subinaequalis* D.C. Int J Curr Microbiol App Sci 8(2):2199–2204

Van Nest BN, Edge AA, Feathers MV, Worley AC, Moore D (2021) Bees provide pollination service to *Campsis radicans* (Bignoniaceae), a primarily ornithophilous trumpet flowering vine. Ecol Entomol 46:117–127

Veereshkumar Kaushik SK, Rajarajan K, Kumarang KM, Uthappa AR, Sridhar KB, Alam B, Handa AK (2020) Pollination biology of *Pongamia pinnata* (L.) Pierre: a potential biodiesel plant. Genet Resour Crop Evol 68:59–67

Xuan TD, Khanh TD (2016) Chemistry and pharmacology of *Bidens pilosa*: an overview. J Pharm Investig 46(2):91–132

Zia-Ul-Haq M, Stanković MS, Rizwan K, Feo VD (2013) *Grewia asiatica* L., a food plant with multiple uses. Molecules 18(3):2663–2682

Zoomi I, Kehri H, Akhtar O, Srivastava P, Pandey D, Narayan R (2019) Medicinal plants of India and their antimicrobial property. In: Ethnopharmacology and diversity of medicinal plant, chapter 10. Apple Academic Press, 205–227pp

Dynamics of Pest Complex, Natural Enemies and Pollinators of *Lycopersicon esculentum* Mill. in Tropics

13

Manoj Kumar Jena, Bikash Mohapatra, and Rajeeb Lochan Moharana

Abstract

Tomato, *Lycopersicon esculentum* (Mill.), is a self-pollinated crop, grown worldwide. It is widely cultivated in the tropics and is susceptible to pests. The most significant pests found on *L. esculentum* are fruit borer, serpentine leaf miner, and whitefly, while flea beetle and aphid are of minor importance. The natural enemies of pests are ladybird beetles, spiders, praying mantis, rove beetles, red ants, dragonflies, and damselflies, and a few parasitoids are found predating on pests. Carpenter bee is the most frequent floral visitor and the main pollinator of tomato. In this chapter, the dynamics of major pests as well as natural enemies and pollinators found in the tropics is reviewed.

Keywords

Dynamics · *Lycopersicon esculentum* · Natural enemies · Pest complex · Pollinators · Tropics

M. K. Jena (✉)
Department of Plant Protection, Institute of Horticultural Sciences, Warsaw University of Life Sciences, Warsaw, Poland
e-mail: manoj_jena@sggw.edu.pl

B. Mohapatra
The Cotton Corporation of India Ltd, Sirsa, Haryana, India

R. L. Moharana
Department of Seed Science and Technology, College of Agriculture, OUAT, Bhawanipatna, Odisha, India

A. K. Chakravarthy (ed.), *The Role of Pollinators in Tropical Landscapes*,
https://doi.org/10.1007/978-981-95-9139-8_13

1 Introduction

Tomatoes, *Lycopersicon esculentum* (Mill.) (Solanaceae), are one of the world's most important vegetables. The insect pests infesting tomato include the fruit borer, *Helicoverpa armigera* (Hubner); whitefly, *Bemisia tabaci* (Genn.); jassids, *Amrasca biguttulla biguttula* (Ishida); thrips, *Thrips tabaci* (Lindeman); and others. Several species of ladybird beetles, *Rodolia cardinalis (Mulsant)*, *Cryptolaemus montrouzieri (Mulsant)*, *Hippodamia variegata (Goeze)*, *Coccinella septempunctata (L.)*, and *Propylea japonica (Thunberg)*, are the dominant predators of sucking pests in tomato (Suresh 2006; Bikash 2013; Singh 2017). Tomato flowers are self-fertile, but pollinators are required for the increase in fruit setting because of the complex floral structure.

2 Pest Complex of *Lycopersicon esculentum*

2.1 Tomato Fruit Borer, *Helicoverpa armigera* (Hubner)

Helicoverpa armigera (Fig. 13.1), leads to a serious reduction in quantity as well as quality of fruits and contributes nearly 40% loss in tomato yield (Tewari and Moorthy 1984). It is a key pest of tomato (Tewari and Moorthy 1984; Gravena 1999; Chaudhuri et al. 2001; Rudenko et al. 2001; Umeh et al. 2002; Reddy and Kumar 2004; Mandal 2012; Sam et al. 2014; Goudia et al. 2023a) and is able to infect the tomato at the time of production (Juma 2015). It has achieved the key pest status because of high polyphagy, mobility, facultative diapause, high fecundity, propensity to develop insecticide resistance, and larval feeding (Wakil et al. 2010). The larva is the damaging stage and is able to infect more than 360 crop plant species, including important cultivated crops worldwide (Duraimurugan and Regupathy 2005a, b). The larva feeds on the floral buds and fruits of tomato

Fig. 13.1 Tomato fruit borer

(Capinera 2000), floral buds and fruits of okra (Dhawan and Sidhu 1984), foliage and pods of chickpea (Atwal and Dhaliwal 1976; Mathur and Upadhyay 1979), and ear heads of wheat (Doval et al. 1972). The larvae may feed on volunteer plants and a variety of weeds before migrating to tomato and pepper plants once they flower (Capinera 2000).

Recently, Goudia et al. (2023a) revealed that the *H. armigera* was recorded between the 5th (1.80 larva/plant) and the 17th SMW (15.60 larva/plant). It attained the peak during the 17th SMW (15.60 larva/plant), when the maximum, minimum, and mean temperature, morning, evening and mean RH, transformed rainfall, and bright sunshine hours (BSH) were 39.95 °C, 24.06 °C, 31.82 °C, 50.57%, 37.71%, 44.41%, 0.71 mm, and 6.76 h, respectively. The pest was prevalent throughout the reproductive cycle of the crop, causing circular or irregular holes within the fruit. Reddy and Kumar (2004) found that the *H. armigera* was active from March to April. Kharpuse and Bajpai (2006) reported a higher *H. armigera* population during the third week of February throughout (Kumar et al. 2004). According to Pandey et al. (2012), *H. armigera* population reached peak in March. The peak period of *H. armigera* larval population occurred in April (Yadav 1980). According to Lal (1996), the tomato fruit borer infestation peaked in late March. *H. armigera* was abundant from March to May with a peak population in April (Pandey et al. 1997). Mahapatra et al. (2007) observed a higher population of the pest in March and April. Ravi and Verma (1997) noticed the pest during the first week of January, and the population peaked in the middle of March. *H. armigera* populations had a significant positive correlation with maximum, minimum, and mean temperature ($r = 0.930$, $r = 0.831$, and $r = 0.898$, respectively).

Growing tall variety of African marigold, *Tagetes erecta (L.)*, as a trap crop was effective in reducing the *H. armigera* damage to fruits by 9.3 and 3.9 times, respectively, by planting rows of marigold every 10 to 20 rows of tomato (Srinivasan et al. 1994; Gajanana et al. 2006). The damage can also be reduced by keeping more spacing between plants (Gajanana et al. 2006) and bagging tomato fruit clusters with organza fabric or TNT (2,4,6-Trinitrotoluene) during organic production that reduced the incidence of borers by 99.7% (Leite et al. 2014). Application of neem oil, karanj oil, and neem seed kernel extract (NSKE) @ 5 or 10 percent was found to be effective in managing *H. armigera* (Mahadevan 1998). Moreover, Jayakumar and Gupta (1999) reported that neem seed kernel extract (NSKE) had an oviposition deterrence effect. On the NSKE-treated substrate, the hatchability of the laid eggs was adversely impacted. Castor, pongamia, and mahua seed extracts have been proven to be effective against *H. armigera* (Vyas et al. 1999; Patil et al. 1997; Gajanana et al. 2006). Root dipping of tomato seedlings in imidacloprid before planting could reduce the infestation of *H. armigera* (Gajanana et al. 2006). Application of Dipel @ 0.5 kg ha^{-1} (Krishnaiah et al. 1981; Kumar and Prasad 2002; Ravita et al. 2012) or Delfin and Halt @ 1.0 kg ha^{-1} (Pawar and Gunjal 1995; Chand and Rabindra 1999; Praveen et al. 2001; Ravi et al. 2008, 2012), nucleopolyhedron virus of *H. armigera, Ha*NPV @ 1.5×10^{12} POB ha^{-1}, spinosad 45 SC @ 75 g a.i. ha^{-1}, and endosulfan 35 EC @ 350 g a.i. ha^{-1}, quinolphos 25 EC @ 250 g a.i. ha^{-1}, and indoxacarb 14.5 SC @ 75 g a.i. ha^{-1} (Ravi et al. 2008; Dhaka et al.

2010; Sreekanth and Seshamahalakshmi 2012) was found to be effective in reducing the incidence. Application of profenophos @ 1000 g a.i. ha^{-1} caused reduction in fruit borer population by 65.20 percent, and lowered the percentage of fruit damage by 28.80% (Kumar and Ravi 2013).

2.2 Serpentine Leaf Miner, *Liriomyza trifolii* (Burg)

Liriomyza trifolii (Fig. 13.2), is a very destructive pest of vegetables, oilseeds, pulses, weeds, ornamentals, and medicinal plants (Wolfenbarger 1954; Oscar et al. 1986; Sharma and Sharma 1997; Meng et al. 1962). It is a key pest of tomato (Parrella et al. 1985; Oscar et al. 1986; Bagmare et al. 1995; Gravena 1999; Marcano and Issa 2000; Chaudhuri et al. 2001; Asalatha 2002; Reddy and Kumar 2004; Ashok and Reddy 2005; Durairaj 2007; Kaur et al. 2010; Mandal 2012; Sam et al. 2014; Goudia et al. 2023a). Adult females can cause stippling on foliage by puncturing it during the feeding and egg laying processes, particularly at the leaf tip and along the leaf margins. The female oviposits by puncturing the leaf tissue and lays eggs on the leaf epidermis (Parrella et al. 1985; Marcano and Issa 2000). The larva after hatching causes mining in leaves, leading to the loss of leaf mesophyll. Three to four days after oviposition, the mine becomes visible and the larva grows in size and weight. The mining pattern is irregular, though it tends to be blotchy near the mine's end. Because the shape is influenced by the plant, it is not a reliable species indicator. Frass becomes more granular at the end of the mine, which is in the case with *Liriomyza bryoniae (Kaltenbach)*. Leaf mining and stippling significantly decrease the rate of photosynthesis in a plant. Leaf fall drop is also a result of mining, which can lead to sun scalding and shading of fruit. The injury to the leaves allows microbial pathogens to enter. Even though leaf mining may decrease plant growth, tomatoes are quite agile and can withstand significant leaf damage. The yield reductions occur when on an average one to three mines per tomato leaf get infested (Levins et al. 1975; Schuster et al. 1976; Parrella et al. 1985).

Fig. 13.2 Leaf miner infested leaf

Planting trap crops like lucerne, cotton, and watermelon in tomato fields can reduce the number of *Liriomyza* adults in the field. To avoid the infestation in the following crop, infested residue from the previous crop should be removed or buried soon after harvest. Shredding crop residues before tillage kills larvae as it reduces the chance of hidden larvae within leaf mines surviving. In the area surrounding the field, the weeds should be eradicated effectively (Altieri and Letourneau 1982; Gajanana et al. 2006). Root dipping of tomato seedlings in imidacloprid and keeping more spacing could reduce the incidence (Gajanana et al. 2006). Application of 4% NSKE can control *L. trifolii* (Vikraktamath et al. 1993). Jayakumar and Uthamasamy (1997) recorded that neem oil (3%), NSKE (5%), and mahua oil (3%) caused 93.3%, 90%, and 90% mortality of *L. trifolii* larvae, respectively. Up to 25 and 45 days after transplanting, imidacloprid 17.8 SL @ 20 g a.i. ha^{-1} proved effective against leaf miner, followed by acetamiprid 20 SP @ 15 g a.i. ha^{-1} and NSKE 5%, whereas emamectin benzoate 5 SG @ 9.5 g a.i. ha^{-1} was the best against leaf miner 65 days after transplanting followed by spinosad 45 SC @ 75 g a.i. ha^{-1} and chlorpyriphos 20 EC @ 525 g a.i. ha^{-1} (Gosalwad et al. 2015). Tarate et al. (2016) observed that emamectin benzoate 5 SG @ 9.5 g a.i. ha^{-1} was the most effective toward *L. trifolii* complemented by spinosad 45 SC @ 75 g a.i. ha^{-1} and lambda cyhalothrin 5 EC @ 50 g a.i. ha^{-1}. Some of the parasitoids were reported to be effective against *L. trifolii.* For example, Eulopids, *Opius pallipes* (Wesmael), *Dacnusa sibirica (Telenga), Diglyphus isaea (Walker), Tetracampe indica* (Math), *Closterocerus insignis* (Waterson), *Closterocerus phytomyzae* (Mani), *Chtysocharis horticola* (Math), *Neochrysocharis formosa* (Westwood), *Diglyphus* sp., *Asecodes* sp., and *Chrysocharis* sp. (Mani 1971; Oscar et al. 1986), and Pteromalid (*Acroclisis melanagromyzae* Math) and braconid, *Opius* sp. (Dubey 1974; Mukharjee 1975).

Fig. 13.3 Flea beetle

2.3 Flea Beetle, *Phyllotreta* sp.

Phyllotreta sp. (Fig. 13.3), are one of the most injurious insect pests of young tomato plants (Gentile and Stoner 1968; Metcalf and Metcalf 1993) but cause minor damage in tomato during fruiting (Goudia et al. 2023a). They also attack on other vegetables, such as okra and cabbage, and oilseed crops, such as mustard and soybean (Burgess 1977; Lamb 1984; Singh and Singh 1993; Nielsen 1994; Lal 1999). They usually overwinter as adults in leaf litter, fence rows, or in the soil in fields. With the arrival of warm and sunny weather, adults become more active. Most of the flea beetles appeared to have a single generation per year (Burgess 1977; Saringer 1984; Cranshaw 2013). Altieri and Gliessman (1983) found that the lowest densities of flea beetles were found in fields that were allowed to remain weedy throughout the season, as opposed to plots with monoculture and polycultures of calciferous crops, which had flea beetle abundances of at least five times higher. Adults cause small, rounded, or irregular shot-hole foliar damage, but the larvae cause damage to tomato seedlings' roots and hypocotyl. The small first-instar larvae and the long later-instar larvae tunnel forward and backward inside the seedlings' roots and hypocotyl. The larvae, aided by their body setae, moved outward and inward within the curved roots and hypocotyls, leaving the vascular bundles and epidermis undamaged. Finally, there is seedling collapse due to larval feeding' was quite striking (Gentile and Stoner 1968; Batiste 1970; Metcalf and Metcalf 1993).

2.4 Whitefly, *Bemisia tabaci* (Genn)

Bemisia tabaci (Fig. 13.4) are economically significant pests in tomato (Yano et al. 1989; Brown et al. 1995; Arnal et al. 1998; Abdel et al. 1998; Gravena 1999; Singh and Korpraditskul 1999; Chaudhuri et al. 2001; Umeh et al. 2002; Reddy and kumar 2004; Waluniba and Ao 2004; Kharpuse 2005; Barde 2006; Lin et al. 2007; Hidayat et al. 2008; Kaur et al. 2010; Mandal 2012; Oda et al. 2012; Srinivasan et al. 2012;

Fig. 13.4 Whitefly

Sam et al. 2014; Juma 2015; Goudia et al. 2023a). It causes physical injury by nourishing on plant phloem sap and has an indirect effect by virus transmission and the growth of sooty mould (Morales 2007). They suck sap from the phloem by thrusting their stylet via the epicuticular surface, epidermis, and sponge tissue (Freeman et al. 2001). The stylet is encased in a gel-like material that suppresses plant defense mechanisms (Kaloshian and Walling 2005). The plants remain stunted, and the leaves clearly curl in upward fashion. In comparison with normal, semi-transparent veins of healthy plants, the leaves of infested plants display dark green veins. A dense whitefly outbreak can reduce plant vigour and yield, as well as trigger early wilting, leaf chlorosis, and defoliation (Summers and Estrada 1996). The fruit growth and establishment may be totally hampered in virus-infected plants (Butter and Kular 1999). They transmit more than 200 plant viruses which include tomato yellow leaf curl virus (TYLCV), tomato chlorotic mottle virus (ToCMoV), tomato crumple mottle virus (TCMV), tomato curly stunt virus (ToCSV), tomato golden mosaic virus (TGMV), tomato golden mottle virus (ToGMoV), tomato mottle virus (ToMoV), tomato yellow leaf curl Sardinia virus (TYLCSV), tomato yellow leaf curl Sardinia virus (TYLCSV), pepper yellow leaf curl virus, etc. (Morales and Jones 2004). Honeydew, which comprises sugars and amino acids, is also produced by whiteflies. Honeydew is a superior constituent for the expansion of sooty mould, which can comprise the plant's entire surface, including leaves, stems, and fruits. Heavy sooty mould growth lowers photosynthesis and causes physiological issues like irregular fruit maturation (McCollum et al. 2004; Morales 2007).

Whitefly populations and TYLCV incidence in tomato were reduced by using white polyethylene mulching, intercropping with maize, *Zea mays* (L.), and crop rotation with maize (Abd-Rabou and Simmons 2012). Muqit and Akanda (2007) investigated the effects of netting on the incidence of TYLCV in the field by raising tomato seedlings and found that netting fine (40 mesh) and coarse (10 mesh) nets, as well as insecticidal spray containing imidacloprid at 0.1% reduced disease incidence by 12 to 37% while increasing yield by 5 to 21%. Fine net (40 mesh) was more effective than coarse net (50 mesh) (10 mesh). Dimetry et al. (1996) demonstrated that the neem seed extracts could decrease the adult populations. Cubillo et al. (1999) revealed that neem derivatives had an insecticide impact on the whitefly. Up to 25 and 45 days after transplanting, imidacloprid 17.8 SL @ 20 g a.i. ha^{-1} was the best against white fly, followed by acetamiprid 20 SP @ 15 g a.i. ha^{-1} and NSKE 5% (Gosalwad et al. 2015). The most well-known whitefly parasitoids found worldwide are in the genera *Encarsia* and *Eretmocerus* of the order Hymenoptera (Shah et al. 2015).

2.5 Tomato Aphid, *Aphis gossypii (Glover)*

,*Aphis gossypii* may attack the tomato crop during early crop stages (Chaudhuri et al. 2001; Umeh et al. 2002; Reddy and kumar 2004; Mandal 2012; Oda et al. 2012; Sam et al. 2014; Juma 2015). It is a minor pest infesting tomato (Kaur et al. 2010).

Intercropping tomato and maize restricted aphid densities while increasing predatory spiders in tomatoes (Umeh and Onukwu 2005).

3 Natural Enemies of *Lycopersicon esculentum*

3.1 Coccinellid

According to Goudia et al. (2023a), a group of predatory coccinellid beetles (Figs. 13.5, 13.6, 13.7, and 13.8), *C. septempunctata*, *Cheilomenes sexmaculatus (Fabricius)*, and *Coccinella transversalis (Fabricius)*, were noticed in the third SMW, with a population of 0.80 beetles per plant, and attained the peak at 3.40 beetles per plant during the tenth SMW, when the maximum and minimum temperature, morning and evening RH were 37.74 °C, 19.06 °C, 46.71%, and 28.29%, respectively.

Fig. 13.5 Coccinellid larva

Fig. 13.6 Coccinellid pupa

Fig. 13.7 Coccinellid adult

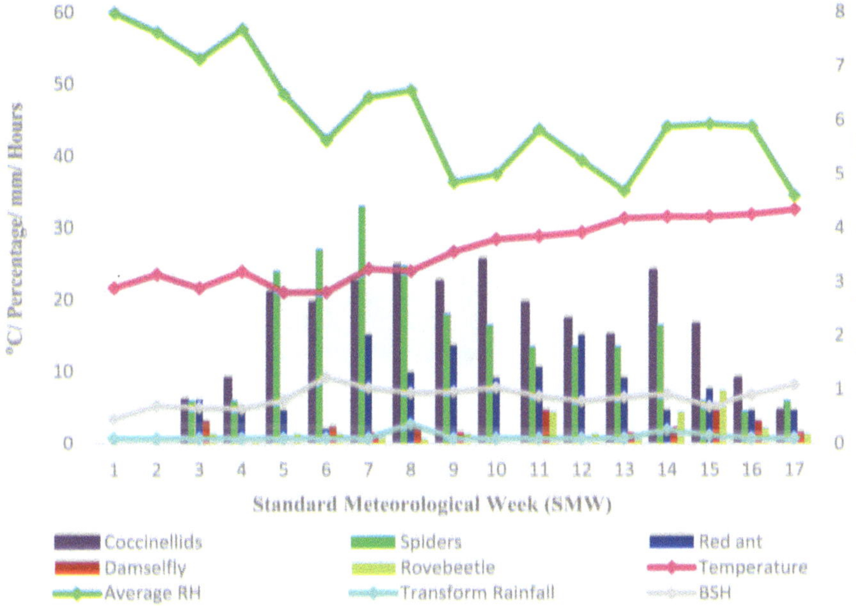

Fig. 13.8 Population fluctuation of natural enemies at different SMW during 2020–21

The coccinellids had a significant positive correlation with BSH ($r = 0.524$) and a significant negative correlation with evening RH ($r = -0.614$). Both the larvae and adults of the ladybird beetle are predators. They ate the nymphs and adult of soft-bodied insects like aphids, jassids, and whiteflies. The coccinellid population was primarily dependent on the presence of soft-bodied insects in the field as well as the RH and BSH. In step-down regression analysis, only morning and evening RH together contributed to 74.6% variation in abundance of coccinellids in tomato ecosystem, whereas regression analysis revealed all six weather parameters were responsible for 76.0% variation (Goudia et al. 2023a). Suresh (2006), Bikash (2013),

Fig. 13.9 Praying mantid

Fig. 13.10 Spider

and Singh (2017) observed that ladybird beetle *C. septumpunctata* was found in tomato field and fed on soft-bodied insects, i.e., aphid, jassid, and whitefly, on 53-day-old crop which lasted up to final fruiting stage (Fig. 13.9).

3.2 Spider

The spiders (Fig. 13.10), lynx, *Oxyopes* sp., jumping, *Phiddipus* sp., and wolf, *Marpissa* sp., were emerged in the third SMW, with a population of 0.80 spiders per plant. The population attained its peak at 4.40 spiders per plant during the eighth SMW, when the maximum and minimum temperature, morning and evening RH were 32.17 °C, 15.69 °C, 64.71%, and 34.29%, respectively (Fig. 13.8). It had a significant positive correlation with BSH ($r = 0.495$) but a significant negative correlation with maximum, minimum, and mean temperature and morning RH ($r = -0.534$, $r = -0.613$, $r = -0.589$, and $r = 0.128$, respectively). The existence of hemipteran and dipteran insects in the field was critical for the spider population. According to regression analysis, all six weather factors were accountable for 72.6%

of the variation, whereas step-down regression analysis revealed that only the minimum temperature was responsible for 37.3% of the variation in spider abundance in the tomato ecosystem (Goudia et al. 2023a). Suresh (2006) and Singh (2017) observed that spiders were predating on aphids during the vegetative growth stage (72-day-old crop).

3.3 Red Ant

The red ant (Fig. 13.11), *Solenopsis* sp., was first emerged in the third SMW with a population of 0.80 per plant. It attained the peak at 2 per plant during the seventh SMW, when the maximum and minimum temperature, morning and evening RH were 32.0 °C, 16.0 °C, 60.86%, and 35.57%, respectively (Fig. 13.8). The red ant population had a statistically significant positive correlation with BSH. In the field, red ants rely primarily on insects, spiders, arthropod eggs, and ticks. According to regression analysis, all six weather factors were accountable for 69.7% of the variation, whereas step-down regression analysis revealed that only BSH contributed to 45.7% of the variation in red ant number in the tomato ecosystem (Goudia et al. 2023a). Singh (2017) observed some red ants, *Oecophylla smaragdina* (Fabricius), in the tomato field.

3.4 Rove Beetle

The rove beetle (Fig. 13.12), *Paederus* sp., first emerged in the third SMW, with a population of 0.20 per plant. The population attained its peak at 1.00 per plant in the fifteenth SMW, when the maximum and minimum temperature, morning and evening RH were 37.74 °C, 19.06 °C, 46.71%, and 26.29%, respectively (Fig. 13.8). It had a significant positive correlation with maximum, minimum, and mean temperature ($r = 0.558$, $r = 0.491$, and $r = 0.532$, respectively). The occurrence of tiny

Fig. 13.11 Red ants

Fig. 13.12 Rove beetle

Fig. 13.13 Damselfly

insects and mites in the field was essential for its survival. In step-down regression analysis, only BSH contributed to 30.1% variation in rove beetle abundance in the tomato ecosystem, whereas regression analysis showed that all six weather parameters were responsible for 45.7% variation (Goudia et al. 2023a).

3.5 Damselfly

The damselfly (Fig. 13.13) first emerged in the third SMW, with a population of 0.40 per plant. The population attained its peak at 0.60 per plant in the eleventh SMW, when the maximum and minimum temperature, morning and evening RH were 37.74 °C, 19.06 °C, 46.71%, and 26.29%, respectively (Fig. 13.8). The damselfly population had a non-significant positive correlation with maximum, minimum, and mean temperature, rainfall and BSH ($r = 0.331$, $r = 0.329$, $r = 0.337$, $r = 0.230$ and $r = 0.197$, respectively) but a non-significant negative correlation with morning, evening and mean RH ($r = -0.263$, $r = -0.073$ and $r = -0.196$, respectively).

Its population was prevalent on the existence of small flying insects in the field. It had no direct effect on weather parameters in the field population. In step-down regression analysis, only BSH contributed to 31.2% variation in damselfly abundance in the tomato ecosystem, whereas regression analysis showed that all six weather parameters accounted for 34.2% variation (Goudia et al. 2023a). Singh (2017) observed damselflies, *Phanerotoma* sp., in the tomato field.

4 Pollinators of *Lycopersicon esculentum*

Honey bees are important pollinators in many of the cross-pollinated crops. They are commonly used to improve pollination, fruit set, and crop yield in a variety of vegetables (Jena et al. 2022; Jena and Sahoo 2022). The population of controlled honey bees has been declining as a consequence of extrajudicial chemical use and climate change (Stanghellini 1997). In Asia, ten species of honeybees (Fig. 13.14) are generally found and engaged for pollination in agro-climatic regions. These species are little bee, *Apis florea* (Fab.), European bee, *Apis mellifera* (Lin.); dammer bee, *Trigona irridipennis* (Smith) (sting less bee); rock bee, *Apis dorsata* (Fab.); Indian bee, *Apis cerana indica* (Fab.), *Apis laboriosa* (Smith) (the largest bee species in the world); red bee, *Apis koschevnikovi* (Enderlein), *Apis andreniformis* (Smith) (world's smallest bee species); Malaysian bee, *Apis nuluensis* (Lin.); and black bee, *Apis nigrocinta* (Smith) (Sakagami et al. 1980). Some non-*Apis* species are also reared for enhancing crop pollination.

Pollination performed by bees generates a considerable increase in fruit and seed set, even in cultures where self-pollination occurs, such as tomato (Silva-Neto et al. 2013; Bartelli et al. 2014, Depra et al. 2014; Santos et al. 2014; Gaglianone et al. 2015). Milking occurs when the bee milks the poricidal anthers with its mandibles, expelling pollen grains. Buzzing or buzz pollination is recognized as the most relevant behaviour for the pollination of poricidal anthers, for native plants and cultivars (Buchmann 1983; Proenca 1992; Harter et al. 2002; De Luca and Vallejo-

Fig. 13.14 Honey bee

Fig. 13.15 Carpenter bee

Marin 2013; Pinheiro et al. 2014), and consists of the vibration of the bees' thorax muscles, shaking the anthers and releasing pollen grains in large quantities.

Tomatoes can be grown in open areas or greenhouses. Although the plants are self-fertile, they have a high dependence on pollinators, to increase seed and fruit set and improve fruit quality. This is because the bee visits to the flowers and improves pollination by increasing pollen deposition on the stigma (Greenleaf and Kremen 2006).

For crops that involve buzz pollination, honey bees are inefficient pollinators because they contract flight muscles steadily than carpenter bees (Averill et al. 2018). Tomato is a major crop pollinated by carpenter bees in a regulated environment. In comparison to non-carpenter bee-pollinated greenhouse tomatoes, carpenter bee-pollinated flowers generate large fruits with higher seed content (Dogterom et al. 1998). Honey bees are less efficient than carpenter bees (Free and Butler 1959; Goudia et al. 2023a, b). Compared to tomato crops pollinated by honey bees, the carpenter bee pollinated fruits had higher number of seeds and bulkier weight (Fig. 13.15) (Yankit et al. 2018).

4.1 Carpenter Bee

The flower-visiting Carpenter bees, *Xylocopa* sp. (Fig. 13.15), emerged in the third SMW, with a population of 0.40 per plot, and peaked at 0.80 per plot in the seventh SMW (Fig. 13.8). It was the most abundant during the peak of blooming. It had a significant negative correlation with maximum, minimum, and mean temperatures ($r = -0.737$, $r = -0.769$, and $r = -0.770$, respectively). In regression analysis, all six weather factors were accountable for 79.9% of the variation, whereas step-down regression analysis revealed that only the minimum temperature contributed to 59.2% of the difference in carpenter bee abundance in the tomato ecosystem (Goudia et al. 2023a).

5 Conclusion

The major pests recorded on tomato were tomato fruit borer, serpentine leaf miner, and whitefly, while aphids and flea beetles were of minor significance. Tomato field was abundant in various natural enemies, ladybird beetles, spiders, praying mantis, rove beetles, and odonators. Among pollinators, carpenter bee was the most frequent floral visitor and considered the main pollinator of tomato.

Acknowledgments The authors are highly grateful to researchers and scientists for providing the source materials provided in the figures.

References

Abdel ML, Hegab MF, Hegazy GM, Kamel MH (1998) Association of certain weather factors with population dynamics of whitefly *Bemisia tabaci* Genn. on tomato plants. Ann Agric Sci Cait 10(1):161–176

Abd-Rabou S, Simmons AM (2012) Some cultural strategies to help manage, *Bemisia tabaci* (Hemiptera: Aleyrodidae) and whitefly-transmitted viruses in vegetable crops. Afr Entomol 20(2):371–379

Altieri MA, Gliessman SR (1983) Effects of plant diversity on the density and herbivory of the flea beetle, *Phyllotreta cruciferae* Goeze, in California collard (*Brassica oleracea*) cropping systems. Crop Prot 2(4):497–501

Altieri MA, Letourneau DK (1982) Vegetation management and biological control in agroecosystems. Crop Prot 1:405–430

Arnal E, Debrot E, Marcano RM, Manlagne A (1998) Population fluctuation of whiteflies and its relation to tomato yellow mosaic in one location in Venezuela. Fitopathol Venezoloana 6(1): 21–26

Asalatha R (2002) Seasonal activity and bio-efficacy of some ecofriendly insecticides against the serpentine leaf miner *Liriomyza trifolii*. M.Sc. (Ag.) Thesis, J.N.K.V.V., Jabalpur, Madhya Pradesh

Ashok Kumar CT, Reddy NA (2005) Influence of weather factors on abundance and management of Serpentine Leaf Miner, *Liriomyza trifolii* (Burgess) on tomato. Ann Plant Prot Sci 13(2): 315–318

Atwal AS, Dhaliwal GS (1976) Agriculture pest of South East Asia and their management, 8th edn. Kalyani Publishers, New Delhi, pp 227–251

Averill AL, Sylvia MM, Hahn N, Couto AV (2018) Bees (Hymenoptera: Apoidea) foraging on American cranberry in Massachusetts. Northeast Nat 25(3):502–512

Bagmare A, Sharma D, Gupta A (1995) Effect of weather parameters on the population build-up of various leaf miner species infesting different host plants. Crop Res 10(3):344–395

Barde SK (2006) Studies on seasonal incidence of pest complex of tomato and management of fruit borer (*Helicoverpa armigera* Hubner) by use of biopesticides. M.Sc. (Ag.) (Ent) Thesis submitted to J.N.K.V.V., Jabalpur, Madhya Pradesh, pp 1–45

Bartelli BF, Santos AOR, Nogueira-Ferreira FH (2014) Colony performance of *Melipona quadrifasciata* (Hymenoptera, *Meliponina*) in a greenhouse of *Lycopersicon esculentum* (Solanaceae). Sociobiol 61(1):60–67

Batiste WC (1970) Insecticidal control of the tobacco flea beetle on seedling tomatoes in Northern California. J Econ Entomol 63:1339–1341

Bikash S (2013) Studies on the pest complex of tomato (*Lycopersicon esculentum* L.). and their sustainable management. M.Sc. (Ag) Thesis. Uttar Banga Krishi Viswavidyalaya, Cooch Behar, West Bengal

Brown JK, Frohlich DR, Rosell RC (1995) The sweet-potato or silverleaf whiteflies - biotypes of *Bemisia tabaci* or a species complex. Annu Rev Entomol 40:511–534
Buchmann SL (1983) Buzz pollination in angiosperms. In: Jones CE, Little RJ (eds) Handbook of experimental pollination biology. Van Nostrand, pp 73–113
Burgess L (1977) Flea Beetles (Coleoptera: Chrysomelidae) attacking rape crops in the Canadian Prairie Provinces-1. Can Entomol 109(1):21–32
Butter NS, Kular JS (1999) Resurgence of whitefly in cotton and its management. Indian J Entomol 61(1):85–90
Capinera JL (2000) Corn Earworm: *Helicoverpa (Heliothis) zea* (Boddie) (Lepidoptera Noctuidae) UF/ IFAS EENY-145 (Retrieved from http://creatures.ifas.ufl.edu/veg/corn_earworm.html)
Chand A, Rabindra RJ (1999) Relative susceptibility of *Helicoverpa armigera* larvae to three commercial formulations of *Bacilus thuringensis*. In: National Symposium on biological control of insects in agriculture, forestry, medicine and veterinary science. Jan.21-22. Department of Zoology, Bharathiar University, Coimbature, Tamil Nadu, India
Chaudhuri N, Deb DC, Senapati SK (2001) Assessment of loss in yield caused by pest complex of tomato under terai region of West Bengal. Res Crop 2(1):71–79
Cranshaw WS (2013) Flea beetles. Insect series: home and garden factsheet no. 5.592. Colorado State University Extension, Fort Collins, Colorado, USA
Cubillo D, Sanabria G, Hilje L (1999) Eficacia de coberturas vivas para el manejo de Bemisia tabaci como vector de geminivirus, en tomate. Man Integ Plagas (Costa Rica) 51:10–20
Das BC (1998) Population dynamics of whitefly, Bemisia tabaci (Genn.) on some vegetable crops. J Entomologic Res 22(3):249–252
De Luca PA, Vallejo-Marin M (2013) What's the'buzz'about? The ecology and evolutionary significance of buzz-pollination. Curr Opin 16(4):429–435
Depra MS, Delaqua GCG, Freitas L, Gaglianone MC (2014) Pollination deficit in open-field tomato crops (*Solanum lycopersicum* L., Solanaceae) in Rio de Janeiro state, Southeast Brazil. J Pollinat Ecol 12:1–8
Dhaka SS, Singh G, Ali N, Yadav A, Yadav A (2010) Field evaluation of insecticides and bio-pesticides against *Helicoverpa armigera* on tomato. Ann Plant Prot Sci 18(1):13–16
Dhawan AK, Sidhu AS (1984) Incidence and relative abundance of different species of spotted bollworms on okra at Ludhiana, Punjab. J Res Punjab Agric Univ 21(4):533–542
Dimetry NZ, Gomaa AA, Salem AA, Abd-El-Moniem ASH (1996) Bioactivity of some formulations of neem seed extracts against the whitefly. Anz Schädlingsk Pflanzenschutz, Umweltschutz 69:140–141
Dogterom MH, Matteoni JA, Plowright RC (1998) Pollination of greenhouse tomatoes by the North American *Bombus vosnesenskii* (Hymenoptera: Apidae). J Econ Entomol 91:71–75
Doval SL, Bohra OP, Sharma SK (1972) New records of *Heliothis armigera* Hubner as a pest of wheat in India and efficacy of some insecticides against its larvae. Indian J Entomol 34:72
Dubey OP (1974) Some new and little-known Chalcidoid parasites (Hymenoptera) of leaf-mining Agromyzidae (Diptera) from India. Orient Insects 8(4):413–418
Duraimurugan P, Regupathy A (2005a) Mitigation of insecticide resistance in *Helicoverpa armigera* (Hübner) (Lepidoptera:Noctuidae) by conjunctive use of trap crops, neem and *Trichogramma chilonis* ishii in cotton. Int J Zool Res, Faisalabad 1(1):53–58
Duraimurugan P, Regupathy A (2005b) Synthetic pyrethroid resistance in field strains of *Helicoverpa armigera* (Hubner) (Lepidoptera: Noctuidae) in Tamil Nadu, South India. Am J Appl Sci 2(7):1146–1149
Durairaj C (2007) Influence of abiotic factors on the incidence of serpentine leaf miner, *Liriomyza trifolii*. Indian J Plant Prot 35(2):232–234
Free JB, Butler CG (1959) Bumblebees. New Naturalist No. 40. London: Collins
Freeman TP, Buckner JS, Nelson DR, Chu CC, Henneberry TJ (2001) Stylet penetration by *Bemisia argentifolii* (Homoptera: Aleyrodidae) into host leaf tissue. Ann Entomol Soc Am 94:761–768

Gaglianone MC, Campos MJO, Franceschinelli E, Depra MS, Silva PN, Montagnana PC, Hautequestt AP, Moraes MCM, Campos LAO (2015) Planode manejo para os polinizadores do tomateiro. Funbio
Gajanana TM, Krishna Moorthy PN, Anupama HL, Raghunatha R, Prasanna Kumar GT (2006) Integrated pest and disease management in tomato: an economic analysis. Agric Econ Res Rev 19:269–280
Gentile AG, Stoner AK (1968) Damage by larvae of the tobacco flea beetle to tomato seedlings. J Econ Entomol 61(1):152–154
Giannini TC, Alves DA, Alves R, Cordeiro GD, Campbell AJ, Awade M, Bento JMS, Saraiva AM, Imperatriz-Fonseca VL (2020) Unveiling the contribution of bee pollinators to Brazilian crops with implications for bee management. Apidologie 51(3):406–421
Gosalwad SS, Toprope VN, Tikotkar AB (2015) Efficacy of insecticides against whitefly and leaf miner in tomato (*Lycopersicon esculentum* Mill). Bioinfolet 12(3A):631–634
Goudia N, Jena MK, Singh RR (2023a) Dynamics of pests, natural enemies, and pollinators of *Lycopersicon esculentum* Mill and correlation with weather parameters. Int J Environ Clim Chang 13(06):128–142
Goudia N, Singh RR, Jena MK (2023b) Evaluation of combined management options for managing pests of *Lycopersicon esculentum* Mill. Biol Forum Int J 15(3):364–369
Gravena S (1999) Integrated management of tomato pests. *XXIV* Congresso Brasilerio de olericultura Reuniaco Lalino-Americana de olericultua, pp 129–149
Greenleaf SS, Kremen C (2006) Wild bee species increase tomato production and respond differently to sur-rounding land use in Northern California. Biol Cons 133(1):81–87
Harter B, Leistikow C, Wilms W, Truylio B, Engels W (2002) Bees collecting pollen from flowers with poricidal anthers in a south Brazilian Araucaria Forest: a community study. J Apic Res 41(1–2):9–16
Hidayat P, Aidawati N, Hidayat SH (2008) Studies on genetic diversity of *Bemisia tabaci* (Hemiptera: Aleyrodididae) in Java and transmission of whitefly-transmitted tomato gemini viruses. Indian J Virol 19:77–77
Jayakumar P, Gupta GP (1999) Effect of neem seed kernal extract (NSKE) on *Helicoverpa armigera*. Pest Res J 11(1):32–36
Jayakumar P, Uthamasamy S (1997) Bio-efficacy of some synthetic insecticides and botanicals against *Liriomyza trifolii*. Indian J Entomol 59(4):347–350
Jena MK, Sahoo S (2022) Blue banded bee, *Amegilla sp.*: a potential pollinator in pollination of crop plants. AgriCos e-Newslett 03(06):4–8
Jena MK, Sahoo S, Parmar PR (2022) A brief review on biodiversity of pollinators. AgriCos e-Newslett 03(02):14–18
Juma VHO (2015) Effects of insecticide treated nets in the management of tomato pests and their impact on natural enemies and yield in Nairobi and Murang'a counties. M.Sc. Thesis submitted to Kenyatta University
Kaloshian I, Walling LL (2005) Hemipterans as plant pathogens. Annu Rev Phytopathol 43:491–521
Kaur S, Kaur S, Srinivasan R, Cheema DS, Lal T, Ghai TR, Chadha ML (2010) Monitoring of major pests on cucumber, sweet pepper and tomato under net-house conditions in Punjab. Pest Mang Hortic Ecosyst 16:50–57
Kharpuse YK (2005) Studies on seasonal incidence and role of botanical against major insect pests of tomato (*Lycopersicon esculentum* M.). M.Sc. (Ag.) (Ent.) Thesis, submitted to J.N.K.V.V., Jabalpur (M.P.), pp 1–53
Kharpuse YK, Bajpai R (2006) Seasonal incidence of major insect pests of tomato *(Lycopersicon esculentum* Mill). Indian J Trop Biodivers 14(2):178–181
Krishnaiah J, Jagan Mohan N, Prasad VG (1981) Efficacy of *Bacilus thuringensis* Ber. for the control of lepidopterous pest of vegetable crops. Entomon 6(2):87–93
Kumar B, Prasad D (2002) Evaluation of neem-based insecticides and biopesticides against *Helicoverpa armigera* infesting chickpea. Indian J Entomol 64(4):411–417

Kumar K, Ravi (2013) Bio-efficacy and residue dynamics of insecticides against fruit borer (Helicoverpa armigera Hub.) in tomato (Lycopersicon esculentum Mill.). M.Sc. Agriculture Entomology Thesis, submitted to Acharya N.G. Ranga Agricultural University
Kumar R, Razvi SMS, Ali S (2004) Seasonal and varietal variation in the population of whitefly (Bemisia tabaci Genn.) and incidence of yellow mosaic virus in urd and mungbean. Indian J Entomol 66(2):155–158
Lal OP (1996) An outbreak of pod borer, *H. armigera* (Hubner) on chickpea in Eastern Uttar Pradesh. Indian J Entomol Res 20:179–181
Lal OP (1999) An unusual high incidence of the flea beetle, *Podagrica bowringi* Baly (Coleoptera: Chrysomelidae) on okra in Uttarkashi district of Bhagirathi valley in central Himalaya (India). J Entomol Res 23(1):85–87
Lamb RJ (1984) Effects of flea beetles, *Phyllotreta* sp. (Chrysomelidae: Coleoptera), on the survival, growth, seed yield and quality of canola, rape and yellow mustard. Can Entomol 116(2):269–280
Leite GLD, Fialho A, Zanuncio JC, Junior RR, Da Costa CA (2014) Bagging tomato fruits: a viable and economical method of preventing diseases and insect damage in organic production. Florida Entomol Soc 97(1):50–60
Levins RA, Poe SL, Littell RC, Jones JP (1975) Effectiveness of a leaf miner control program for Florida tomato production. J Econ Entomol 68:772–774
Lin K, Wu K, Zhang Y, Guo Y (2007) Over wintering and population dynamics of *Bemisia tabaci* biotype B in greenhouse during the spring in northern China. Crop Prot 26(12):1831–1838
Mahadevan NR (1998) Azadirachtin to control tomato fruit borer. Insect Environ 3(4):10–14
Mahapatra SD, Aswal JS, Mishra PN (2007) Monitoring population dynamics of tomato fruit borer, *Helicoverpa armigera* Hubner moths through pheromone traps in Uttaranchal Hills. Indian J Entomol 69(2):172–173
Mandal SK (2012) Bio-efficacy of cyazypyr 10% OD, a new anthranilic diamide insecticide, against the insect pests of tomato and its impact on natural enemies and crop health. Acta Phytopathol Entomol Hung 47(2):233–249
Mani MS (1971) Some chalcidoid parasites (Hymenoptera) of leaf-mining Agromyzidae (Diptera) from India. J Nat Hist 5(5):591–598
Marcano R, Issa S (2000) Spatial and vertical distribution of *Liriomyza trifolii* on tomato. Boletin de Entomologica venezolana 8(1):115–122
Mathur YK, Upadhyay KD (1979) A textbook of entomology. Arran Publishing House, Kanpur, p 158
McCollum TG, Stoffella PJ, Powell CA, Cantliffe DJ, Hanif-Khan S (2004) Effects of silverleaf whitefly feeding on tomato fruit ripening. Postharvest Biol Technol 31:183–190
Meng HL, Chang GS, Rens Z (1962) Further studies on cotton bollworm, *Heliothis armigera*. Acta Entomol Sim 11:82
Metcalf RL, Metcalf RA (1993) Destructive and useful insects, 5th edition. McGraw-Hill Book Co., New York, pp 146–149
Morales FJ (2007) Tropical whitefly IPM project. Adv Virus Res 69:249–311
Morales FJ, Jones PG (2004) The ecology and epidemiology of whitefly-transmitted viruses in Latin America. Virus Res 100:57–65
Mukherjee MK (1975) Records of some known and descriptions of some new species of Chalcids (Hymenoptera) from India. Memorial School Entomol Agra 4:60
Muqit A, Akanda AM (2007) Management of tomato yellow leaf curl virus through netting. The Agriculturists 5(1 and 2):1–5
Nielsen GC (1994) Cabbage stem flea beetle in oil seed rape [thresholds]. SP-rapport (Denmark)
Oda M, Hanboonsong Y, Jamjanya T, Srichompoo K, Kotaki T (2012) Occurrence of insect pests in a tomato field under a pesticide-free dry season water-saving cultivation in Northeast Thailand. JARQ, Jpn Agric Res 46(1):59–64
Oscar P, Minkenberg JM, Lenteren JCV (1986) The leafminers *Liriomyza bryoniae* and *L. trifolii* (Diptera: Agromyzidae), their parasites and host plants: a review. Agricultural University Wageningen Papers, pp 86–82

Pandey AK, Singh SS, Singh R (1997) Incidence of whitefly (Bemisia tabaci) and yellow mosaic virus in relation to date of sowing in mungbean. Indian J Entomol 59(4):361–365

Pandey BM, Tripathi MK, Lakshmi V (2012) Seasonal incidence of gram pod borer, *Helicoverpa armigera* (Hub.) on chickpea in Varanasi area. J Exp Zool India 15(2):667–669

Parrella MP, Jones VP, Youngman RR, Lebeck LM (1985) Effect of leaf mining and leaf stippling of *Liriomyza* sp. on photosynthetic rates of chrysanthemum. Ann Entomol Soc Am 78:90–93

Patil S, Kotikal YK, Revanappa PDR (1997) Effects of intercropping tomatoes (*Lycopersicon esculentum* mill) on the infestation of tomato fruit borer, *Helicoverpa armigera* Hubner. Adv Agric Res India 8:141–146

Pawar VM, Gunjal SB (1995) Field evaluation of *Bacillus thuringensis* (Halt) against lepidopterous pod borers of pigeon pea. Pestology 19:27–29

Pinheiro M, Gaglianone MC, Nunes CEP, Sigrist MR, Santos I (2014) Polinizacao por abelhas. In AR Rech, K Agostini, PE Oliveira, IC Machado (Eds.), Biologia daPolinizac ao (cap. 9, p. 211). Editora Projeto Cultural. Portman, ZM, Orr MC, Griswold T (2019) a review and updated classification of pollen gathering behavior in bees (Hymenoptera, Apoidea). J Hym Res 71:171–208

Praveen PM, Dhandapani N, Kennedy JS (2001) Efficacy of *Bacillus thuringensis var kurstaki* (Berliner) formulations for the management of tomato fruit borer, *Helicoverpa armigera* (Hubner). Pestology 15(9):58–61

Proenca CEB (1992) Buzz pollination-older and more widespread than we think? J Trop Ecol 8(01): 115–120

Ravi G, Verma S (1997) Seasonal incidence of chickpea pod borer, *Helicoverpa armigera* and its larval parasitoid on chickpea crop. Indian J Ent 59(4):359–361

Ravi M, Santharam G, Sathiah N (2008) Eco-friendly management of tomato fruit borer, *Helicoverpa armigera* (Hubner). J Biopest 1(2):134–137

Ravita, Chaudhary K, Kumar S (2012) Bio-efficacy of selected isolated *Bacillus thuringiensis* against *Helicoverpa armigera*. J Cotton Res Dev 26(2):155–158

Reddy NA, Kumar CTA (2004) Insect pest of tomato, *Lycopersicon esculentum* Mill. in eastern dry zone of Karnataka. Insect Environ 10(1):40–42

Rudenko NE, Zubanov AP, Sherbinin BH, Chalenko VV (2001) Pest control by the pneumatic method. Zeschita Rastenii:10–16

Sakagami SF, Matsumura T, Ito K (1980) *Apis laboriosa* in Himalayas, the little know world largest honeybee (Hymenoptera: Apidae). Insect Matsumsura 19:47–77

Sam GA, Osekre EA, Mochiah MB, Kwoseh C (2014) Evaluation of insecticides for the management of insect pests of tomato, *Solanum Lycopersicon* L. J Biol Agric Health 4(5):49–57

Santos AOR, Bartelli BF, Nogueira-Ferreira FH (2014) Potential pollinators of tomato, *Lycopersicon esculentum* (Solanaceae), in open crops and the effect of a solitary bee in fruit set and quality. J Econ Entomol 107(3):987–994

Saringer G (1984) Summer diapause of cabbage stem flea beetle, *Psylliodes chrysocephala* L. (Col, Chrysomelidae). Zeitschrift für Angewandte Entomologie 98(1–5):50–54

Schuster DJ, Jones JP, Everett PH (1976) Effect of leaf miner control on tomato yield. Proc Florida State Hortic Soc 89:154–156

Shah MMR, Zhang S, Liu T (2015) Whitefly, host plant and parasitoid: a review on their whitefly, host plant and parasitoid: a review on their interactions. Asian J Appl Sci Eng 4:48–61

Sharma D, Sharma S (1997) Status of *Liriomyza trifolii* (Burgess) and its host plants in Jabalpur district in Madhya Pradesh. Crop Res (Hissar) 14(2):351–355

Silva-Neto CM, Lima FG, Goncalves BB, Bergamini LL, Bergamini BAR, Elias MAS, Franceschinelli EV (2013) Native bees pollinate tomato flowers and increase fruit production. J Pollinat Ecol 11:41–45

Singh S (2017) Studies on insect pests of tomato with special reference to seasonal incidence and management of serpentine leaf miner *Liriomyza trifolii* Burgess. M.Sc. (Ag.) Thesis Indira Gandhi Krishi Vishwavidyalaya, Raipur (Chhattisgarh), India

Singh H, Korpraditskul V (1999) Evaluation of some plant extracts against aphids, key vectors of chilly-virus complex. *Azadirachta indica* A. Juss:139–146

Singh OP, Singh KJ (1993) New record of flea beetle, *Chaetocnema sp.* on soybean in Madhya Pradesh. J Insect Sci 6(1):128–129

Sreekanth M, Seshamahalakshmi M (2012) Studies on relative toxicity of biopesticides to *Helicoverpa armigera* (Hubner) and *Maruca vitrata* (Geyer) on pigeonpea (*Cajanus cajan* L.). J Biopest 5(2):191–195

Srinivasan K, Krishna Moorthy PN, Raviprasad TN (1994) African marigold as a trap crop for the management of the fruit borer *Helicoverpa armigera* on tomato. Int J Pest Mang 40:56–63

Srinivasan R, Riley D, Diffie S, Sparks A, Adkins S (2012) Whitefly population dynamics and evaluation of whitefly transmitted tomato yellow leaf curl virus (TYLCV)-resistant tomato genotypes as whitefly and TYLCV reservoirs. J Econ Entomol 105(4):1447–1456

Stanghellini MS (1997) The effects of honey bee and bumble-bees pollination of fruit set and abortion of cucumber and watermelon. Am Bee J 137:386–391

Summers CG, Estrada D (1996) Chlorotic streak of bell pepper: a new toxicogenic disorder induced by feeding of the silverleaf whitefly, *Bemisia argentifolii*. Plant Dis 80:822

Suresh Y (2006) Study on the occurrence of pest complex of tomato (*Lycopersicon esculentum* Mill) and their management. M.Sc (Ag.) thesis. Acharya N. G. Ranga Agricultural University, Hyderabad, Andhra Pradesh, India

Tarate R, Mohite P, Dhumal S (2016) Efficacy of new molecules of insecticides against leaf miner infesting tomato. Indian J Appl Res 6(2):456–458

Tewari GC, Moorthy PNK (1984) Yield loss in tomato caused by fruit borer. Indian J Agric Sci 54(4):341–343

Umeh VC, Onukwu D (2005) Development of environmentally-friendly tomato insect pest control options under tropical conditions. J Veg Sci 11:73–84

Umeh VC, Kuku FO, Nawanguma EL, Adebayo OS (2002) A survey of the insect pests and farmers' practices in the cropping of tomato in Nigeria. International Information System for the Agricultural Science and Technology. Directory of Open Access Journals- Lund University Libraries (FAO)

Vikraktamath CA, Tiwari GC, Srinivasan K, Gupta M (1993) American serpentine leaf miner is a new threat to crops. Indian Fmg 43:10–12

Vyas BN, Ganesan S, Raman K, Godrej NB, Mistry KB, Singh PP, Saxena RC (1999) Effect of three plant extracts on growth and development of three noctuid pests. Azadirachta indica A. Juss. Singh RP and Saxena RC edition. Oxford and IBH Publishing Co. 17(5):103–109

Wakil W, Ghazanfar MU, Kwon YJ, Qayyum MA, Nasir F (2010) Distribution of *Helicoverpa armigera* Hubner (Lepidoptera: Noctuidae) in tomato fields and its relationship to weather factors. Entomol Res 40:290–297

Waluniba, Ao A (2004) Incidence of insect pest complex on cauliflower (Brassica oleracea var. botrytis L.) in Medziphema, Nagaland. Indian J Entomol 66(4):312–315

Wolfenbarger DO (1954) Potato yields associated with control of aphids and the serpentine leaf miner. Florida Entomol 37:7–12

Yadav DN (1980) Studies on the natural enemies of *Helicoverpa armigera* (Hubner) and its biological control using an eggparasite, *Trichogramma australicum* Giranit. (Hymenoptera, Trichogramatidae). G.A.U. Res J 6(1):62–63

Yankit P, Rana K, Sharma HK, Thakur M, Thakur RK (2018) Effect of bumble-bees pollination on quality and yield of tomato (*Solanum lycopersicum* Mill.) grown under protected conditions. Int J Curr Microbiol Appl Sci 7(01):257–263

Yano E, Lenleren JC, Rabbinge Van, Hylsparl R, Jordoan PM, Van JC, Kenteren L (1989) The parasite host relationship between *Encarsia formosa* and *Trialeurodis vaporarium*. XXX. Simulation studies of population growth of green-house white fly on tomato. Agric Univ Wageningen, Papers 89(2):55–73

Pollinators in Tropical Forests

14

B. Tambat, A. N. Sringeswara, G. N. Chaithra, and N. Deeksha Raj

Abstract

Pollinators play a vital role in sustaining biodiversity and ensuring ecological balance, particularly in forests of tropical region. They involve in reproduction of approximately 75% of global crop plants and about 94% of flowering plant species in tropical ecosystems. Among them, bees are the most significant group, responsible for around 66% of pollination in tropical regions, facilitating the reproduction of numerous tree species across various families and genera. This chapter explores the diversity of pollinators in tropical forests, their ecological and economic importance, the threats they face, and the implications for food security. Despite their critical role, over recent decades due to habitat loss, climate change, the spread of pests and diseases, and the widespread use of chemical pesticides, pollinators are declining rapidly. However, both scientific research and traditional ecological knowledge offer a range of strategies for pollinator conservation and the restoration of pollination services essential to ecosystem resilience and sustainable agriculture.

Keywords

Tropical forest · Pollination · Pollinators · Honeybees · Threats

B. Tambat (✉) · A. N. Sringeswara · N. Deeksha Raj
Department of Forestry and Environmental Science, UAS, GKVK, Bangalore, Karnataka, India

G. N. Chaithra
Department of Biotechnology, College of Agriculture, UASM, Hassan, Karnataka, India

A. K. Chakravarthy (ed.), *The Role of Pollinators in Tropical Landscapes*,
https://doi.org/10.1007/978-981-95-9139-8_14

1 Introduction

In tropical forest ecosystems, pollination plays a crucial role in plant reproduction, influencing around 95% of canopy tree species. Within the sub-canopy and understory layers, approximately 20–25% of plants depend on vertebrates—such as bats and other animals—for pollination, while the remaining species are primarily pollinated by insects. As pollinators move from flower to flower in search of food sources like nectar and pollen, they inadvertently or intentionally make contact with the reproductive structures of the plants. This interaction allows pollen from one flower to be transferred to another flower of stigma, enabling fertilization and ultimately leading to the *growth of fruit and seed.*

A wide variety of organisms—including bees, birds, bats, butterflies, moths, ants, ladybugs, and other insects—are essential pollinators (Fig. 14.1). Many food crops rely on these organisms for successful reproduction, as they cannot produce fruits without pollination. For instance, crops such as tomatoes and eggplants require wild bees to achieve proper fruit development. Bees alone are estimated to be responsible for about two-thirds of all pollination activity globally (Khalifa et al. 2021).

The Brazilian Cerrado, a highly biodiverse and threatened savanna region, is home to over 7000 plant species, about 40% of them are endemic to the region. Many of these plants have significant economic, medicinal, or ornamental value (Klink and Machado 2005). Around 86% of trees and shrubs in this ecosystem rely on animals for pollination (Gottsberger et al. 1988). Therefore, preserving pollinator populations is vital to the health of the ecosystem. A decline in pollinators can

Fig. 14.1 Pollinators diversity exits in the tropics

significantly reduce fruit production and even influencing self-pollinating species (Arroyo 1976; Aizen et al. 2009; De Deus and Oliveira 2016).

2 Pollinators in Tropical Ecosystems

Roughly 90% of flowering plants rely on biological vectors to transfer their pollen, making pollination a key ecological interaction. This relationship is mutually beneficial: plants enhance their reproductive success, and in return, animals receive rewards—typically nectar or pollen. The range of pollinators is vast and includes both wild and domesticated species. Globally, wild pollinators include around 20,000 bee species, along with other insects, bats, birds, and various vertebrates. A few species of bee have been domesticated, such as the European honeybee (*Apis mellifera*), Asian honeybee (*Apis cerana*), some bumblebees (*Bombus* spp.), and solitary bees. While bees are the most common pollinators, other insect groups such as flies, wasps, beetles, moths, ants, thrips, and even some reptiles and mammals also contribute significantly (IPBES et al. 2016a).

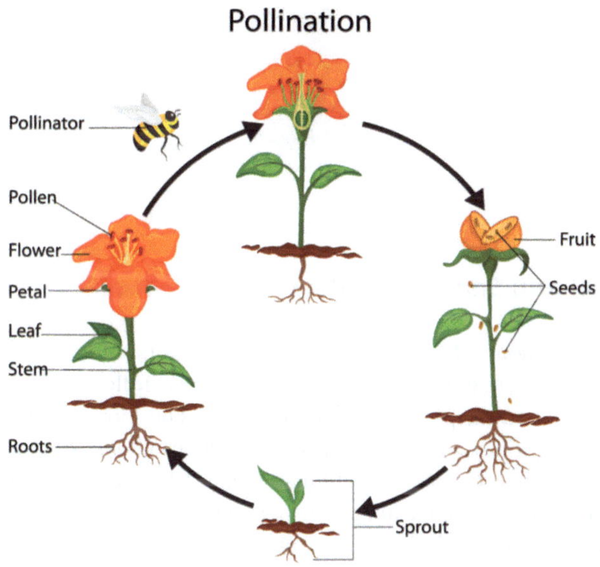

2.1 Bees

Bees are considered the most important pollinators in agricultural systems worldwide. Both domesticated and wild bees play critical roles. Notably, *Apis mellifera* and stingless bees (Meliponines) are prominent for their ecological and agricultural importance (Potts et al. 2016a, b; Slaa et al. 2006; Giannini et al. 2015, 2020). In

Brazil, a study of 23 cultivated crops identified 144 bee species as pollinators. Among them, 63 were social bees, with the genera *Trigona* and *Melipona* most frequently observed (Giannini et al. 2020). Non-bee insects like beetles (e.g., Curculionidae, Chrysomelidae) and flies (e.g., Syrphidae) also significantly contribute to pollination and may offer resilience in the face of bee population declines (Rader et al. 2016).

The Amazon is particularly rich in stingless bee species, harboring around 190 species (Pedro 2014). These bees have broad dietary preferences (Ramalho 2004; Lichtenberg et al. 2017) and are considered essential pollinators in forest ecosystems (Bawa 1990). In a study of 161 Amazonian species, 54% were pollinated by bees, followed by beetles (14%), and bats and moths (7% each) (Paz et al. 2021). Such information is vital to conserving ecosystem services and ensuring fruit tree productivity in the Amazon.

Coin of Ephesus with a bee

2.2 Social and Semi-social Bees

Social bees are among the most effective pollinators of trees, contributing to pollination in approximately 18% of studied species, followed closely by beetles (17%), flies (6%), and thrips (5%) (Ganesh 1995). The most utilized pollinators in agriculture include honeybees (*Apis* spp.), bumblebees (*Bombus* spp.), and stingless bees. Solitary bees like *Osmia* spp. are also managed for crops such as blueberries and alfalfa. The impacts of farming practices on pollinators and biodiversity vary across agroecosystems, underlining the need for pollinator-friendly management strategies (IPBES et al. 2016a).

2.3 Bats

In tropical and subtropical regions, nectar-feeding bats serve as vital pollinators and seed dispersers. In the Neotropics, over 500 plant species across 27 families rely on bats for pollination, particularly those visited by nectarivorous bats belong to the subfamily *Glossophaginae* (Altringham 1996). Bat-pollinated flowers are typically pale-colored, nocturnal, and produce large quantities of nectar. As bats feed, flowers deposit pollen onto the bat's head and upper surface of body for transfer to the next flower. Fourteen bat species, including some nectar specialists, have been identified as cactus pollinators. The tree bat (*Ardops nichollsi* subsp. *annectens*), found in forested regions of Antigua and Guadeloupe, exemplifies species reliant on intact forest habitats. In one study, *Syzygium mundagam* was found to be exclusively pollinated by bats (Ganesh 1995).

2.4 Insects and Birds

Some species show specialized traits that attract both insects and birds. For instance, *Gmelina arborea* is pollinated by both insect groups (e.g., *Xylocopa violacea*, *Amegilla cingulata*, *Mylabris pustulata*) and birds from the Passeriformes order. *Xylocopa violacea* had the highest visitation frequency and demonstrated strong compatibility with the floral structure, identifying it as a key pollinator. The species of birds that possess long bills are more effective pollinators, and *G. arborea* appears to have evolved traits favoring bird as well as insect pollination. This species supports both insects and birds as pollinators offering floral resources and play a role in enhancing biodiversity in forest and agroforestry settings (Kumar et al. 2021).

2.5 Mammals

Pollination by non-flying mammals is relatively rare in tropical forests but does occur under certain ecological conditions. In the Western Ghats, for instance, only 2 out of 89 surveyed tree species were pollinated by arboreal mammals. These cases often arise in areas where other animal pollinators like bats are absent, or where flowers become crucial food resources during times of scarcity. Squirrels like *Callosciurus pygerythrus* is also known for foraging on flowers. Such pollination systems, typically uncommon, are ecologically significant and often found in regions which are less favorable to more typical pollinators (Ganesh and Devy 2006).

2.6 Animal Pollinators

In the tropics, species dependent on animal pollinators for reproduction are comparatively more, particularly due to the complex and specialized pollination mechanisms seen in those species (Machado and Sazima 2008). A study of 161 plant species in the Amazon forests demonstrates a significant reliance on animals as pollinators rather than other means of abiotic agents wind or water for pollination (Paz et al. 2021). This implies animal-mediated pollination is vital for food security, especially in the traditional Amazonian communities.

3 Current Status and Trends in Pollinator Populations

Despite their critical role in agriculture and forest ecosystems, pollinators suffer significantly and their populations are being declining drastically in recent decades (IPBES et al. 2016b). This decrease is closely associated with reduced yields in pollinator-dependent crops. Although comprehensive data is lacking, available evidence suggests a global trend of pollinator population decline. Currently, 16.5% of animal pollinators are reported as threatened species (IUCN Red List), with the number rising to 30% for species native to islands. In the Caribbean, five nectar-feeding bat species from Cuba are listed as "least concern," while the Jamaican nectar-feeding bat (*Phyllonycteris aphylla*) is classified as "critically endangered." Unfortunately, legal protections for bats in Caribbean nation are often lacking, and due to misconceptions about their ecological role, they are sometimes actively persecuted (IPBES et al. 2016a).

The Food and Agriculture Organization (FAO) estimates that about one-third of global food production depends in some way on animal pollination, primarily by bees (Klein et al. 2007). Early estimates valued global pollination services at around US$117 billion (Costanza et al. 1997), later updated to between US$235 billion and US$577 billion per year as of 2015 (Costanza et al. 2014; IPBES et al. 2016a). In Brazil, pollination services were valued at US$12 billion annually in 2013 (Giannini et al. 2015), and in Pará, a key Amazonian state, the value was estimated at nearly US$983 million in 2016 (Borges et al. 2020). Beyond boosting yields, pollinators also improve crop quality (Giannini et al. 2015).

4 Importance of Pollinators

Pollinators provide a wide range of essential ecological, economic, and cultural services:

- The Americas hold a vast wealth of natural resources that exceed global averages in their contribution to human well-being.
- These resources, particularly pollination services, support food, water, and energy security.
- About 90% of flowering plant species rely on animals for pollination (IPBES et al. 2016b).
- This process benefits both plants and pollinators—plants increase their reproductive success, while pollinators receive nectar or pollen as food.
- Roughly 75% of global crop types and one-third of agricultural production volume depend, at least partially, on animal pollinators (IPBES et al. 2016a).
- Pollination boosts both the quantity and quality of fruits and seeds in 39 of the 57 leading global crops.
- Crops such as cocoa and coffee—important to the Caribbean economy—require wild or managed pollinators for optimal production in fruit yield.

- Pollination is vital for nutritional security, particularly through crops that provide essential vitamins and micronutrients (e.g., fruits, vegetables, nuts, seeds, and oils).
- For rural communities, honeybee products, including honey, wax, and propolis, serve as source of income.
- Stingless bees also produce goods with medicinal and cosmetic applications.
- Some pollinators, such as butterflies, have aesthetic and economic value in ecotourism and craft industries.
- Pollination supports ecosystem resilience and genetic diversity in plant populations, helping ecosystems adapt to environmental changes.
- Animal pollination sustains wildlife by enabling fruit and seed production in natural habitats.

5 Additional Benefits

Pollinators contribute to cultural and traditional practices as well:

- Bat guano is traditionally used as fertilizer in many farming systems.
- Cultural and indigenous traditions often include the protection of pollinators and their host plants.
- During droughts, communities frequently safeguard nectar-producing plants for their value.

6 Ecological Connectivity

Pollinators such as birds, butterflies, and bats migrate across island and continental landscapes. Hence, to develop operative conservation strategies, it is important to understand these animals migratory patterns. Moreover, maintaining connectivity between wild and agricultural habitats helps to sustain pollinator populations, especially during extreme climatic events. Natural and semi-natural areas can act as refuges or sources for pollinators that support crop production (IPBES et al. 2016a).

7 Pollination as an Ecosystem Service

Pollination is a foundational ecosystem service critical for agriculture, biodiversity, and human well-being. It:

- Enhances plant genetic variability, allowing ecosystems to recover from disturbances and better cope with climate change.
- Maintains biodiversity in wildlife populations by supporting plant reproduction.
- Facilitates the production of fruits and seeds, feeding both humans and wildlife.

- Underpins other ecosystem services, such as raw material supply, nutrient recycling, and cultural services.
- Crops that depend on animal pollination tend to have higher economic value than non-pollinated crops.
- The global value of pollination was estimated between US$235–577 billion in 2015 (IPBES et al. 2016b).
- Pollinator loss could raise production costs and market prices for many staple and cash crops.

8 Major Threats to Pollinators

Pollinator populations have declined sharply since the mid-twentieth century (Carson 1962; Nabhan and Buchmann 1997), posing serious risks to global food security (Potts et al. 2016a, b). Key threats include:

- Habitat loss and fragmentation: Driven primarily by agricultural expansion, deforestation reduces food and nesting sites for pollinators and restricts their movement between habitats.
- Pesticide use: Agrochemicals have sub-lethal and sometimes lethal effects on bees and other pollinators (IPBES et al. 2016b).
- Invasive species: Non-native pollinators may outcompete or displace native species, altering pollination networks.
- Exotic pollinators: These may spread invasive weeds or transmit diseases to native bees.
- Disease and pathogens: Managed bees, especially *Apis mellifera*, are vulnerable to parasites and viruses, which may spread to wild populations through movement and trade.
- Climate change: Altered temperature and precipitation patterns, plus extreme events like hurricanes, impact pollinator behavior and survival. For example, hurricanes have damaged bee colonies in Mexico and across the Caribbean.
- Genetically modified crops: The use of herbicide-resistant GMOs may reduce floral diversity, limiting food for pollinators and possibly affecting wild populations.

In Brazil, climate projections suggest a 13% reduction in pollinator presence by 2050, particularly in the eastern Amazon (Giannini et al. 2017; Giannini et al. 2020). Knowledge gaps remain a major hurdle for effective conservation of pollinators in tropical regions.

9 Causes of Pollinator Decline

Multiple interacting drivers contribute to pollinator declines:

- Land use change: Conversion of forests to farmland disrupts natural plant-pollinator relationships and reduces habitat connectivity.
- Agricultural intensification: Increased reliance on chemicals and monocultures harms pollinators directly and indirectly.
- Competition and aggression from invasive species: Introduced pollinators can displace or harm native species.
- Disease transmission: Movement of managed bees contributes to the spread of pests and pathogens.
- Extreme weather events: Hurricanes and climate shifts alter the distribution, seasonality, and abundance of pollinators.
- Loss of floral resources: Herbicides and GM crops reduce the diversity and availability of plants that provide nectar and pollen.

10 Conservation

India, one of the world's mega biodiversity country and with four biodiversity hotspots, is home to over 55,000 plant species (Mao et al. 2023) and more than 100,000 animal species (Banerjee et al. 2023). However, 91% of its plant species are threatened by habitat degradation (Gardenfors et al. 2001), and around 70% of medicinal plants are found in tropical forests (Negi et al. 2019). Protecting this biodiversity is crucial for food security, human well-being and sustainable development.

Key conservation strategies include:

- Education and training: Raising awareness and building capacity among communities and stakeholders is essential to protect pollinators.
- Legislation and governance: Stronger legal frameworks and enforcement mechanisms are needed to safeguard pollinator habitats and populations.
- Science and innovation: Data-driven policies and accessible farming practices should integrate pollinator conservation into broader agricultural goals.
- Climate resilience: Given their vulnerability to climate change, especially in the tropics, actions are needed urgently to mitigate future risks to pollinator populations.

References

Aizen MA, Garibaldi LA, Cunningham SA, Klein AM (2009) How much does agriculture depend on pollinators? Lessons from long-term trends in crop production. Ann Bot 103(9):1579–1588

Altringham JD (1996) Bats: biology and behaviour. Oxford University Press, Oxford

Altringham JD, Senior P, Ruckstuhl KE, Neuhaus P (2005) Social systems and ecology of bats. In: Sexual segregation in vertebrates, pp 280–302. Cambridge University Press, Cambridge

Arroyo MTK (1976) Geitonogamy in animal pollinated tropical angiosperms: a stimulus for the evolution of self-incompatibility. Taxon 25(5–6):543–548

Banerjee D, Raghunathan C, Rizvi, Das D (2023) Animal discoveries 2022: new species and new records. Zoological Survey of India, Kolkata

Bawa KS (1990) Plant-pollinator interactions in tropical rain forests. Annu Rev Ecol Syst 21(1): 399–422

Borges RC, Brito RM, Imperatriz-Fonseca VL, Giannini TC (2020) The value of crop production and pollination services in the eastern Amazon. Neotrop Entomol 49(4):545–556

Carson R (1962) Silent Spring. Houghton Mifflin

Costanza R, Darge R, De Groot R, Farber S, Grasso M, Hannon B, Limburg K, Naeem S, Oneill RV, Paruelo J, Raskin RG (1997) The value of the world's ecosystem services and natural capital. Nature 387(6630):253–260

Costanza, R., De Groot, R., Sutton, P., Van Der Ploeg, S., Anderson, S.J., Kubiszewski, I., Farber, S. And Turner, R.K (2014) Changes in the global value of ecosystem services. Glob Environ Chang, 26:152–158

De Deus, F.F. And Oliveira, P.E (2016) Changes in floristic composition and pollination systems in a "Cerrado" community after 20 years of fire suppression. Rev Bras Bot, 39: 1051–1063

Ganesh T (1995) Fruiting patterns among canopy trees and fruits use by vertebrates in a wet evergreen forest of the southern western Ghats India. Doctoral dissertation, Dept of Ecology, Pondicherry University

Ganesh T, Devy MS (2006) Interactions between non-flying mammals and flowers of *Culleniaexarillata*Robyns (Bombacaceae), a canopy tree from the wet forests of Western Ghats, India. Curr Sci:1674–1679

Gardenfors U, Hilton-Taylor C, Mace GM, Rodríguez JP (2001) The application of IUCN Red List criteria at regional levels. Conserv Biol 15(5):1206–1212

Giannini TC, Cordeiro GD, Freitas BM, Saraiva AM, Imperatriz-Fonseca VL (2015) The dependence of crops for pollinators and the economic value of pollination in Brazil. J Econ Entomol 108(3):849–857

Giannini TC, Costa WF, Cordeiro GD, Imperatriz-Fonseca VL, Saraiva AM, Biesmeijer J, Garibaldi LA (2017) Projected climate change threatens pollinators and crop production in Brazil. PLoS One 12(8):e0182274

Giannini TC, Alves DA, Alves R, Cordeiro GD, Campbell AJ, Awade M, Bento JMS, Saraiva AM, Imperatriz-Fonseca VL (2020) Unveiling the contribution of bee pollinators to Brazilian crops with implications for bee management. Apidologie 51:406–421

Gottsberger, G., Camargo, J.M. And Silberbauer-Gottsberger, I (1988) A bee pollinated tropical community: the beach dune vegetation of Ilha de Sao Luis, Maranhao, Brazil, Bot Jahrb Syst, 109(4): 469–500

https://agritech.tnau.ac.in/crop_improvement/crop_imprv_pollmode.html#:~:text=The%20process%20by%20which%20pollen,2)%20Allogamy%20or%20cross%20pollination.&text=Transfer%20of%20pollen%20grains%20from,as%20autogamy%20or%20self%20pollination

IPBES, S.G. Potts, V. L. Imperatriz-Fonseca, H. T. Ngo, (eds) (2016a) The assessment report of the intergovernmental science-policy platform on biodiversity and ecosystem services on pollinators, pollination and food production. Secretariat of the Intergovernmental Science-Policy Platform on Biodiversity and Ecosystem Services, Bonn, Germany., 552 pages

IPBES, SG Potts, V.L. Imperatriz-Fonseca, Ngo, H. T., Biesmeijer, J. C., Breeze, T. D., Dicks, L. V., Garibaldi, A., Hill, R., Settele, J., Vanbergen, A. J., Aizen, M. A., Cunningham, S. A., Eardley, C., Freitas, B. M.Gallai, N., Kevan, P. G., Kovács-Hostyánszki, A., Kwapong, P. K., Li, J., Li, X., Martins, D. J., Nates-Parra, G., Pettis, J. S., Rader, R., & Viana, B. F. (Eds.) (2016b) Summary for policymakers of the assessment report of the intergovernmental science-policy platform on biodiversity and ecosystem services on pollinators, pollination and food production. Secretariat of the Intergovernmental Science-Policy Platform on Biodiversity and Ecosystem Services, Bonn, Germany, 36 pages

Khalifa SA, Elshafiey EH, Shetaia AA, El-Wahed AAA, Algethami AF, Musharraf SG, AlAjmi MF, Zhao C, Masry SH, Abdel-Daim MM, Halabi MF (2021) Overview of bee pollination and its economic value for crop production. Insects 12(8):688

Klein AM, Vaissiere BE, Cane JH, Steffan-Dewenter I, Cunningham SA, Kreme C, Tscharntke T (2007) Importance of pollinators in changing landscapes for world crops. Proc R Soc B Biol Sci 274(1608):303–313

Klink, C.A. And Machado, R.B (2005) Conservation of the Brazilian cerrado. Conserv Biol, 19(3): 707–713

Kumar KS, Khanduri VP, Tripathi SK (2021) Reproductive adaptations and the availability of pollinating vectors in white Indian teak (GmelinaarboreaRoxb.) in tropical rain forest of Indo-Burma Hotspot. Trees, Forests People 3:100058

Lichtenberg EM, Mendenhall CD, Brosi B (2017) Foraging traits modulate stingless bee community disassembly under forest loss. J Anim Ecol 86(6):1404–1416

Machado IC, Sazima M (2008) Pollination and breeding system of MelochiatomentosaL. (Malvaceae), a keystone floral resource in the Brazilian Caatinga. Flora-Morphol Distrib Funct Ecol Plants 203(6):484–490

Mao AA, Agrawala DK, Mukherjee S (2023) Plant discoveries 2023. Botanical Survey of India, Kolkata

Nabhan GP, Buchmann SL (1997) Services provided by pollinators. In: Nature's Services: societal dependence on natural ecosystems, pp 133–150. Cambridge University Press, Cambridge

Negi VS, Maikhuri RK, Maletha A, Phondani PC (2019) Ethnobotanical knowledge and population density of threatened medicinal plants of Nanda Devi Biosphere Reserve, Western Himalaya, India. Iran J Sci Technol Trans A: Sci 43:63–73

Paz FS, Pinto CE, de Brito RM, Imperatriz-Fonseca VL, Giannini TC (2021) Edible fruit plant species in the Amazon forest rely mostly on bees and beetles as pollinators. J Econ Entomol 114(2):710–722

Pedro SRM (2014) The stingless bee fauna in Brazil (Hymenoptera: Apidae). Sociobiol 61(4): 348–354. https://doi.org/10.13102/sociobiology.v61i4.348-354

Potts SG, Imperatriz-Fonseca V, Ngo HT, Aizen MA, Biesmeijer JC, Breeze TD, Dicks LV, Garibaldi LA, Hill R, Settele J, Vanbergen AJ (2016a) Safeguarding pollinators and their values to human well-being. Nature 540(7632):220–229

Potts SG, Imperatriz-Fonseca V, Ngo HT, Biesmeijer JC, Breeze TD, Dicks LV, Garibaldi LA, Hill R, Settele J, Vanbergen AJ (2016b) The assessment report on pollinators, pollination and food production: summary for policymakers. Secretariat of the Intergovernmental Science-Policy Platform on Biodiversity and Ecosystem Services

Rader R, Bartomeus I, Garibaldi LA, Garratt MP, Howlett BG, Winfree R, Cunningham SA, Mayfield MM, Arthur AD, Andersson GK, Bommarco R (2016) Non-bee insects are important contributors to global crop pollination. Proc Natl Acad Sci 113(1):146–151

Ramalho M (2004) Stingless bees and mass flowering trees in the canopy of Atlantic Forest: a tight relationship. ActaBotanicaBrasilica 18:37–47

Slaa EJ, Chaves LAS, Malagodi-Braga KS, Hofstede FE (2006) Stingless bees in applied pollination: practice and perspectives. Apidologie 37(2):293–315

Part III

Pollinator Conservation Strategies

Utilization and Conservation of Insect Pollinators in Commercial Flower Crops in Tropics and Subtropics

15

Nilanjana Chaudhuri, Rajiv Kumar, and A. K. Chakravarthy

Abstract

Pollination is a major phenomenon of dynamic ecological interaction in flower crops. The interaction between flower crops and pollinators is mutualistic. Flower adaptations over evolutionary period are designed to cater to the pollinators. Select species of flower crops are commercially cultivated under polyhouse/controlled and hi-tech mechanization conditions. These require a different set-up and species of pollinators for commercial flower production. The arrangement of petals, color, shape, size, and dispensation of flowers on a plant play a key role in attracting pollinators to plants. Select species of flowers have evolved even to deceive the pollinators! Human-induced changes in land use and environment have led to global pollinator crisis. Emphasis on pollinator research, impact of climate change on pollinators, sustainable agriculture, pollinator conservation, assessing impacts of invasives and reducing application of chemicals will facilitate food security and contribute to biodiversity conservation.

Keywords

Flower crops · Tropics · Insect pollinators · Cross pollination

N. Chaudhuri (✉)
Agricultural Entomology, Regional Research Station, Terai Zone, Uttar Banga Krishi Viswavidyalaya, Pundibari, Cooch Behar, West Bengal, India

R. Kumar
Division of Horticulture and Medicinal plants, ICAR-Indian Institute of Horticultural Research, Bengaluru, Karnataka, India

A. K. Chakravarthy
Society for Science and Technology Applications (SSTA), Bangalore, Karnataka, India

A. K. Chakravarthy (ed.), *The Role of Pollinators in Tropical Landscapes*,
https://doi.org/10.1007/978-981-95-9139-8_15

1 Introduction

Pollination has a long history that extended way back to the Greek philosopher, Theophrastus (around 320 BC). The contribution by Charles Darwin on pollination in orchids (Darwin 1862) and reproduction in plants (Darwin 1876) was perhaps the pioneering study that led to the present-day scientific studies. The first fossil record of pollination by biological entities was evidenced in early Triassic period from fern plant of Gymnosperms, plant-pollinator mutualisms date back to the Cretaceous period, when insects acquired food from flowers, and flowers attained higher reproductive rates via the movement of pollen by insects. The relationship between beetles and angiosperms was documented in the early Cretaceous period and the mutualism between hymenopterans and angiosperms was witnessed in late Cretaceous period. Darwin (1859) was the first also to describe the origin of plant-pollinator interactions.

Floriculture comprises traditional and modern flower crops production. Traditional flowers are cultivated in fields/open-air conditions. Tropical flowers usually refer to crops like roses, jasmines, china roses, lavender, chrysanthemums, and carnations. Basically this category or ornamentals also include several genera and species of orchids. The flowers are frequented by a wide variety of insect pollinators besides bats, birds, and rodents like squirrels. Documented information on pollinators of commercial flowers in tropics and subtropics has remained scanty. For instance, contribution on pollinations is barely 5% of the world. Hence, a chapter on utilization and conservation of insect pollinators in commercial flower crops was thought of paramount importance.

2 Insect Pollinators

Pollination is a process in flowering plants, which is a pre-requisite for fruit and seed formation and production. The phenomenon demands effective integration of pollinator and pollinizer. A pollinizer is a plant that is a pollen source for effective fertilization. An insect cannot be considered a pollinator if it visits a flower and does not intentionally or accidentally touch the reproductive parts (Dafni et al. 2005). Insects are the dominant class of pollinators, contributing over 80% to pollination services. Honeybees are critically important for crop pollination (Thapa 2006; Klein et al. 2007; Belavadi and Ganeshaiah 2013).

Flowers are a magic! The interaction between flowering plants and pollinating insects is a mutualism. Adaptation in flowers is intended to attract specific pollinators, and the process is called the pollination syndrome. Common pollinators of flower crops include bees (Melittophily), butterflies (Psychophily), moths (Phalaenophily), flies (Myophily), carrion flies (Sapromyophyily), beetles (Canthorophily), birds (Ornitophily), and bats (Chiropterophily). Flowers adaptation related to pollination syndromes include color, shape, fragrance, food, and rewards.

Flower adaptations are designed to attract general pollinators or be highly adapted to attract a limited group or taxa of pollinators. Insect pollinated flowers are generally large and brightly colored; stigmas are small, compact, and do not protrude outside flower; stamens are not pendulous and located within flowers; and nectar guides are present to guide pollinators.

3 Floriculture

Flowering plants play a crucial role in sustaining populations of pollinators and ecosystem services. Floriculture includes commercial flowers, bedding plants, flowering plants, foliage plants or houseplants, cut cultivated greens, and cut flowers. Cut flowers are usually sold in bunches or as bouquets with cut foliage. Hi-tech floriculture refers to growing of high-quality cut-flowers under controlled conditions in poly-house through tissue cultured plants. In this systems of flower production, water soluble fertilizers, part-mechanization, cold chain, packaging, and post-harvest technology for export and domestic markets are utilized. The traditional flowers like chrysanthemum, jasmine, crossandra, rose, tuberose, aster, marigold, champaka, and others are cultivated in open fields. In contrast, modern flower plants are cultivated under controlled and hi-tech mechanization, but in developing countries like India and South East Asia the open-field, low-cost shade house, open poly-house, or naturally ventilated houses are generally utilized for commercial flower productions. Flower crops are also grown in polyhouses and common species include anthuriums, orchids, carnation, lilium, and others. Common cultivated commercial flowers in tropics and sub-tropics are rose, carnation, gerbera, gladiolus, marigold, hibiscus, jasmine, and orchids.

4 Pollination in Commercial Flowers

4.1 Carnation, *Dianthus caryophyllus* L.

The carnation *Dianthus carthuslanorum* (Caryophyllaceae) is an example for a plant species that demands a specialized pollination. The most obvious characteristic features of *D. carthuslanorum* are its conspicuous, tubular shaped flowers, providing a mechanical barrier to nectar consumption for visitors with short proboscides. The flowers have two stigma lobes and ten stamens and mainly visited and pollinated by butterflies that feed on the nectar secreted from nectaries (Knuth 1898). Carnation flowers can have pollination either by self-pollination or by bees or by visual appeal and scent that stamens from flowers.

Two closely related carnation species, *Dianthus carthustanorum* and *D. sylvestris*, represent ideal systems for studying plant-pollinator interactions. Two butterfly species, *Satyrus ferula* (Fabricius) and *Melangargia galalthea* L. (Satyridae), are the main pollinators of *D. carthusianorum* (Bloch et al. 2016). Butterflies viz. hesperids (e.g. *Ochlodes venatus* (Turati). *Thymelicus sylvestris* (Poda)) and others too visit carnation flowers. In contrast, *D. sylvestris* is mainly pollinated by the hummingbird hawkmoth (*Macroglossum stellatarum* L.) and the noctuid moth *Hadena compta* (Schiff) (Erhardt 1988). Wind, an abiotic factor, plays a limited role in carnation flower pollination.

4.2 Gerbera

Gerbera (Asteraceae) is a flowering plant belonging to daisy family. It is fifth most important cut flower plant and is being increasingly used in gardens and landscapes. Gerbera is grown for large bright colored flowers (sometimes with combination of multiple colors in different florets) that are available in wide range of colors mono or multiple. Once the flowering begins, these plants profusely flower over a long period of time. Both self-pollinated and cross pollinated varieties are found in Gerbera. In open field, the flowers are attractive to insects like bees and butterflies. Bumblebees, seat bees, and miner bees that nest on the ground also are attracted to the flowers of Gerbera. Wasps, hornets, and yellow jackets may also help spread pollen. A honeybee typically pollinates several female flowers on a disk with each visit. Butterflies do not take much pollen with each landing, but they flit from flower to flower, collecting pollen each time.

Flowers of gerbera are composite with three types of flowers viz., disc florets, trans florets, and day florets. Gerbera flowers show protogyny, i.e., female stigma matures before male anthers.

4.2.1 Flower-Seed Production

In cross-pollinated flower crops like gladiolus, roses, gerbera, and others, honeybees are avoided during crossing of flower-crops germplasm to maintain purity of the cross. Honeybees carry pollen in legs and wings. Honeybees carrying pollen of other varieties, when visit crosses flowers, they get contaminated with foreign pollen. Honeybees are avoided in advancing of generations of flower plants from first to seventh generation. The advancing generation of flower plants are served with enclosing the plants under mosquito nylon net or iron mesh net to avoid visits of honeybees. This is because there may be chances of cross-pollination by bees. In Asteraceae family, flowers have two distinct floret, via disc floret, and ray florets. Ray florets are pistillate or female flowers.

4.3 Gladiolus

The Gladiolus is a monocot (Tridaceae) and has flowers attractive to an assembly of pollinators like bees, flies, beetles, butterflies, and birds. In Gladiolus, moth-pollinated flowers are large, long-tubed, and pale-colored. Pollinators forage on Gladiolus primarily for nectar. Flowers are aromatic, open fully at night, and produce relatively sucrose concentrated nectar. Variations in the tube length in flowers of Gladiolus have been the subject of investigation (Bruce Anderson et al. 2010). Pollinators, mainly hawkmoths belonging to the family sphingidae under different genera Gladiolus plants to have ecotypes in South Africa based on the tube lengths, either long, short, or intermediate pollinators driven mediation and radiation, have also been reflected in flower morphology, color, and flower aroma. Butterfly flowers, in contrast, open fully during the day, close partially or completely at night, are often bright crimson to scarlet and produce dilute nectar that may be sucrose-rich or hexose-rich (Goldblatt and Manning 2002).

Bee-pollinated Gladiolus flowers have shorter tubes and insect-guides that are distinctly marked by bright colors. The bright-red colored flowers are attractive to birds like the Sunbirds (*Nectariana* spp.) that are rewarded by the copious amounts of nectar.

4.4 Hibiscus

The genus Hibiscus comprises 200–300 species, distributed chiefly in tropical and subtropical regions (Pfeil et al. 2002). Hibiscus (Perry 2004) belongs to Malvaceae family with flowers of varied colors. Hibiscus is native to warm temperate, subtropical, and tropical regions. Species are known for large and showy flowers. The genus includes annual and perennial herbaceous plants, and woody shrubs and small trees. Each Hibiscus species can breed with another species. Hibiscus flowers are polyandrous in that they receive pollen from different plants to get seeds.

The pollinators use petals as a cue to locate the flowers as they are large and showy and it is pollinated by anthophorid bees and bumblebees (Snow and Spira 1993; Kodoh and Wigham 1998). During feeding on Hibiscus flowers, pollen get glued to various parts of the insect body, which is then carried to other flowers and deposited. *H. tiliaceus* is pollinated by an endemic Galapagus carpenter bee, *Xylocopa darwini*, which also nests in this tree (Williams 1926).

4.5 Jasmine

Jasmine is a shrub or a vine in Oleaceae family. Jasmines are native to tropical and subtropical regions of Eurasia, Africa, Australasia, and Oceania. Their center of diversity is in South and South East Asia (Christenhusz and Byng 2016). In Jasmine, the pollination was established after the visit of ants and scarab beetles (Dafni and Dukas 1986). Of all the pollinators, ants are the key major pollinators of jasmine (Gomez 1999). The pollinators of jasmine also Hummingbird hawk moth *Macroglossum stellarum* (Shrivastava et al. 2011). Interestingly in Jasmine, there are diurnal as well as nocturnal pollinators. Diurnal pollinators are drawn to flowers by nectar, while nocturnal flowers by fragrance (Fig. 15.1).

4.6 Orchids

With over 9% of the angiosperm flora, orchids (Orchidaceae) form the largest group in India (Adit et al. 2021). The flowers of orchids are pollinated by various vertebrate

Fig. 15.1 Butterfly pollinating flowers

and invertebrate taxa. Pollinators of orchids include insects, rodents, birds, and other vertebrates. Orchids use nectar as a main source of attraction. Nectar is stored in nectaries. Not all orchids have nectaries, certain orchids attract pollinators by their pseudo pollens.

Wasps (Vespidae) pollinate about 60% of the orchids. These flowers have well-developed landing platform and have nectar glands or marks of contrasting color pointing the way to nectar. The nectar is present in minimal quantities and more or less concealed. Butterfly pollinated orchids are bright and bear sweet fragrance. Such flowers may or may not have a nectar gland. In certain species nectar remains concealed much below in the spurs. In India, *Habenaria foliosa* var. *foetida* is pollinated by blue tiger butterfly (*Tirumala limniace*) during day time and by moth *Dysgonia* during the night. During day time, the fetid odor attracts butterfly, whereas during the night, whitish green color and for nectar attracts the moths (Dangat and Gurav 2014). The flowers of orchids pollinated by flies are placed horizontally, fringed and some have landing platform. The flowers may open during day or night and emit odor. *Epipactis veratrifolia* is pollinated by hoverfly (*Ischidon scutellaris*) while it rests in its flowers for laying the eggs (Kumar and Rawat 2011).

4.7 Rose, *Rosa* sp.

Flowers of rose species are mostly hermaphroditic and homogamous, meaning that they open with both male (stamens number from 50 to 250, depending on species) and female (pistils) maturing simultaneously. A fertilized ovary forms a berry like structure called a hip. However, most locally cultivated certain varieties of roses are self-pollinating and self-fertilizing and can survive without cultivars of roses. Some do not produce a hip insect pollination, whereas others cannot. However, little is known about most species in the genus Rosa. Some plants are devoid of pollen, do not set fruit, and must be propagated vegetatively.

Bumblebees (*Bombus*) primarily use the pollen scent of *R. rugosa* flowers to assess the availability of forage, but visual stimuli from the anthers are also

Fig. 15.2 Pollinator on rose flower

important. Petal odors and colors are of secondary use in the bees' selection of rewarding flowers. It has been suggested that dish-shaped flowers, such as those of roses, could act as parabolic sonic reflectors that amplify the sound of the bees buzzing in the flowers to release pollen (Kevan 2017) (Fig. 15.2).

5 Threats

Habitat modifications like agriculture, grazing, division of native landscapes, and development of regions supporting wild vegetation cause loss of native bee flora and nesting areas utilized by pollinators. Introductions of pollinators may have beneficial as well as harmful impacts. Honeybees have become competitors with local bees, birds, and other pollinators. Bumblebees have moved to islands and continents where they did not occur earlier. These introductions may have benefited agriculture, but their impacts on native plants and animals are deleterious. Pesticides are another major factor against pollinators. Monoculture, lower density of weed plants, lack of flower rich grasslands, and declining pastures add to declining populations of pollinators.

6 Conservation

The crops that need pollination service will be in flower only in certain months or seasons. During rest of the year, it is necessary to understand on what plants the pollinators forage. Hence, it is necessary to develop a floral calendar for the given region. If the target crops are closer to natural habitats like forests, the plant species closer to cropped area should be identified, and the flowering phenology of each species should be recorded. In addition, the common weeds and other species of plants in the vicinity of the crop are also to be identified and their flowering times recorded. All these information can be used to analyze their importance in conserving populations of pollinators in the fields. Regulation of pollination systems is a recent and new strategy for conservation. To conserve pollinator populations, the role of conserving non-cropped areas/refugia in and around cultivated areas is crucial. Pollinators' populations may be transported to deficit areas. Resistant stock of bees against parasitic mites and viruses can be developed. According to Adit et al. (2024), a collaborative initiative among scientists, society, and administrators is crucial for the conservation of pollinators' populations.

References

Adit A, Jalal JS, Koul M, Tandon R (2021) A conspectus of orchid studies in India. Rheedea 31(3): 218–233. https://doi.org/10.22244/rheedea.2021.31.03.14

Adit A, Sngh VK, Tandon R, Shivanna KR (2024) Pollination research in India: current status, future vision and conservation implications. Intgr Conserv 3(4):330–341. https://doi.org/10.1002/inc3-74

Anderson B, Alexandersson R, Johnson SD (2010) Evolution and co-existence of pollination Ecotypes in an African Gladiolus (Iridaceae). Evolution 64(4):960–972

Belavadi VV, Ganeshaiah KN (2013) Insect pollination manual, NICRA Project. University of Agricultural Sciences, Bangalore, p 44

Bloch D, Werdenberg N, Erhardt A (2016) Pollination crisis in the butterfly-pollinated wild carnation Dianthus carthusianorum? New Phytol 169(4):699–706

Christenhusz MJM, Byng JW (2016) The number of known plants species in the world and its annual increase. Phytotaxa 261(3):201–217. https://doi.org/10.11646/phytotaxa.261.3.1

Dafni A, Dukas R (1986) Insect and wind pollination in *Urginea maritime* (Liliaceae). J Plant System Evol 154:1–10. https://doi.org/10.1007/BF00984864

Dafni A, Kevan PG, Husband BC (2005) Practical pollination biology. Enviroquest, Cambridge, 590 pp

Dangat TD, Gurav RV (2014) Pollination in *Habenaria foliosa* var. foetida (Orchidaceae). Richardiana 14:219–228

Darwin C (1859) On the origin of species by means of natural selection, or the preservation of favoured races in the struggle for life. London: John Murray-Royal Society Publishing.org

Darwin CR (1862) On the various contrivances by which British and foreign orchids are fertilised by insects, and on the good effects of intercrossing. John Murray, London

Darwin CR (1876) The effects of cross and self fertilisation in the vegetable kingdom. John Murray, London

Erhardt A (1988) Pollination and reproduction in *Dianthus silvester* Wulf. — In Cresti M, Gori P, Pacini E (Eds.): Sexual reproduction in higher plants, pp 351–356. Berlin, Heidelberg, New York, Tokyo: Springer.

Goldblatt P, Manning JC (2002) Evidence for moth and butterfly pollination in Gladiolus (Iridaceae: Crocoideae). Ann Mo Bot Gard 89:110–124

Gomez JM (1999) Effectiveness or ants as pollinators of *Lobularia maritime*: effects on main sequential fitness components of the host plant. J Oecalogia 122:90–97. https://doi.org/10.1007/PL00008840

Kevan PG (2017) Pollination in roses. In Reference Module in Life Sciences. Elsevier. https://doi.org/10.1016/B978-0-12-809633-8.05070-6

Klein AM, Vaissiere BE, Cane JH, Steffan Dewenter I, Cunningham SA, Kremen C, Tscharntke T (2007) Importance of pollinators in changing landscapes for world crops. Proc R Soc Ser B 274: 303–313

Knuth P (1898) Handbuch der Bluetenbiologie. Engelmann, Leipzig

Kodoh H, Wigham T (1998) The effect of petal size manipulation on pollinator/seed predator mediated female reproductive success of *Hibiscus moscheutos*. Oecologia 117:70–79

Kumar P, Rawat GS (2011) Nature Precedings: hdl: https://doi.org/10.1001/npre.2011.6295.1

Perry LB (2004) Hibiscus: hardy and tropical plants for the garden. Timber Press, p 36. ISBN 978-0-88192-6545

Pfeil BE, Brubaker CL, Craven LE, Crips MD (2002) Phylogeny of Hibiscus and the tribe Hibiscaceae (Malvaceae) using chloroplast DNA sequences of ndhF and the rpl16 intron. Syst Bot 27:333–350

Shrivastava S, Rajak N, Shrivastava L (2011) The pollination template in *Jasminum*. Elixir Horm Signal 35(2011):2754–2759

Snow AA, Spira TP (1993) Individual variation in the vigor of self pollen and selfed progeny in *Hibiscus moscheutos* (Malvaceae). Am J Bot 80:160–164

Thapa RB (2006) Honeybees and other insect pollinators of cultivated plants: a review. J Instit Agric Anim Sci 27:1–23

Williams FX (1926) Expedition of the California Academy of Sciences to the Galapagos Islands, 1905-1906, The bees and aculeate wasps of the Galapagos Islands. Proc California Acad Sci Fourth Ser 2:347–357

Pollinators in Fruit and Vegetable Cultivated Ecosystems: Diversity, Climate-Driven Shifts, and Conservation Strategies

16

P. V. Rami Reddy, Kolla Sreedevi, and P. Sree Chandana

Abstract

Pollination is a crucial and an economically important ecosystem service offered by certain group of insects majorly bees, flies, beetles, butterflies, and moths. Pollinators form an integral component of natural and man-made ecosystems. It is estimated that 35% of the global crop production, comprising 87 of the prominent crops worldwide, is benefited by insect pollination. Though agriculture in general is benefitted to a great extent by insect pollination, the proportion of commercially important crops essentially dependent on entomophily is very high in horticulture including fruits, vegetables, nuts, and spices. Besides factors like habitat loss, pesticide, pollution, climate change, etc., are an emerging threat to pollination process by disrupting the evolved synchrony of plant pollinator interactions. Behavioral responses of pollinator insects to overcome higher temperatures significantly reduce pollination process. This chapter presents the pollinator complex of major fruit and vegetable crops, their contribution to yield enhancement, and their vulnerability to climate change and conservation strategies.

Keywords

Pollinators · Foraging behavior · Fruits · Vegetables · Entomophily · Fruit set · Climate Change · Honey bees · Conservation

P. V. R. Reddy (✉)
ICAR-Indian Institute of Horticultural Research, Bengaluru, Karnataka, India

K. Sreedevi
ICAR-National Bureau of Agricultural Insect Resources, Bengaluru, Karnataka, India

P. Sree Chandana
ICAR-Indian Institute of Rice Research, Hyderabad, Telangana, India

A. K. Chakravarthy (ed.), *The Role of Pollinators in Tropical Landscapes*,
https://doi.org/10.1007/978-981-95-9139-8_16

1 Introduction

Flowering plants and their insect foragers have co-evolved over thousands of years, and their mutualism resulting in pollination has helped sustaining the plant biodiversity and the existence of several life forms, including man on the earth. The role of insect pollinators in agroecosystems is widely acknowledged. Though agriculture in general is benefitted to a great extent by insect pollination, the proportion of key crops essentially dependent on entomophily is high in horticulture including fruits, vegetables, nuts, and spices. Major horticultural crops that rely on insect pollination are apple, apricot, avocado, citrus, litchi, mango, grapes, peach, cherry, olive, strawberry, passion fruit, melons, gourds, cauliflower, radish, crucifers, tomato, onion, pumpkin, oil palm, coffee, cardamom, and various nuts (Abrol 1990; Free 1993). The diversity of insect pollinators is high in horticultural ecosystems including bees (*Apis* spp., and non-*Apis* sp.), wasps, butterflies, flies, and beetles. European bee (*Apis mellifera*) and Indian honeybee (*Apis cerana*) are the widely used managed pollinators followed by bumble bees. Approximately 73% of crops globally are pollinated by bees, 10% by flies, and rest by bats, wasps, beetles, birds, butterflies, and moths (Abrol 2009). Bee diversity at global level includes about 19,000 described species (Linsley 1958). Except *Apis mellifera* L., commonly called European bee, all other species of bees are grouped as "wild bees." India is rich in honey bee diversity and is home to species, viz., *A. cerana*, *A. florea*, *A. dorsata*, *A. andreniformis*, and *A. laboriosa*. Within species *A. cerana*, there are sub-species like *A. ceranacerana*, *A. cerana indica*, etc. In addition, many other pollinators like bumble bees, stingless bees, megachilids, and other pollinators occur in different agro-ecosystems (Thakur 2012). Of late, the numbers of native pollinator fauna have been dwindling due to several factors like loss of habitats, use of dangerous agrochemicals, monoculture, displacements by invasive species, climate change, etc. This decline will have drastic adverse effect on the biodiversity in general and horticultural production in particular. According to an estimate, inadequate pollination translate into more than 50% reduction in fruit and vegetable production, and the loss would be 100% in case of monoecious crops like cucurbits. In recent past, area under horticultural crops is expanding fast and hence the importance of pollinators. The essentiality of pollinators is underscored by an estimate that there was threefold enhancement in the cultivated area under pollinator-dependent crops in the past 50 years (Aizen and Harder 2009). In this contribution, we precisely compile the literature on the diversity of pollinators associated with important horticultural crops.

2 Pollinators of Fruit Crops

2.1 Apple (*Malus pumila*)

The apple flowers produce copious amounts of nectar and pollen and the blossom period lasts for about nine days. Although apple trees produce numerous flowers, fruit set is a mere 5%. Wind pollination has been ruled out in apple (Free 1966). Wild bees of the genera *Andrena*, *Bombus*, *Halictus*, and *Osmia* were considered main pollinators of apple (Free 1966). In contrast, honey bees (*Apis dorsata* and *A. mellifera*) are recorded as the major pollinators of apple and other temperate fruits (Singh and Misra 2007). *A. cerana* and *Eristalis* spp. are also often important pollinators of apple and pear (Verma and Chauhan 1985). Introduction of three to eight *E. tenax* flies into cages containing an apple tree and appropriate pollenizer branches helped in enhancing fruit set. Pollen collecting honey bees are generally regarded as more efficient pollinators than nectar collectors (Solomon and Kendall 1970). Placement of three to four colonies/ha of *A. mellifera* in apple and pear orchards had a significant effect on fruit set and yield. However, *A. cerana* was reported a reliable pollinator, particularly under adverse climatic conditions (Sharma 2000).

2.2 Citrus (*Citrus* spp.)

The extensively cultivated genus *Citrus*, which includes orange (*Citrus sinensis*), mandarin (*C. reticulata*), grapefruit (*C. paradisi*), and lemon (*C. limon*), exhibits wide variations in pollination requirements (Roubik 1995). There can be parthenocarpic fruit development on selfing, but in some cases there is a self-sterility. Sweet lime and mandarin oranges are greatly benefited from cross pollination. Most seeded varieties are self-fruitful but need pollination. For instance, 'Clementine' tangerine and sweet oranges are completely or partially self-incompatible (Sanford 1992) and require insects for pollination.

Chacoff and Aizen (2007) recorded 50 insects species visiting grapefruit flowers. The most repeated visitor was *Apis mellifera* followed by stingless bees (*Trigona argentina* and *Tetragonisca angustula*) and solitary bees (*Xylocopa eximia* and *Megachile* sp.). They found that hand- and open-pollinated flowers result in about six times higher fruit production than emasculated and bagged (insect excluded) flowers. The fruit set and mean fruit weight were 35.30% greater in uncovered sweet orange compared to the covered flowers. Some compounds like Bee-Here, eugenol, geraniol, citral, and lemon grass extract were effective in attracting honey bees (Malerbo-Souza et al. 2004). In India, Phartiyal et al. (2012) recorded 12 insect visitors on flowers of sweet orange in *Terai* region. Among the pollinators, the Hymenoptera constituted major group of insects, including *A. dorsata*, *A. cerana indica*, *Trigona* spp., *Campsomeriella megachalis* F., and *C. collaris*. Among the Dipterans, syrphid flies were common foragers including *Syrphus corollae* Fab., *Episyrphus balteatus* De Geer, *Spherophoria* spp., and *Melanostoma* spp. The percent fruit setting was higher (8.4–12%) in panicles under open pollination compared to completely bagged panicles (1.2–8.0%). Bhuyan and Bhattacharyya (2002) studied the foraging activity of honeybee on Assam lemon and recorded 77% fruit set in honeybee pollinated trees compared to 46% in naturally pollinated ones. They recommended four *A. cerana* colonies per ha for optimum yield. Similar observations on the positive impact of bee pollination in citrus were reported by Gogoi et al. (2007).

2.3 Custard Apple (*Annona* spp.)

Annonaceous flowers are protogynously dichogamous where stigma receptivity is over by the time pollen shedding occurs, thus limiting the scope of pollination without external factors (Gottsberger 1970). The flowers attract pollinators through strong odors. Some species have thermogenic flowers that attain a temperature higher than that of surrounding air, so as to enhance the intensity of volatilization of the odoriferous chemicals in the petals. Annonaceous flowers produce no nectar, but pollinators are rewarded with fleshy edible petals. The flowers also serve as mating sites for pollinators (Gottsberger 1988).

A majority of *Annonaceae* are pollinated by nitidulid beetles although some are pollinated by thrips, bugs (Farre et al. 1997), and cockroaches (Nagamitsu and Inoue 1997). Reiss (1971) was the first to document pollination by nitidulid beetles, in cultivated atemoyas and cherimoyas. The species complex of beetles and the performance of each species vary geographically. The four major pollinators in Israel, viz., *Carpophilus humeralis*, *C. heemipterus*, *C. mutilates*, and *Haptoncus luteolus*, are equally effective as pollinators. In Florida, USA, out of nine species of nitidulids, *C. mutilatus* was the most important pollinator in terms of efficacy and abundance (Nadel and Pena 1994). The number of species visiting flowers in Ecuador, Colombia, and the Caribbean region was similar to that found in Florida (Pena and Bennett 1995). The increased number of beetles was reported to positively influence the fruit set and also the quality of the fruit (Nadel and Pena 1994; Lopez and Uquillas 1997).

There are also instances when augmentation of nitidulid populations did not result in increased fruit set in atemoya orchards. In Israel, one-time additions of decaying apples as an attractant failed to increase the number of beetles in the flowers (Galon et al. 1982). However in Australia, adding pineapples in an atemoya orchard throughout the flowering season enhanced beetle population (George et al. 1992). The effect of nitidulid-pheromone bait station on sugar apple and atemoya fruit set was determined in southern Florida (Pena et al. 1999).

2.4 Guava (*Psidium guajava* L.)

Guava produces all bisexual flowers but about 35% cross pollination occurs due to insect foraging, especially honey bees (Kumar et al. 1996b). Guava is a highly cross pollinated crop and pollination by honey bees result in higher fruit yield. The different modes of pollination in guava showed significant flower drop in self-pollination indicating high degree of self-incompatibility. The fruit set, fruit weight, number of seeds per fruit, and test weight were statistically significant in guava orchard provided with five colonies of *A. cerana* compared to orchard without bee hives (Kuberappa et al. 2002) in India. In Brazil, flowers excluded from flower visitors set only 15.09%, while it was 39.5% in open pollinated ones (Alves and Freitas 2007).

Honey bees are the main pollinators of guava (Morton 1987) and accounted for 92.86% of total pollinators. Pollinator fauna of guava included *A. dorsata*, *A. mellifera*, *A. cerana indica*, *A. florea*, *Trigona iridipennis*, *Megachile lanata*, *Chalicodoma lanatum*, and *Xylocopa fenestrata* (Kuberappa et al. 2002). Reddy and Jacob (1999) studied the foraging behavior of honey bees on guava and found that the peak activity of all species was between 9 and 11 AM. Among the three species studied, *Apis dorsata* was the dominant forager followed by *A. mellifera* and *A. cerana*. Anitha et al. (2012) tested certain bee attractants for efficacy and found that a commercial formulation, viz., Bee Scent @1.5 ml/L was most effective in attracting the highest number of bees. Plants sprayed with attractants had fruits of higher weight and size compared to control.

2.5 Litchi (*Litchi chinensis*)

Litchi produces three basic types of flowers, i.e., male with no pistil (M_1), female with rudimentary stamen (F), and male with incomplete pistil (M_2) on the same panicle in succession (Costes 1988). As litchi flowers are functionally unisexual, fruit set occur only if pollen from male flowers (M1 and M2) is transferred to the stigma and hence they need an external agent (Stern and Gazit 1996). Wind pollination occur but fruit set is higher by insect pollination, and little or no fruit set is obtained under selfing. Litchi flowers are entomophilic, and pollination is performed by insects, mainly the honey bees (Free 1993; King et al. 1989). Pollination of the litchi by *Apis mellifera* was studied by Stern and Gazit (1996). The density of bees found on inflorescence was high during female flowering. Pollen density collected on bees was low during M1 flowering and increased to high values during M2 flowering. They also found a highly significant positive correlation between bee density and sugar concentration in the nectar.

Honeybees and stingless bees comprise more than 90% of the flower visitors on litchi. In India, 16 species of bees, flies, wasps, and other insects were recorded visiting litchi flowers by Singh and Chopra (1998). Most common were the honey bees, viz., *A. cerana*, *A. dorsata*, *A. mellifera*, and *A. florea* and syrphid flies, viz., *Melanostoma univittatum* and *Episyrphus balteatus*. Pollination by these insects resulted in 38.7% increase in fruit set and 50% increase in retention. In north India and in Thailand, *A. mellifera* was the most significant pollinator of litchi (Kumar et al. 1996a; Wongsiri and Chen 1995). Besides, stingless bee, *Trigona iridipennis* is also reported to help in litchi pollination (Kumar et al. 1996b). *Apis cerana* is a reliable pollinator litchi in Thailand, and at least one hive per acre is required to get desired level of fruit set (Wongsiri and Chen 1995; Du Toit and Swart 1995). There was 88 % reduction in fruit set when litchi flowers were excluded from insect foragers (Kumar et al. 1996a) and South Africa (Du Toit 1994). Chaudhary (2004) studied the influence of pollination by bees on the morpho-physiological characters of litchi fruit. The highest volume of fruits (1860 ml), fruit size (3.15 × 2.87 cm), fruit weight (19.96 g), pulp weight (12.47 g), skin weight (3.05 g), seed weight (4.34 g), and fruit-seed ratio (1:0.22) were observed in bee-pollinated litchi trees.

Apis dorsata on guava flower	Honey bees *A. cerana* foraging on passionfruit	*Apis dorsata* on litchi flowers

Fig. 16.1 Honey bees on the flowers

Mahanta and Rahman (1997) found five standard *A. cerana* colonies per hectare resulted in 124% increase in fruit yield over self-pollination, while the efficacy of *A. mellifera* in improving litchi fruit yield was reported by Kumar et al. (1996a, b) and Rai and Srivastav (2012). Details on litchi pollination are in another chapter in this volume (Fig. 16.1).

2.6 Mango (*Mangifera indica* L.)

Mango produces both male and hermaphrodite flowers, and the ratio of perfect to male flowers varies from 1% to 35% (Mukherjee 1953) depending on the cultivar and the time of emergence and its position on the tree (Singh 1960). Though mango trees produce millions of flowers, the fruit set is too low. The scanty productiveness of many mango varieties has been attributed by several workers to inadequate pollination. In spite of having perfect flowers, cross pollination by insects is essential to achieve adequate fruit set and research results proved the essentiality of insects in mango pollination. About 60–100% reduction in fruit set was observed when panicles were completely excluded from insect foraging (Bhatia et al. 1995; Reddy et al. 2011). Hermaphrodite flowers produce nectar around the ovary. Insects foraging for either nectar or pollen contribute to the transfer pollen to the stigma. Anthesis occurs early in the morning when the stigma also becomes instantly receptive. The nectar secreted from the opened flowers attracts a large number of insects (Free and Williams 1976).

In India, mango flowers are reported to be visited by more than 20 species of insects. Dipterans (flies) that constitute the major chunk of total species followed by hymenopterans (bees, ants, wasps), hemipterans (bugs), and lepidopterans (butterflies, moths). However, four species, viz. *Apis florea* (Hymenoptera:Apidae), *Chrysomya megacephala* and *Stomorhina discolor* (Diptera:Calliphoridae), *Eristalinus arvorum* (Dipteral:Syrphidae), and *Tetragonulairi dipennis* are the most frequent and dominant visitors significantly contributing to mango pollination (Reddy et al. 2012). In Israel, of 46 distinct species, 2 blow flies (*Chrysomya albiceps* and *Lucilia sericata*), *Apis mellifera*, and the housefly (*Musca domestica*) were found important in mango pollination (Dag and Gazit 2000). In Taiwan, 126 individual insects belonging to 39 species in 23 families, and 5 orders were recorded as visitor or pollinator on mango (Sung et al. 2006). Bees were recorded extracting extra-floral nectar from the bases of mango panicles which according to Du Toit and Swart (1993) would help to sustain honey bee colonies in South Africa.

Pollinator abundance is also influenced by varietal differences. For instance, cultivar 'Ratna' attracted highest number of insect foragers (3.24/10 panicles) while it was lowest in 'Sindhu' (0.63). Dipterans and bees showed varied preferences to varieties as was evident in case of 'Alphonso', which recorded the highest number of bees (2.6) while dipteran activity was the lowest (0.02) (Reddy et al. 2013b).

2.6.1 Pollinator Enhancement

As mango is a highly seasonal crop with a relatively short span of flowering, it is desirable to augment pollinators during that period to ensure sufficient pollination. Placing four to six bee colonies per acre of mango orchard had not only enhanced floral visitation but also stabilized fruit set. Studies from Israel also report the positive impact of introduction of honey bee and the bumblebee (*Bombus terrestris*) on mango yield (Dag and Gazit 2000). Besides honey bees, the activity of other pollinators, especially that of *C. megacephala*, can be enhanced by placing simple fish meal based traps. These traps were found to attract adult flies which breed on fish-based diet. These traps should be erected just before flowering so that population can be enhanced by the time full blossom occurs (Reddy et al. 2013a).

2.7 Passion Fruit (*Passiflora edulis*)

Passion fruit bears bisexual flowers that arise singly from the leaf axils. These blossoms are large, showy, fragrant, and rich in both nectar and pollen, which encourages visits from insects and enhances cross-pollination. The species is protandrous, as the anthers release pollen before the stigmas become receptive; however, stigma receptivity lasts throughout the flower's opening period (Cox 1957). The strong scent, abundant nectar secretion, and sticky pollen grains confirm the entomophilous nature of the crop (Free 1993).

In India, Kishore et al. (2010) examined the floral biology of purple, giant, yellow, and *P. foetida* types in Sikkim. Nishida (1963) documented a wide range of floral visitors in Hawaii, including insects from Diptera, Hymenoptera, Coleoptera, Thysanoptera, and Orthoptera, with *Xylocopa varipuncta* and the hoverfly *Eristalis arvorum* (Syrphidae) being the most frequent. In El Salvador, bumble bees (*Bombus* spp.), stingless bees (*Trigona* spp.), and carpenter bees (*Xylocopa* spp.) were the main pollinators (Free 1993). In Brazil, *Xylocopa* spp., *Epicharis* spp., and Africanized honey bees (*Apis mellifera scutellata*) dominated floral visitation (Ruggiero et al. 1976).

Other observations include *Eristalis arvorum* and the grasshopper *Conocephalus saltator* feeding on pollen in Hawaii (Akamine and Girolami 1957), while in Malaysia, *Patinopoda latipes* was considered the key pollinator (Mardan et al. 1991). In Venezuela, three carpenter bee species—*X. mordax*, *X. scutellata*, and *X. (Megaxylocopa) fimbriata*—were most effective. Free (1993) also noted the importance of honey bees, especially in lowland and humid upland conditions where they can be easily managed. In Florida, *Apis mellifera* serves as the sole pollinator of *P. edulis*, with up to 25% fruit set achieved (Hardin 1986).

In Indian conditions, *Apis mellifera* was the dominant pollinator of purple, giant, and yellow passion fruit, while *A. cerana* was more common in *P. foetida*. Peak bee activity occurred between 07:00 and 08:00 h in purple and giant varieties, and between 13:00 and 14:00 h in the yellow type (Kishore et al. 2010).

2.8 Other Fruit Crops

Contribution of bees to pollination and fruit set in underutilized fruit crops like Aonla, Avocado, Jackfruit, Macadamia, Rambutan, Strawberry, and Tamarind is also documented. Sphinx moths and honey bees are reported as major pollinators of Papaya. Pollination in figs by tiny wasps of the genus *Blastophaga* is a well-documented ecological interaction. These insects complete their life cycle within the gall flowers of wild caprifigs. In commercial orchards, growers traditionally gather caprifigs containing developing wasps and suspend them in the branches of Smyrna fig trees, ensuring pollination and fruit set. However, several modern fig cultivars are able to produce edible fruits without this intervention (Free 1993).

3 Pollinators of Vegetable Crops

3.1 Cucurbits

The Cucurbitaceae comprises vine-crops such as cucumber, ridge gourd, pumpkin, and watermelon. All melons and gourds are monoecious in nature (male and female flowers are borne at different positions on the same plant), and hence self-pollination is not possible. In addition, the sticky and heavy nature of pollen grains restrict wind pollination, and hence, these crops depend mainly on insects for pollination.

3.1.1 Bitter Gourd (*Momordica charantia* L.)

Pollination in bitter gourd is largely dependent on insect pollinators especially by honey bees. Among 17 species of floral insect visitors of bitter gourd, Hymenoptera alone constituted 87.56% of the total population of the pollinators followed by Diptera (5.81%), Lepidoptera (4.68%), and Coleoptera (1.95%). Among Hymenopterans, *Apis florea, Trigona iridipennis, Amegilla zonata* (Apidae), *Halictus gutturosus* and *Halictus* sp., (Halictidae) and *Megachile lanata* (Megachilidae) were more frequently found species. Lepidopterans were next in the order of diversity including *Pieris brassicae*, *Delias eucharis* and *Colotis eucharis* (Pieridae), *Pelopidas mathias* (Hesperidae), and *Danaus chrysippus* (Nymphalidae). The rest belonged to Diptera and Coleoptera, viz., *Eristalinusobscuritarsis*, *Asarkina* sp. (Syrphidae), *Rhinia* sp. (Rhiniidae) and *Illeiscincta*, *Coccinella transversalis* (Coccinellidae), *Aulacophorafoveicollis* (Chrysomelidae), respectively. Among these insects, commonly occurring foragers were *T. iridipennis* (77.52%) was the highest followed by *H. gutturosus* (7.60%) and *A. florea* (3.36%) on bright colored flowers of bitter gourd (Subhakar et al. 2011). Observations on foraging speed reflected that *T. iridipennis* spent the longer time (40.62 s/ flower), while *H. gutturosus* spent only 10.80 s. *A. florea* spent minimum time on flowers (Subhakar et al. 2011).

Basavaraj and Sattagi (2005) documented *Apis florea* as the most predominant forager on bitter gourd constituting 43% and the rest included *A. cerana* (26%), *A. dorsata* (13%), and others (18%). The activity of *Apis* pollinators was at peak between 10 am and 12 noon. The observations on various insect visitors of bitter gourd flowers revealed the presence of only diurnals. As flowers close completely during night, there was no activity of The absence of nocturnal pollinators (Subhakar et al. 2011). Bee pollination resulted in higher fruit set (84.14%) compared to open pollination (74.76%) (Dorjay et al. 2017).

3.1.2 Bottle Gourd (*Lagenaria siceraria*)

Bottle gourd flowers start opening in the evening time, so nocturnal flower visitors become important. Shrivastava and Shrivastava (1991) studied on bottle gourd (*L. siceraria*) crop and revealed that flower visitors mainly recorded during night time included lepidopterans and coleopterans. Sihag (1990) found *Xylocopa fenestrate* as a good pollinator of bottle gourd. Bottle gourd flowers are also frequented and foraged by *Cyrtopeltis tenuis*, sphingid moths, the beetle *Epilachna punctata*, and pyralid moths (Shrivastava 1990).

Since bottle gourd flowers open in the late evening, they are more foraged by nocturnal insects, mainly lepidopterans including sphingid/hawk moths, pyralids, pierids followed by coleopterans, hymenopterans, orthopterans, and dictyopterans. Ten insect foragers belonging to nine families of five Orders were recorded visiting flowers of bottle gourd. The lepidopterans were the major floral visitors comprising of six species, viz., *Arthoscistahilarialis* from Pyralidae and *Diaphania indica* from Crambidae, *Hippotioncelerio* from Sphingidae, *Delias eucharis* and *Pieris brassicae* from Pieridae, and *Anadevidiapeponis* from Noctuidae. These were followed by hymenopteran, viz., *Oecophyllasmaragdina* (Formicidae), coleopteran, *Aulacophorafoevicollis* (Chrysomelidae), orthopteran, *Hieroglyphus banian*, and dictyopteran, viz., *Mantis religiosa* L. (Subhakar et al. 2011).

3.1.3 Ridge Gourd (*Luffa* sp.)

Ridge gourd also bloom at late evenings like bottle gourd flowers. Hence, the composition of flower visitors was more or less similar to bottle gourd as in both the cases, nocturnal visitors play major role. Ramesh (2007) recorded 21 species of insect foragers on ridge gourd flowers. They comprised 12 Hymenopterans (12),

Dipterans (4), Coleopterans (3), and Lepidopterans (2). Of them, three honey bee species, viz., *Apis cerana*, *A. florea*, and *A. dorsata* were the dominant ones constituting 78% of all insect visitors.

In Andhra Pradesh, lepidopterans were recorded the major floral visitors of ridge gourd comprising six species, viz., *Diaphania indica* and *Glyphodesbivitralis* from Crambidae, *Hippotioncelerio* from Sphingidae, and *Delias eucharis* and *Pieris brassicae* from Pieridae (Subhakar et al. 2011). In Karnataka, 33 species of insect pollinators were recorded on ridge gourd flowers, of which 22 species belonged to Order Hymenoptera, five to Diptera, four to Coleoptera, and two to Lepidoptera. The foraging activity of *A. cerana*, *A. florea*, and *T. iridipennis* was at its peak in the morning hours, whereas other pollinators were observed between 1200 and 1600 h. Bee visitation had significantly improved the fruit parameters, viz., fruit set, weight, volume, and number of seeds per fruit. Similar observations were also reported by Lakshmi and Kuberappa (2015).

3.1.4 Cucumber (*Cucumis sativus*)

More than 25 species of insects are reported to forage on flowers of cucumber. Of them, about 60% were Hymenopterans, and rest belonged to Diptera, Lepidoptera and Coleoptera. *A. dorsata*, *A. cerana*, *A. florea*, and *T. iridipennis* comprised more than 82% of the total insect pollinators of cucumber crop (Prakash 2002; Sajjanar et al. 2004). Honey bee, viz., *A. cerana*, *A. florea*, and *A. dorsata*, constituted more than 84% of the total insect pollinators (Khaja Rubina 2010). In Assam also, *A. cerana* was found the predominant pollinator comprising 51.4% of the total pollinators population (Islam 2007). The peak foraging of A. cerana was recorded in morning hours (8–9 am). Cervancia and Bergonia (1991) documented that *Xylocopa chlorina*, *Xylocopa philippinensis*, *Megachilea trata*, and *A. dorsata* were the main pollinators of cucumber flowers in Philippines.

Meena and Rana (2008) reported significant increase in fruit set with bee pollination in cucumber (74.96%). Similar way, Rana et al. (2006) also recorded a significant positive impact of bee pollination on the seed weight. The cucumber crop pollinated by *A. cerana* enhanced 25.21% and 48.6% higher fruit yield than self-pollinated and pollinated by natural insects, respectively, and recommended seven *A. cerana* colonies per ha for optimum yield. Fruit yield was highest (74 t/ha) when *A. cerana* colonies were introduced at 5/ha compared to yield (24 t/ha) in field which had no bee colonies (Islam and Deka 2009).

3.1.5 Cole Crops

In cole crops, though economic yield is constituted by vegetative parts like leaves and underground bulbs, pollination is essential for seed production. Since they are of highly cross pollinated group, they depend on external agents for pollen transfer and insects play greater role. Honey bee, *A. cerana*, is a major pollinator of cauliflower seed crop. Open pollination is found to significantly increase seed set, seed weight, and yield over exclusion of insect pollinators. Similarly, carrot *Apis mellifera* and *A. cerana* are the main visitors, and an intensive bee pollination under caged conditions reported to have enhanced seed yield by 60–74% (Jhajj et al. 1996). In radish, all the *Apis* species are foragers, and there was 84.5% increase in seed yield in the bee pollinated crop over 22.49% in the pollination exclusion (Brar et al. 2010) (Table 16.1).

4 Pollinators and Climate Change: Potential Shifts and Implications

Plant–pollinator interaction is a highly evolved mutualism that has contributed immensely not only to the strengthening of ecosystems and biodiversity conservation but also to meet food and nutrition requirement of mankind. Climate change is one of major threats to pollination services (Potts et al. 2010). Since bees and other insect pollinators are ectothermic, the temperature of their surroundings determines their activity. Behavioral responses of pollinator insects to avoid extreme temperatures have the potential to significantly reduce pollination services. Shifts in climate cause changes in behavior of pollinators. The time taken for thermoregulation at higher temperatures comes at the cost of foraging (Memmott et al. 2007; Hegland et al. 2009).

A large proportion of crops are potentially vulnerable to declines in honey bee and other pollinator insects. Despite widespread concern about declines in pollination services, not much data are available on the patterns of change in most pollinator assemblages. Pollinator declines are presumed to be more frequent and of severe nature in specialist, univoltine, and non-migrant species (Abrol 2009). In the UK and the Netherlands, a 70% drop of wild flowers dependent on insect pollination since the 1980s has been recorded (Biesmeijer et al. 2006). A trend of 50–60% decline in honey bee numbers world over during last 15–25 years is forcasted (Gallai et al. 2009).

4.1 Impact of Climate Change on Plant-Pollinator Interactions

As per the estimates of the Intergovernmental Panel on Climate Change (IPCC), the average increase in world earth surface temperatures during the twenty-first century ranges from 1.1 °C under low emission scenario to 6.4 °C under high emission scenario (IPCC 2007). The likely impact of this rise in temperatures on the organisms depends on thermal sensitivity, and it is reported that the elevated temperature even at smaller level will have more serious implications in tropics than at higher latitudes (Deutsch et al. (2008). The rising atmospheric global temperature and CO_2 could disrupt the synchrony between plants and insects by advancing or delaying phenology leading to asynchrony (De Lucia et al. 2012). Successful pollination of crops relies on the synchronization of plant flowering and pollinator activity. Many crops, such as litchi and mango in tropical climates and almonds or apples in temperate regions, bloom in concentrated bursts over short timeframes, requiring a large number of pollinators at that particular time. Climate change threatens this balance by altering both flowering schedules and pollinator life

Table 16.1 Status of entomophily dependency of certain horticultural crops

Crop	Scientific name	Extent of dependency on entomophily	Major pollinators involved
Almond	*Prunus amygdalus*	Essential	Honey bees
Anise	*Pimpinella anisum*	Essential	Honey bee
Apple	*Malus pumila*	Essential	Honey bees, blue mason orchard bees
Apricot	*Prunus armenica*	Essential	Bees
Avocado	*Persea americana*	Essential	Bees, flies, bats
Banana	*Musa* sp.		Birds, fruit bats
Blueberry	*Vaccinium cyanococcus*	Essential	Over 115 kinds of bees, including bumblebees, mason bees, mining bees, and leafcutter bees
Cardamom	*Elettaria cardamomum*	Essential	Honey bees, solitary bees
Carrot	*Daucus carota* subsp. *sativus*	Essential	
Cashew	*Anacardium occidentale*	Moderate	Bees, moths, fruit bats
Citrus	*Citrus* spp.	Moderate	
Cauliflower	*Brassica oleracea var. botrytis*	Essential	
Cherry	*Prunus* spp.	Essential	Honey bees, solitary bees, flies
Coconut	*Cocos nucifera*		Insects and fruit bats
Coffee	*Coffeea arabica, C. robusta*	Essential of *robusta* coffee. However, bee pollination increases fruit yield	Stingless bees, other bees or flies
Coriander	*Coriandrum sativum*	Essential	Honey bees, solitary bees
Eggplant	*Solanum melongena*	Not essential but cross pollination increases fruit set	Solitary bees (bumble bees when grown under polyhouses)
Fennel	*Foeniculum vulgare*	Essential	Honey bees
Guava	*Pasidium guajava*	Essential	Honey bees
Grape	*Vitis vitifera*	Low	Bees
Grapefruit	*Citrus paradisi*	Essential	Bees
Jamun	*Syzygiumcumini*	Essential	Honey bees, flies
Kiwifruit	*Actinidia deliciosa*	Essential	Honey bees, solitary bees

(continued)

Table 16.1 (continued)

Crop	Scientific name	Extent of dependency on entomophily	Major pollinators involved
Litchi	*Litchi chinensis*	Essential	Honey bees,
Macadamia Nut	*Macadamia integrifolia*		Bees, beetles, wasps
Mango	*Mangifera indica*	Essential	Flies, bees
Nutmeg	*Myristica fragrans*	Essential	Honey bees, bird
Okra	*Abelmoschus esculentus*	Low	Honey bees
Oil Palm	*Elaeisguineensis*	Essential	Cameroon weevil
Onion	*Allium cepa*	Essential	Honey bees
Papaya	*Carica papaya*	Essential	Moths, birds, bees
Peach	*Prunus persica*	Moderate	Honey bees
Pear	*Pyrus* spp.	Essential	Honey bees, flies, mason bees
Pumpkin	*Cucirbita moschata*	Essential	Squash and gourd bees, bumblebees
Radish	*Raphanus raphanistrum* subsp. *sativus*	Essential	Honey bees
Strawberry	*Fragaria x ananasa*	Essential	Bees
Tomato	*Solanum lycopersicum*	Required under protected cultivation	Bumble bees for buzz pollination
Vanilla	*Vanilla planifolia*	Essential	Bees
Watermelon	*Citrullus lanatus*	Essential	Bees

With inputs from Free (1993), Roubik (1995) and different other sources cited under references

cycles. Insects and plants often respond differently to temperature shifts, creating temporal (phenological) and spatial (distributional) mismatches that can have serious demographic consequences.

For plants, these mismatches may reduce insect visitation and pollen transfer, while pollinators face diminished food sources. Global observations reveal that spring events are advancing by about 2.3 days per decade and species are shifting their ranges poleward by approximately 6.1 km per decade (Parmesan and Yohe 2003). Modelling by Memmott et al. (2007) demonstrated that temperature-driven shifts in flowering times reduced floral resources for 17–50% of pollinator species. Similarly, Williams et al. (2007) linked declines in British bumblebee populations to climatic niche changes and projected widespread losses of bee diversity across Europe.

Case studies highlight these risks: Kehrberger and AndreaHolzschuh (2019) reported that rising temperatures trigger earlier blooming of *Pulsatilla vulgaris*, potentially before the emergence of its solitary bee pollinators, threatening both the plant (a red-list species) and its pollinators. However, Kudo et al. (2004) observed that the emergence of bumblebee queens in Japan was not disrupted despite advances in crop flowering, indicating that synchrony outcomes differ by species and region.

The impacts are generally more severe for specialist pollinators, which depend on specific plants, whereas generalists—with broader diets and adaptive capacity—are less affected by asynchrony. Beyond timing mismatches, warming also alters plant physiology: water-stressed crops produce fewer, less attractive flowers, and changes in nectar and pollen quality have been observed under warming scenarios (Bartomeus et al. 2011).

4.2 Effects of Heightened CO_2 Levels on Pollinators

The consequences of elevated atmospheric carbon dioxide (CO_2) levels for pollinators and their mutualistic plant partners remain highly uncertain. Increased CO_2 concentrations can alter the carbon-to-nitrogen balance in plants, which in turn affects herbivory patterns, such as those observed in butterfly larvae (Rusterholz and Erhardt 1998). How such shifts cascade to influence pollinator communities is not yet well understood. Rising CO_2 also drives changes in plant community composition, particularly by influencing the relative abundance of C_3 and C_4 species within ecosystems. These changes interact with broader climate-related stresses, which often act synergistically with human-induced disturbances (Myers 1992).

Higher CO_2 availability typically enhances carbohydrate concentrations in nectar and stimulates greater reproductive investment in many plants (Erhardt et al. 2005). Sweeter nectar can be advantageous for certain pollinator groups, including bees and flies. However, the interaction between elevated CO_2 and temperature complicates predictions of plant–pollinator dynamics. For example, increased CO_2 has been shown to accelerate flowering in entomophilous C_3 plants (Springer and Ward 2007), whereas rising temperatures may constrain nectar production through water limitation or shortened floral longevity. Despite their importance, the combined effects of CO_2 enrichment and warming on plant–pollinator relationships remain poorly studied and warrant urgent investigation.

Climate change influences honey bees at multiple levels (Fig. 16.2). It may directly affect their biology, physiology, and even behavior, while also modifying the floral environment, thereby altering resource availability and impacting colony growth and foraging efficiency (Le Conte and Navajas 2008). Temperature fluctuations can further influence the developmental cycle, which varies across honey bee races. The colony's ability to store energy and regulate its development is a key adaptive trait. For instance, in early spring, rising temperatures stimulate the queen to resume egg-laying, leading to rapid worker population growth. However, a sudden cold spell during this period may force bees to forage despite limited resources, increasing the risk of starvation (Figs. 16.3 and 16.4).

Apis florea

Apis cerana

Apis dorsata

Chrysomya megacephala

Eristalinus arvorum

Fig. 16.2 Pollinator diversity on mango

4.3 Climate Change: Relevance of Unconventional Native Species

Studies on mango, onion, and watermelon conducted at ICAR-Indian Institute of Horticultural Research, Bengaluru, clearly indicated temperature as the crucial abiotic factor that affects pollinator activity. It was also evident that species had

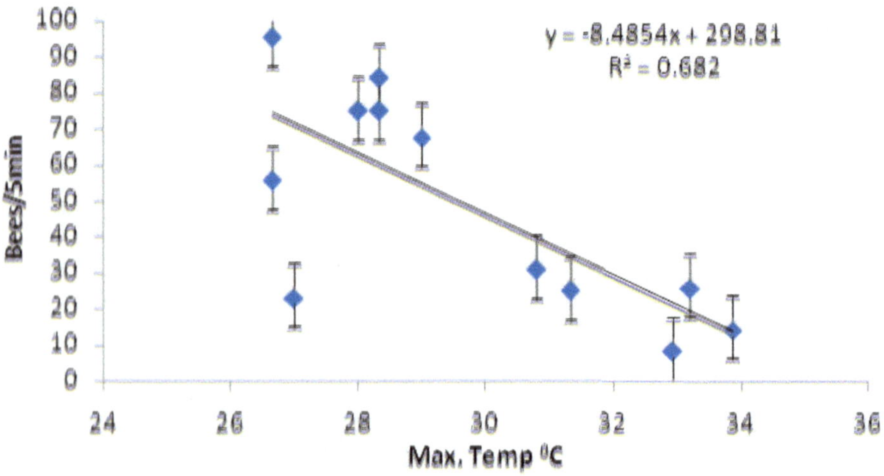

Fig. 16.3 Variability in worker bee outward movement as influenced by the maximum temperature

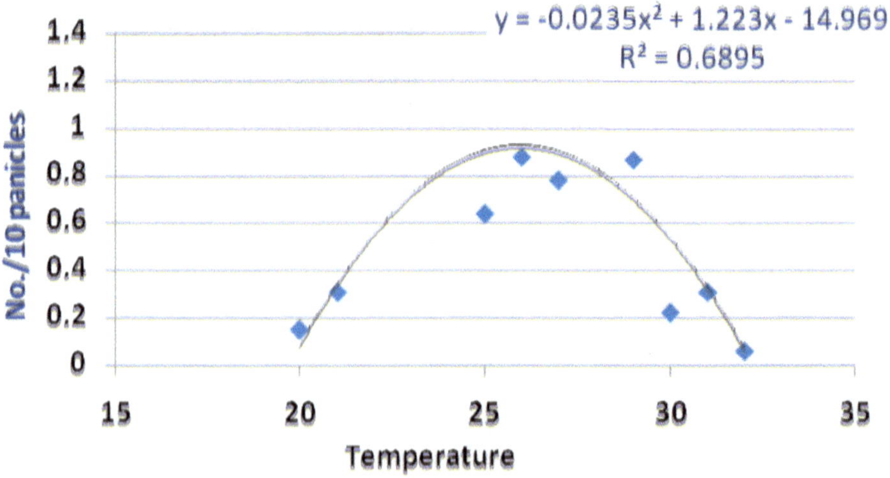

Fig. 16.4 Foraging behavior of *Apis florea* as influenced by temperature on mango

shown variability in their optimum thermal niche breadth. For instance, little bee *Apis florea* exhibited a decreasing trend in its foraging efficiency beyond 30 °C, while that of Calliphorid (*Chrysomya megacephala*) and Syrphids remained unaffected. Accordingly Levin's thermal niche index of these species was superior to that of *Apis* spp. It was also observed that the density of *C. megacephala* remained stable throughout the blossom period unlike bees whose maximum activity coincided with peak blossom. Specially designed experiments also proved the pollination efficiency of *C. megacephala* in mango and onion crops resulting in fruit comparable to *Apis* spp. Considering its merits, an artificial diet and mass rearing protocol has been developed to enhance pollinator populations through mass releases. Details on climate change and pollinators are furnished in another chapter in this volume.

5 Conservation of Pollinators

- Conserving native pollinators: Importance of native species in buffering the adverse effects of climate change has been established. Under the most extreme IPCC scenario (A1F1), pollination services by domesticated honey bees are reported to reduce by 14.5%, whereas pollination services provided by most native, wild taxa are predicted to increase, resulting in an estimated aggregate change in pollination services of +4.5% by 2099 (Rader et al. 2013). Hence, care should be taken to conserve native species.
- Avoid broad spectrum insecticide sprays especially during blossom period.
- Allow off-season flora like flowering weeds in and around the field, on hedges, bunds, etc.
- Do not burn natural habitats around the orchards.
- Augment natural populations through attractants (Varun Rajan and Reddy 2018) and low-cost fish meal traps (for Dipteran pollinators) (Reddy et al. 2017).

6 Conclusion

Research on pollinators in relation to climate change in different agricultural ecosystems remains limited. Knowledge gaps are particularly evident regarding the thermal sensitivity of crops and their pollinators, a concern that is especially relevant in tropical regions where insect-pollinated crops are highly diverse. In addition to honey bees, native and non-*Apis* pollinators deserve greater attention, as they contribute significantly to maintaining pollination services and mitigating the adverse effects of climate change. To bridge these gaps, comprehensive baseline studies are essential for understanding how climate change may reshape crop–pollinator interactions and ecological dynamics.

Acknowledgments The financial assistance from ICAR, New Delhi, through NICRA (National Innovations on Climate Resilient Agriculture) project is gratefully acknowledged. Thanks are also due to Director, ICAR-IIHR, Bengaluru, for facilitating and encouraging studies on pollinators. Authors acknowledges all original contributors whose works are cited in this paper. Special thanks to Ms. N. P. Jishna for her editorial assistance.

References

Abrol DP (1990) Conservation of pollinators for promotion of agricultural production in India. J Anim Morphol Physiol 38(1&2):123–139

Abrol DP (2009) Plant-pollinator interactions in the context of climate change - an endangered mutualism. J Palynol 45:1–25

Aizen MA, Harder LD (2009) The global stock of domesticated honey bees is growing slower than agricultural demand for pollination. Curr Biol 19:915–918. https://doi.org/10.1016/j.cub.2009.03.071

Akamine EK, Girolami DG (1957) Problems in fruit set in yellow passion fruit. Hawaii Farm Sci 5: 3–5

Alves JE, Freitas B (2007) Pollination requirements of guava. Ciência Rural 37(5):1281–1286

Anitha M, Sivaram V, Jayaramappa KV (2012) Influence of bee attractants on pollination and yield parameters in Guava (*Psidium guajava* L.). Int J Plant Reprod Biol 4(1):41–46

Bartomeus I, Ascher JS, Wagner D, Danforth BN, Colla S, Kornbluth S, Winfree R (2011) Climate-associated phenological advances in bee pollinators and bee-pollinated plants. Proc Natl Acad Sci U S A 108:20645–20649

Basavaraj R, Nidagundi, Sattagi HN (2005) Pollinator fauna and foraging activity of bees in bitter gourd. Karnataka J Agric Sci 18(4):982–985

Bhatia R, Gupta D, Chandel JS, Sharma NK (1995) Relative abundance of insect visitors on flowers of major subtropical fruits in Himachal Pradesh and their effect on fruit set. Indian J Agric Sci 65(12):907–912

Bhuyan M, Bhattacharyya PR (2002) Foraging of honeybee upon Assam lemon (*Citrus acida* Roxb.) under agro-climatic condition of Jorhat, Assam, India. Geobios 29:33–36

Biesmeijer JC, Roberts SPM et al (2006) Parallel declines in pollinators and insect-pollinated plants in Britain and The Netherlands. Science 313(5785):351–354

Brar HS, Gatoria GS, Chhuneja PK (2010) Requirement of Apis mellifera Linn. Colonies for maximizing seed yield of radish, *Rhaphanus sativus* Linn. J Insect Sci 23(4):366–374

Cervancia CR, Bergonia EA (1991) Insect pollination of cucumber (*Cucumis sativus* L.) in the Philippines. In: Proceedings of the Sixth International Symposium on Pollination, Tilburg, Netherlands, 27–31 August 1990, pp. 278–282

Chacoff NP, Aizen MA (2007) Pollination requirements of pigmented grapefruit (*Citrus paradisi* Macf.) from Northwestern Argentina. Crop Sci 47:1143–1150

Chaudhary N (2004) Factors influencing inter and intra specific preferences of honey bee for three concurrently flowering vegetable/ spice crops. Ph.D Thesis. CCSHAU, Hisar. 96+xxviii pp

Costes E (1988) Analyzearchitecturate et modelisation du litchi (*Litchi chinensis* Sonn.). PhD diss. University of Montpelier, France

Cox JE (1957) Flowering and pollination of passion fruit. Agric Gaz NS Wales 68:573–576

Dag A, Gazit S (2000) Mango pollinators in Israel. J Appl Hortic 2(1):39–43

De Lucia E, Nabity PD, Zavala J, Berenbaum MR (2012) Climate change: resetting plant-insect interactions. Plant Physiol 160(4):1677–1685

Deutsch CA, Tewksbury JJ, Huey RB, Sheldon KS, Ghalambor CK, Haak DC, Martin PR (2008) Impacts of climate warming on terrestrial ectothermsacross latitude. Proc Natl Acad Sci USA 105:6668–6672

Dorjay N, Abrol DP, Uma Shankar (2017) Insect visitors on cucumber and bitter gourd flowers and impact on quantity of crop production by different pollination treatment. J Apicult 32(2):77–88

Du Toit A (1994) Pollination of avocados, mangoes and litchis. Inligtingsbulletin Institut vir Tropiese en Subtropiese Gewasse 262:7–8

Du Toit AP, Swart DJ (1993) Yearbook South African Mango Growers' Association 13:129–130

Du Toit AP, Swart DJ (1995) Pollination and fruit set in Bengal litchi during 1993 at Politsi: second report. Year Book South Afr Litchi Growers Assoc 7:31–32

Erhardt A, Rusterholz HP, Stöcklin J (2005) Elevated carbon dioxide increases nectar production in *Epilobium angustifolium* L. Oecologia 146:311–317

Farre M, Perez JM, De los Hermoso MA, Orgata R (1997) Estudio sobre polinizacion de chirimoyo (*Annona cherimola* Mill) en Espana. Memorias Congreso International de Annonaceas. Chapingo, Mexico, pp 43–55

Free JB (1966) The pollinating efficiency of honey bee visits to apple flowers. Jour Hort Sci 41:91–94

Free JB (1993) Insect pollination of crops. Academic Press, London. 684 pp

Free JB Williams IH (1976) Insect pollination of *Anacardium occidentale* L., *Mangifera indica* L., *Blighiasapida*, Koenig

Gallai N, Salles JM, Settele J, Vaissi BE (2009) Economic valuation of the vulnerability of world agriculture confronted with pollinator decline. Ecol Econ 68:810–821

Galon, I., Gazit, S. And Podoler, H (1982) Improvement of natural fruit set of Annona by increasing the population of pollinating inscets. Alon Halontea 9, 611–614

George AP, Nissen RJ, Campbell JA (1992) Pollination and selection in *Annona* species (cherimoya, atemoya and sugar apple). Acta Horticult 321:178–185

Gogoi B, Rahman A, Rahman S, Deka MK (2007) Foraging behavior and effect of *Apis cerana* pollination on fruit set and yield of Assam lemon (*citrus lemon* L. burn). Indian J Agric Sci 77(2):120–122

Gottsberger G (1970) BeitragezurBiologie von Annonaceen-Bluten. Osterreichische Botanische Zeitschrift 118:237–279

Gottsberger G (1988) The reproductive biology of primitive angiosperms. Taxon 37:630–643

Hardin LC (1986) Floral biology and breeding systems of the yellow passion-fruit, *Passiflora edulis* f. *flavicarpa*. Proceedings of the Interamerican Society for Tropical Horticulture 30:35–44

Hegland SJ, Nielsen A, Lázaro A, Bjerknes AL, Totland (2009) How does climate warming affect plant-pollinator interactions? Ecol Lett 12:184–195

IPCC (2007) Climate change 2007: synthesis report - contribution of Working Groups1, 2 and 3 to the Fourth Assessment Report of the Intergovernmental Panel on Climate Change. In: Change IPoC, Geneva

Islam N (2007) Role of Apis cerana (Hymenoptera, Apidae) on fruit set and yield of cucumber, M. Sc. thesis. Assam Agricultural University, Jorhat, p 47

Islam N, Deka MK (2009) Role of *Apis cerana* (Hymenoptera: Apidae) on fruit set and yield of cucumber. Geobios 36:161–164

Jhajj MS, Gatoria GS, Bakhetia DRC (1996) Three decades of bee keeping in Punjab. In: National Conference on Agriculture and Technology, Ludhiana, pp 23–27

Kehrberger S, AndreaHolzschuh A (2019) Warmer temperatures advance flowering in a spring plant more strongly than emergence of two solitary spring bee species. PLoS One 14(6): e0218824. https://doi.org/10.1371/journal.pone.0218824

Khaja Rubina S (2010) Pollinators diversity with special reference to role of honey bees in quantitative and qualitative improvement of cucumber, *Cucumis sativus* L. M.Sc. Thesis. University of Agricultural Sciences, Bangalore

King J, Exley EM, Vithange V (1989) Insect pollination for yield increases in lychee. In: Batten DJ (ed) Proceedings on 4th Australasian Conference on Tree and Nut Crops. Exotic Fruit Growers Assn, Lismore, pp 142–145

Kishore K, Pathak KA, Shukla R, Bharali R (2010) Studies on floral biology of passion fruit (*Passiflora* Spp.). Pak J Bot 42(1):21–29

Kuberappa GC, Eswarappa G, Vishweshwaraiah CH, Sajjanar SM (2002) Activity and abundance of insects visiting guava (*Psidium guajava* L.) flowers. Pest Manage Econ Zool 11:59–63

Kudo G, Nishikawa Y, Kasagi T, Kosuge S (2004) Does seed production of spring ephemerals decrease when spring comes early? Ecol Res 19:255–259

Kumar R, Chaudhary OP, Lenin JK (1996a) Studies on pollination of litchi in Dehradun, India. Indian Bee J 58:9–12

Kumar R, Lenin JK, Chandran K (1996b) Honey bee and pollination of *Psidium guajava* L. Indian Bee J 56(4):206–207

Lakshmi KS, Kuberappa GC (2015) Impact of honey bees pollination on ridge gourd. Indian J Entomol 77(2):198–199. https://doi.org/10.5958/0974-8172.2015.00040.1

Le Conte Y, Navajas M (2008) Climate change: impact on honey bee populations and diseases. Rev Sci Tech 27(2):499–510

Linsley EG (1958) The ecology of solitary bees. Hilgardia 27:543–599

Lopez E, Uquillas C (1997) *Carpophilushemipterus* (Coleoptera: Nitidilidae) as cherimoya (*Annona cherimola* Mill.) pollination agent under controlled conditions. Acta Entomol Chilena 21:89–99
Mahanta M, Rahman A (1997) Studies on pollination of litchi (*Litchi chinensis* Sonn.) by honey bee (*Apis cerana* F.). Indian Bee J 59(2):74–75
Malerbo-Souza DT, Nogueira-Couto RH, Couto LA (2004) Honey bee attractants and pollination in sweet orange, *Citrus sinensis* (L.) Osbeck, var. Pera-Rio. J Venom Anim Toxins Incl Trop Dent Discourse 10(2):144–153
Mardan M, Yatim IM, Khalid MR (1991) Nest biology and foraging activity of carpenter bee on passion fruit. In: The 6th International Symposium on Pollination. Tilburg, pp 127–132
Memmott J, Craze PG, Waser NM, Price MV (2007) Global warming and the disruption of plant–pollinator interactions. Ecol Lett 10:710–717
Meena T, Rana RS (2008) Studies on the role of insect pollination in cucumber yield. Pest Technol 2(2):130–133
Morton J (1987) Guava. In: Morton JF (ed) Fruits of warm climates. Julia F. Morton, Maimi
Mukherjee SK (1953) The mango, its botany, cultivation, uses and future improvement. Econ Bot 7(2):130–162
Myers N (1992) Synergisms: joint effects of climate change and other forms of habitat destruction. In: Peters RL, Lovejoy TE (eds) Global warming and biological diversity. Yale University Press, Connecticut, pp 344–354
Nadel H, Pena JE (1994) Identity, behavior and efficacy of nitidulid beetles (Coleoptera: Nitidulidae) pollinating commercial Annona species in Florida. Environ Entomol 23:878–886
Nagamitsu T, Inoue T (1997) Cockroach pollination and breeding system of Uvariaelmeri (Annonaceae) in a lowland mixed- dipterocarp forest in Sarawak. Am J Bot 84:208–213
Nishida T (1963) Ecology of the pollinators of the passion fruit. Hawaii Agric Exper Station, Honolulu, Tech Bull 55:356–363
Parmesan C, Yohe G (2003) A globally coherentfingerprint of climate change impacts acrossnatural systems. Nature 421:37–42
Pena JE, Bennett FD (1995) Arthropods associated with Annona spp. in the neotropics. Fla Entomol 78:329–349
Pena JE, Castineiras A, Bartelt R, Duncan R (1999) Effect pf pheromone bait stations for sap beetles (Coleoptera: Nitidulidea) on Annona fruit set. Fla Entomol 82:475–480
Phartiyal T, Srivastava P, Khan MS, Srivastava RM (2012) Abundance of insect pollinators in citrus crop and their impact on fruit setting under tarai Agroclimatic condition. J Entomol Res 36(3): 211–213
Potts SG, Jacobus CB, Kremen C, Neumann P, Schweiger O, William EK (2010) Global pollinator declines: trends, impacts and drivers. Trends Ecol Evol 25(6):345–353
Prakash KB (2002) Pollination potentiality of Indian honeybee viz. *A. cerana* on cucumber Ucumis sativus (Linn.) S.W. (Cucurbitaceae). M. Sc. (Agri.) Thesis, University of Agricultural Sciences, Bangalore
Rader R, Reilly J, Bartomeus I, Winfree R (2013) Native bees buffer the negative impact of climate warming on honey bee pollination of watermelon crops. Glob Chang Biol 9(10):3103–3110
Rai V, Srivastav P (2012) Studies on the impact of bee pollination on yield and quality of litchi (*Litchi chinensis* Sonn.). Prog Hort 44(2):262–264
Ramesh RG (2007) Pollination potentiality of honey bee species in ridge gourd (*Luffa acuntangula* L.). M.Sc. Thesis. University of Agricultural Sciences, Bangalore
Rana RS, Rana BS, Joshi AK (2006) Pollinator fauna in *Cucumis sativus* Linn. and their effect on seed productivity in Solan district of Himachal Pradesh. Proc. Natl. Seminar Sustainable Beekeeping Development and Honey Festival, RAU, Pusa, Bihar, pp 36
Reddy PVR, Jacob TK (1999) Foraging behaviour of honeybees in guava. Insect Environ 5(2):96
Reddy PVR, Dinesh MR, Verghese A, Vasugi C (2011) Insect Pollinators: A key component to increase fruit set in mango. Presented in the Global conference on Augmenting Mango Productivity and Utilization. 21–24 June 2011, Lucknow

Reddy PVR, Varun Rajan V, Verghese A (2012) Potential impact of climate change on honey bees (*Apis* spp.) and their pollination services. Pest Manage Horticult Ecosyst 18(2):121–127
Reddy PVR, Varun Rajan V, Verghese A (2013a) A semi-synthetic diet based trap to attract and monitor the calliphorid pollinator, *Chrysomya megacephala* (F.). 100th Indian Science Congress, Kolkata, 3–7 January 2013
Reddy PVR, Varun Rajan V, Vasugi C, Verghese A (2013b) Pollinator activity in relation to varietal differences in mango (*Mangifera indica* L.) International Conference on Insect Science, Bangalore, 14–17 February 2013
Reddy PVR, Varun Rajan V, Vasugi C (2017) Conservation of mango pollinators under climate change scenario. Technical Folder. Published by ICAR-IIHR, Bengaluru
Reiss A (1971) Pollination and fruit set of *Annona cherimola* and *Annona squamosa*. M.Sc. Thesis, the Hebrew University of Jersalem. Israel Report (1970):101–102
Roubik DW (1995) Pollination of cultivated plants in the tropics. FAO Agric Serv Bull 118, FAO, Rome
Ruggiero C, Banzatto DA, Lam-Sanchez A (1976) Studies on natural and controlled pollination in yellow passion fruit (Passiflora edulis f. flavicarpa Deg.). Acta Horticult 57:121–124
Rusterholz HP, Erhardt A (1998) Effects of elevated CO2 on flowering phenology and nectar production of nectar plants important for butterflies of calcareous grasslands. Oecologia 113: 341–349
Sajjanar SM, Kuberappa GC, Prabhuswamy HP (2004) Insect visitors of cucumber (*Cucumis sativus* L.) and the role of honey bee *Apiscerana* F., in its pollination. Pest Manage Econ Zool 12(1):23–31
Sanford MT (1992) Pollination of citrus by honey bees. RFAA092. Florida Coop. Ext. Serv., Inst. of Food and Agric. Sci., University of Florida
Sharma HK (2000) Pollination studies on apple and pear. In: Seventh IBRA conference on Tropical Bee and Fifth Asian Apiculture Association Conference, Ching Mai, Thailand, pp 56–13
Shrivastava U (1990) Insect pollination in some cucurbits. In: The Sixth International Symposium on Pollination. Tilburg, pp 27–31
Shrivastava GP, Shrivastava U (1991) Coevolution of stamens and carpels in cucurbits and of their insect pollinators. In: Proceedings of The Sixth International Symposium on Pollination, Tilburg, pp 347–353
Sihag RC (1990) Behaviour and ecology of sub-tropical carpenter bee, *Xylocopa fenestrata* F. 2. Host plant association. Indian Bee Jounal 52(1–4):35–40
Singh LB (1960) The Mango. Leonard Hill, London
Singh R, Chopra SK (1998) Flower visitors of litchi and their role in pollination. Pest Manage Econ Zool 6:1–5
Singh VP, Misra KK (2007) Pollination management in apple for sustainable production. Progress Horticult 39(2):139–148
Solomon ME, Kendall DA (1970) Pollination by the Syrphid fly, Eristalistenax. Long Ashton Research Station. Annual
Springer CJ, Ward JK (2007) Flowering time and elevated atmospheric CO_2. New Phytol 176:243–255
Stern RA, Gazit S (1996) Lychee pollination by the honey bee. J Amer Soc Hort Sci 120(1): 152–157
Subhakar G, Sreedevi K, Manjula K, Reddy NPE (2011) Pollinator diversity and abundance in bitter gourd, *Momordica charantia* Linn. Pest Manage Hortic Ecosyst 17:23–27
Sung IH, Lin M-Y, Chang C-H, Cheng A-S, Chen W-S (2006) Pollinators and their behaviors on mango flowers in southern Taiwan. Formosan Entomol 26:161–170
Thakur M (2012) Bees as pollinators—biodiversity and conservation. Int Res J Agric Sci Soil Sci 2(1):001–007
Varun Rajan V, Reddy PVR (2018) Natural attractants enhance pollinator visitation frequency to mango flowers. Pest Management in Horticultural Ecosystems 24(2):101–104

Verma LR, Chauhan P (1985) Distribution, abundance and diversity of insect pollinators in apple orchard of Shimla hills. Indian J Ecol 12:286–292

Williams PH et al (2007) Can vulnerability among British bumblebee (Bombus) species be explained by niche position and breadth? Biol Conserv 138:493–505

Wongsiri S, Chen P (1995) The effects of agricultural development on honey bees in Thailand. Bee World 76:3–5

Non-Apis Pollinators: Role and Conservation 17

U. Amala, H. S. Venu, A. N. Shylesha, and T. M. Shivalingaswamy

Abstract

Non-Apis pollinators among insects play a vital role in pollination process. Non-Apis bees like leaf cutters bees, sweat bees, carpenter bees, minor bees, furrow bees and others play a part in pollination. Besides hymenoptera, anthophilous, dipterous flies, beetles, butterflies, moths, thrips and other insects play a role in pollination. This chapter outcomes and conservation importance of non-Apis pollinators. These pollinators constitute an important component of pollinators in the wild and significantly contribute to pollination process.

Keywords

Non-Apis bees · Management · Conservation · Ecosystem services

1 Contribution of Non-Apis Pollinators to Agriculture

1.1 Pollination Is a Vital Ecosystem Service

Pollination by animals was estimated to account for 75% of the crop plants cultivated worldwide. The pollination service provided by the non-Apis bees accounts for a monetary value of $219 billion per year, contributing to 9.5% of global crop value (Morse and Calderone 2011). Potts et al. (2011) documented that native non-Apis bees provide pollination services to an economic value of $937 million and $2.4 billion per year. Wild non-Apis bees possess wide adaptability, polylectic in foraging behaviour, less prone to pests and disease attack and greater survival ability. On

U. Amala (✉) · H. S. Venu · A. N. Shylesha · T. M. Shivalingaswamy
ICAR-National Bureau of Agricultural Insect Resources, Bengaluru, Karnataka, India

A. K. Chakravarthy (ed.), *The Role of Pollinators in Tropical Landscapes*,
https://doi.org/10.1007/978-981-95-9139-8_17

the contrary, honeybees require greater hive maintenance with frequent hive inspection for pests and diseases, viral, mite and fungal pathogens that pose a threat to bees survival and development.

In India, the quantitative contribution to the pollination and yield enhancement in agricultural and horticultural crops by the native non-Apis bees lacks documentary evidence. Chaudhary et al. (2013) reported that native pollinators contribute to a worth of $40 billion worth of crops produces annually in the US alone. The major insects that serve as crop pollinators are bees (66%), 19% by flies, 6.5% by bats, 5% by wasps and beetles, 3.5% by birds, butterflies and moths. Free (1993) reported that of the total crops pollinated, only 15% were pollinated by domesticated honeybees and 80% by native non-Apis bees.

1.2 Need of Non-Apis Pollinators

The key family taxon executing pollination service is Apidae (Klein et al. 2007), but a worldwide decline in managed colonies of honeybees has increased reliance on diverse communities of wild bees for crop pollination services. Honeybees, mainly *Apis mellifera*, remain the most economically valuable pollinators of crop monocultures worldwide also shown for several single crops and yields of horticultural crops decrease by more than 90% without pollinators (Watanabe 1994).

Certain crops like almond, blueberry, raspberry, tomato, cherry and apples directly depend upon solitary bees for effective pollination and fruit set. Non-Apis bees also serve as efficient pollinators of flowers with distinct floral morphology, like poricidal anthers of Solanaceae flowers. Buzzing activity by native buzz pollinating bees is vital to dislodge the pollen from the anther cone to fall on the stigma to cause pollination. Honeybees are incapable of buzz pollination; flower visitation by the native buzz pollinating bees has a definite role in the pollination and fruit set in Solanaceae flowers. Honeybees are more prone to several threat factors, viz.: (i) the spread of pests like parasitic mites (*Varroa jacobsoni*, *V. destructor* and *Acarapis woodi*) the small hive beetle (*Aethina tumida*) and the microsporidian parasite *Nosema ceranae*; (ii) improper pesticide and herbicide use; and (iii) loss of honeybees due to Colony Collapse Disorder (Kevan and Viana 2003; Garibaldi et al. 2009). Moreover, maintenance of honeybee colonies demands trained manpower with weekly hive monitoring for pests and diseases to protect the colonies from swarming and absconding. Native non-Apis bees are gaining attention in recent days as they have diverse foraging preferences and versatile nesting behaviour. Conservation of native non-Apis bees is gaining attention for utilisation for managed pollination services.

2 Non-Apis Bees

2.1 Leafcutter Bees

Leafcutter bees are solitary in habit; construct their nests in hollow cavities using leaf bits, resin, soil and flower petals; and were reported to be pollinators of papilionaceous flowers (Amala and Shivalingaswamy 2022) (Fig. 17.1). They prefer to construct their nests in hollow pre-existing cavities of size 5–10 mm in diameter (Amala et al. 2017, 2019). The mated female bees forage for the pollen, collect the pollen grains in their abdominal scopae and bring it back to their nest. She will construct linear cells (4–5 cells per nest). She lays down one pollen ball per cell and oviposits on the top of the pollen ball to ensure food availability for her hatching broods till pupation. The larvae, upon hatching, feed on the pollen ball and pupate inside the cell. After the pupal stage, the emerging adult bee will chew open the leaf bits and fly out from the nest.

2.2 Carpenter Bees

Carpenter bees belonging to the subfamily Xylocopinae (Figs. 17.2 and 17.3) construct nests in hardwood or soft pithy stems by tunnelling the pithy material, constructing cells and provisioning with pollen balls in each cell. The bee makes partitions in the cells using the chewed-up material. The larvae feed on the pollen mass, pupate inside the cell and the adult chews apart the partition layer and emerges out from the nest. Large carpenter bees of the genus *Xylocopa* constructs unbranched or linear nests in hardwood stems/branches and were important pollinators of leguminous crops (Ali et al. 2016). Small carpenter bees of the.

Fig. 17.1 Leaf cutter bee, *Megachile* sp.

Fig. 17.2 Large carpenter bee, *Xylocopa* sp.

Fig. 17.3 Small carpenter bee, *Ceratina hieroglyphica*

genus *Ceratina* constructs nests in soft pithy stems by making a series of cells in the stems and provisioning pollen ball in the cells ovipositing on the pollen ball (Amala et al. 2019).

Fig. 17.4 *Hoplonomia westwoodi*

2.3 Miner Bees

Miner bees are small, medium-sized bees' solitary in habit that nests underground during the onset of early spring season. Mated adult female bees burrow the soil for nest construction, forage for the pollen, and carry the pollen in the hind legs for provisioning the brood. They prefer to nest in well-drained soil that eases the burrowing activity.

2.4 Sweat Bees

They are solitary bees that belong to the family Halictidae that construct nests in barren soil. The adult female bees burrow into the soil, constructing branched cells in clusters underground with pollen provision in each cell. Pearl banded bees belong to the subfamily Nomiinae (Fig. 17.4) and are efficient pollinators of Solanaceous vegetables. These bees construct their nests in flower pots or barren grounds, with the presence of bees sighted at the nest entrance with a chimney-shaped turret. The chimney-shaped turret collapsed during rainy days and the nest-founding bees were involved in the repair process. The nest architecture of *H. westwoodi* has a main shaft that ends in laterals ending in brood cells with a pollen provision for the immature stages (Amala and Shivalingaswamy 2018).

3 Non-Apis Bees: A Potential Group of Buzz Pollinators

Native non-Apis bees are potential buzz pollinators of several solanaceous vegetables. Honeybees seldom visit the flowers of Solanaceae crops as they are incapable of buzz pollination. The flowers of Solanaceae have poricidal anthers wherein the pollen grains are trapped inside the anther cone, which needs a forceful wind to dislodge the pollen grains that fall on the stigma to effect the pollination. During the flower visitation, the buzz-pollinating bees vibrate their thoracic wing muscles creating a buzzing vibration and helping to dislodge the pollen from the anther cone. Amala and Shivalingaswamy (2019) reported that flower visitation by two buzz-pollinating native bee species, viz., blue banded bee, *Amegilla zonata* (Anthophorinae: Apidae) and sweat bee, *Hoplonomia westwoodi* (Nominae: Halictidae) significantly increased the fruit and seed set in tomatoes under field conditions. The solitary bee *Amegilla holmesi* is an ideal candidate pollinator for greenhouse tomatoes as evidenced by increased fruit yields and quality (Bell et al. 2006).

4 Wasps as Pollinators

Wasps belongs to the group "Aculeata," which possess a feature of modification of ovipositors to stingers. These insects mimic the bees except for lack of fuzzy hairs in their body (Fig. 17.5).

5 Fig Wasps

Fig wasp, *Blaptophaga psenes* (Family: Agaonidae), and fig trees have a unique mutualistic relationship where the wasps pollinate the figs while simultaneously using the fig as a breeding ground. Capri figs are wild, inedible figs that house the fig

Fig. 17.5 Wasps foraging on flowers

wasp and are crucial for pollinating Smyrna figs, which are the edible figs with superior flavour. Smyrna figs require the caprifig and its wasp for pollination to develop properly. The lifecycle of the wasp is typified in the wild edible caprifig, *Ficus carica sylvestris*. Figs are inverted flowers (syconia) with the flowers growing on the inside of a hollow structure. Female fig wasps enter figs (syconia) through a tiny opening, depositing eggs and pollinating the flowers inside. Male wasps, wingless and smaller, emerge first; mate with the females; and then die. The fertilised female wasps, now carrying pollen, exit the fig and fly to another to repeat the process, ensuring fig pollination and wasp reproduction. In a shorter life span of adult female wasp (as short as two days), she flies out to fertilise another fig and deposit another generation of fig wasps (Suhs et al. 2009). This relationship is obligate, meaning neither the fig nor the wasp can reproduce successfully without the other.

The female wasps pollinating the Smyrna fig (*Ficus carica*) are crucial for growers, as proper fertilisation is vital for the uniform ripening of the fruits (Liu et al. 2011). The wasps carry pollen from the caprifig to the Smyrna fig, enabling fertilization and fruit development. Growers plant Capri fig adjacent to Smyrna fig to aid in pollination. The process of using caprifigs to pollinate Smyrna figs is called caprification.

6 Myrmecophily

6.1 Ants as Pollinators

Ants visit the floral parts to collect nectar, and pollen gets attached to their body parts incidentally, resulting in pollination (Fig. 17.6). It was reported that the secretions of the metapleural gland contain an antimicrobial secretion that decreases the pollen viability in different plants.

Fig. 17.6 Ants visiting different flowers

7 Myophily

7.1 Flies as Pollinators

Dipteran flies are the most abundant among the insect orders and were reported to be incidental pollinators. These true flies have a single pair of wings, a soft body with short, bristly antennae. The major families of Diptera reported to be efficient pollinators were Syrphidae, Bombyliidae, Tachinidae, Calliphoridae and Empididae (Sajad and Saeed 2012) (Fig. 17.7).

Flies prefer to forage in the flowers with the following attributes:

- Pale and dull to dark brown or purple
- Sometimes flecked with translucent patches
- Putrid order, like rotting meat, carrion, dung, humus, sap and blood
- Nectar guides not present
- Produce pollen and funnel-like flowers

7.2 Syrphid Flies as Pollinators

Dipteran flies are a vital group of pollinators after the order Hymenoptera, with a diverse range of species (Mitra 2010). They are commonly called "hover flies" or "syrphid flies" hover around the flowers to collect pollen and nectar for their egg maturation and oviposition. Adult flies after emergence visit the flowers to meet their energy requirements for further flights and mating events. Hover flies efficiently handle the smaller-sized flowers and are reported to have pollen carrying ability for longer distances compared to honeybees (Meyer et al. 2009). These flies were reported to be efficient pollinators of crops like onion, mustard and other crops. The flies capable of making active hovering flights with hairy body parts enable easier pick up of the pollen grains from the flowers and efficient pollen transfer.

Fig. 17.7 Visitation of flies in different flowers

Rotheray and Gilbert (2011) reported that syrphid fly, *Episyrphus balteatus*, possess a tongue length of 2.9 mm with a shorter proboscis and larger labella was reported to be an efficient pollinator of shallow open flowers. Adult flies actively hover around the flowers foraging for the nectar to meet the energy requirements for flight, dispersal and seeking mates for oviposition (Larson et al. 2001). A species of hover fly, *Ischiodon scutellaris*, was reported to pollinate the orchid species, *Epipactis veratrifolia*, in the Western Himalayas while laying eggs inside the flowers (Kumar and Rawat 2011).

7.3 Pollination in Mango by Dipteran Flies

The pollination and fruit set in mango were reported to be facilitated by wind to some extent, but insects were the prime pollinators in mango. Floral visitation by the insects during the flowering period increases the percent fruit set in mango (Fig. 17.8). Fajardo et al. (2008) reported 53% increase in fruit set in mango through visitation by insects during the anthesis period. Honeybees, *Apis* sp., flies viz., *Musca domestica* and *Chrysomya megacephala* were the major pollinators in mango that increased the fruit set and yield in Taiwan (Sung et al. 2006). Per cent fruit set in open-pollinated flowers was the highest (13.5 ± 6.5 / panicle), followed by the flowers pollinated by the calliphorid fly, *Chrysomya megacephala* (11.3 ± 5.4/ panicle) and pollination by the little bee, *Apis florea* (10.09 ± 4.7/ panicle) (Rajan and Reddy 2019). Among the different pollinators foraging in the flowers of mango, *Apis florea* was the most abundant (3.86/panicle/minute) followed by a calliphorid, *Chrysomya megacephala* (1.62/panicle/minute) and a syrphid, *Eristalinus arvorum* (0.70/panicle/minute) (Reddy et al. 2010).

Fig. 17.8 Flies in mango flowers

7.4 Biting Midges

Midge, Forcipomyia hardyi (Ceratopogonidae: Diptera) was reported to be economically important pollinators of cocoa bean, *Theobroma cacao* L. To encourage the natural breeding of the midges, piling of damaged pseudostems of banana and harvested cocoa pods was advocated. Winder (1978) reported that immature stages of the midges survive in different parts of the plant and emerged adults visits the flowers resulting in pollination exhibiting a co-evolved mechanism of pollination. Insect visitation during anthesis is vital for the pollination of cocoa flowers (Doherty and Zoll 2012). Groeneveld et al. (2010) reported that crawling movements and vigorous wing movements of the midges aid in efficient pollination and fruit set in cocoa.

8 Cantharophilly

8.1 Beetles as Pollinators

Sap beetles belonging to the family Nitidulidae were the efficient pollinators in custard apple, *Annona* sp. These beetles breed in the decomposing fruits and are attracted toward the fruit trees during the bloom period. They were capable of transferring the pollen from the stamens to the stigma during the period of anther dehiscence (George et al. 1989). The beetles were observed to transfer the pollen from the male to female flowers while they feed on pollen and nectar from the flowers. Different species of beetles like *Carpophilus hemipterus, C. mutilates* and *Haptoncus luteolus* were the commonly reported pollinators.

The osmophores located at the internal base of the petals release the odours (Gottsberger and Silberbauer-Gottsberger 2006). The temperature of the flower buds increases during the anthesis, which causes the volatilisation of the flower odours that attract the pollinating beetles. The flower petals provide a concealed environment for the beetles. The beetles gain entry into the flower, gradually feed on the pollen and nutritive tissue of the petals. The petals sheltering the beetles ensure frequent movement of the beetles touching the stamen and stigma, causing the pollination in custard apple (Paulino-Neto 2014). The flowers of the custard apple help the pollen-carrying beetles by serving as a protective chamber against the adverse physical/biotic factors, pollen as a food and a site of mating for the adult beetles (Fig. 17.9).

9 Role of Oil Palm Weevil (*Elaeidobius kamerunicus*) in Pollinating Oil Palms

Oil palm is an entomophilous crop. Pollinating weevil, *Elaeidobius kamerunicus* (Coleoptera: Curculionidae; Derelominae), is an efficient pollinator in oil palm. Insect pollination plays a vital role in in transfer of pollen during the wet periods

Fig. 17.9 Beetle in flowers of coriander

at times of greater atmospheric humidity and low pollen density. *E. kamerunicus* can effectively handle the flowers that were deeply embedded inside the female inflorescence (Sambathkumar and Ranjith 2011). Other insects like *Prosoestus* spp., nitidulid beetles, *Microporium* spp. and Staphylinid beetles *Atheta* spp. were reported the flower visitors in oil palm. The weevils used their body hairs, wings and specialized hair-like structures located in the periphery of the abdomen, making frequent visits to the female inflorescence, making it ideal for pollen transfer and fertilisation. The larger-sized insects with more hairiness in the body contribute to the pollination and adequate fruit production.

The weevil was reported to be well adapted for the habitat of dense inflorescence of oil palm, and it equally pollinates younger and older palms. The male inflorescence was reported to release the volatiles that attract the weevils. The weevils breed on the spent male inflorescence and visit the female inflorescence during the flowering period, resulting in pollen transfer. These weevils will never damage the inflorescence. The weevil inhabits and completes its development in the male inflorescence. During the development period, the weevils feed on the pollen and oviposits inside the male inflorescence. The hatched larvae develop inside the inflorescence, pupates inside within the spikelets of the flowers (Tuo et al. 2011).

Ponnamma et al. (2002) reported that the release of adult weevils during the flowering period increased the fruit set to an extent of 48.89% with 66.09% increase in bunch weight. Due to efficient pollen transfer ability, the weevil was introduced into Malaysia to increase the pollination and yield in oil palm (Syed et al. 1982). After the introduction of the weevil, the insect established at a rapid rate and hand pollination was completely discontinued in the country. There was an increase in fruit set from 36% to 80% with 12% increase in fruit to bunch ratio and 28–54% increase in bunch weight in oil palm (Appiah and Dwarko 2013). The yield increase in oil palm by 20% accounted for a monetary benefit of $115 million per year has been achieved in Indonesian oil palm orchards. Caudwell et al. (2003) reported that the release of the pollinating weevils at 50 per acre increased oil palm productivity from 1660.90 to 1897.83 kg/acre.

10 Psychophily

10.1 Butterflies as Pollinators

Butterflies are vital bioindicators of a healthy ecosystem; any vital change in the ecosystem will negatively impact the abundance and diversity of the localized populations (Fig. 17.10). The butterfly habitat needs vary with the life stages. The immature stages of the butterflies need the preferred host plants for feeding, whereas the adult stages require nectar/pollen yielding plants to meet their energy requirements for flight and mating process. They prefer to alite in broad statured colourful, fragrant flowers. Other plant parts like leaves, stems and branches of the host plants serve as perching or landing sites for butterflies that aids in settling and collecting the floral rewards. They prefer to forage in the flowers in clusters that are diurnal in blooming habit, with ample nectar deep hidden inside the flowers.

Fig. 17.10 Butterflies visitation in different flowers

11 Sphingophily

11.1 Moths as Pollinators

Moths are nocturnal pollinators that prefer to pollinate the white or pale coloured flowers with heavy fragrances with diluted nectar. They are active fliers with a long tongue that can fly upwind tracking the airborne fragrances towards the source point (Tahir et al. 1985). Moths prefer to forage in the white or dull coloured flowers in clusters with a proper landing platform, which blooms during late evening or night hours, produce ample deeply hidden nectar inside the floral parts.

12 Pollination of Papaya

The major floral visitor in papaya is the hawk moth, which visits both the 'female pistillate' and 'male staminate' flowers during anthesis. Fuentes and Santamaría (2014) reported that wild papaya are dioecious in nature that depends on floral visit by pollinators for the fruit set. Recently cultivated varieties of papaya with self-compatible hermaphrodite flowers demands pollen deposition by insects for effective pollination and fruit set. Xenogamous pollination, or cross-pollination, is essential for papaya due to its dioecious nature, where male and female flowers are on separate plants. This ensures genetic diversity and prevents self-pollination, which can lead to inbreeding depression. Hawkmoths account for >95% of legitimate visits in papaya, which aids in xenogamous pollination in papaya. Martins and Johnson (2009) reported that species of hawkmoths, viz., *Hippotion celerio*, *Nephele comma*, and *Agrius convolvuli*, were the main pollinators of papaya. Planting the host plants preferred by the hawkmoths as hedgerows around the papaya orchards will help in the natural breeding of the moth species that aids in pollination and effective fruit set.

13 Moths as Pollinators in Avocado

The flowering type of avocado is protogynous, i.e., opening of functional female flowers initially followed by male flowers on the second day, inhibiting self-pollination. Extreme climatic conditions like extreme low temperatures will delay the female flower phase that resulting in open flowers overnight that necessitating nocturnal pollination. Moths are effective pollinators of nocturnal flowers due to their crepuscular nature (Pattemore et al. 2018). A moth species, *Ichneutica steropastis*, was reported to carry 10–33% of avocado pollen in its body parts in avocado (Buxton et al. 2021).

14 Role of Yucca Moths in Pollination

Yucca moth, *Tegeticula* sp. (Prodoxidae: Lepidoptera), is a small white coloured moth with a body colour similar to the flowers of Yucca. The adult moth is devoid of a long tongue, which is a characteristic feature of moths and butterflies. The adult possess tentacles around its mouth that aid in pollination. Adult female moth uses their tentacles to gather the pollen grains and hold the grains under their chins. After mating, the female moth oviposits in the ovary located at the bottom of the flower. After the oviposition, female moth collects a pollen ball from other flowers and places in the stigma of the floral surface, resulting in fertilisation. This mechanism of pollination was reported to be co-evolved, and the yucca moth were reported to be the principal pollinators of Yucca flowers (Bogler et al. 1995).

After fertilization, the flower stalks get elongated and develop capsular fruits. The capsular fruits will scatter the seeds from the capsules during the heavy wind hours. The moth's eggs attached to the seeds hatch, and the hatched larvae feed on the seeds through the seed columns during the fruit maturation. After the completion of the larval period, the mature larvae fall to the soil, pupate, and new generation adult moths emerge during the following season. The mutualistic relationship between the plants pollinated by the adult moths to produce viable seed, and the larvae of the moth feeding on the seeds ensures the survival of both the plant and the yucca moth.

15 Thripophily

15.1 Thrips as Pollinators

Thrips inhabiting the flowers are tiny, hairy, occurring in large numbers, congregating in the flowers, feeding on the pollen and nectar, serving as accidental pollinators. The flowers serve as an ideal congenial site for the foraging of the pollen, breeding and ovipositional site for the thrips. The cuticle setae of the thrips aid in picking up the pollen grains from the flowers. The flower thrips feed generally on both the pollen and nectar of the flowers.

The stickiness and surface ornamentation of the pollen grains were the vital physical factors of the pollen grains that play a role in the firm attachment of the grains to the body parts of the thrips. The movement pattern of thrips, viz., frequent spreading, congregation in the floral parts, and grooming behaviour, helps in the attachment of the pollen grains in the body parts of the thrips. The attractive petals of the flowers invite the mated adult thrips for oviposition. The larvae of the thrips upon emergence move upwards and downwards the flowers, resulting in pollen transfer and effective pollination (Ananthakrishnan 1982).

The mouthparts of the thrips have a functional left mandible that is used to pierce the pollen and nectar glands in the flowers (Varatharajan et al. 2016). The higher abundance of the thrips coincides with the flower dehiscence, and secluded flower organs serve as protection sites for the immature stages of the thrips. The larvae of thrips inhabit the keel petal of groundnut flowers to feed on the nectar accidentally transfers the pollen grains that lead to self-pollination in groundnut (Velayudhan et al. 1986). In Mahogany, three species of flower thrips, viz., *Frankliniella bispinosa*, *F. insularis*, *Pseudothrips inequalis*, inhabit the staminal tube with lots of pollen and leave the staminal tube through the stigma through a marginal cleft, which forces them to contact the anthers and stigma to cause the pollination (Howard et al. 1995).

16 Conservation

Pollinator garden is a concept of growing mixture of annual, biennial and perennial flowering herbs and shrubs in a protected garden to ensure year-round supply of pollen and nectar to bees with the aim of in-situ conservation of bees (Shivalingaswamy et al. 2020). Conservation of native pollinators result in enhanced pollination of agricultural and horticultural crops with increased yield. Maintenance of native biodiversity of plants and insects also result in a balanced ecosystem.

Abandoned wooden structures attract carpenter and leaf cutter bees for nesting. This concept of construction of "bee homes" in pollinator garden helps to sustain population. Bees need nectar and pollen source for survival and buildup of population. Use of locally available natural nesting structures for buildup of the population could be replicated in the urban landscapes which would create awareness among common public and school children.

Pollinator Garden

- Develop the pollinator gardens in a sunny site to sustain an active population of pollinators.
- Select diverse species of plants with flowers of different shapes and sizes to suit the mouthparts (both long tongued and short tongued bees) of the pollinators.
- Avoid use of chemical pesticides or use minimum in the garden.

References

Ali H, Shebl M, Alqarni AS, Owayss AA, Ansari MJ (2016) Nesting biology of two species of the large carpenter bees *Xylocopa pubescens* and *Xylocopa fenestrata* (Hymenoptera: Apidae) in north-western Pakistan. Oriental Ins. https://doi.org/10.1080/00305316.2016.1272500

Amala U, Shivalingaswamy TM (2018) Nesting biology, seasonality and host range of sweat bee, *Hoplonomia westwoodi* (Gribodo) (Hymenoptera: Halictidae: Nomiinae). Sociobiology 65(3): 491–496

Amala U, Shivalingaswamy TM (2019) Nest architecture and life cycle of Small Carpenter bee, *Ceratina binghami* Cockerell (Xylocopinae: Apidae: Hymenoptera). Sociobiology 66(1):29–33

Amala U, Shivalingaswamy TM (2022) Leafcutter Bees (Hymenoptera: Megachilidae) as pollinators of pigeon pea (*Cajanus cajan* (L.) Millsp., Fabaceae): artificial trap nests as a strategy for their conservation. Sociobiology 69(1). https://doi.org/10.13102/sociobiology.v69i1.7202

Amala U, Shivalingaswamy TM, Kumar V (2017) An unusual nesting site by leafcutter bee *Megachile* (*Aethomegachile*) *laticeps* Smith. J Kansas Entomol Soc 90:77–81

Amala U, Shivalingaswamy TM, Kumar V, Pratheepa M (2019) Pruned petioles of papaya as nesting sites of the leafcutting bee, *Megachile laticeps* Smith and its pollen fidelity. J Apic Res 58:660–664

Ananthakrishnan TN (1982) Thrips and pollination biology. Curr Sci 51(4):168–172

Appiah SO, Dwarko AD (2013) Studies on entomophilous pollination towards sustainable production and increased profitability in the oil palm: a review. Elx Agric 55:12878–12883

Bell MC, Spooner-Hart RN, Haigh AM (2006) Pollination of greenhouse tomatoes by the Australian blue banded bee *Amegilla (Zonamegilla) holmesi* (Hymenoptera: Apidae). J Econ Entomol 99(2):437–442

Bogler DJ, Neff JL, Simpson BB (1995) Multiple origins of the yucca-yucca moth association. Proc Natl Acad Sci 98:6864–6867

Bosch J, Blas M (1994) Foraging behaviour and pollinating efficiency of *Osmia cornuta* and *Apis mellifera* on almond (Hymenoptera, Megachilidae and Apidae). Appl Entomol Zool 29:1–9

Buxton MN, Hoare RJB, Broussard MA, Noort TV, Fale GRT, Nathan T, Pattemore DE (2021) Moths as potential pollinators in avocado (*Persea americana*) orchards in temperate regions. New Zealand J Crop Hort Sci 51(1):27–38

Caudwell RW, Hunt D, Reid A, Mensah BA, Chinchilla C (2003) Insect pollination of oil palm—a comparison of the long-term viability and sustainability of *Elaeidobious kamerunicus* in Papua New Guinea, Indonesia, Costa Rica and Ghana. ASD Oil Palm Papers 25:1–16

Chaudhary N, Saharawat YS, Sivaram V (2013) Conservation agriculture: an option to increase pollinators and sustainability. World J Agric Sci 9(3):210–213

Doherty DCO, Zoll JJK (2012) *Forcipomyia hardyi* (Diptera: Ceratopogonidae), a potential pollinator of Cacao (*Theobroma cacao*) flowers in Hawaii. Proc Hawaiian Ento Soc 44:79–81

Fajardo JAC, Medina JR, Oscar S, Opina, Cervancia CR (2008) Insect pollinators and floral visitors of mango (*Mangifera indica* L. cv. Carabao). Philippine Agric Sci 91(4):372–382

Free JB (1993) Insect pollination of crops. Academic Press, London, p 99

Fuentes G, Santamaría JM (2014) Papaya (*Carica papaya* L.): origin, domestication, and production. In: Ming R, Moore PH (eds) Genetics and genomics of papaya. Springer, New York, pp 3–15

Garibaldi LA, Aizen MA, Cunningham S, Klein AM (2009) Pollinator shortage and global crop yield: looking at the whole spectrum of pollinator dependency. Commun Integr Biol 2:37–39

George AP, Nissen RJ, Ironside DA, Anderson P (1989) Effects of nitidulid beetles on pollination and fruit set of *Annona* spp. hybrids. Sci Hortic 39(4):289–299

Gottsberger G, Silberbauer-Gottsberger I (2006) Life in the Cerrado: a South American tropical seasonal ecosystem. Vol. II. Pollination and seed dispersal. Reta Verlag, Ulm

Groeneveld JH, Tscharntke T, Moser G, Clough Y (2010) Experimental evidence for stronger cacao yield limitation by pollination than by plant resources. Persp Plt Ecol Evol Sys 12:183–191

Howard FW, Nakahara S, Williams DS (1995) Thysanoptera as apparent pollinators of West Indies mahogany, *Swietenia mahagoni* (Meliaceae). Ann Sci For 52:283–286

Kevan PG, Viana BF (2003) The global decline of pollination services. Biodiversity 4:3–8

Klein AM, Vaissiere BJH, Cane I, Dewenter SA, Cunningham C, Kremen E (2007) Importance of crop pollinators in changing landscapes for world crops. Proc R Soc London B, Biol Sci 274: 303–313

Kumar P, Rawat GS (2011) Vermiculiphily: larvae pollinating orchids nature proceedings 10101/ npre.2011.6295.1

Larson BMH, Kevan PO, Inouye DW (2001) Flies and flowers: taxonomic diversity of anthophiles and pollinators. Can Entomol 133:439–465

Liu C, Yong DR, Peng YQ (2011) Body size in a pollinating fig wasp and implications for stability in a fig-pollinator mutualism. Ent Exp Appl 138:249–255

Martins DJ, Johnson SD (2009) Pollination ecology of papaya (*Carica papaya*) on small-holder farms in Kenya. In: Proc African Poll Initiative, Kenya, pp 1–4

Meyer B, Jauker F, Steffan-Dewenter I (2009) Contrasting resource-dependent responses of hoverfly richness and density to landscape structure. Basic Appl Ecol 10:178–186

Mitra B (2010) Diversity of flower visiting flies (Insecta: Diptera) in India and their role in pollination. Rec Zool Surv India 11:95–107

Morse RA, Calderone NW (2011) The value of wild bees as pollinators of US crops in 2000. Bee Cult 128(3):1–15

Pattemore DE, Buxton MN, Cutting BT, McBrydie H, Goodwin M, Dag A (2018) Low overnight temperatures associated with a delay in 'Hass' avocado (*Persea americana*) female flower opening, leading to nocturnal flowering. J Pollinat Ecol 23:127–135

Paulino-Neto HF (2014) Escaravelhos e as suas flores no cerrado. In: Silva TRS, Moura CWN, Lima LCL, Santos FAR (eds) Botânica na América Latina: conhecimento, interação e difusão. Sociedade Botânica do Brasil, Salvador, pp 462–473

Ponnamma KN, Dhileepan K, Sasidharan VG (2002) Record of the pollinating weevil *Elaeidobius kamerunicus* (Faust) (Coleoptera: Curculionidae) in oil palm plantations of Kerala. Curr Sci 55(19):992–994

Potts SG, Biesmeijer JC, Kremen O, Schweiger R, Kunin WE (2011) Global pollinator declines: trends, impacts and drivers. Trends Ecol Evol 25(6):345–353

Rajan VV, Reddy PVR (2019) A dead heat in pollination race: a comparative evaluation of the efficiency of a fly (*Chrysomya megacephala*) and a bee (*Apis florea*) in mango pollination. J Entomol Zool Stud 7(1):1087–1091

Reddy PVR, Verghese A, Dinesh A (2010) Diversity and foraging behaviour of pollinator fauna in mango, National Symposium on Perspectives and Challenges of Integrated pest management for Sustainable Agriculture, Dr. Y. S. Parmar University of Horticulture and Forestry, Nauni, Solan, Himachal Pradesh, November 19–21

Rotheray GE, Gilbert SF (2011) The natural history of hoverflies. Forrest Text, Tresaith

Sajad A, Saeed S (2012) Floral host plant range of Syrphid flies (Syrphidae: Diptera) under natural conditions in Southern Punjab, Pakistan. Pak J Bot 42(2):1187–1200

Sambathkumar S, Ranjith AM (2011) Insect pollinators of oil palm in Kerala with special reference to African weevil, *Elaeidobius kamerunicus* Faust. Pest Mgmt Hort Ecosys 17(1):14–18

Shivalingaswamy TM, Amala U, Gupta A, Raghavendra A (2020) Non-Apis bee diversity in an experimental pollinator garden in Bengaluru—a Silicon Valley of India. Sociobiology 67(4): 593–598

Suhs AB, Somavilla A, Putzke J, Kohler A (2009) Pollen vector wasps (Vespidae; Hymenoptera) of *Schinustere binthefolius* Raddi (Anacardiaceae) Santa Cruz do Sul, RS, Brazil. Brazilian J Bio Sci 7(2):138–143

Sung IH, Lin MY, Chang CH, Cheng AS, Chen WS, Ho KK (2006) Pollinators and their behaviors on mango flowers in southern Taiwan. For Entomol 26:161–170

Syed RA, Law IH, Corley RHV (1982) Insect pollination of oil palm: introduction, establishment and pollinating efficiency of *Elaeidobius kamerunicus* in Malaysia. Planter 58:547–561

Tahir I, Farooq S, Bhat MR (1985) Insect pollinators and pests associated with cultivated buck-wheat in Kashmir (India). Fagopyrum 5:3–5

Tuo Y, Koua HK, Hala N (2011) Biology of *Elaeidobius kamerunicus* and *Elaeidobius plagiatus* (Coleoptera: Curculionidae) main pollinators of oil palm in West Africa. Eur J Sci Res 3:426–432

Velayudhan R, Gopinathan K, Bhakthavathsalam N (1986) Pollination potential, population dynamics and dispersal of Thrips species (Thysanoptera; Insecta) infesting flowers of *Dolichus lablab* L. (Fabaceae). Proc Indian Nat Sci Aca 51(5):574–580

Varatharajan R, Maisnam S, Shimray CV, Rachana RR (2016) Pollination potential of thrips (Insecta: Thysanoptera) – an overview. Zoo's Print 31(4):6–12

Watanabe ME (1994) Pollination worries rise as honey bees decline. Science 265:1170

Winder JA (1978) Cocoa Flower Diptera: their identity, pollinating activity and breeding sites. Pest Artic News Summ 24:5–18

Pollination and Pollinators of Cultivated and Wild Flowering Plants

18

Gaurav Singh, Shivani Khokhar, and Sindhu Sheoran

Abstract

Pollination is one of the most important ecosystem services that facilitates the reproduction and survival of angiosperms. A majority of the flowering plants are animal-pollinated, while some are water- or wind-pollinated. Insects are undoubtedly the most important animal pollinators and play an important role in sustaining global food production and biodiversity. Among insects, bees and flies are the most well-known pollinators; however, butterflies, moths, beetles, wasps, and ants are less known. Besides, vertebrate pollinators such as bats and birds are also important. This chapter discusses the importance of pollination, the modes of pollination, and the different types of pollinators of cultivated and wild flowering plants, and emphasizes the worldwide importance of different types of insect pollinators and their role in the pollination of flowering plants.

Keywords

Anemophily · Animal pollination · Bats and birds · Chiropterophily · Insect pollination · Nocturnal pollination · Ornithophily

Authors Gaurav Singh, Shivani Khokhar and Sindhu Sheoran have equally contributed to this chapter.

G. Singh (✉)
Maharana Pratap Horticultural University, Karnal, Haryana, India
e-mail: gs.apento@mhu.ac.in

S. Khokhar
Maharana Pratap Horticultural University, Karnal, Haryana, India

S. Sheoran
Hawkesbury Institute for the Environment, Western Sydney University, Penrith, NSW, Australia

A. K. Chakravarthy (ed.), *The Role of Pollinators in Tropical Landscapes*,
https://doi.org/10.1007/978-981-95-9139-8_18

1 Introduction

Pollination is a pivotal ecological process that underpins the reproduction and survival of flowering plants, maintaining ecosystem stability and supporting global agricultural productivity. The success of pollination depends on the diverse array of pollination agents, broadly categorized into two groups: abiotic (mainly wind and water) and biotic (animal pollinators). Animal pollinators, especially insects, facilitate the survival of many flowering plants and play a key part in crop production worldwide. Several cultivated and wild plants are pollinated by insects. In some crops, insect pollination is essential, while in others, the crop yield increases considerably in the presence of insect pollinators. This chapter highlights major pollination agents, including insect pollinators and the crops pollinated by them. This chapter briefly discusses the threats faced by pollinators and ways to minimize those threats.

2 Pollination

Sexual reproduction in angiosperms involve three stages: pollination, fertilization, and seed maturation (Lyons et al. 1989). Pollination is a fundamental ecological process facilitating the reproduction, survival, and diversification of flowering plants (Ollerton et al. 2019). It involves the transfer of pollen from the male reproductive structures (anthers) of a flower to the female reproductive structures (stigma) of the same or a different flower, leading to fertilization and the production of seeds. The pollination process is crucial for maintaining biodiversity and ecosystem stability. Typically, the process of pollination process in angiosperms develops in three critical phases: (i) the pollen grains are released from the anthers (male part of a flower); (ii) the pollen grains are transferred from the male part to the female part (stigma) by pollination vectors; and (iii) the pollen grains are successfully placed on the surface of stigma (maternal part of the flower) followed by successful pollen germination (Faegri and Pijl 2013).

There are two main types of pollination: self-pollination and cross-pollination. Self-pollination occurs when pollen is transferred from the anthers to the stigma of the same flower or to another flower on the same plant (Abrol 2012). It often ensures reproductive success and increases genetic uniformity, as it does not rely on external agents for pollen transfer. Cross-pollination, on the other hand, involves the transfer of pollen between flowers of different plants (Abrol 2012). This type of pollination promotes genetic diversity, reduces the risk of inbreeding, and enhances adaptability to changing environmental conditions.

The effectiveness of pollination relies on a wide variety of abiotic and biotic agents (Faegri and Pijl 1979; Abrol 2012; Faegri and Pijl 2013; Ollerton 2021). Abiotic pollination is aided by natural carriers like water and wind, while biotic pollination is aided by animals (Ackerman 2000; Ollerton 2021). Abiotic pollination, observed in ~20% of all angiosperm families (Ackerman 2000), can be anemophily and hydrophily when pollen grains are transported by wind (Culley et al. 2002) and water, respectively (Faegri and Pijl 2013). Wind pollination mostly occurs in grasses, most conifers and many deciduous plants, while hydrophily occurs in aquatic plants that release pollen directly into water (Faegri and Pijl 1979; Ackerman 2000; Abrol 2012; Faegri and Pijl 2013; Ollerton 2021). Anemophily is much more common than hydrophily, and around 98% of plant species requiring abiotic pollination are wind-pollinated (Faegri and Pijl 1979, 2013). However, hydrophily is rarely observed in a few aquatic plants (Ollerton 2021).

Biotic or animal pollination is aided by pollinating agents such as insects, birds, bats, and mammals, and is by far the most common pollination strategy among plants (Faegri and Pijl 1979, 2013; Ollerton 2017, 2021). Plants relying on wind/water pollination produce a large number of pollen grains to ensure the successful transfer of released pollen grains to the stigma. However, animal pollination is a mutualistic relationship between plants and pollinating animals where animals visit plants for resources and incidentally help in the transfer of pollen, resulting in fertilization (Abrol 2012). Understanding the roles and interactions of diverse pollination agents is of paramount importance in the face of ongoing global pollinator declines. The delicate balance between plants and their pollinators can be disrupted by environmental changes, habitat destruction, and pesticides. Consequently, this disruption poses a significant threat to both natural and agricultural systems that rely on pollinators for crop production.

3 Wind Pollination

Anemophily is a remarkable reproductive strategy observed in flowering plants that evolved unique adaptations to facilitate pollination through wind currents. Wind pollination takes advantage of the natural movement of air to transport pollen. While it may lack the visual spectacle of biotic pollination, the remarkable adaptations and wide diversity of wind-pollinated flowering plants highlight the incredible array of evolutionary solutions in nature. Anemophily is highly incidental and less precise compared to animal pollination (Faegri and Pijl 1979). Consequently, to achieve successful pollination, anemophilous plants exhibit various anatomical and reproductive adaptations to maximize the efficiency of pollen transfer (Faegri and Pijl 1979; Ackerman 2000). Key adaptations include smaller and lighter pollen grains, reduced petals and scent, abundant pollen production, exposed and feathery stigmas, long stamen filaments, etc. (Harder 1998; Ackerman 2000). Wind-pollinated flowers have reduced petals or lack showy floral parts and often lack a strong scent, as there is no need to attract pollinators with colourful displays or enticing fragrances. Since wind dispersal of pollen is relatively inefficient compared to insect pollination, the sheer abundance of pollen increases the likelihood of successful pollination. The stamens (male reproductive structures) of wind-pollinated flowers such as *Plantago* and *Corylus* often have long, flexible filaments which help to position the anthers and pollen outward, where they can be caught by the wind (Faegri and Pijl 1979). For example, the stigmas of grasses are usually exposed and feathery or branched to increase the surface area and capture airborne pollen efficiently (Faegri and Pijl 1979).

Numerous plant families have species that rely on wind pollination as their primary reproductive strategy. Some of the most well-known examples of wind-pollinated plants include cattails (Typhaceae), conifers (Pinaceae), grasses (Gramineae), rushes (Juncaceae), willows (Salicaceae), sedges (Cyperaceae), and trees like pines and oaks (Faegri and Pijl 1979; Ackerman 2000; Ollerton 2021). Some of the world's most key food crops such as rice, wheat, rye, barley, oats, etc. are wind-pollinated, which explains the characteristic "tassels" or "inflorescence" structures seen in reproductive parts (Faegri and Pijl 1979; Ackerman 2000). Anemophily has also been observed in plants that are primarily insect-pollinated (Dafni and Dukas 1986; Berry and Calvo 1989; Vroege and Stelleman 1990; Bullock 1994; Gomez and Zamora 1996; Goodwillie 1999). This mode of pollination is particularly prevalent in environments where insect populations are limited, such as open grasslands, deserts, and aquatic habitats. Although in some plants anemophily and entomophily can happen together, their relative importance in pollination varies among plant species, the same species grown in different habitats, different environmental conditions, etc. (Faegri and Pijl 1979). For example, primarily wind-pollinated plants such as *Plantago media* and *P. lanceolata* are also pollinated by insects (Clifford 1962; Faegri and Pijl 1979). Similarly, *Calluna* flowers nectar attracts insects for pollination, while a great deal of pollen is also transferred by wind (Faegri and Pijl 1979). Also, wind pollination might be happening in a number of important entomophilous plant species as the additional transfer of pollen by the action of wind current is inevitable.

Understanding the mechanisms and ecological significance of wind pollination is essential for studying plant reproductive ecology and aids in conservation efforts, especially in areas where certain wind-pollinated species play critical roles in ecosystem function and biodiversity maintenance. Recognizing the importance of wind-pollinated plants can inform land management practices and help safeguard the ecological integrity of habitats.

4 Water Pollination

Water pollination (hydrophily) represents an essential but often overlooked aspect of plant ecology, particularly in freshwater and marine ecosystems. While relatively rare, water pollination is a fascinating process of plant reproduction occurring in a small number of aquatic plants (Cox 1988; Ollerton 2021). Water pollination can be ephydrophily or hyphydrophily based on the occurrence of pollination on the surface of water or in the water, respectively (Faegri and Pijl 1979; Cox 1988). Unlike the more prevalent biotic and abiotic pollination methods, hydrophilous plants release pollen directly onto the water's surface, and currents carry the pollen to the female flowers, ensuring successful pollination (Cox 1988). Water-pollinated plants often possess unique adaptations to prevent pollen from sinking or being washed away. Some of the most well-known examples of water-pollinated flowering plants include Vallisneria (Eelgrass), Hydrilla, Ceratophyllum (Hornwort), Zostera (Seagrass), Najas (Naiads), etc. (Faegri and Pijl 1979; Cox 1988).

Hydrophilous plants typically produce large quantities of buoyant, non-sticky pollen that can easily float on the water's surface. For example, in plants such as *Ruppia* and *Callitriche autummlis*, pollen grains are released into the water, but the pollen floats up on the surface to reach the exposed stigmas where pollen spreads like yellow oil droplets (Faegri and Pijl 1979). In flowers such as *Neptunia* and *Aeschynomene* spp., the released pollen forms a sac or foam on the water surface, into which the female flowers emerge and receive pollen (Mahabale 1968). These female flowers often have elongated, feathery, or branched stigmas that can capture floating pollen grains from the water's surface. Interestingly, in some plants such as *Vallisneria spiralis*, *Hydrilla*, and *Lemma trisulca*, the whole male flower is released onto the water surface where it touches the shallow depressions on the stigmas of emerged female flowers (Faegri and Pijl 1979; Cox 1988). Besides, plants like *Callitriche hamulata, Ceratophyllum, Halophila*, and *Najas* are hyphydrophilous (Faegri and Pijl 1979; Cox 1988). For instance, sinking pollen grains are captured by stigmas in *Najas*, whereas tube-like pollen grains wrap around the stigmas in *Zostera* (Faegri and Pijl 1979). It is worth noticing that ephydrophilous pollen has water-repellent properties which help pollen grains to float on the water; however, hyphydrophilous pollen can withstand wetting which helps in the survival of pollen grains under the water (Faegri and Pijl 1979).

5 Biotic Pollination or Animal Pollination

Although abiotic pollination is common, many plant species need animal pollinators for successful fertilization (Ollerton et al. 2011; Abrol 2012; Ollerton 2021). One might argue that staple food crops such as rice, wheat, and maize account for ~65% of food production worldwide and do not require animal pollination; however, the majority of fruits, vegetables, spices, and nuts require animal pollination (Klein et al. 2007). Animal pollination, a vital ecological process, serves as a linchpin in maintaining agricultural productivity and securing economic stability. It significantly enhances crop yield, quality, and market value, ensuring food security, rural livelihoods, and international trade. There is a mutualistic relationship between plants and animals, which is recent in evolutionary terms, as angiosperms have diversified considerably over the last 100 million years, and animals (insects, birds, and mammals) have radiated roughly in an identical way (Waser and Forrest 2014). This intricate relationship sustains biodiversity and contributes significantly to economies worldwide.

Globally, 87.5% of flowering plant species (~352,000 species) depend on animal pollination for successful reproduction (Ollerton et al. 2011). If we talk specifically about the crop species, 75% of the estimated 1330 species (food, fruits, fibres, vegetables, beverages, spices, and medicinal plants) grown worldwide are animal-pollinated (Raven 1987). To further narrow it down, 87 of the 115 studied crops in a global synthesis require pollination by animals, and this contribution accounts for 35% of world food output (Klein et al. 2007). A previous assessment also suggests that 84% of 264 European crops depend upon some sort of animal pollination

(Williams 1994). Animal pollination services can be mediated by bats (Chiropterophily), birds (Ornithophily), small mammals (Therophily), and insects (Entomophily) (Faegri and Pijl 1979, 2013; Abrol 2012). Animal pollination services are crucial for ensuring food supply for the increasing population (Klein et al. 2007; Ollerton et al. 2011; Potts et al. 2016; Ollerton 2017).

6 Bat Pollination

Most of the bat species are nocturnal and have an important role in arthropod suppression (Kunz et al. 2011). In addition, pollination and seed dispersion are among the important ecosystem services provided by bats, especially in the tropics and subtropics (Kunz et al. 2011). Bats are considered niche pollinators because their unique adaptations allow them to pollinate certain plants that other pollinators may not effectively service. Pollination services are predominantly provided by bats belonging to the Pteropodidae and Phyllostomidae families (Fleming et al. 2009; Kunz et al. 2011). Being strong fliers, bats cover large distances and facilitate pollen transfer over long distances, promoting genetic diversity within plant populations. Mexican long-nosed bats, lesser long-nosed bats, and fruit bats are key pollinators of flowering plants.

Many bat-pollinated plants have large, night-blooming flowers with copious nectar that attract bats. Bat-assisted pollination has been observed in ~528 plant species spanning 67 families and 28 orders of flowering plants worldwide (Fleming et al. 2009). *Pteropodid bats are accountable for pollinating ~168 plant species from 100 genera and 41 families, while phyllostomid bats facilitate the pollination of ~360 plant species from 159 genera and 44 families (Fleming et al. 2009). The Agavaceae and Cactaceae are the two most important families of plants pollinated by bats (Kunz et al. 2011). Many species of columnar cacti and paniculate *Agave* are primarily pollinated by phyllostomid bats (Fleming and Valiente-Banuet 2002). Particularly in tropical and arid environments, bats can be critical pollinators for plants such as agave (*Agave tequilana, A. sisalana*), saguaro cactus (*Carnegiea gigantea*), and balsa tree (*Ochroma pyramidale*) (Rocha et al. 2006; Kunz et al. 2011). Bats occasionally pollinate important fruit plants such as durians, guavas, mangoes, petai, star apple, some bananas (*Musa balbisiana, M. maclayi*), and velvet beans in their respective native regions (Kunz et al. 2011). Pollination of Mahua tree or Indian butter tree (*Madhuca latifolia* and *M. indica*: Sapotaceae) by Indian short-nosed fruit bat (*Cynopterus sphinx*), Fulvous fruit bat (*Rousettus leschenaultia*), and Indian flying fox (*Pteropus giganteus*) is a classic example of high-value ecosystem service provided by bats (Nathan et al. 2009; Isaac et al. 2010; Stephenraj et al. 2010; Kunz et al. 2011).

7 Bird Pollination

Birds are probably the most important group of vertebrate pollinators and play a crucial role in many ecosystems. There are around 1000 known species of birds that pollinate plants (Whelan et al. 2015), and among them, hummingbirds (Trochilidae), sunbirds (Nectariniidae), false-sunbirds (Philepittidae), lories (Loridae), honeyeaters (Meliphagidae), toucans, tanagers are notable examples (Whelan et al. 2008; Whelan et al. 2015; Anderson et al. 2016; Serrano-Serrano et al. 2017; Ollerton 2021). As they forage for nectar or fruits, birds inadvertently transfer pollen between flowers, facilitating plant reproduction. Birds, with specialized beaks and affinity for brightly coloured flowers, are excellent pollinators for plants producing copious amounts of diluted nectar (Faegri and Pijl 1979; Anderson et al. 2016). Birds are attracted to specific types of flowers based on characteristics, such as shape, colour, and diluted nectar production (Cronk and Ojeda 2008; Johnson and Nicolson 2008). This specialization leads to efficient and precise pollination of many wild flowering plants (Yumoto 2000; Van Dulmen 2001; Cronk and Ojeda 2008; Ollerton et al. 2008; Kelly et al. 2010; Hoffmann et al. 2011; Serrano-Serrano et al. 2017; Andrews et al. 2022) and some commercial crops such as Arabica coffee (Martínez-Salinas et al. 2022), blueberries (Ramirez-Mejia et al. 2023), loquat (Fang et al. 2012), etc.

It is necessary to know that the reliance on vertebrate pollinators, especially bats and birds, varies depending on the specific geographic region and the availability of other pollinators. While bats and birds are essential for the pollination of many wild flowering plant species, the majority of commercially important crops are primarily pollinated by insect pollinators. However, it does not necessarily mean that vertebrate pollinators are not important. According to a global synthesis of 126 studies on animal-pollinated plants, the exclusion of respective vertebrate pollinators, but not insects, reduced seed/fruit production by 63% (Ratto et al. 2018). Bat-pollinated plants exhibit a stronger reliance on specific vertebrate pollinators compared to plants pollinated by birds (Ratto et al. 2018). When bats were removed, there was an average decline of 83% in fruit and seed production, whereas the exclusion of birds led to a 46% reduction (Ratto et al. 2018). The degree of plant dependence on respective vertebrate pollinators was greater in tropical regions compared to regions located at higher latitudes (Ratto et al. 2018). Considering the significant adverse effects linked to the decline of vertebrate pollinators, there is a need to initiate efficient conservation measures to safeguard threatened vertebrate pollinators. Habitat loss, climate change, and anthropogenic factors impact bat and bird populations, leading to potential disruptions in plant-pollinator interactions. Conserving bat and bird populations and their natural habitats is essential for sustaining their vital role as pollinators of cultivated and wild flowering plants. Recognizing the significance of bat and bird pollination helps drive conservation initiatives to protect unique and valuable pollinators.

8 Insect Pollination

There is a long evolutionary history between insects and angiosperms (Stephens et al. 2023), as the most common ancestor of angiosperms is believed to be insect-pollinated (Hu et al. 2008; Labandeira and Currano 2013; Asar et al. 2022). This notion is supported by the prevalence of insect pollination in the early-diverging angiosperms, a significant number of which are reportedly fly- and beetle-pollinated (Luo et al. 2018). Insect-pollinated plants have evolved a remarkable array of adaptations to attract and utilize insect partners efficiently, such as vibrant colours, alluring fragrances, and unique nectar guides, all designed to attract insects (Faegri and Pijl 1979). These visual and olfactory cues serve as beacons, guiding pollinators towards the flower's reproductive structures. Many insect-pollinated flowers have evolved specialized structures, such as tubular or spurred shapes, to accommodate specific pollinator species (Herrera 1996). These structures ensure effective pollen transfer as insects probe for nectar or pollen. Plants often exhibit pollination syndromes, where specific traits align with particular groups of pollinators (Faegri and Pijl 1979). For instance, flowers with long tubular corollas may cater to long-tongued insects like butterflies, while flat, open flowers are favoured by bees and flies (Johnson and Steiner 2000). Some plants synchronize their flowering periods with the activity patterns of their target pollinators (Heinrich 1975). For example, certain flowers bloom at specific times of the day or during particular seasons when their pollinators are most active.

9 Insect Pollinators

Insects are the most important pollinators accounting for 9.5% of agricultural production worldwide and contributing an estimated amount of €153 billion towards the global economy (Gallai et al. 2009). Pollination services provided by insects are valuable for crop growers as they increase their output and help maintain ecosystem stability (Gallai et al. 2009). Approximately 70% of the world's most-produced crop species (Gallai et al. 2009) and 84% of crop species grown in Europe benefit from some sort of insect pollination (Williams 1994). Insects and plants share a tremendous mutualistic relationship as plants provide resources in the form of nectar and pollen, and insects passively provide pollination services by mediating the transfer of pollen grains, required for the successful reproduction (Ollerton et al. 2011; Waser and Forrest 2014). The attractiveness of flowers, with vibrant colours and aromatic scents, lures insects to seek nectar and pollen. In the process of foraging, pollen adheres to their bodies, and as they move between flowers, cross-pollination occurs. Bees, flies, butterflies, moths, beetles, and wasps are the major insect pollinators (Waser and Forrest 2014) (Table 18.1). Insect pollination not only increases crop yield but also improves their quality (Aizen et al. 2009; Bommarco et al. 2012; Klatt et al. 2014). Pollinated crops tend to have more uniform size and shape, better taste, higher shelf life, and increased commercial and nutritional value (Garratt et al. 2014; Klatt et al. 2014; Wietzke et al. 2018). Insects may be specific or general pollinators

Table 18.1 Different subclasses of entomophily

Pollination subclasses	Types of insects
Melitophily	Bees
Cantharophily	Beetles
Myophily	Syrphids and Bombylid flies
Sphigophily	Hawk moths
Psychophily	Butterflies
Phalaeophily	Moths

depending on the crops (Waser and Forrest 2014); however, bees are probably the most important insect pollinators of cultivated and wild plants (Ollerton 2021).

9.1 Bees

Bees, with specialized body structures, are the most efficient and celebrated insect pollinators. They are characterized by branched body hairs, specialized structures for pollen collection, and adaptations for precision flight. Bees are the largest group of pollinators in tropical and temperate biomes (Ollerton 2017), with around 20,000 described species (Waser and Forrest 2014; Ascher and Pickering 2023), that contribute to pollinating vast numbers of wild plants. There are far more species of, e.g., wasps, flies, beetles, etc., but bees are important to far more plant species. Essentially, almost all bees are pollinators, whereas other groups have a mix of pollinator and non-pollinator species. In addition, according to the UN Food and Agriculture Organization, 71% of the major world food crops are pollinated by bees (Kluser et al. 2010). Bees contribute significantly to the pollination of fruits, vegetables, and oilseed crops, including apples, almonds, blueberries, canola, and sunflowers (Khalifa et al. 2021). The quantity and quality of crop yields are shaped by both the number of bee visits and the combined impacts of different bee species, and better crop deliverables are documented when more than one bee species is involved in pollination (Greenleaf and Kremen 2006; Osterman et al. 2023). Honeybees (*Apis* spp.), bumble bees (*Bombus* spp.), stingless bees (Meliponini), and solitary bees (e.g., *Osmia* spp. and Megachilidae) are responsible for pollinating crops.

9.1.1 Honey Bees

Honey bees (*Apis* genus: Apidae) are considered the most important and generalist pollinators worldwide (Free 1993; Klein et al. 2007; Winfree et al. 2011). Being generalist pollinators, honey bees are effective pollinators for diverse crop species, including fruits, vegetables, nuts, oilseeds, and some field crops (Hung et al. 2018; Khalifa et al. 2021) (Table 18.2). They have specialized structures such as pollen baskets on hind legs, allowing them to collect and transport large amounts of pollen from flower to flower. Their foraging behaviour and ability to visit multiple flowers in a single flight maximize the chances of successful pollination. Their efficient pollination services improve crop production and quality and contribute significantly

Table 18.2 Honey bees as crop pollinators of economically important crops

Crop	Bee species	Locality	References
	Apis mellifera L.	Argentina	Geslin et al. (2017)
	A. mellifera	Australia	Bernauer et al. (2022)
	A. mellifera scutellata	Brazil	Viana et al. (2014)
	A. mellifera	Canada	Bonneau et al. (2021)
Apple	*A. mellifera*	China	Wu et al. (2021)
	A. mellifera	England	Garratt et al. (2016)
	Apis cerana, A. mellifera	India	Sharma and Gupta (2001) Sharma et al. (2003)
	A. mellifera	Pakistan	Khan and Khan (2004)
	A. mellifera	USA	Vicens and Bosch (2000)
	A. mellifera	India	Abrol (1993)
Almond	*A. mellifera*	Spain	Bosch and Blas (1994)
	A. mellifera	USA	Pitts-Singer et al. (2018) Sáez et al. (2020)
Avocado	*A. mellifera*	Central America	Peña and Carabalí (2018)
Brassica(Broccoli, Chinese cabbage and Knol-khol)	*A. mellifera*	India	Sushil et al. (2013)
Blueberry	*A. mellifera*	USA	Grant et al. (2021)
Cherry	*A. mellifera*	USA	Boyle and Pitts-Singer (2019)
Chinese Cabbage	*A. mellifera*	India	Sushil et al. (2013)
Citrus	*A. mellifera*	Brazil	Malerbo-Souza et al. (2004)
Coconut	*A. mellifera*	Mexcio	Meléndez-Ramírez et al. (2004)
		Japan	Free et al. (1975)
Coriander	*A. cerana*	India	Patil and Pastagia (2016)
Cranberry	*A. mellifera*	USA	Morse and Calderone (2000)
Cucumber	*A. mellifera*	USA	Morse and Calderone (2000)
Cumin	*Apis dorsata, Apis florea, A. mellifera*	India	Meena et al. (2018)
Guava	*A. cerana, A. dorsata, A. florea, A. mellifera*	India	Rajagopal and Eswarappa (2005)
Kiwifruit	*A. mellifera*	Australia	Howpage et al. (2001)

(continued)

Table 18.2 (continued)

Crop	Bee species	Locality	References
Litchi	*A. mellifera*	India	Kumar and Kumar (2014) Kumar (2012)
Mango	*A. mellifera*	Australia	Willcox et al. (2019)
	A. mellifera	Brazil	Siqueira et al. (2008)
	A. cerana	India	Deuri et al. (2018)
	A. florea	India	Rajan and Reddy (2019)
	A. cerana, A. mellifera	Taiwan	Sung et al. (2006)
Onion	*A. mellifera*	India	Pathak (1999)
	A. florea	India	Abrol (2012)
Pear	*A. mellifera*	Argentina	Geslin et al. (2017)
	A. mellifera	USA	Naumann et al. (1994)
Peach	*A. cerana*	Nepal	Partap et al. (2000)
Plum	*A. cerana*	Nepal	Partap et al. (2000)
Strawberry	*A. cerana, A. dorsata, A. florea, A. mellifera*	India	Abrol et al. (2019)
	A. mellifera	Australia	Radford et al. (1979)
Sunflower	*A. mellifera*	Brazil	Chambó et al. (2011)
	A. mellifera	Pakistan	Idrees et al. (2023)
Sweet cherry	*A. mellifera*	USA	Naumann et al. (1994)
Watermelon	*A. cerana cerana* *A. mellifera ligustica*	China	Su et al. (2017)
	A. mellifera	Egypt	Taha and Bayoumi (2009)
	A. mellifera	India	Layek et al. (2021)

to the production of essential food items (Aizen and Harder 2009; Hung et al. 2018; Khalifa et al. 2021). The increased fruit set, uniform fruit size, and increased seed production are among the benefits of honey bee pollination (Klatt et al. 2014; Hung et al. 2018). For instance, in crops like canola (rapeseed) (Sabbahi et al. 2005), sunflowers (Greenleaf and Kremen 2006), and white clovers (Cecen et al. 2007), pollination by honey bees is critical for seed formation, crucial for future crop cultivation. Studies showed that many crop species experience yield improvements ranging from 30% to 100% due to honey bee pollination (Oz et al. 2009; Klatt et al. 2014; Rollin and Garibaldi 2019; Khalifa et al. 2021) (Table 18.2). According to an estimate, honey bee pollination services contributed USD 11.68 billion to pollination-dependent crops in the USA in 2009 (Calderone 2012). Thus, ensuring

Fig. 18.1 Four different species of honey bees ((**a**) *A. cerana,* (**b**) *A. dorsata,* (**c**) *A. florea,* and (**d**) *A. mellifera*) foraging in a mustard field in Haryana (India). (Photos by Sindhu Sheoran)

the conservation and well-being of honey bee populations is crucial to maintaining their vital role in supporting natural and agricultural systems.

A relatively recent molecular phylogeny analysis suggests 12 recognized species in the *Apis* genus (Lo et al. 2010), and most of these species are confined to specific regions within Asia. These species include *Apis andreniformis, A. binghami, A. cerana* (Fig. 18.1a), *A. dorsata* (Fig. 18.1b), *A. florea* (Fig. 18.1c), *A. koschevnikovi, A. laboriosa, A. mellifera* (Fig. 18.1d), *A. nigrocincta,* and *A. nuluensis* (Crane 2009; Lo et al. 2010). Eusocial in nature, all honey bee species exhibit a reproductive division of labour within their large, overlapping generational colonies (Wilson and Hölldobler 2005). Although all species aid in pollination in different parts of the world, the managed honey bee, *Apis mellifera* (L.), is native to Europe and is considered the foremost and most widely distributed pollinator species (Klein et al. 2007; Kluser et al. 2010; Winfree et al. 2011).

Apis mellifera is the most extensively researched bee species and offers indispensable pollination benefits to animal-pollinated crops across the world (Klein et al. 2007). Humans have intentionally introduced *A. mellifera* to all inhabited continents (Free 1993; Klein et al. 2007; Winfree et al. 2011). Most of the agricultural crops largely depend upon *A. mellifera* pollination as they increase the yield in several animal-pollinated crops (Klein et al. 2007). It is the most frequent flower visitor worldwide and provides high-valued pollination services to many cultivated plant species (Calderone 2012; Gallai et al. 2009; Hung et al. 2018). These pollination services are vital to agricultural crop production (Klein et al. 2007; Ollerton et al. 2011; Potts et al. 2016).

In many agricultural and horticultural scenarios, it has become a standard practice to deploy *A. mellifera* colonies during the flowering period of crops. It is done to enhance the effectiveness of pollination, ultimately leading to improved crop yield and quality (Monck et al. 2008; Rucker et al. 2012; Lorenz 2016; Phankaew 2016; Hoover and Ovinge 2018). Many major crops such as almonds, apples, berries, cherries, cucumbers, melons, pumpkins, squash, etc. benefit from *A. mellifera* managed pollination services (Klein et al. 2007; Aizen and Harder 2009; Partap 2011; Bernauer et al. 2022; David et al. 2022; Kamo et al. 2022). This growing demand for regulated pollination services has concurrently led to a worldwide rise in the number of managed honey bee hives (Aizen et al. 2008; Shahbandeh 2022). Apart from *A. mellifera,* other *Apis* species such as *A. cerana, A. dorsata,* and *A. florea* are also important for crop pollination in specific regions (Abrol 2012).

However, multiple stress factors like climate change, deforestation, fragmentation duc to anthropogenic factors, habitat destruction, pest pressure, poor nutrition, and pesticide exposure are affecting honey bee health (Potts et al. 2010; Rader et al. 2013; Staveley et al. 2014; Goulson et al. 2015; Potts et al. 2016). While the reasons are complex and interacting, Varroa mites, *Varroa destructor*, and the viruses they carry are some of the major drivers (Steinhauer et al. 2018). There has been considerable decline in honey bee population owing to *Varroa* mite infestation (Kluser et al. 2010; Staveley et al. 2014) and Colony Collapse Disorder (CCD) (James and Pitts-Singer 2008; Vanengelsdorp et al. 2009; Staveley et al. 2014), particularly in the USA but also other countries. Additionally, the increasing application of pesticides in agriculture and apiculture is a matter of grave concern for honey bee pollination. Pesticides directly kill honey bees in the field, indirectly effect their foraging behaviour, and result in residual toxicity in hive products (Sanchez-Bayo and Goka 2014; Johnson 2015).

A. mellifera is the main pollinator of agricultural crops globally; a decline in its pollination services will have severe consequences on crop production (Klein et al. 2007; Rader et al. 2013). It is high time to explore the importance of non-*Apis* pollinators as the over-reliance of agricultural and horticultural crops on a single pollinator species is risky (Winfree 2008; Potts et al. 2010; Rader et al. 2013). As a response, the focus of research has already shifted towards other pollinators like native bees and non-bee insects. Efforts have been made to conserve or restore wild pollinators in native ranges (Nicholson et al. 2019) as they have the potential to compensate for the losses and perform pollination (Rader et al. 2009; Garibaldi et al. 2013; Rader et al. 2013, 2016; Willcox et al. 2019).

9.1.2 Honey Bee Versus Alternative Pollinators

Increasing the number of honey bee hives may not necessarily boost crop production because, for several crops, honey bees may not be the most efficient pollinators compared to other pollinators (McGrady et al. 2020; Willis Chan and Raine 2021; Mashilingi et al. 2022). This emphasizes the significance of protecting alternative pollinators and wild pollinator assemblages to meet global pollination demand. As a result, research into the roles of alternative managed or wild pollinators is essential to explicate the pollination services they provide. Several species of non-*Apis* bees, native bees, ants, wasps, flies, butterflies, moths, and beetles make a significant contribution to the pollination of crops (Delaplane et al. 2000; James and Pitts-Singer 2008). Several researchers have compared the pollination services of different pollinators and highlighted the importance of wild bees and non-bee pollinators (Rader et al. 2009, 2013, 2016; Garibaldi et al. 2013; Hung et al. 2018; Willcox et al. 2019).

Despite being the primary pollinator and most frequent flower visitor in most of the crops (Winfree et al. 2011; Hung et al. 2018), pollen deposition (single visit) by *A. mellifera* did not differ significantly from the alternative pollinators (Rader et al. 2009; Hung et al. 2018). Additionally, *A. mellifera* visitations were not even observed in 33% and 49% of the studied plant-pollinator interaction networks and plant species, respectively (Hung et al. 2018). Moreover, its role in the pollination of wild plants is still not clear and not well supported by empirical data (Klein et al. 2007; Potts et al. 2010). Even in cultivated crops, alternative pollinator taxa are equally efficient as honey bees in terms of their pollen deposition but are not equally abundant (Rader et al. 2009). Although honey bees visit flowers at significantly higher frequencies, there is non-significant difference in flowers visited per minute compared to alternative pollinators (Rader et al. 2009). Individual visits by honey bees are less effective, but higher abundance increases their overall pollinating activity and makes them more effective as a single taxon (Rader et al. 2009). Their overall pollinating activity and abundance are higher as they are maintained in a larger number (managed pollination) during the flowering season while most of the other pollinator taxa are unmanaged. This does not necessarily mean that unmanaged or wild pollinators are less effective (Rader et al. 2009, 2016; Garibaldi et al. 2013), and their vital role in maintaining ecosystem stability cannot be neglected and can in no way be replaced by *A. mellifera*. Integrating honey bees and wild bees in crop pollination can be an improved strategy as they complement each other (Greenleaf and Kremen 2006; Winfree et al. 2007; Garibaldi et al. 2011) and crops benefit from wild bee pollination services even when honey bees are abundant (Garibaldi et al. 2013; Button and Elle 2014).

9.1.3 Non-*Apis* Bees

A wide array of non-*Apis* bees such as stingless bees, bumble bees, blue-banded bees, allodapine bees, sweat bees, and many other solitary bees are present worldwide which are more efficient pollinators of specific crops compared to honey bees (Delaplane et al. 2000; McGrady et al. 2020; Willis Chan and Raine 2021). Non-*Apis* bees may be evolutionary more suited to those specific crops and thus, may be more beneficial for crop production compared to honey bees. For instance, wild bees are likely efficient pollinators of crops like alfalfa, cranberry, and red clover, not well suited for honey bee pollination (Delaplane et al. 2000; James and Pitts-Singer 2008). They can even compliment honey bee pollination by increasing the seed set and yield in crops like sunflower and blueberry (Greenleaf and Kremen 2006; Nicholson and Ricketts 2019).

9.1.4 Stingless Bees

Stingless bees (Meliponini) are an overlooked group of pollinators, but they play a vital role in crop pollination and ecosystem health. Stingless bees form another group of eusocial bees (Michener 2000). They belong to the Meliponini tribe (Hymenoptera: Apidae) and are closely related to honey bees and orchid bees (Roubik 1989). Contrary to the name, stingless bees do have a sting, but a non-functional one that cannot be used for any kind of colony defence (Michener 2000). Instead, they have different behavioural defence mechanisms and even use mandibles to bite and ward off the enemies. Generally, the biting intensity of stingless bees is very low (like an ant), but some of them like *Trigona hyalinata* have powerful mandibles and suicidal biting as a nest defence strategy (Shackleton et al. 2015). Globally, there are ~516 estimated species (>600 according to some estimates) of stingless bees belonging to 60 genera (Rasmussen and Cameron 2010) and they are majorly found in most tropical and subtropical parts of the world like Africa, Australia, Southeast Asia, and tropical America (Michener 2000; Roubik 2023). There are around 400 identified species of stingless bees in the neotropical region of the world with colony sizes ranging from a few hundred to thousands of workers (Pedro and De Camargo 2013; Roubik 2023).

Stingless bees play a significant role in crop pollination and biodiversity conservation, especially in tropical and subtropical regions of the world (Lima et al. 2016). They are useful for local and small-scale agricultural systems, especially in regions where honey bees might not be abundant. Their gentle nature, diverse foraging behaviour, and adaptability make them valuable allies in supporting sustainable agriculture and enhancing food security. Their agility and small size allow stingless bees to access flowers with challenging morphologies, promoting effective pollination. According to a recent review of global studies, stingless bees reportedly visit flowers of 1746 plant genera from 221 families (Bueno et al. 2023). Among these, stingless bees have mostly frequently been observed visiting flowers of 15 important families: Acanthaceae, Anacardiaceae, Apocynaceae, Arecaceae, Asteraceae, Bignoniaceae, Euphorbiaceae, Fabaceae, Lamiaceae, Malvaceae, Melastomataceae, Myrtaceae, Poaceae, Orchidaceae, Rubiaceae, and Sapindaceae (Bueno et al. 2023).

Stingless bees have earlier been reported as the most frequent flower visitors in ~90 crop species and effective pollinators in nine cultivated food and fruit crops (Heard 1999). Stingless bees are considered effective pollinators in many economically important crops like apples, avocado, blueberries, macadamia, mango, tomatoes, etc. (Heard 1999; Slaa et al. 2006; Giannini et al. 2015) (Fig. 18.2 and Table 18.3). Studies have revealed that fruit set and size are positively influenced by stingless bee pollination in crops (Table 18.3). Due to smaller size, stingless bees perfectly fit on the flowers and can be efficient pollinators of crops like mangoes (Singh et al. 2021, 2024, 2025). Their short foraging range makes them good agents for pollination under protected cultivation (Khan and Yogi 2017). Species of stingless bees can efficiently buzz-pollinate: *Heterotrigona itama* in chillies (Azmi et al. 2016) and cucumber (Azmi et al. 2017), *Melipona fasciculata* in tomatoes and eggplants (Nunes-Silva et al. 2013), and *Trigona* (now *Tetragonula*) *carbonaria* in capsicums (Greco et al. 2011). Stingless bees also provide shared pollination

Fig. 18.2 A native stingless bee (*Tetragonula* sp.) foraging on mango flowers in the Northern Territory of Australia. (Photo credit: James Makinson)

services among important fruit crops like avocado, macadamia, and mango (Willcox et al. 2019) and help supplement the pollination services provided by other pollinators (Viana et al. 2014). Studies are needed to ascertain the pollination efficiency of stingless bees in economically important crops in key tropical and subtropical regions of the world. Recognizing and promoting the importance of stingless bees as crop pollinators is essential for conservation and for maintaining the ecological balance in agricultural landscapes.

9.1.5 Bumble Bees

Bumble bees (Apidae: Bombini) are crucial crop pollinators, providing essential pollination services in agricultural crops and wild plant species (Wahengbam et al. 2019; Khalifa et al. 2021) (Table 18.4). A few bumble bee species are kleptoparasitic, but most of the species are eusocial like honey bees, with comparatively smaller colonies. There are >250 species of bumble bees (Williams et al. 2008), widely distributed in almost all biogeographic regions of the world, particularly in temperate regions (Velthuis and Van Doorn 2006). Among them, *Bombus occidentalis* Greene, *B. ignites* Linn, *B. impatiens* Cresson, *B. lucorum*, and *B. terrestris* Linn are used in commercial pollination globally (Velthuis and Van Doorn 2006). Besides this, *B. pascuorum* Scopoli and *B. hortorum* Linn are considered good pollinators of red clover, field beans, and wild flowering plants (Bohart 1972; Fussell and Corbet 1991).

Bumble bees possess several morphological and behavioural characteristics and can be better pollinators than honey bees in certain crop settings. Their larger body size, longer tongue, fuzzy appearance, ability to buzz-pollinate, extended foraging period, and generalist behaviour make them highly efficient and effective pollinators of crops like red clover, tomatoes, bell peppers, blueberries, cranberries, kiwifruits, etc. (Rao and Stephen 2009; Yankit et al. 2018; Nayak et al. 2019; Kumar and Khan 2022). Their robust and fuzzy body helps them to forage for longer duration and remain active at lower temperatures compared to honey bees (Corbet et al. 1993).

Table 18.3 Stingless bees as important pollinators of economically important crops

Crop	Bee species	Locality	References
Annatto, achiote	*Melipona melanoventer*	Amazonas	Maues and Venturieri (1995)
	Melipona fuliginosa	Many regions	Wille (1976)
	Hypotrigona pothieri	Ivory Coast	Lobreau-Callen et al. (1990)
	Melipona seminigra merrillae	Amazonas	Absy and Kerr (1977)
	Melipona rufiventris	Amazonas	Absy et al. (1984)
Apple	*Melipona quadrifasciata anthidioides*	Brazil	Viana et al. (2014)
	Tetragonula	Australia	Tierney et al. (2023)
Avocado	*Trigona nigra*, *Nannotrigona perilampoides*	Mexico	Can-Alonzo et al. (2005)
	Geotrigona acapulconis, *Trigona nigerrima*, *Partamona bilineata*, *Scaptotrigona pectoralis*, *Trigona nigra*, *Scaptotrigona mexicana*, *Trigona fulviventris*, *Plebeia frontalis*	Mexico	Ish-Am et al. (1999)
Camu camu	*Melipona* sp., *Scaptotrigona postica*	Peru	Peters and Vasquez (1987)
Carambola, starfruit	*Melipona favosa*	Suriname	Engel and Dingemans-Bakels (1980)
Chayote, choko	*Trigona corvina* and *Partamona cupira*	Costa Rica	Wille (1983)
Cherry	*Trigona*	Brazil	Giannini et al. (2015)
Chilli	*Trigona laeviceps*	Indonesia	Putra et al. (2014)
Coconut	*Melipona* spp. stingless bees	Costa Rica, Suriname, Mexico	Hedström (1986) Engel and Dingemans-Bakels (1980) Meléndez-Ramírez et al. (2004)
Coffee	*Lepidotrigona terminate*	Indonesia	Klein et al. (2003)
Cupuassu	*Plebeia minima*	Brazil, Belem	Venturieri (1994)
	Trigona clavipes and *Trigona lurida*	Brazil	Heard (1999)
Macadamia	*Tetragonula carbonaria*	Australia Eastern	Heard and Exley (1994)
Mango	Meliponini and Diptera	Brazil	Cortopassi-Laurino et al. (1991)
	Meliponini	India	Singh (1989)
	Trigona	Australia	Anderson et al. (1982)
	Tetragonula	Australia	Willcox et al. (2019)

(continued)

Table 18.3 (continued)

Crop	Bee species	Locality	References
	Tetragonula	Australia	Singh et al. (2021), Singh et al. (2024), Singh et al. (2025)
	Tetragonisca angustula	Mexico	Sosa-Najera et al. (1994)
Pomegranate	*Triogna*	Brazil	Giannini et al. (2015)
Pumpkin	*Partamona bilineata*	Mexico	Meléndez-Ramírez et al. (2002)
Rambutan	*Scaptotrigona Mexicana Tetragonisca*	Mexico	Rincón-Rabanales et al. (2015)
Strawberry	*Trigona,*	Brazil	Giannini et al. (2015)
	Tetragonisca angustula	Brazil	Malagodi-Braga and Kleinert (2004), Slaa et al. (2006)
Squash	*Trigona*	Brazil	Giannini et al. (2015)
Tomato	*Tetragonula pagdeni*	Thailand	Wongsa et al. (2023)
Watermelon	*Tetragonula iridipennis*	India	Layek et al. (2021)

Modified Meléndez Ramírez et al. (2018)

Table 18.4 Crops pollinated by different bees

Bees	Crops
Bumble bees	Apple, apricot, blackberry, blueberry, cherry, cranberry, kiwifruit, peach, pear, plum, raspberry, strawberry, watermelon, capsicum, cucumber, eggplant, hot pepper, tomato
Leaf-cutter bees	Alfalfa, blueberry, canola, clover, carrot
Sweat bees	Cucumber, ginger, hot chilli, lettuce, melon, tomato, watermelon
Mason bees	Apple, almond, apricot, blackberry, cherry, pear, raspberry, tart

Bumble bees start their activity earlier than honey bees and continue foraging in adverse weather conditions, like cold and rainy days (Willmer et al. 1994; Peat et al. 2005). For instance, bumble bees reportedly start their foraging activity ~15 to 40 min earlier than honey bees in cucumber and watermelon fields (Stanghellini et al. 2002). Their ability to buzz-pollinate deep-throated flowers such as tomatoes and bell peppers sets them apart from honey bees (Goulson 2010; Nayak et al. 2019; Kumar and Khan 2022). Additionally, bumble bees are effective cross-pollinators due to their large foraging range, making them crucial for seed production in cucumber, pumpkin, squash, and watermelon (Stanghellini et al. 1997). Their efficient pollination services help ensure high yields and quality in controlled environments (Yankit et al. 2018). Additionally, bumble bees are efficient pollinators of plant communities with tubular flowers (Fontaine et al. 2006). However, native bumble bee populations are threatened by intensive agriculture, habitat loss, destruction of nesting habitats, decrease in wild flowering plants' abundance and diversity, indiscriminate use of pesticides, and pests and diseases (Goulson et al. 2008). By promoting the conservation of bumble bee populations, the sustainability of agriculture, preserve biodiversity, and secure global food production can be ensured.

9.1.6 Leafcutter Bees

The leafcutter bees (*Megachile* genus) are solitary bees extensively utilized for pollinating cultivated and wild plants, especially alfalfa (Pitts-Singer and Cane 2011) (Table 18.4). Leafcutter bees cut the bits from the leaves of plants, preferably belonging to the Leguminosae family, for nest construction (Michener 2000). These bees, particularly *Megachile rotundata*, are widely managed for pollination purposes, especially in crops like alfalfa and canola (Pitts-Singer and Cane 2011). According to a North American study, the use of leafcutter bees in alfalfa pollination produced 46,000 metric tons of alfalfa seeds in 2004, which was roughly two-thirds of the global production of alfalfa (Pitts-Singer 2008). *Megachile rotunda* also significantly contributes to hybrid seed production in Canadian canola (Pitts-Singer 2008). They are effective pollinators for blueberries (Mackenzie 2008) and have shown potential as substitutes for managed honey bees in certain conditions for crops like canola (Soroka et al. 2001) and carrots (Tepedino 1996). *Megachile addenda* can be an effective pollinator of cranberries in New Jersey, USA (Cane et al. 1996). Interestingly, despite their successful use in North America, these bees are not utilized as managed pollinators in other parts of the world, particularly Europe, where they are native (Pitts-Singer and Cane 2011). Their presence remains minimal in European settings, as indicated by low population counts in Hungary (Móczár 1961) and Spain (Ortiz-Sanchez and Aguirre-Segura 1991), and challenges in sustaining their numbers in French alfalfa fields (Tasei 1975).

9.1.7 Allodapine Bees

Allodapine bees (tribe Allodapini: subfamily Xylocopinae) are a group of native bees found primarily in tropical regions, particularly in Africa, Southeast Asia, and the Pacific Islands (Michener 2007). Although Allodapine bees are found extensively across the Old World, the greatest diversity is within Africa (Michener 2007). More than 125 species of allodapine bees belonging to 8 genera are documented from Africa, followed by ~100 species (5 genera) from Australia (Michener 2007; Engel and Dathe 2011). *Braunsapis* is the largest genus in the tribe Allodapini which showed behaviour ranging from social behaviour to social parasitism (Reyes and Michener 1990; Aenmay et al. 2006; Michener 2007). Allodapine bees exhibit interesting nesting behaviour, with some species creating nests in pre-existing cavities like beetle tunnels in wood or bamboo (Michener 2007). Allodapine bees showed a high degree of specialization in foraging behaviour. Some species have

Fig. 18.3 A native allodapine bee from the *Braunsapis* genus (left) and a sweat bee from the Halictidae family (right) foraging on mango flowers in the Northern Territory of Australia

specific preferences for certain types of flowers, making them efficient and effective pollinators for those plants.

They are often rare but efficient pollinators for many cultivated and wild plants, especially in natural habitats. For instance, allodapine bees from the genus *Exoneura* are efficient pollinators of apples in Australia; however, their low abundance results in near negligible contribution to apple pollination services (Bernauer et al. 2022). These bees often nest in stems of native vegetation (Bernauer et al. 2021) and visit apple orchards in different regions of Australia (Prendergast et al. 2021; Bernauer et al. 2022; Brown et al. 2022). Other species of allodapine bees such as *Braunsapis hewitti* and *Braunsapis* spp. are efficient pollinators of mangoes in Taiwan (Sung et al. 2006) and Australia (Singh et al. 2024) (Fig. 18.3), and raspberries in Australia (Coates et al. 2022). While allodapine bees are not major pollinators for widely cultivated crops, they can still play a role in pollinating some tropical and subtropical plants, such as leafless vanilla complex (Andriamihaja et al. 2020) and aloe (Hargreaves et al. 2012). As research on allodapine bees continues, more plant associations may be discovered, shedding light on their unique role as pollinators in ecosystems. Many allodapine bee species are endemic to specific regions and vulnerable to habitat loss and other threats. Recognizing their role as pollinators contributes to conservation and the preservation of their natural habitats. Supporting the conservation of allodapine bees and their habitats is essential to maintaining the diversity and resilience of pollinator populations and the ecosystems they support.

9.1.8 Sweat Bees

The family Halictidae, often referred to as halictid bees or sweat bees, comprises more than 4500 species distributed across the globe (Ascher and Pickering 2023). Halictidae majorly includes ground-nesting bees (Michener 2007). These short-tongue bees are recognized for their vibrant colours, relatively small to medium size, and often iridescent integument (Michener 2007). Among halictids, the genus *Lasioglossum* is the most diverse and contains 1881 species (Ascher and Pickering

2023); however, only a few species are recognized and utilized for their pollination ability (Nelson et al. 2023). Halictid bees are essential pollinators in both natural and agricultural ecosystems, contributing significantly to the pollination of various flowering plants (Table 18.4). Some common examples include pollination of cucumbers by *Lasioglossum* spp. (Lowenstein et al. 2015), gingers by *Nomia* spp. and *Thrinchostoma afaciatum* (Sakai et al. 1999), lettuces by *L. villosulum* (Goubara and Takasaki 2004), hot chillies by *Lasioglossum* spp. (Landaverde-González et al. 2017), melons by *L. malachurum* (Rodrigo Gómez et al. 2016), tomatoes by *Hoplonomia westwoodi* (Amala and Shivalingaswamy 2017), watermelons by *Halictus* spp. and *Lasioglossum* spp. (Campbell et al. 2019), etc. Sweat bees are significant pollinators for various wildflower species such as Brazilian orchids (*Campylocentrum, Cyclopogon, Prescottia*) (Singer and Cocucci 1999), *Spiranthes* (Catling 1983), *Isotria* (Mehrhoff III 1983), *Calopogon* and *Cephalanthera* (Pijl and Dodson 1966), *Caladenia, Prasophyllum, Thelymitra*, and *Dendrobium* (Adams and Lawson 1993).

Halictid bees supplement the pollination services provided by honeybees and bumble bees. When managed pollinators are not available, halictid bees ensure adequate pollination. Halictid bees are known for adaptability to urban environments. They can thrive in gardens, parks, and other green spaces in cities, contributing to pollination services for urban agriculture and ornamental plants (Frankie et al. 2009; Lowenstein et al. 2015, 2019). Halictids visit and pollinate various ornamental plants cultivated in gardens and landscapes, including flowers like yarrows, daisies, poppies, asters, and sunflowers (Frankie et al. 2009). Some halictid bee species have shown resilience to certain environmental stressors, such as habitat disturbances and pesticide exposure. Their ability to withstand such challenges can be advantageous for pollination in agricultural settings. Promoting the conservation and protection of halictid bee populations and natural habitats is crucial for maintaining pollination services, ensuring crop productivity, and supporting biodiversity in agricultural and wild landscapes.

9.1.9 Mason Bees (*Osmia* spp.)

In recent years, Mason bees (Megachilidae; *Osmia*) have gained attention for their potential as alternative pollinators to supplement honeybees, which have faced population declines due to factors like habitat loss, pesticides, and diseases. This has led to efforts to promote the conservation and management of *Osmia* bees, for ecological importance and their role in supporting agriculture. *Osmia* bees are solitary bees that do not form colonies like honeybees or bumble bees (Michener 2007). They are called "mason bees" because they construct nests using mud or other materials, sealing individual chambers where they lay their eggs. *Osmia* bees are well-known alternative pollinators, especially for their role in fruit crop pollination (Sedivy and Dorn 2014).

Several rosaceous fruit crops, including apples, apricots, almonds, cherries, peaches, pears, and plums, are efficiently pollinated by *Osmia* bees (Torchio 1976; Bosch et al. 2000; Bosch and Kemp 2002; Biddinger et al. 2013; Sedivy and Dorn 2014) (Table 18.4). Among *Osmia* bees, *Osmia bicornis* (Latreille), *O. cornifrons*

(Radoszkowski), *O. cornuta,* and *O. lignaria* Say are the four most widely studied species for crop pollination (Sedivy and Dorn 2014). While *O. corniforns* has been widely used to pollinate three-fourths of Japan's apple orchards since the 1940s (Bosch and Kemp 2002), *O. cornuta* and *O. ribifloris*, respectively, are excellent pollinators of almonds (Bosch et al. 2000) and blueberries (Sampson et al. 2004a, b) in the USA. Additionally, *O. bicornis* (formerly *O. rufa*) is an efficient pollinator of apples, oilseed rapes, and strawberries (Wilkaniec and Radajewska 1996; Gruber et al. 2011). Other *Osmia* species: *O. excavate, O. jacoti, O. pedicornis, O. tersula,* and *O. taurus* have the potential to be good pollinators of fruit orchards (Wei et al. 2002; Sheffield et al. 2008; Sedivy and Dorn 2014).

9.1.10 Blue-Banded Bees

Blue-banded bees (Anthophorini: *Amegilla*), native to Australia, are characterized by pale-yellowish thorax, black-coloured body, and bluish abdominal bands (Leijs et al. 2017). Blue-banded bees are widespread in Australia and often seen hovering on cultivated, wild flowering (Fig. 18.4a), and ornamental plants in urban settings (Leijs et al. 2017). Blue-banded bees are excellent pollinators of crops requiring buzz pollination within greenhouse environments. They are efficient pollinators of capsicums, chilli peppers, and tomatoes throughout Australia, excluding Tasmania (Dollin 2020). Blue-banded bees also pollinate tomatoes in India (Amala and Shivalingaswamy 2017) and 11 species of Bornean gingers in Malaysia (Sakai et al. 1999). An adequate level of pollination services in tomatoes can be achieved by having a total of 282 nesting female *Amegilla chlorocyanea* bees per hectare within greenhouse settings (Hogendoorn et al. 2007). Buzz pollination activity of blue-banded bees, *A. chlorocyanea* and *A. holmesi*, results in a 20–24% yield increase in greenhouse tomatoes and eliminates the need for mechanical pollination

Fig. 18.4 A blue-banded bee, *Amegila* sp. (**a**) and a cuckoo bee, *Thyreus* sp. (**b**) visiting flowers of Pink Turkey Bush (*Calytrix exstipulate*) in the Northern Territory of Australia. (Photo credit: James Makinson)

(Bell et al. 2006; Hogendoorn et al. 2006). Similarly, *A. zonata* pollination enhanced tomato seed setting and fruit weight in India (Amala and Shivalingaswamy 2017).

Blue-banded bees offer several benefits as crop pollinators, and their importance cannot be overstated. As bumble bee populations face challenges and decline in some areas, blue-banded bees offer an alternative to buzz pollination services. This helps maintain consistent pollination, particularly for crops that rely on buzz pollination for optimal fruit production. The efficient buzz pollination provided by blue-banded bees eliminates the need for labour-intensive mechanical or hand pollination. This reduces costs for farmers and promotes sustainable farming practices.

9.1.11 Other Bee Pollinators

Apart from the abovementioned bee pollinators, large carpenter bees (*Xylocopa* spp.), small carpenter bees (*Ceratina* spp.), squash bees (*Peponapis* and *Xenoglossa* spp.), mining bees (*Andrena* and *Habropoda* spp.), and cuckoo bees (*Thyreus* spp.) (Fig. 18.4b) are efficient pollinators in different crop settings. Large carpenter bees (*Xylocopa varipuncta* and *X. pubescens*), recognized for their ability to excavate nest tunnels in wood, can be important pollinators for a variety of flowering plants, including ocotillo (Enz 2001), orchids (Sugiura 2013), Rabbiteye blueberries (Sampson et al. 2004a, b), sunflowers, Virginia bluebells (Enz 2001), and honey melons (in protected conditions) (Sadeh et al. 2007). Small carpenter bees (*Ceratina* spp.) efficiently pollinate apples, beans, coffees, and cowpeas (Eardley et al. 2009); Acacia (Sornsathapornkul and Owens 1998); and teak trees (Tangmitcharoen and Owens 1997).

Hoary squash bees, *Eucera pruinose* or *Peponapis pruinose*, are reliable pollinators of *Cucurbita* crops (Tepedino 1981). These specialized pollinators can be better pollinators than the European honey bee, *A. mellifera*, on specific crops. For example, *P. limitaris* can deposit significantly more pollen grains compared to *A. mellifera*, and therefore, are more efficient pollinators of *Cucurbita moschata* (Canto-Aguilar and Parra-Tabla 2000). Similarly, mining bees (*Melandrena* spp.) are key pollinators of apple in New York, USA, and have the potential to provide reliable pollination services in the wake of future declines in *A. mellifera* populations (Park et al. 2016). The alkali bee (*Nomia melandri*), which is perhaps the most intensively managed species of ground-nesting bees, can efficiently pollinate alfalfa (Cane 2023).

9.2 Non-bee Pollinators

Non-bee pollinators, including flies, butterflies, moths, beetles, wasps, and ants, provide valuable pollination services and potential insurance against bee decline (Rader et al. 2016). Among them, flies seem more important as they are active pollinators of more than 100 crops (Ssymank et al. 2008). A global synthesis of 37 crop studies from five different continents underlined the importance of non-bee pollinators (Rader et al. 2016). Non-bee insects made 25–50% of total floral visits and performed more visits compared to bee pollinators to compensate for less efficient pollen transfer. The pollination services provided by non-bee insects were more or less similar to those rendered by honey and wild bees (Rader et al. 2016). They increase food sets irrespective of bee visitations and are vital to global crop production (Garibaldi et al. 2013; Rader et al. 2016).

9.2.1 Flies

Flies encompass a diverse species inhabiting habitats, many of which possess the desirable traits for crop pollination (Cook et al. 2020a, b). Several species of flies have been observed visiting flowers of more than 1100 species of plants, demonstrating potential to enhance crop yield in numerous cases (Heath 1982; Cook et al. 2020a, b). Globally, flies rank as the second most abundant floral visitors to crops and, in certain instances, exhibit comparable pollination efficiency to bees (Free 1993; Larson et al. 2001; Ollerton et al. 2011; Rader et al. 2020). Despite this, potential research on the role of flies as crop pollinators remains relatively limited. Among the 86 families of flies that have been documented as flower visitors, certain groups such as Syrphidae (hover flies), Calliphoridae (blow flies), and Anthomyiidae (flower flies) have shown promise as effective pollinators in horticultural settings (Cook et al. 2020a, b). Fly species from genera such as *Calliphora, Chrysomya, Eristalinus, Lucilia, Musca, Stomorhina*, and *Sarcophaga* have been identified as pollinators in mangoes across different regions of the world (Ssymank et al. 2008; Saeed et al. 2016; Willcox et al. 2019; Cook et al. 2020a, b; Bora et al. 2022).

The most important fly pollinators include hover flies and blow flies, documented to carry and deposit substantial pollen loads under specific conditions (Rader et al. 2009; Orford et al. 2015). While fly species may deposit fewer pollen grains per visit, their high visit frequency and foraging behaviour render them efficient pollinators than bees (Kavazos and Wallman 2012; Cook et al. 2020a, b). Other fly species that may pollinate specific plants include bee flies (Bombyliidae), flesh flies (Sarcophagidae), fruit flies (Tephritidae and Drosophilidae), robber flies (Asilidae), moth flies (Psychodidae), march flies (Bibionidae), etc. (Cook et al. 2020a, b). Despite these intriguing findings, fly pollination has not received proportional attention, particularly concerning its significance in crop pollination, especially within horticultural contexts (Cook et al. 2020a, b). The diversity and adaptability of flies as pollinators highlight the complexity and interdependence of natural ecosystems, where multiple species work together to ensure successful plant reproduction and the maintenance of biodiversity. It is essential to recognize that while flies are not well-studied or celebrated as pollinators as bees and butterflies, they are still crucial contributors to plant reproduction and ecosystem functioning. Further exploration is warranted to fully understand the contribution of flies to crop pollination and to integrate their importance into pollinator conservation and management strategies.

9.2.2 Hover Flies

Hover flies or syrphid flies (Syrphidae) are the most well-known group of pollinating flies (Klecka et al. 2018; Doyle et al. 2020; Rader et al. 2020). Among the three subfamilies of Syrphidae, Eristalinae and Syrphinae subfamilies are important from a pollination perspective with ~3800 and 1800 species, respectively (Rotheray and Gilbert 2011). *Eristalis, Eristalinus, Ischiodon, Mesembrius, Syritta,* etc. are some of the important genera of hover flies (Fig. 18.5a–c). Syrphid flies are important for agroecosystems as they provide two important services—biological control of insect pests and crop pollination (Doyle et al. 2020; Dunn et al. 2020). They are easily recognizable due to their distinctive appearance, resembling bees or wasps. Hover flies are classified as generalist foragers because they visit a diverse range of flowering plants for nectar and pollen (Lucas et al. 2018). This adaptability allows them to support diverse crops, making them valuable in mixed cropping systems (Schirmel et al. 2018). Hover flies are active early in the morning, providing essential pollination services when other pollinators might still be less active (Karbassioon and Stanley 2023; Singh 2024 unpublished data).

In a global synthesis of reports of 105 crop plants, hover flies were reported to visit 52% of crops (Rader et al. 2020), and the economic evaluation of visitation was estimated at around US$300 billion per year (Doyle et al. 2020). Hover flies were the

Fig. 18.5 Hover flies (**a**–**c**) and blow fly (**d**) visiting crop flowers. Figure (**a**–**c**) show *Eristalinus* sp. (**a**), *Mesembrius* sp. (**b**), and *Ischiodon* sp. (**c**) hover flies and *Chrysomaya* sp. blow fly (**d**) foraging on mango flowers in the Northern Territory of Australia. (Photo credit: James Makinson (**a** and **c**))

most important non-bee floral visitors to the studied crops (Rader et al. 2020). Among hover flies, *Eristalis tenax* and *E. balteatus* were the two most abundant hover fly species reported visiting 28 and 24 different crop plants, respectively (Doyle et al. 2020; Rader et al. 2020). Hover flies are key pollinators for several crops (Ssymank et al. 2008; Hodgkiss et al. 2018; Moquet et al. 2018), due to their ability to carry a large amount of pollen on their bodies per visit to the flowers (Inouye et al. 2015; Doyle et al. 2020). Hover fly pollination led to improved fruit set and increased crop yields, like in rapeseed/canola (*Brassica napus*) (Jauker and Wolters 2008), strawberries (*Fragaria* spp.) (Hodgkiss et al. 2018), cranberries (Gervais et al. 2018), sweet pepper (Pekas et al. 2020), mangoes (Sánchez et al. 2022a; Singh 2024 unpublished data), watermelon (Sánchez et al. 2022b), etc. Hover flies (*Mesembrius* sp.) are frequent visitors of mangoes in the Northern Territory of Australia (Singh 2024 unpublished data) (Fig. 18.5c). They carry and deposit a considerable number of mango pollen grains and help in mango pollination (Singh 2024 unpublished data). They are well suited for greenhouse and indoor crop cultivation since they can tolerate controlled environments and fluctuating temperatures (Jarlan et al. 1997; Ssymank et al. 2008). Hover flies are also pollinators for wildflower species. For instance, over 70% of the animal-pollinated wild flowering species of Europe are visited by hover flies (Rotheray and Gilbert 2011). Plant communities with open flowers are efficiently pollinated by hover flies; however, hover flies are inefficient pollinators of plants with tubular flowers (Fontaine et al. 2006).

Hover flies also carry or deposit a considerable number of crop pollen grains. A meta-analysis of 71 pollinator visitation networks and 30 pollen transport networks revealed that there is no significant difference in the pollen loads of honey bees and hover flies (Orford et al. 2015). Additionally, pollen grains transferred by hover flies to the stigmas of pak choi and onions were statistically similar to those of honey bees (Howlett et al. 2009; Rader et al. 2009); however, hover flies were less effective on kiwifruits (Stavert et al. 2016; Howlett et al. 2017). They often complement the pollination services of other pollinators (Albrecht et al. 2012). When bee populations are low or weather conditions are unfavourable, hover flies can step in and ensure adequate pollination. Incorporating hover flies into integrated pest management (IPM) strategies helps reduce the reliance on chemical pesticides by reducing the pest population while enhancing crop pollination (Doyle et al. 2020; Rodríguez-Gasol et al. 2020). Their ability to control pest populations and provide pollination makes them valuable allies in sustainable agriculture (Dunn et al. 2020; Li et al. 2023). Promoting hover fly populations and protecting their natural habitats are crucial for maintaining biodiversity and ecological balance. Hover flies are not only beneficial as pollinators and predators, but play crucial roles in supporting wild plant populations and wildlife. Therefore, promoting habitat conservation and adopting sustainable agricultural practices that support hover fly populations can have positive impacts on crop production and ecosystem health.

9.2.3 Blow Flies

Blow flies (Calliphoridae) are often overlooked in discussions of crop pollination, but they form an important group of flower-visiting flies and play a vital role in facilitating plant reproduction (Cook et al. 2020a, b). Blow flies are regularly observed in flower-visitor communities, including economically important crop plants (Fig. 18.5d) (Ssymank et al. 2008; Rajan and Reddy 2019; Willcox et al. 2019; Rader et al. 2020; Singh et al. 2021; Finch et al. 2023). Their diverse ecological traits, adaptable behaviour, and interactions with other pollinators highlight their significance as essential contributors to sustainable agriculture. Blow flies are known for their attraction to odorous compounds (Finch et al. 2023), often leading them to visit a variety of flower types. While their foraging behaviour appears less specialized compared to other pollinators, blow flies have demonstrated effective pollen transfer capabilities in crops (Heath 1982; Pérez-Balam et al. 2012; Rajan and Reddy 2019; Willcox et al. 2019). Their visitation to flowers results in pollen deposition, contributing to cross-pollination and enhancing reproductive success in crops (Willcox et al. 2019).

Their adaptability to different habitats allows them to access and pollinate a wide range of plants. This ecological plasticity enhances their role as pollinators, especially in regions where other pollinator populations may be limited. Blow flies often interact with other pollinator species in complex ways. Co-pollination involving blow flies and other pollinators leads to increased pollination efficiency and crop yield (Willcox et al. 2019). These interactions underscore the importance of considering blow flies as integral components of pollination networks, particularly in scenarios where multiple pollinator species coexist (Willcox et al. 2019). Crops that benefit from blow fly pollination include blueberries (Cook et al. 2020a, b), avocados (Cook et al. 2023), hybrid carrots (Howlett 2012), onions (Munawar et al. 2011), and mangoes (Rajan and Reddy 2019; Willcox et al. 2019; Finch et al. 2023). Blow flies from the genus *Chrysomaya* aid the transfer of pollen between mango flowers, enhancing fruit set and contributing to higher mango yields (Willcox et al. 2019; Singh 2024 unpublished data). Blow flies also help in higher fruit production and improved fruit quality of strawberries (Porter 1977). While blow flies might not be the sole or primary pollinators for crops, their role as supplementary pollinators can still contribute to overall crop productivity and yield.

9.2.4 Other Dipteran Pollinators

Although syrphid flies are the major group of pollinating flies, other true flies are equally important at least in some plant pollination networks (Orford et al. 2015). Their importance is largely overshadowed by bees, butterflies, and hover flies. Non-syrphid fly dipterans were more frequently observed in 33 farms in England. They made up 82% of the total abundance and 73% of the species richness of the dipteran floral visitors (Orford et al. 2015). Besides this, non-syrphid fly dipterans carried 84% of the total pollen carried by dipteran pollinators, and there was no significant difference in the pollen loads of syrphid flies and other flies (Orford et al. 2015). Fly families such as Bombyliidae (bee flies), Muscidae (house flies), Rhiniidae (nose flies), and Sarcophagidae (flesh flies) are frequently observed on flowers and may be important. For example, nose flies from the genus *Stormorhina* pollinate mangoes in Australia (Willcox et al. 2019) and Malaysia (Huda et al. 2015).

9.3 Butterflies and Moths

The adult moths of the majority of the Lepidopteran species are nectar feeders (except, for example, fruit-sucking moths). They exhibit diverse behaviours and morphological adaptations that make them effective and efficient pollinators of flowering plants with open structures and shallow nectar sources (Reddi and Bai 1984). Almost all the butterflies use visual cues (i.e., flower colours) to locate flowers and use long and slender proboscis to draw nectar from deep within the flowers (Reddi and Bai 1984). While feeding, pollen grains get adhered to their proboscis, head, wings, and legs, which are eventually transferred to other conspecific flowers (Reddi and Bai 1984). Some butterfly species are strong fliers, and their ability to cover large areas contributes to pollen transfer between distant plants, promoting genetic diversity within plant populations.

Butterflies are essential pollinators of wildflowers and cultivated plants, including species of milkweed, lantana, and zinnia (Schemske 1976; Faegri and Pijl 1979; Ward and Johnson 2013). Skipper and metalmark butterflies are important pollinators of wildflowers, grasses, and some agricultural crops, especially in tropical regions. The small skipper butterflies (*Thymelicus flavus*) show strong floral constancy, which is an important character from a pollination perspective (Goulson et al. 1997). An interesting example of butterfly pollination involves a rare specific interaction between the *Meneris tulbaghia* butterfly and red-flowered Fynbos plants, where *M. tulbaghia* exclusively pollinates red-coloured flowers (Johnson and Bond 1994). Butterflies are often occasional pollinators; however, in certain crops, such as the peacock flowers (*Caesalpinia pulcherrima*), they are reported to play a significant role (Cruden and Hermann-Parker 1979). Papilionid butterflies are important floral visitors of *Caesalpinia* flowers and carry pollen grains on their large and scaled wings, which facilitates pollination (Cruden and Hermann-Parker 1979). Two species of papilionid butterflies, *Papilio helenus* and *P. Memnon*, are effective pollinators of *Habenaria* orchids in China (Tan et al. 2023). Similarly, the pipevine swallowtail, *Battus philenor*, is the most important pollinator of wildflower *Phlox drummondii* (Burgin et al. 2023).

Moths are predominantly nocturnal pollinators, but selected species are crepuscular or even diurnal (Faegri and Pijl 1979). The significance of moth pollination is overshadowed by the fact that their immature stages (larvae) are harmful and cause significant crop losses (Macgregor et al. 2015). Moths typically have less vibrant colours compared to butterflies and have stout bodies. Moths often visit flowers with strong, sweet fragrances and pale or white flowers that are visible at night. Their long proboscis allows them to access nectar from deep, tubular flowers. Moths are pollinators of wildflowers, fruit trees like apple and cherry, and agricultural crops like tobacco and cotton. Among moths, crepuscular hawkmoth (*Eupanacra mydon*) is an effective pollinator of orchids (*Habenaria dentata*) with two pailionid butterflies in China (Tan et al. 2023). Hawk moths or sphinx moths are very nocturnal pollinators for plants with long tubular flowers, such as honeysuckles (*Lonicera* spp.), evening primrose (*Oenothera biennis*), tobacco flowers (*Nicotiana* spp.), petunias, and orchids (Abrol 2012). Further, clearwing moths are adapted to feed on nectar from shallow flowers and pollinate flowers with short corollas. Even ornamental plants such as gladiolus are pollinated by noctuid and sphingid moths (Goldblatt and Manning 2002). Pollination of *Epidendrum densiflorum* orchids in neotropics is an interesting example where the flowers of *Epidendrum* are exclusively pollinated by males of diurnal lepidopterans (Silveira et al. 2023). Males of four species of Ithomiinae butterflies (*Episcada hymenaea* Prittwitz, *Hypothyris euclea* Doubleday, *Methona themisto* Hübner, and *Placidina euryanassa* C.& R. Felder) and five species of Arctiinae tiger moths (*Antichloris eriphia* Fabricius, *Calodesma collaris* Dury, *Cyanopepla jucunda* Walker, *Philoros rubriceps* Walker, and *Phoenicoprocta teda* Walker) help in the pollination of *Epidendrum* orchids (Silveira et al. 2023).

Butterflies and moths only visit a few flowers on each plant and tend to keep moving between plants. They can be efficient cross-pollinators due to this foraging behaviour. Therefore, there is a need to study lepidopteran pollination in different geographical regions to understand their role in pollen transfer. Implementing practices that support butterfly and moth populations and promote pollinator-friendly landscapes can have significant benefits for human well-being and the environment.

9.4 Beetles

Beetles and weevils are not as well-known for pollination services as bees or flies, but they are generalist pollinators and visit a wide variety of flowering plants (Henderson 1986). Their foraging behaviour supports pollination of crops, wildflowers, and native plants, contributing to overall plant diversity in ecosystems. Beetles are active during low-light conditions, such as dusk and dawn. They support pollination in the early morning or late evening when other pollinators, like bees, may not be as active. Beetles are particularly effective pollinators for crops with simple, bowl-shaped, or open flowers (Dafni et al. 1990). These types of flowers are easy for beetles to access, and their body structure allows them to carry a substantial amount of pollen. Beetles can be abundant in certain landscapes, especially in regions where other pollinators are less prevalent. Beetles contribute significantly to large-scale pollination. For instance, *Goniothalamus wynaadensis* (Annonaceae) is effectively pollinated by nitidulid beetles (*Caprophilus, Colopterus, Urophorus*) and curculionid weevils in the Western Ghats of India (Anjali and Sreekala 2023). Many water lily species, including the famous lotus (*Nelumbo* spp.) and various water lilies (*Nymphaea* spp.), are pollinated by beetles.

Scarabaeidae is amongst the most important family of beetle pollinators. Dynastid scarab beetles are exclusive pollinators of *Annona* spp. (Annonaceae) (Gottsberger

1989). *Cyclocephala caelestis* and *Myrmecocephalus* sp. are significant pollinators for Magnolia trees, including species like *Magnolia tamaulipana, Magnolia grandiflora*, and *M. soulangeana* (Dieringer et al. 1999). These large, showy flowers are often pollinated by beetles attracted to sweet fragrance. Another *Cyclocephala* species (*C. variolosa*) is the sole pollinator of *Philodendron* (Araceae) in São Paulo, Brazil (Maia et al. 2023). Additionally, a tropical rainforest herb, *Dieffenbachia longispatha* is also pollinated by scarab beetles from *Cyclocephala* and *Eriosceli* genus (Young 1986). *Cyclocephala* beetles are also good pollinators of *Cyclanthus bipartitus* plants (Beach 1982). Another genus of scarabaeid beetles, *Amphicoma* includes the primary pollinators of poppy (*Anemone coronaria*), Flanders poppy (*Papaver rhoeas*), the Persian buttercup (*Ranunculus asiaticus*), and tulips (*Tulipa agenensis*) (Dafni et al. 1990). Other examples of beetle pollination include pollination of cycads by boganiid beetle (*Cretoparacucujus cycadophilus*) (Cai et al. 2018), *Orchidantha inouei* by *Onthophagus* dung beetles (Sakai and Inoue 1999), peacock moraeas by hopliine beetles (*Anisochelus, Anisonyx, Monochelus*, and *Peritrichia*) (Steiner 1998), *Philodendron solimoesens C. colasi* (Gibernau et al. 1999), etc. Beetles visit and pollinate flowers of several other plants such as daisy, sunflower, yellow tulip tree (*Liriodendron tulipifera*), Magnolia vine (*Schisandra chinensis*), wild ginger (*Asarum* spp.), dutchman's pipe (*Aristolochia* spp.), pawpaw (*Asimina triloba*), carrion flowers, etc.

Among Coleoptera, weevil pollinators have received very little attention. Due to the almost exclusively phytophagous nature of weevils, they are generally considered antagonistic to plants. However, recent studies show that weevils could be the most diverse group of specialized brood site pollinators (Haran et al. 2023). Brood site pollination is a mutualistic mechanism where immature stages of insects develop inside the plant tissues in exchange for plant pollination (Sakai 2002). Fig wasps are the classic example of this mutualism; however, weevils could be the most diverse group showing brood site pollination mutualism (Haran et al. 2023). Some of the reported interactions include pollination of *Annona senegalensis* by Ochyromerini, *Carludovica palmata* by Derelomini, *Cecropia peltate* by Eugnomini, etc. (Haran et al. 2023). Pollination of oil palm by oil palm weevil, *Elaeidobius kamerunicus*, is one of the interesting examples of weevil pollination (Gintoron et al. 2023).

Beetles and weevils may not be as efficient as bees. Nonetheless, their generalist behaviour and adaptability to various habitats allow them to play an essential role in the pollination of plants. Promoting the conservation and habitat preservation of beetles and weevils, with other pollinators, is crucial for maintaining the health of agricultural and natural ecosystems. Future research should focus on understanding the beetle pollination mechanism in economically important crops, and wherever possible, efforts should be made to ascertain the contribution of beetle pollination to fruit or seed production. A diverse and resilient community of pollinators, including beetles, helps ensure the productivity and sustainability of global food production and supports biodiversity conservation efforts.

9.5 Wasps

Wasps are not as well-known for their role in pollination as bees and flies, but some wasp species do play a role in pollinating plants. It is important to note that not all wasp species are effective pollinators, and some may have a reputation for being pests. Fig wasps (Agaonidae), potter wasps (Eumeninae), and scoliid wasps (Scoliidae) are some examples of wasp pollinators (Abrol 2012). Fig wasps have a unique mutualistic relationship with fig trees (*Ficus* spp.). Female fig wasps are essential for pollinating fig flowers, and fig trees are entirely dependent on specific species of fig wasps for pollination and reproduction (Weiblen 2002; Cook and Rasplus 2003; Abrol 2012). The wasps crawl into the fig's receptacle, where they pollinate the flowers and lay eggs. In return, the fig provides a protected environment and nourishment for the developing wasp larvae (Weiblen 2002). Other notable examples include pollination of *Ficus macrophylla* by female *Pleistodontes froggatti* (Abrol 2012), *Calochilus holtzei* orchids by male scoliid wasps and male flower wasps (Thynnidae) (Gumprecht 1977), and *Caladenia nobilis* orchids by male *Rhagigaster discrepans* flower wasp (Phillips et al. 2020). Besides this, spider wasps and potter wasps are potential pollinators. Spider wasps are solitary wasps known for their hunting behaviour, as they capture and paralyse spiders to provision their nests. While they primarily hunt spiders for their offspring, some species may inadvertently transfer pollen between flowers as they forage for nectar. Potter wasps are solitary wasps that construct pot-like nests from mud or clay. They are known to visit flowers for nectar and may contribute to pollination as they move between flowers.

9.6 Ants

The primary ecological role of ants is not pollination, but they can have secondary interactions with certain plants that involve pollen transfer. In many cases, ants are more important for their role in seed dispersal and as predators or scavengers in ecosystems; however, they can act as pollen vectors. Ant pollination is less prevalent, and ants are reported to have limited pollination ability (Beattie et al. 1984). However, a recent study highlights the significance of ants in pollination (Das and Das 2023). This global synthesis identified 70 ant species as pollinators of 41 plant species from 23 families, and these species belong to four subfamilies, viz., Formicinae ($n = 34$ species), Myrmicinae ($n = 18$), Dolichoderinae ($n = 15$), and Pseudomyrmecinae ($n = 3$) (Das and Das 2023). Formicinae is the largest subfamily of pollinating ants and includes specific and generalist ant-pollinators. Only one species pollinates ≤ 4 plant species, and the majority of the species ($n = 56$) act as pollinators of only one plant species (Das and Das 2023). Further, Euphorbiaceae, Cytinaceae, Orchidaceae, Apiaceae, Roseaceae, Cucurbitaceae, Arecaceae, Brassicaceae, Anacardiaceae, Liliaceae, Fabaceae, and Myrtaceae are the major plant families pollinated by ants (Das and Das 2023). *Proformica longiseta* from Europe and *Camponotus crassus* from South America are the two most frequent ant-pollinators, respectively (Das and Das 2023). The role of ant pollination varies among different geographical regions. For instance, *Camponotus* and *Monomorium* species are pollinators of mangoes in South Africa (Carvalheiro et al. 2010); however, *Camponotus* and *Iridomyrmex* ants carry very few or no mango pollen grains on their bodies (Huda et al. 2015; Singh 2024 unpublished data). However, ants are mainly secondary or occasional pollinators/visitors for plants such as myrmecophytes, bird's foot violet (*Viola pedate*), *Coffea mauritiana, Acacia,* and *Euphorbia* species (Philpott et al. 2006).

9.7 Nocturnal Pollinators

Previous evaluations of animal pollination services have predominantly focused on diurnal pollinators, particularly emphasizing the role of managed and wild bees; however, the economic value of nocturnal pollination services is unclear. However, pollinators that forage at night are understudied (Macgregor and Scott-Brown 2020; Buxton et al. 2022). Many plants have adapted to pollination by nocturnal animals. Several species of bats, bees, beetles, butterflies, flies, moths, and thrips and species of rodents and marsupials have been observed as nocturnal floral visitors (Fig. 18.6); however, little is known about nocturnal pollinators (Macgregor and Scott-Brown 2020). Some nocturnal flowers have specialized structures to accommodate their nocturnal pollinators. Floral scents, night vision, and thermogenic sensitivity are the three key mechanisms of nocturnal pollinator-plant interactions (Macgregor and Scott-Brown 2020). Bat-pollinated flowers are often large and pale-coloured, making them more visible in low-light conditions, and bats use their keen sense of smell and echolocation to locate and navigate to flowers.

The pollination of *A. tequilana* and durians by bats is one of the best examples of nocturnal pollination services provided by animals (Kunz et al. 2011). Thrips are generally considered pests of major agricultural and horticultural crops; however, thrips can contribute to pollination in *Solanum melongena* (Velayudhan and Annadurai 1987), *Elaeis guineensis* (Syed 1979), and *Coffea canephora* (Infante et al. 2017) in the tropics and subtropics (Ananthakrishnan 1993; Macgregor and Scott-Brown 2020). The presence of thrips is positively correlated with the fruit set

Fig. 18.6 Butterfly (**a**), moth (**b**), and green lacewing (**c**) as nocturnal floral visitors of mangoes in the Northern Territory of Australia. (Photo credit: James Makinson)

in *Sambucus nigra* (Scott-Brown et al. 2019). Fruit crops from the family Myrtaceae are pollinated by nocturnal bees (Cordeiro et al. 2019). Other examples of nocturnal pollinators include sphinx moths (Sphingidae), hawkmoths (*Hemaris*), also known as hummingbird moths, and scarab beetles. Lesser long-nosed bat, Mexican long-tongued bat, small marsupial-sugar glider (*Petaurus breviceps*), long-tailed rat (*Uromys caudimaculatus*), and Virginia opossum (*Didelphis virginiana*) are also observed on flowers, which potentially contribute to nocturnal pollination.

Nocturnal pollinators contribute significantly to apple (Robertson et al. 2021) and bushberry fruit set (Cutler et al. 2012). Nocturnal pollinators also complement the services provided by diurnal pollinators like bees and butterflies, ensuring continuous pollination and reproductive success for plants (Luo et al. 2011; Robertson et al. 2021). Understanding and conserving nocturnal pollination is crucial for the health of ecosystems and the success of various plant species. As human activities continue to impact the environment, preserving the habitats and resources that support nocturnal pollinators becomes increasingly important to ensure the sustainability of ecosystems.

10 Conclusion

The diversity of pollination agents highlights the co-evolutionary relationships between flowering plants and pollinators. These interactions have shaped the remarkable diversity of floral forms, colours, and scents observed in nature. However, the delicate balance between plants and pollinators faces unprecedented challenges due to habitat loss, climate change, and other anthropogenic disturbances (Dicks et al. 2021). Many animal pollinators, particularly bees and butterflies, are experiencing population declines worldwide due to habitat loss, pesticide use, diseases, climate change, and other factors (Dicks et al. 2021). These declines threaten the stability of natural and agricultural ecosystems reliant on pollinators. The fragmentation and degradation of natural habitats reduce the availability of suitable foraging and nesting sites for pollinators, leading to reduced pollinator diversity and abundance. Certain pesticides harm pollinators and disrupt foraging behaviour, navigation, and reproductive capabilities. Altered weather patterns and shifts in flowering and pollinator activity affect the synchronization between plants and their pollinators, potentially leading to reduced pollination success. Understanding the roles and preferences of these factors is crucial to formulating effective conservation strategies to safeguard wild plant populations and crops.

Insect pollinators are indispensable components of ecological systems, contributing to biodiversity, food production, and environmental health. Bees, butterflies, moths, flies, and beetles are instrumental in the pollination of numerous crop species. Their efficient foraging behaviours and specialized adaptations have resulted in a remarkable diversity of pollination strategies; consequently, their decline or loss could have severe consequences for ecosystem stability and human well-being. Protecting and supporting insect pollinators is not only necessary for the survival of countless plant species but also crucial for ensuring sustainable agriculture and food security in the face of a changing world. To safeguard insect pollination and address the challenges encountered by pollinators, conservation efforts are crucial. It includes: (i) protecting and restoring natural habitats; (ii) implementing sustainable agricultural practices that minimize the use of harmful pesticides and promote integrated pest management; (iii) raising awareness on pollinators and their conservation among the general public, farmers, and policymakers; (iv) encouraging the creation of pollinator-friendly landscapes, gardens, and urban green spaces; and (v) conducting research to better understand the factors impacting pollinators.

References

Abrol DP (1993) Insect pollination and crop production in Jammu and Kashmir. Curr Sci 65:265–269

Abrol DP (2012) Pollination biology: biodiversity conservation and agricultural production. Springer, New York

Abrol DP, Gorka AK, Ansari MJ, Al-Ghamdi A, Al-Kahtani S (2019) Impact of insect pollinators on yield and fruit quality of strawberry. Saudi J Biol Sci 26(3):524–530

Absy ML, Kerr WE (1977) Algumas plantas visitadas para obtenção de pólen por operárias de Melipona seminigra merrillae em Manaus. Acta Amaz 7(3):309–315

Absy ML, Camargo JM, Kerr WE, Miranda ID (1984) Plant species visited by pollen-collecting Meliponinae in Central Amazonas. Braz J Biol 44(2):227–237

Ackerman JD (2000) Abiotic pollen and pollination: ecological, functional, and evolutionary perspectives. Plant Syst Evol 222(1–4):167–185

Adams PB, Lawson SD (1993) Pollination in Australian orchids: a critical assessment of the literature 1882-1992. Aust J Bot 41(5):553–575

Aenmay TK, Tierney SM, Pillay N, Schwarz MP (2006) Nesting biology of an African allodapine bee, *Braunsapis vitrea*: female biased sex allocation in the absence of worker-like behavioural castes. Ethol Ecol Evol 18(3):205–220

Aizen MA, Harder LD (2009) The global stock of domesticated honey bees is growing slower than agricultural demand for pollination. Curr Biol 19(11):915–918

Aizen MA, Garibaldi LA, Cunningham SA, Klein AM (2008) Long-term global trends in crop yield and production reveal no current pollination shortage but increasing pollinator dependency. Curr Boil 18(20):1572–1575

Aizen MA, Garibaldi LA, Cunningham SA, Klein AM (2009) How much does agriculture depend on pollinators? Lessons from long-term trends in crop production. Ann Bot 103(9):1579–1588

Albrecht M, Schmid B, Hautier Y, Müller CB (2012) Diverse pollinator communities enhance plant reproductive success. Proc R Soc B Biol Sci 279(1748):4845–4852

Amala U, Shivalingaswamy T (2017) Role of native buzz pollinator bees in enhancing fruit and seed set in tomatoes under open field conditions. J Entomol Zool 5(3):1742–1744

Ananthakrishnan TN (1993) The role of thrips in pollination. Curr Sci 65:262–264

Anderson DL, Sedgley M, Short JRT, Allwood AJ (1982) Insect pollination of mango in Northern Australia. Aust J Agric Res 33(3):541–548

Anderson S, Kelly D, Robertson A, Ladley J (2016) Pollination by birds: a functional evaluation. In: Sekercioglu Ç, Wenny D, Whelan C (eds) Why birds matter: avian ecological function and ecosystem services. University of Chicago Press, Chicago, pp 73–106

Andrews CE, Anderson SH, Van der Walt K, Thorogood R, Ewen JG (2022) Evaluating the success of functional restoration after reintroduction of a lost avian pollinator. Conserv Biol 36(4): e13892

Andriamihaja CF, Ramarosandratana AV, Grisoni M, Jeannoda V, Besse P (2020) The leafless Vanilla species-complex from the South-West Indian Ocean Region: a taxonomic puzzle and a model for orchid evolution and conservation research. Diversity 12(12):443

Anjali J, Sreekala AK (2023) Floral biology and pollinator captivating semiochemicals in *Goniothalamus wynaadensis* (Bedd.) (Annonaceae) an endemic tree species of the Western Ghats. Plant Sci Today 10(3):399–408

Asar Y, Ho SYW, Sauquet H (2022) Early diversifications of angiosperms and their insect pollinators: were they unlinked? Trends Plant Sci 27:858–869

Ascher J, Pickering J (2023) Discover life bee species guide and world checklist (Hymenoptera: Apoidea: Anthophila). www.discoverlife.org/mp/20q?guide=Apoidea_species

Azmi WA, Seng CT, Solihin NS (2016) Pollination efficiency of the stingless bee, *Heterotrigona itama* (Hymenoptera: Apidae) on Chili (*Capsicum annuum*) in Greenhouse. J Trop Plant Physiol 8:1–11

Azmi WA, Samsuri N, Hatta MFM, Ghazi ROZIAH, Seng CT (2017) Effects of stingless bee (*Heterotrigona itama*) pollination on greenhouse cucumber (*Cucumis sativus*). Malays Appl Biol 46(1):51–55

Beach JH (1982) Beetle pollination of *Cyclanthus bipartitus* (Cyclanthaceae). Am J Bot 69(7): 1074–1081

Beattie AJ, Turnbull C, Knox RB, Williams EG (1984) Ant inhibition of pollen function: a possible reason why ant pollination is rare. Am J Bot 71(3):421–426

Bell MC, Spooner-Hart RN, Haigh AM (2006) Pollination of greenhouse tomatoes by the Australian bluebanded bee *Amegilla (Zonamegilla) holmesi* (Hymenoptera: Apidae). J Econ Entomol 99(2):437–442

Bernauer OM, Cook JM, Tierney SM (2021) Nesting biology and social organisation of the allodapine bee, *Exoneura angophorae* (Hymenoptera: Apidae): montane environmental constraints yield biased sex allocation yet phenology is unhindered. Insect Soc 68(4):337–349

Bernauer OM, Tierney SM, Cook JM (2022) Efficiency and effectiveness of native bees and honey bees as pollinators of apples in New South Wales orchards. Agric Ecosyst Environ 337:108063

Berry PE, Calvo RN (1989) Wind pollination, self-incompatibility, and altitudinal shifts in pollination systems in the high Andean genus *Espeletia* (Asteraceae). Am J Bot 76:1602–1614

Biddinger DJ, Joshi NK, Rajotte EG, Halbrendt NO, Pulig C, Naithani KJ, Vaughan M (2013) An immunomarking method to determine the foraging patterns of *Osmia cornifrons* and resulting fruit set in a cherry orchard. Apidologie 44:738–749

Bohart GE (1972) Management of wild bees for the pollination of crops. Annu Rev Entomol 17(1): 287–312

Bommarco R, Marini L, Vaissière BE (2012) Insect pollination enhances seed yield, quality, and market value in oilseed rape. Oecologia 169:1025–1032

Bonneau MN, Samson-Robert O, Fournier V, Chouinard G (2021) Commercial bumble bee (*Bombus impatiens*) hives under exclusion netting systems for apple pollination in orchards. Renew Agric Food Syst 36(3):234–244

Bora MS, Deka MK, Bora SS (2022) Foraging behaviour of some syrphid flies on mango. Indian J Entomol 84(2):418–420

Bosch J, Blas M (1994) Foraging behaviour and pollinating efficiency of *Osmia cornuta* and *Apis mellifera* on almond (Hymenoptera, Megachilidae and Apidae). Appl Entomol Zool 29(1):1–9

Bosch J, Kemp WP (2002) Developing and establishing bee species as crop pollinators: the example of *Osmia* spp. (Hymenoptera: Megachilidae) and fruit trees. Bull Entomol Res 92(1): 3–16

Bosch J, Kemp WP, Peterson SS (2000) Management of *Osmia lignaria* (Hymenoptera: Megachilidae) populations for almond pollination: methods to advance bee emergence. Environ Entomol 29(5):874–883

Boyle NK, Pitts-Singer TL (2019) Assessing blue orchard bee (*Osmia lignaria*) propagation and pollination services in the presence of honey bees (*Apis mellifera*) in Utah tart cherries. PeerJ 7: e7639

Brown J, Groom SV, Rader R, Hogendoorn K, Cunningham SA (2022) Land cover associations of wild bees visiting flowers in apple orchards across three geographic regions of Southeast Australia. Agric Ecosyst Environ 324:107717

Bueno FGB, Kendall L, Alves DA, Tamara ML, Heard T, Latty T, Gloag R (2023) Stingless bee floral visitation in the global tropics and subtropics. Glob Ecol Conserv 43:e02454

Bullock SH (1994) Wind pollination of neotropical deciduous trees. Biotropica 26:172–179

Burgin GA, Bronzo-Munich O, Garner AG, Acevedo IA, Hopkins R (2023) Characterizing each step of pollination in *Phlox drummondii* reveals that a single butterfly species predominates in the pollinator assemblage. Am J Bot 110(5):e16172

Button L, Elle E (2014) Wild bumble bees reduce pollination deficits in a crop mostly visited by managed honey bees. Agric Ecosyst Environ 197:255–263

Buxton MN, Gaskett AC, Lord JM, Pattemore DE (2022) A global review demonstrating the importance of nocturnal pollinators for crop plants. J Appl Ecol 59(12):2890–2901

Cai C, Escalona HE, Li L, Yin Z, Huang D, Engel MS (2018) Beetle pollination of cycads in the Mesozoic. Curr Biol 28(17):2806–2812

Calderone NW (2012) Insect pollinated crops, insect pollinators and US agriculture: trend analysis of aggregate data for the period 1992–2009. PLoS One 7(5):e37235

Campbell JW, Stanley-Stahr C, Bammer M, Daniels JC, Ellis JD (2019) Contribution of bees and other pollinators to watermelon (*Citrullus lanatus* Thunb.) pollination. J Apic Res 58(4): 597–603

Can-Alonzo C, Quezada-Euán JJ, Xiu-Ancona P, Moo-Valle H, Valdovinos-Nunez GR, Medina-Peralta S (2005) Pollination of 'criollo' avocados (*Persea americana*) and the behaviour of associated bees in subtropical Mexico. J Apic Res 44(1):3–8

Cane JH (2023) The extraordinary alkali bee, *Nomia melanderi* (Halictidae), the world's only intensively managed ground-nesting bee. Annu Rev Entomol 69:99

Cane JH, Schiffhauer D, Kervin LJ (1996) Pollination, foraging, and nesting ecology of the leaf-cutting bee Megachile (Delomegachile) addenda (Hymenoptera: Megachilidae) on cranberry beds. Ann Entomol Soc Am 89(3):361–367

Canto-Aguilar MA, Parra-Tabla V (2000) Importance of conserving alternative pollinators: assessing the pollination efficiency of the squash bee, *Peponapis limitaris* in *Cucurbita moschata* (Cucurbitaceae). J Insect Conserv 4:201–208

Carvalheiro LG, Seymour CL, Veldtman R, Nicolson SW (2010) Pollination services decline with distance from natural habitat even in biodiversity-rich areas. J Appl Ecol 47(4):810–820

Catling PM (1983) Pollination of northeastern North American Spiranthes (Orchidaceae). Can J Bot 61(4):1080–1093

Cecen S, Gosterit A, Gurel F (2007) Pollination effects of the bumble bee and honey bee on white clover (Trifolium repens L.) seed production. J Apic Res 46(2):69–72

Chambó ED, Garcia RC, Oliveira NTED, Duarte-Júnior JB (2011) Honey bee visitation to sunflower: effects on pollination and plant genotype. Sci Agric 68:647–651

Clifford HT (1962) Insect pollinators of *Plantago lanceolata* L. Nature 193:196

Coates JM, Brown J, Cunningham SA (2022) Wild bees nest in the stems of cultivated Rubus plants and act as effective crop pollinators. Agric Ecosyst Environ 325:107741

Cook JM, Rasplus JY (2003) Mutualists with attitude: coevolving fig wasps and figs. Trends Ecol Evol 18(5):241–248

Cook DF, Deyl RA, Mickan BS, Howse ET (2020a) Yield of southern highbush blueberry (*Vaccinium corymbosum*) using the fly *Calliphora albifrontalis* (Diptera: Calliphoridae) as a pollinator. Austral Entomol 59(2):345–352

Cook DF, Voss SC, Finch JT, Rader RC, Cook JM, Spurr CJ (2020b) The role of flies as pollinators of horticultural crops: an Australian case study with worldwide relevance. Insects 11(6):341

Cook DF, Tufail MS, Voss SC, Deyl RA, Howse ET, Foley J, Norrish B, Delroy N, Shivananjappa SL (2023) Blow flies (Diptera: Calliphoridae) ability to pollinate Hass avocado trees within paired tree enclosures. J Appl Entomol 147(8):577–591

Corbet SA, Fussell M, Ake R, Fraser A, Gunson C, Savage A, Smith K (1993) Temperature and the pollinating activity of social bees. Ecol Entomol 18(1):17–30

Cordeiro GD, Dos Santos IGF, da Silva CI, Schlindwein C, Alves-dos-Santos I, Dötterl S (2019) Nocturnal floral scent profiles of Myrtaceae fruit crops. Phytochemistry 162:193–198

Cortopassi-Laurino M, Knoll FRN, Ribeiro MF (1991) Food plant preferences of *Friesella schrottkyi*. Acta Hortic 288:382–385

Cox PA (1988) Hydrophilous pollination. Annu Rev Ecol Evol Syst 19(1):261–279

Crane E (2009) *Apis* species:(honey bees). In: Encyclopedia of insects. Academic, pp 31–32

Cronk Q, Ojeda I (2008) Bird-pollinated flowers in an evolutionary and molecular context. J Exp Bot 59(4):715–727

Cruden RW, Hermann-Parker SM (1979) Butterfly pollination of *Caesalpinia pulcherrima*, with observations on a psychophilous syndrome. J Ecol 67:155–168

Culley TM, Weller SG, Sakai AK (2002) The evolution of wind pollination in angiosperms. Trends Ecol Evol 17(8):361–369

Cutler GC, Reeh KW, Sproule JM, Ramanaidu K (2012) Berry unexpected: nocturnal pollination of lowbush blueberry. Can J Plant Sci 92(4):707–711

Dafni A, Dukas R (1986) Insect and wind pollination in *Urginea martima* (Liliaceae). Plant Syst Evol 154:1–10

Dafni A, Bernhardt P, Shmida A, Ivri Y, Greenbaum S, O'Toole C, Losito L (1990) Red bowl-shaped flowers: convergence for beetle pollination in the Mediterranean region. Isr J Plant Sci 39(1–2):81–92

Das S, Das A (2023) Ants are more than just curious bystanders to some flowers—they act as significant pollinators. Front Insect Sci 3:1145761

David MA, Yommi A, Sánchez E, Martinez A, Murillo N, Marcellán O, Atela O, Palacio MA (2022) Strategic use of honey bees (*Apis mellifera* L.) to increase the number and size of fruits in kiwifruit (*Actinidia chinensis* var. *deliciosa*). Eur J Agron 133:126420

Delaplane KS, Mayer DR, Mayer DF (2000) Crop pollination by bees. CABI Publishing, Oxon

Deuri A, Rahman A, Gogoi J, Borah P, Bathari M (2018) Pollinator diversity and effect of *Apis cerana* F. pollination on yield of mango (*Mangifera indica* L.). J Entomol Zool Stud 6(5): 957–961

Dicks LV, Breeze TD, Ngo HT, Senapathi D, An J, Aizen MA, Basu P, Buchori D, Galetto L, Garibaldi LA, Gemmill-Herren B (2021) A global-scale expert assessment of drivers and risks associated with pollinator decline. Nat Ecol Evol 5(10):1453–1461

Dieringer G, Cabrera RL, Lara M, Loya L, Reyes-Castillo P (1999) Beetle pollination and floral thermogenicity in *Magnolia tamaulipana* (Magnoliaceae). Int J Plant Sci 160(1):64–71

Dollin A (2020) Blue banded bees. Aussie Bee Online, 27, 01–06. https://www.aussiebee.com.au/aussiebeeonline027.pdf

Doyle T, Hawkes WL, Massy R, Powney GD, Menz MH, Wotton KR (2020) Pollination by hoverflies in the Anthropocene. Proc R Soc B 287(1927):20200508

Dunn L, Lequerica M, Reid CR, Latty T (2020) Dual ecosystem services of syrphid flies (Diptera: Syrphidae): pollinators and biological control agents. Pest Manag Sci 76(6):1973–1979

Eardley CD, Gikungu M, Schwarz MP (2009) Bee conservation in Sub-Saharan Africa and Madagascar: diversity, status and threats. Apidologie 40(3):355–366

Engel MS, Dathe HH (2011) Order Hymenoptera, family Apidae: tribe Allodapini, genus *Braunsapis*. Arthropod Fauna UAE 4:630–635

Engel MS, Dingemans-Bakels F (1980) Nectar and pollen resources for stingless bees (Meliponinae, Hymenoptera) in Surinam (South America). Apidologie 11(4):341–350

Enz J (2001) Nectar robbers and non-nectar robbers: their visitation rates and effects on seed set in Virginia bluebells, *Mertensia virginica* (L.). In: Pers. Botany 2001 conference, plants and people

Faegri K, Pijl VDL (1979) Principles of pollination ecology. Pergamon Press, Oxford/New York

Faegri K, Pijl VDL (2013) Principles of pollination ecology (III). Pergamon Press, Oxford/New York

Fang Q, Chen YZ, Huang SQ (2012) Generalist passerine pollination of a winter-flowering fruit tree in Central China. Ann Bot 109(2):379–384
Finch J, Gilpin AM, Cook J (2023) Fishing for flies: testing the efficacy of "stink stations" for promoting blow flies as pollinators in mango orchards. J Pollinat Ecol 33:79–100
Fleming TH, Valiente-Banuet A (2002) Columnar cacti and their mutualists: evolution, ecology, and conservation. University of Arizona Press, Tucson
Fleming TH, Geiselman C, Kress WJ (2009) The evolution of bat pollination: a phylogenetic perspective. Ann Bot 104:1017–1043
Fontaine C, Dajoz I, Meriguet J, Loreau M (2006) Functional diversity of plant–pollinator interaction webs enhances the persistence of plant communities. PLoS Biol 4(1):e1
Frankie G, Thorp R, Hernandez J, Rizzardi M, Ertter B, Pawelek J, Witt S, Schindler M, Coville R, Wojcik V (2009) Native bees are a rich natural resource in urban California gardens. Calif Agric 63(3):113–120
Free JB (1993) Insect pollination of crops. Academic, London
Free JB, Raw A, Williams IH (1975) Pollination of coconut (*Cocos nucifera* L.) in Jamaica by honeybees and wasps. Appl Anim Ethol 1(3):213–223
Fussell M, Corbet SA (1991) Forage for bumble bees and honey bees in farmland: a case study. J Apic Res 30(2):87–97
Gallai N, Salles JM, Settele J, Vaissière BE (2009) Economic valuation of the vulnerability of world agriculture confronted with pollinator decline. Ecol Econ 68(3):810–821
Garibaldi LA, Steffan-Dewenter I, Kremen C, Morales JM, Bommarco R, Cunningham SA, Carvalheiro LG, Chacoff NP, Dudenhöffer JH, Greenleaf SS, Holzschuh A (2011) Stability of pollination services decreases with isolation from natural areas despite honey bee visits. Ecol Lett 14(10):1062–1072
Garibaldi LA, Steffan-Dewenter I, Winfree R, Aizen MA, Bommarco R, Cunningham SA, Kremen C, Carvalheiro LG, Harder LD, Afik O, Bartomeus I (2013) Wild pollinators enhance fruit set of crops regardless of honey bee abundance. Sci 339(6127):1608–1611
Garratt MP, Breeze TD, Jenner N, Polce C, Biesmeijer JC, Potts SG (2014) Avoiding a bad apple: insect pollination enhances fruit quality and economic value. Agric Ecosyst Environ 184:34–40
Garratt MP, Breeze TD, Boreux V, Fountain MT, McKerchar M, Webber SM, Coston DJ, Jenner N, Dean R, Westbury DB, Biesmeijer JC (2016) Apple pollination: demand depends on variety and supply depends on pollinator identity. PLoS One 11(5):e0153889
Gervais A, Chagnon M, Fournier V (2018) Diversity and pollen loads of flower flies (Diptera: Syrphidae) in cranberry crops. Ann Entomol Soc Am 111(6):326–334
Geslin B, Aizen MA, Garcia N, Pereira AJ, Vaissière BE, Garibaldi LA (2017) The impact of honey bee colony quality on crop yield and farmers' profit in apples and pears. Agric Ecosyst Environ 248:153–161
Giannini TC, Boff S, Cordeiro GD, Cartolano EA, Veiga AK, Imperatriz-Fonseca VL, Saraiva AM (2015) Crop pollinators in Brazil: a review of reported interactions. Apidologie 46:209–223
Gibernau M, Barabé D, Cerdan P, Dejean A (1999) Beetle pollination of *philodendron solimoesense* (Araceae) in French Guiana. Int J Plant Sci 160(6):1135–1143
Gintoron CS, Mohammed MA, Sazali SN, Deka EQ, Ong KH, Shamsi IH, King PJH (2023) Factors affecting pollination and pollinators in oil palm plantations: a review with an emphasis on the *Elaeidobius kamerunicus* Weevil (Coleoptera: Curculionidae). Insects 14(5):454
Goldblatt P, Manning JC (2002) Evidence for moth and butterfly pollination in Gladiolus (Iridaceae- Crocoideae). Ann Mo Bot Gard 89:110–124
Gomez JM, Zamora R (1996) Wind pollination in high-mountain populations of *Hormathophylla spinosa* (Cruciferae). Am J Bot 83:580–585
Goodwillie C (1999) Wind pollination and reproductive assurance in *Linanthus parviflora* (Polemoniaceae), a self-incompatible annual. Am J Bot 86:948–954
Gottsberger G (1989) Beetle pollination and flowering rhythm of *Annona* spp. (Annonaceae) in Brazil. Plant Syst Evol 167:165–187

Goubara M, Takasaki T (2004) Pollination effects of the sweat bee *Lasioglossum villosulum* trichopse (Hymenoptera: Halictidae) on genic male-sterile lettuce. Appl Entomol Zool 39(1): 163–169

Goulson D (2010) Bumblebees: behaviour, ecology and conservation, 2nd edn. Oxford University Press

Goulson D, Ollerton J, Sluman C (1997) Foraging strategies in the small skipper butterfly, *Thymelicus flavus*: when to switch? Anim Behav 53(5):1009–1016

Goulson D, Lye GC, Darvill B (2008) Decline and conservation of bumble bees. Annu Rev Entomol 53:191–208

Goulson D, Nicholls E, Botias C, Rotheray EL (2015) Bee declines driven by combined stress from parasites, pesticides, and lack of flowers. Sci 347(6229):1255957

Grant KJ, DeVetter L, Melathopoulos A (2021) Honey bee (*Apis mellifera*) colony strength and its effects on pollination and yield in highbush blueberries (*Vaccinium corymbosum*). PeerJ 9: e11634

Greco MK, Spooner-Hart RN, Beattie AG, Barchia I, Holford P (2011) Australian stingless bees improve greenhouse capsicum production. J Apic Res 50(2):102–115

Greenleaf SS, Kremen C (2006) Wild bees enhance honey bees' pollination of hybrid sunflower. Proc Natl Acad Sci USA 103(37):13890–13895

Gruber B, Eckel K, Everaars J, Dormann CF (2011) On managing the red mason bee (*Osmia bicornis*) in apple orchards. Apidologie 42:564–576

Gumprecht R (1977) Seltsame Bestaubungsvorgange bei Orchideen. Die Orchid 3(28):5–23

Haran J, Kergoat GJ, de Medeiros BA (2023) Most diverse, most neglected: weevils (Coleoptera: Curculionoidea) are ubiquitous specialized brood-site pollinators of tropical flora. Peer Commun J 3:e49

Harder LD (1998) Pollen-size comparisons among animal-pollinated angiosperms with different pollination characteristics. Biol J Linnean Soc 64:513–525

Hargreaves AL, Harder LD, Johnson SD (2012) Floral traits mediate the vulnerability of aloes to pollen theft and inefficient pollination by bees. Ann Bot 109(4):761–772

Heard T (1999) The role of stingless bees in crop pollination. Annu Rev Entomol 44(1):183–206

Heard TA, Exley EM (1994) Diversity, abundance, and distribution of insect visitors to macadamia flowers. Environ Entomol 23(1):91–100

Heath ACG (1982) Beneficial aspects of blowflies (Diptera: Calliphoridae). N Z Entomol 7(3): 343–348

Hedström I (1986) Pollen carriers of *Cocos nucifera* L. (Palmae) in Costa Rica and Ecuador (Neotropical region). Rev Biol Trop 34(2):297–301

Heinrich B (1975) Energetics of pollination. Annu Rev Ecol Evol Syst 6(1):139–170

Henderson A (1986) A review of pollination studies in the Palmae. Bot Rev 52:221–259

Herrera CM (1996) Floral traits and plant adaptation to insect pollinators: a devil's advocate approach. In: Floral biology: studies on floral evolution in animal-pollinated plants. Springer US, Boston

Hodgkiss D, Brown MJ, Fountain MT (2018) Syrphine hoverflies are effective pollinators of commercial strawberry. J Pollinat Ecol 22(6):55–66

Hoffmann F, Daniel F, Fortier A, Hoffmann-Tsay SS (2011) Efficient avian pollination of *Strelitzia reginae* outside of South Africa. S Afr J Bot 77(2):503–505

Hogendoorn K, Gross CL, Sedgley M, Keller MA (2006) Increased tomato yield through pollination by native Australian *Amegilla chlorocyanea* (Hymenoptera: Anthophoridae). J Econ Entomol 99(3):828–833

Hogendoorn K, Coventry S, Keller MA (2007) Foraging behaviour of a blue banded bee, *Amegilla chlorocyanea* in greenhouses: implications for use as tomato pollinators. Apidologie 38(1): 86–92

Hoover SE, Ovinge LP (2018) Pollen collection, honey production, and pollination services: managing honey bees in an agricultural setting. J Econ Entomol 111(4):1509–1516

Howlett BG (2012) Hybrid carrot seed crop pollination by the fly *Calliphora vicina* (Diptera: Calliphoridae). J Appl Entomol 136(6):421–430

Howlett BG, Evans LJ, Pattemore DE, Nelson WR (2017) Stigmatic pollen delivery by flies and bees: methods comparing multiple species within a pollinator community. Basic Appl Ecol 19:19–25

Howlett BG, Walker MK, Newstrom-Lloyd LE, Donovan BJ, Teulon DA (2009) Window traps and direct observations record similar arthropod flower visitor assemblages in two mass flowering crops. J Appl Ento 133(7):553–564

Howpage D, Spooner-Hart RN, Vithanage V (2001) Influence of honey bee (Apis mellifera) on kiwifruit pollination and fruit quality under Australian conditions. NZ J Crop Hortic Sci 29 (1):51–59

Hu S, Dilcher DL, Jarzen DM, Winship TD (2008) Early steps of angiosperm pollinator coevolution. Proc Natl Acad Sci USA 105:240–245

Huda AN, Salmah MC, Hassan AA, Hamdan A, Razak MA (2015) Pollination services of mango flower pollinators. J Insect Sci 15(1):113

Hung KLJ, Kingston JM, Albrecht M, Holway DA, Kohn JR (2018) The worldwide importance of honey bees as pollinators in natural habitats. Proc R Soc B Biol Sci 285(1870):20172140

Idrees A, Qasir ZA, Hasnat AU, Ahmad A, Ahmad S, Aqueel A, Li Z, Rady A, Ali S, Li J (2023) Effectiveness of honeybee (*Apis mellifera*) visit on the pollination of different sunflower cultivars. J King Saud Univ Sci 35:102837

Infante F, Ortíz JA, Solis-Montero L, Mound LA, Vega FE (2017) Thrips (Thysanoptera) of coffee flowers. Ann Entomol Soc Am 110(3):329–336

Inouye D, Larson BM, Ssymank A, Kevan PG (2015) Flies and flowers III: ecology of foraging and pollination. J Pollinat Ecol 16(16):115–133

Isaac SS, Stephenraj D, Kunz TH (2010) Florivory and ecosystem services provided by three species of Megachiropteran bats (Chiroptera: Pteropodidae) in India. Unpublished manuscript. Boston University, Boston

Ish-Am G, Barrientos-Priego F, Castañeda-Vildozola A, Gazit S (1999) Avocado (*Persea americana* mill.) pollinators in its region of origin. Rev Chapingo Ser Hortic 5:137–143

James RR, Pitts-Singer TL (2008) Bee pollination in agricultural ecosystems. Oxford University Press Inc

Jarlan A, De Oliveiha D, Gingras J (1997) Effects of *Eristalis tenax* (Diptera: Syrphidae) pollination on characteristics of greenhouse sweet pepper fruits. J Econ Entomol 90:1650–1654

Jauker F, Wolters V (2008) Hover flies are efficient pollinators of oilseed rape. Oecologia 156:819–823

Johnson RM (2015) Honey bee toxicology. Annu Rev Entomol 60(1):415–434

Johnson SD, Bond WJ (1994) Red flowers and butterfly pollination in the fynbos of South Africa. In: *Plant-animal interactions in Mediterranean-type ecosystems*, pp 137–148

Johnson SD, Nicolson SW (2008) Evolutionary associations between nectar properties and specificity in bird pollination systems. Biol Lett 4(1):49–52

Johnson SD, Steiner KE (2000) Generalization versus specialization in plant pollination systems. Trends Ecol Evol 15(4):140–143

Kamo T, Nikkeshi A, Tawaratsumida T, Tanaka Y, Nakamura S, Kishi S (2022) Pollination efficiency of bumblebee, honeybee, and hawkmoth in kabocha squash, *Cucurbita maxima*, production in Kagoshima, Japan. Appl Entomol Zool 57(2):119–129

Karbassioon A, Stanley DA (2023) Exploring relationships between time of day and pollinator activity in the context of pesticide use. Basic Appl Ecol 72:74

Kavazos CRJ, Wallman JF (2012) Community composition of carrion-breeding blowflies (Diptera: Calliphoridae) along an urban gradient in South-Eastern Australia. Landsc Urban Plan 106(2): 183–190

Kelly D, Ladley JJ, Robertson AW, Anderson SH, Wotton DM, Wiser SK (2010) Mutualisms with the wreckage of an avifauna: the status of bird pollination and fruit-dispersal in New Zealand. N Z J Ecol 34(1):66

Khalifa SA, Elshafiey EH, Shetaia AA, El-Wahed AAA, Algethami AF, Musharraf SG et al (2021) Overview of bee pollination and its economic value for crop production. Insects 12(8):688

Khan MR, Khan MR (2004) The role of honey bees *Apis mellifera* L. (Hymenoptera: Apidae) in pollination of apple. Pak J Biol Sci 7(3):359–362

Khan MS, Yogi MK (2017) Insect crop pollinators. In: *Industrial entomology*. Springer, Singapore

Klatt BK, Holzschuh A, Westphal C, Clough Y, Smit I, Pawelzik E, Tscharntke T (2014) Bee pollination improves crop quality, shelf life and commercial value. Proc R Soc B Biol Sci 281(1775):20132440

Klecka J, Hadrava J, Biella P, Akter A (2018) Flower visitation by hoverflies (Diptera: Syrphidae) in a temperate plant-pollinator network. PeerJ 6:e6025

Klein AM, Steffan-Dewenter I, Tscharntke T (2003) Pollination of *Coffea canephora* in relation to local and regional agroforestry management. J Appl Ecol 40(5):837–845

Klein AM, Vaissière BE, Cane JH, Steffan-Dewenter I, Cunningham SA, Kremen C, Tscharntke T (2007) Importance of pollinators in changing landscapes for world crops. Proc R Soc B Biol Sci 274(1608):303–313

Kluser S, Neumann P, Chauzat M, Pettis J (2010) Global honey bee colony disorders and other threats to insect pollinators. https://archive-ouverte.unige.ch/unige:32251

Kumar R (2012) Planned honey bee pollination in litchi (*Litchi chinensis* Sonn.)-a new production strategy for enhancing fruit yield and quality in India. In: *IV international symposium on lychee, Longan and other Sapindaceae fruits*

Kumar G, Khan MS (2022) Insects as crop pollinators. In: *Insects as service providers*. Springer Nature, Singapore

Kumar R, Kumar V (2014) Impact of pollination by European honey bee, *Apis mellifera* L. on the yield and quality of litchi (*Litchi chinensis* Sonn.) fruits in India. Pest Manag Hortic Ecsyst 20(2):127–132

Kunz TH, Braun TE, Bauer D, Lobova T, Fleming TH (2011) Ecosystem services provided by bats. Ann N Y Acad Sci 1223(1):1–38

Labandeira CC, Currano ED (2013) The fossil record of plant–insect dynamics. Annu Rev Earth Planet Sci 41:287–311

Landaverde-González P, Quezada-Euán JJG, Theodorou P, Murray TE, Husemann M, Ayala R, Moo- Valle H, Vandame R, Paxton RJ (2017) Sweat bees on hot chillies: provision of pollination services by native bees in traditional slash-and-burn agriculture in the Yucatán Peninsula of tropical Mexico. J Appl Ecol 54(6):1814–1824

Larson BMH, Kevan PG, Inouye DW (2001) Flies and flowers: taxonomic diversity of anthophiles and pollinators. Can Entomol 133(4):439–465

Layek U, Kundu A, Bisui S, Karmakar P (2021) Impact of managed stingless bee and western honey bee colonies on native pollinators and yield of watermelon: a comparative study. Ann Agric Sci 66(1):38–45

Leijs R, Batley M, Hogendoorn K (2017) The genus *Amegilla* (Hymenoptera, Apidae, Anthophorini) in Australia: a revision of the subgenera *Notomegilla* and *Zonamegilla*. ZooKeys 653:79–140

Li H, Wyckhuys KA, Wu K (2023) Hoverflies provide pollination and biological pest control in greenhouse-grown horticultural crops. Front Plant Sci 14:1118388

Lima MAP, Martins GF, Oliveira EE, Guedes RNC (2016) Agrochemical-induced stress in stingless bees: peculiarities, underlying basis, and challenges. J Comp Physiol 202(9–10): 733–747

Lo N, Gloag RS, Anderson DL, Oldroyd BP (2010) A molecular phylogeny of the genus *Apis* suggests that the giant honey bee of The Philippines, *A. breviligula* Maa, and the plains honey bee of southern India, *A. indica* Fabricius, are valid species. Syst Entomol 35(2):226–233

Lobreau-Callen D, Le Thomas A, Darchen B, Darchen R, d'Amico N (1990) Some factors determining foraging behaviour of *Hypotrigona pothieri* (Hymenoptera: Apidae) in the vegetation of Ivory Coast. Apidologie 21(1):69–83

Lorenz S (2016) The endangerment of bees and new developments in beekeeping: a social science perspective using the example of Germany. Int J Environ Stud 73(6):988–1005
Lowenstein DM, Matteson KC, Minor ES (2015) Diversity of wild bees supports pollination services in an urbanized landscape. Oecologia 179:811–821
Lowenstein DM, Matteson KC, Minor ES (2019) Evaluating the dependence of urban pollinators on ornamental, non-native, and 'weedy' floral resources. Urban Ecosyst 22:293–302
Lucas A, Bodger O, Brosi BJ, Ford CR, Forman DW, Greig C, Hegarty M, Jones L, Neyland PJ, De Vere N (2018) Floral resource partitioning by individuals within generalised hoverfly pollination networks revealed by DNA metabarcoding. Sci Rep 8:5133
Luo CW, Huang ZY, Chen XM, Li K, Chen Y, Sun YY (2011) Contribution of diurnal and nocturnal insects to the pollination of *Jatropha curcas* (Euphorbiaceae) in southwestern China. J Econ Entomol 104(1):149–154
Luo SX, Zhang LJ, Yuan S, Ma ZH, Zhang DX, Renner SS (2018) The largest early-diverging angiosperm family is mostly pollinated by ovipositing insects and so are most surviving lineages of early angiosperms. Proc R Soc B Biol Sci 285:20172365
Lyons EE, Waser NM, Price MV, Antonovics J, Motten AF (1989) Sources of variation in plant reproductive success and implications for concepts of sexual selection. Am Nat 134(3):409–433
Macgregor CJ, Scott-Brown AS (2020) Nocturnal pollination: an overlooked ecosystem service vulnerable to environmental change. Emerg Top Life Sci 4(1):19–32
MacGregor CJ, Pocock MJ, Fox R, Evans DM (2015) Pollination by nocturnal Lepidoptera, and the effects of light pollution: a review. Ecol Entomol 40(3):187–198
MacKenzie K (2008) Pollination practices and the use of bees in Vaccinium crops. In: IX international vaccinium symposium, vol 810, pp 527–538
Mahabale TS (1968) Spores and pollen of water plants and their dispersal. Rev Palaeobot Palynol 7: 285–296
Maia ACD, Dötterl S, Gonçalves EG, Silberbauer-Gottsberger I, Gottsberger G (2023) Floral scent chemistry of sympatric species of Philodendron (Araceae) sharing a common pollinator in the fragmented coastal Atlantic Forest of southeastern Brazil. Flora 300:152224
Malagodi-Braga KS, Kleinert ADMP (2004) Could *Tetragonisca angustula* Latreille (Apinae, Meliponini) be effective as strawberry pollinator in greenhouses? Aust J Agric Res 55(7): 771–773
Malerbo-Souza DT, Nogueira-Couto RH, Couto LA (2004) Honey bee attractants and pollination in sweet orange, *Citrus sinensis* (L.) Osbeck, var. Pera-Rio. J Venom Anim Toxins Incl Trop Dis 10:144–153
Martínez-Salinas A, Chain-Guadarrama A, Aristizábal N, Vilchez-Mendoza S, Cerda R, Ricketts TH (2022) Interacting pest control and pollination services in coffee systems. Proc Nat Acad Sci 119(15):e2119959119
Mashilingi SK, Zhang H, Garibaldi LA, An J (2022) Honeybees are far too insufficient to supply optimum pollination services in agricultural systems worldwide. Agric Ecosyst Environ 335: 108003
Maues MM, Venturieri GC (1995) Pollination biology of anatto and its pollinators in Amazon area. Honey Bee Sci 16:27–30
McGrady CM, Troyer R, Fleischer SJ (2020) Wild bee visitation rates exceed pollination thresholds in commercial Cucurbita agroecosystems. J Econ Entomol 113(2):562–574
Meena NK, Lal G, Meena RS, Meena BM, Meena RD (2018) Pollinator's diversity and abundance on cumin (*Cuminum cyminum* L.) and their impact on yield enhancement at semi-arid regions. J Entomol Zool Stud 6:1017–1021
Mehrhoff LA III (1983) Pollination in the genus Isotria (Orchidaceae). Am J Bot 70(10):1444–1453
Meléndez-Ramírez V, Magaña-Rueda S, Parra-Tabla V, Ayala R, Navarro J (2002) Diversity of native bee visitors of cucurbit crops (Cucurbitaceae) in Yucatán, México. J Insect Conserv 6:135–147
Meléndez-Ramírez V, Parra-Tabla V, Kevan PG, Ramírez-Morillo I, Harries H, Fernández-Barrera M, Zizumbo-Villareal D (2004) Mixed mating strategies and pollination by insects

and wind in coconut palm (*Cocos nucifera* L. (Arecaceae)): importance in production and selection. Agricul For Entomol 6(2):155–163

Meléndez Ramírez V, Ayala R, Delfín González H (2018) Crop pollination by stingless bees. In: Pot-pollen in stingless bee melittology. Cham: Springer Internation Publishing

Michener CD (2000) The bees of the world. The Johns Hopkins University Press, pp. 139–153

Michener CD (2007) The bees of the world (II). The Johns Hopkins University Press, Baltimore

Móczár L (1961) The distribution of wild bees in the lucerne fields of Hungary (Hymenoptera: Apoidea). Ann Mus Hist-Nat Hung 53:451–461

Monck M, Gordon J, Hanslow K (2008) Analysis of the market for pollination services in Australia. http://www.rirdc.gov.au

Moquet L, Laurent E, Bacchetta R, Jacquemart AL (2018) Conservation of hoverflies (Diptera, Syrphidae) requires complementary resources at the landscape and local scales. Insect Conserv Divers 11(1):72–87

Morse RA, Calderone NW (2000) The value of honey bees as pollinators of U.S. crops in 2000. Bee Cult 128:1–15

Mullin CA, Frazier M, Frazier JL, Ashcraft S, Simonds R, Van Engelsdorp D, Pettis JS (2010) High levels of miticides and agrochemicals in North American apiaries: implications for honey bee health. PLoS One 5(3):e9754

Munawar MS, Raja S, Niaz S, Sarwar G (2011) Comparative performance of honeybees (*Apis mellifera* L.) and blow flies (*Phormia terronovae*) in onion (*Allium cepa* L.) seed setting. J Agric Res 49(1):49–56

Nathan PT, Karuppudurai T, Raghuram H, Marimuthu G (2009) Bat foraging strategies and pollination of *Madhuca latifolia* (Sapotaceae) in southern India. Acta Chiropt 11(2):435–441

Naumann K, Winston ML, Slessor KN, Smirle MJ (1994) Synthetic honey bee (Hymenoptera: Apidae) queen mandibular gland pheromone applications affect pear and sweet cherry pollination. J Econ Entomol 87(6):1595–1599

Nayak RK, Rana K, Sharma HK, Rana VS, Thakur M (2019) Influence of bumble bee pollination on quantitative and qualitative parameters of kiwifruit. Indian J Hortic 76(2):294–299

Nelson W, Evans L, Donovan B, Howlett B (2023) Lasioglossum bees–the forgotten pollinators. J Apic Res 62(1):39–46

Nicholson CC, Ricketts TH (2019) Wild pollinators improve production, uniformity, and timing of blueberry crops. Agric Ecosyst Environ 272:29–37

Nicholson CC, Ricketts TH, Koh I, Smith HG, Lonsdorf EV, Olsson O (2019) Flowering resources distract pollinators from crops: model predictions from landscape simulations. J Appl Ecol 56: 618–628

Nunes-Silva P, Hrncir M, da Silva CI, Roldão YS, Imperatriz-Fonseca VL (2013) Stingless bees, Melipona fasciculata, as efficient pollinators of eggplant (*Solanum melongena*) in greenhouses. Apidologie 44(5):537–546

Ollerton J (2017) Pollinator diversity: distribution, ecological function, and conservation. Annu Rev Ecol Evol Syst 48(1):353–376

Ollerton J (2021) Pollinators and pollination: nature and society. Pelagic Publishing Ltd

Ollerton J, Cranmer L, Stelzer R, Sullivan S, Chittka L (2008) Bird pollination of Canary Island endemic plants. Nat Prec. https://doi.org/10.1038/npre.2008.1977.1

Ollerton J, Winfree R, Tarrant S (2011) How many flowering plants are pollinated by animals? Oikos 120(3):321–326

Ollerton J, Liede-Schumann S, Endress ME, Meve U, Rech AR, Shuttleworth A, Keller HA, Fishbein M, Alvarado-Cardenas LO, Amorim FW et al (2019) The diversity and evolution of pollination systems in large plant clades: apocynaceae as a case study. Ann Bot 123:311–325

Orford KA, Vaughan IP, Memmott J (2015) The forgotten flies: the importance of non-syrphid Diptera as pollinators. Proc R Soc B Biol Sci 282(1805):20142934

Ortiz-Sanchez FJ, Aguirre-Segura A (1991) Composition and seasonal patterns of a community of Apoidea in Almeria. Rev Esp Entomol 67:3–22

Osterman J, Benton F, Hellström S, Luderer-Pflimpfl M, Pöpel-Eisenbrandt AK, Wild BS, Theodorou P, Ulbricht C, Paxton RJ (2023) Mason bees and honey bees synergistically enhance fruit set in sweet cherry orchards. Ecol Evol 13(7):e10289

Oz M, Karasu A, Cakmak I, Goksoy AT, Turan ZM (2009) Effects of honeybee (*Apis mellifera*) pollination on seed set in hybrid sunflower (*Helianthus annuus* L.). Afr J Biotechnol 8(6). https://doi.org/10.4314/AJB.V8I6.60012

Park MG, Raguso RA, Losey JE, Danforth BN (2016) Per-visit pollinator performance and regional importance of wild Bombus and Andrena (Melandrena) compared to the managed honey bee in New York apple orchards. Apidologie 47:145–160

Partap U, Shukla AN, Verma LR (2000) Pollination of peach and plum by *Apis cerana*. In: Asian bees and beekeeping: progress of research and development. IBH, Oxford/New Delhi

Partap U (2011) The pollination role of honeybees. In Honeybees of Asia. Springer, Berlin, Heidelberg

Pathak CS (1999) Hybrid seed production in onion. J New Seeds 1(3–4):89–108

Patil PN, Pastagia JJ (2016) Effect of bee pollination on yield of coriander, *Coriandrum sativum* Linnaeus. Int J Plant Prot 9(1):79–83

Peat J, Darvill B, Ellis J, Goulson D (2005) Effects of climate on intra-and interspecific size variation in bumble-bees. Funct Ecol 19(1):145–151

Pedro SRM, de Camargo JMF (2013) Stingless bees from Venezuela. In: *Pot-honey: a legacy of stingless bees*. Springer, New York

Pekas A, De Craecker I, Boonen S, Wäckers FL, Moerkens R (2020) One stone; two birds: concurrent pest control and pollination services provided by aphidophagous hoverflies. Biol Control 149:104328

Peña JF, Carabalí A (2018) Effect of honey bee (*Apis mellifera* L.) density on pollination and fruit set of avocado (*Persea americana* Mill.) Cv. Hass J Apic Sci 62:5–14

Pérez-Balam J, Quezada-Euán J, Alfaro-Bates R, Medina S, McKendrick L, Soro A, Paxton RJ (2012) The contribution of honey bees, flies and wasps to avocado (*Persea americana*) pollination in southern Mexico. J Pollinat Ecol 8:42–47

Peters CM, Vasquez A (1987) Estudios ecológicos de Camu-Camu (*Myrciaria dubia*). I. Producción de frutos en poblaciones naturales. Acta Amaz 17(0):161–188

Phankaew C (2016) Apiculture and pollination industry survey in Thailand. Int J Agric Ext 4(2): 95–103

Phillips RD, Bohman B, Brown GR, Tomlinson S, Peakall R (2020) A specialised pollination system using nectar-seeking thynnine wasps in *Caladenia nobilis* (Orchidaceae). Plant Biol 22(2):157–166

Philpott SM, Uno S, Maldonado J (2006) The importance of ants and high-shade management to coffee pollination and fruit weight in Chiapas, Mexico. Arthropod Divers Conserv 15(1): 473–487

Pijl VDL, Dodson CH (1966) Orchid flowers: their pollination and evolution. Coral Gables. University of Miami Press, Florida

Pitts-Singer TL (2008) Past and present management of alfalfa bees. In: James RR, Pitts-Singer J (eds) Bee pollination in agricultural ecosystems. Oxford University Press, New York

Pitts-Singer TL, Cane JH (2011) The alfalfa leaf cutting bee, *Megachile rotundata*: the world's most intensively managed solitary bee. Annu Rev Entomol 56:221–237

Pitts-Singer TL, Artz DR, Peterson SS, Boyle NK, Wardell GI (2018) Examination of a managed pollinator strategy for almond production using *Apis mellifera* (Hymenoptera: Apidae) and *Osmia lignaria* (Hymenoptera: Megachilidae). Environ Entomol 47(2):364–377

Porter J (1977) Strawberry yields increased by better pollination. In: Horticulture industry, pp 375–376

Potts SG, Biesmeijer JC, Kremen C, Neumann P, Schweiger O, Kunin WE (2010) Global pollinator declines: trends, impacts and drivers. Trends Ecol Evol 25(6):345–353

Potts SG, Imperatriz-Fonseca V, Ngo HT, Aizen MA, Biesmeijer JC, Breeze TD, Dicks LV, Garibaldi LA, Hill R, Settele J, Vanbergen AJ (2016) Safeguarding pollinators and their values to human Well-being. Nature 540(7632):220–229

Prendergast KS, Leclercq N, Vereecken NJ (2021) Honey bees (Hymenoptera: Apidae) outnumber native bees in Tasmanian apple orchards: perspectives for balancing crop production and native bee conservation. Austral Entomol 60(2):422–435

Putra RE, Permana AD, Kinasih I (2014) Application of Asiatic honey bees (*Apis cerana*) and stingless bees (*Trigona laeviceps*) as pollinator agents of hot pepper (*Capsicum annuum* L.) at local Indonesia farm system. Psyche. https://doi.org/10.1155/2014/687979

Rader R, Howlett BG, Cunningham SA, Westcott DA, Newstrom-Lloyd LE, Walker MK, Teulon DAJ, Edwards W (2009) Alternative pollinator taxa are equally efficient but not as effective as the honeybee in a mass flowering crop. J Appl Ecol 46(5):1080–1087

Rader R, Reilly J, Bartomeus I, Winfree R (2013) Native bees buffer the negative impact of climate warming on honey bee pollination of watermelon crops. Glob Change Biol 19(10):3103–3110

Rader R, Bartomeus I, Garibaldi LA, Garratt MP, Howlett BG, Winfree R, Cunningham SA, Mayfield MM, Arthur AD, Andersson GK, Bommarco R (2016) Non-bee insects are important contributors to global crop pollination. Proc Natl Acad Sci USA 113(1):146–151

Rader R, Cunningham SA, Howlett BG, Inouye DW (2020) Non-bee insects as visitors and pollinators of crops: biology, ecology, and management. Annu Rev Entomol 65:391–407

Radford BJ, Nielsen RGH, Rhodes JW (1979) Agents of pollination in sunflower crops on the central Darling Downs. Qld Aust J Exp Agric 19(100):565–569

Rajagopal D, Eswarappa G (2005) Pollination potentiality of honeybees in increasing productivity of guava in Karnataka. In: Changing trends in pollen spore research, pp 131–141

Rajan VV, Reddy PVR (2019) A dead heat in pollination race: a comparative evaluation of the efficiency of a fly (*Chrysomya megacephala*) and a bee (*Apis florea*) in mango pollination. J Entomol Zool 7(1):1087–1091

Ramirez-Mejia AF, Lomáscolo S, Blendinger PG (2023) Hummingbirds, honeybees, and wild insect pollinators affect yield and berry quality of blueberries depending on cultivar and farm's spatial context. Agric Ecosyst Environ 342:108229

Rao S, Stephen WP (2009) Bumble bee pollinators in red clover seed production. Crop Sci 49(6): 2207–2214

Rasmussen C, Cameron SA (2010) Global stingless bee phylogeny supports ancient divergence, vicariance, and long-distance dispersal. Biol J Linn Soc 99(1):206–232

Ratto F, Simmons BI, Spake R, Zamora-Gutierrez V, MacDonald MA, Merriman JC, Tremlett CJ, Poppy GM, Peh KSH, Dicks LV (2018) Global importance of vertebrate pollinators for plant reproductive success: a meta-analysis. Front Ecol Environ 16(2):82–90

Raven PH (1987) Scope of the plant conservation problem world-wide. In: Botanic gardens and the world conservation strategy: proceedings of an International Conference 1985 held at Las Palmas de Gran Canaria. Academic, London

Reddi CS, Bai GM (1984) Butterflies and pollination biology. Proc Anim Sci 93:391–396

Reyes SG, Michener CD (1990) Observations on a parasitic allodapine bee and its hosts in Java and Malaysia (hymenoptera Anthophoridae Allodapine). Trop Zool 3(2):139–149

Rincón-Rabanales M, Roubik DW, Guzmán MA, Salvador-Figueroa M, Adriano-Anaya L, Ovando I (2015) High yields and bee pollination of hermaphroditic rambutan (*Nephelium lappaceum* L.) in Chiapas, Mexico. Fruits 70(1):23–27

Robertson SM, Dowling APG, Wiedenmann RN, Joshi NK, Westerman EL (2021) Nocturnal pollinators significantly contribute to apple production. J Econ Entomol 14(5):2155–2161

Rocha M, Good-Ávila SV, Molina-Freaner F, Arita HT, Castillo A, García-Mendoza A, Silva-Montellano A, Gaut BS, Souza V, Eguiarte LE (2006) Pollination biology and adaptive radiation of Agavaceae, with special emphasis on the genus Agave. Aliso: J Syst Floristic Bot 22(1):329–344

Rodrigo Gómez S, Ornosa C, Selfa J, Guara M, Polidori C (2016) Small sweat bees (Hymenoptera: Halictidae) as potential major pollinators of melon (*Cucumis melo*) in the Mediterranean. Entomol Sci 19(1):55–66

Rodríguez-Gasol N, Alins G, Veronesi ER, Wratten S (2020) The ecology of predatory hoverflies as ecosystem-service providers in agricultural systems. Biol Control 151:104405

Rollin O, Garibaldi LA (2019) Impacts of honeybee density on crop yield: a meta-analysis. J Appl Ecol 56(5):1152–1163

Rotheray GE, Gilbert F (2011) The natural history of hoverflies. Forest Text, Tresaith

Roubik DW (1989) Ecology and natural history of tropical bees. Cambridge University Press

Roubik DW (2023) Stingless bee (Apidae: Apinae: Meliponini) ecology. Annu Rev Entomol 68(1): 231–256

Rucker RR, Thurman WN, Burgett M (2012) Honey bee pollination markets and the internalization of reciprocal benefits. Am J Agric Econ 94(4):956–977

Sabbahi R, DeOliveira D, Marceau J (2005) Influence of honey bee (Hymenoptera: Apidae) density on the production of canola (Crucifera: Brassicacae). J Econ Entomol 98(2):367–372

Sadeh A, Shmida A, Keasar T (2007) The carpenter bee, *Xylocopa pubescens* as an agricultural pollinator in greenhouses. Apidologie 38(6):508–517

Saeed S, Naqqash MN, Jaleel W, Saeed Q, Ghouri F (2016) The effect of blow flies (Diptera: Calliphoridae) on the size and weight of mangos (*Mangifera indica* L.). PeerJ 4:e2076

Sáez A, Aizen MA, Medici S, Viel M, Villalobos E, Negri P (2020) Bees increase crop yield in an alleged pollinator-independent almond variety. Sci Rep 10(1):3177

Sakai S (2002) A review of brood-site pollination mutualism: plants providing breeding sites for their pollinators. J Plant Res 115:0161–0168

Sakai S, Inoue T (1999) A new pollination system: dung-beetle pollination discovered in *Orchidantha inouei* (Lowiaceae, Zingiberales) in Sarawak, Malaysia. Am J Bot 86(1):56–61

Sakai S, Kato M, Inoue T (1999) Three pollination guilds and variation in floral characteristics of Bornean gingers (Zingiberaceae and Costaceae). Am J Bot 86(5):646–658

Sampson BJ, Danka RG, Stringer SJ (2004a) Nectar robbery by bees *Xylocopa virginica* and *Apis mellifera* contributes to the pollination of rabbiteye blueberry. J Econ Entomol 97(3):735–740

Sampson BJ, Stringer SJ, Cane JH, Spiers JM (2004b) Screenhouse evaluations of a mason bee *Osmiaribifloris* (Hymenoptera: Megachilidae) as a pollinator for blueberries in the southeastern United States. Small Fruits Rev 3(3–4):381–392

Sánchez M, Velásquez Y, González M, Cuevas J (2022a) Hoverfly pollination enhances yield and fruit quality in mango under protected cultivation. Sci Hortic 304:111320

Sánchez M, Velásquez Y, González M, Cuevas J (2022b) Pollination effectiveness of the hoverfly, *Eristalinus aeneus* (Scopoli, 1763) in diploid and triploid associated watermelon crop. Insects 13(11):1021

Sanchez-Bayo F, Goka K (2014) Pesticide residues and bees—a risk assessment. PLoS One 9(4): e94482

Schemske DW (1976) Pollinator specificity in *Lantana camara* and *L. trifolia* (Verbenaceae). Biotropica 8:260–264

Schirmel J, Albrecht M, Bauer PM, Sutter L, Pfister SC, Entling MH (2018) Landscape complexity promotes hoverflies across different types of semi-natural habitats in farmland. J Appl Ecol 55: 1747–1758

Scott-Brown AS, Arnold SE, Kite GC, Farrell IW, Farman DI, Collins DW, Stevenson PC (2019) Mechanisms in mutualisms: a chemically mediated thrips pollination strategy in common elder. Planta 250:367–379

Sedivy C, Dorn S (2014) Towards a sustainable management of bees of the subgenus Osmia (Megachilidae; Osmia) as fruit tree pollinators. Apidologie 45:88–105

Serrano-Serrano ML, Rolland J, Clark JL, Salamin N, Perret M (2017) Hummingbird pollination and the diversification of angiosperms: an old and successful association in Gesneriaceae. Proc R Soc B 284(1852):20162816

Shackleton K, Al Toufailia H, Balfour NJ, Nascimento FS, Alves DA, Ratnieks FL (2015) Appetite for self-destruction: suicidal biting as a nest defense strategy in Trigona stingless bees. Behav Ecol Sociobiol 69:273–281

Shahbandeh M (2022) Number of beehives worldwide 2010–2020. https://www.statista.com/statistics/818286/number-of-beehives-worldwide

Sharma HK, Gupta JK (2001) Impact of bee pollination on sustainable production in apple orchards. Pest Manag Econ Zool 9(1):45–48

Sharma HK, Gupta JK, Thakur JR (2003) Effect of bee pollination and polliniser proportion on apple productivity. In: VII international symposium on temperate zone fruits in the tropics and subtropics, vol 662, pp 451–454

Sheffield CS, Kevan PG, Westby SM, Smith RF (2008) Diversity of cavity-nesting bees (Hymenoptera: Apoidea) within apple orchards and wild habitats in the Annapolis Valley, Nova Scotia, Canada. Can Entomol 140(2):235–249

Silveira RS, Singer RB, Ferro VG (2023) Pollination in *Epidendrum densiflorum* hook (Orchidaceae: Laeliinae): fraudulent trap-flowers, self-incompatibility, and a possible new type of mimicry. Plants 12(3):679

Singer RB, Cocucci AA (1999) Pollination mechanism in southern Brazilian orchids which are exclusively or mainly pollinated by halictid bees. Plant Syst Evol 217:101–117

Singh G (1989) Insect pollinators of mango and their role in fruit setting. Acta Hortic 231:629–632

Singh G, Cook JM, Spooner-Hart RN, Makinson JC (2025) Insect floral visitors vary spatiotemporally and influence fruit production in mango orchards. Ecol Entomol 50(6):1121–1136

Singh G, Makinson JC, Gilpin AM, Spooner-Hart RN, Cook JM (2024) Wild native insects are efficient pollinators of mangoes in the Northern Territory of Australia. Agri Ecosyst Env 15(374):109161

Singh G, Cook JM, Spooner-Hart RN, Makinson JC (2021) What are the potential mango pollinators in the NT? Tree Crop News. https://www.treecrop.com.au/news/what-are-potential-mango-pollinators-nt/

Siqueira KMM, de Kiill LHP, Martins CF, Lemos IB, Monteiro SP, Feitoza EA (2008) Estudo comparativo da polinização de Mangifera indica L. em cultivo convencional e orgânico na região do Vale do Submédio do Sáo Francisco. Rev Bra Frut 30(2):303–310

Slaa EJ, Chaves LAS, Malagodi-Braga KS, Hofstede FE (2006) Stingless bees in applied pollination: practice and perspectives. Apidologie 37(2):293–315

Stephens RE, Gallagher RV, Dun L, Cornwell W, Sauquet H (2023) Insect pollination for most of angiosperm evolutionary history. New Phytol 240(2):880–891

Sornsathapornkul P, Owens JN (1998) Pollination biology in a tropical acacia hybrid (*A. mangium* Willd. × *A. auriculiformis* A. Cunn. ex Benth). Ann Bot 81(5):631–645

Soroka JJ, Goerzen DW, Falk KC, Bett KE (2001) Alfalfa leafcutting bee (Hymenoptera: Megachilidae) pollination of oilseed rape (*Brassica napus* L.) under isolation tents for hybrid seed production. Can J Plant Sci 81(1):199–204

Sosa-Najera MS, Martínez-HernÁndez E, Lozano-García MDS, Cuadriello-Aguilar JI (1994) Nectaropolliniferous sources used by *Trigona* (*Tetragonisca*) *angustula* in Chiapas, southern México. Grana 33(4–5):225–230

Ssymank A, Kearns CA, Pape T, Thompson FC (2008) Pollinating flies (Diptera): a major contribution to plant diversity and agricultural production. Biodiversity 9(1–2):86–89

Stanghellini MS, Ambrose JT, Schultheis JR (1997) The effects of honey bee and bumble bee pollination on fruit set and abortion of cucumber and watermelon. Am Bee J 137:386–391

Stanghellini MS, Ambrose JT, Schultheis JR (2002) Diurnal activity, floral visitation and pollen deposition by honey bees and bumble bees on field-grown cucumber and watermelon. J Apic Res 41(1–2):27–34

Staveley JP, Law SA, Fairbrother A, Menzie CA (2014) A causal analysis of observed declines in managed honey bees (*Apis mellifera*). HERA 20(2):566–591

Stavert JR, Liñán-Cembrano G, Beggs JR, Howlett BG, Pattemore DE, Bartomeus I (2016) Hairiness: the missing link between pollinators and pollination. PeerJ 4:e2779

Steiner KE (1998) Beetle pollination of peacock moraeas (Iridaceae) in South Africa. Plant Syst Evol 209:47–65

Steinhauer N, Kulhanek K, Antúnez K, Human H, Chantawannakul P, Chauzat MP, VanEngelsdorp D (2018) Drivers of colony losses. Curr Opin Insect Sci 26:142–148

Stephenraj D, Isaac SS, Kunz TH, Stanley JA (2010) Foraging behaviour of Megachiropteran Bats (Chiroptera: Pteropodidae) in Courtallam, Tamilnadu, South India. Bioinfolet 7(2):175–180

Su X, Hua Q, Chen Y, Jiang X, Xi W, Zhao D, Bie Z (2017) Behavior of *Apis cerana cerana* and *Apis mellifera ligustica* as pollinator for long-season cultivated watermelon in tunnel greenhouse under summer high temperature condition. Environ Entomol 39(1):104–110

Sugiura N (2013) Specialized pollination by carpenter bees in Calanthe striata (Orchidaceae), with a review of carpenter bee pollination in orchids. Bot J Linn Soc 171(4):730–743

Sung IH, Lin MY, Chang CH, Cheng AS, Chen WS, Ho KK (2006) Pollinators and their behaviours on mango flowers in southern Taiwan. Formosan Entomol 26:161–170

Syed RA (1979) Studies on oil palm pollination by insects. Bull Entomol Res 69(2):213–224

Sushil SN, Stanley J, Hedau NK, Bhatt JC (2013) Enhancing seed production of three Brassica vegetables by honey bee pollination in north-western Himalayas of India. Univ J Agri Res 1 (3):49–53

Taha EKA, Bayoumi YA (2009) The value of honey bees (*Apis mellifera*, L.) as pollinators of summer seed watermelon (*Citrullus lanatus colothynthoides* L.) in Egypt. Acta Biol Szeged 53(1):33–37

Tan SL, Chen XH, Liao HH, Huang L, Xiao HW, Luo HL, Xiong DJ, Yang BY, Ren ZX (2023) Swallowtail butterflies and hawkmoths contribute equally to the pollination of *Habenaria dentata* (Orchidaceae). Flora 300:152230

Tangmitcharoen S, Owens JN (1997) Floral biology, pollination, pistil receptivity, and pollen tube growth of teak (*Tectona grandis* Linn f.). Ann Bot 79(3):227–241

Tasei JN (1975) Le probleme de l'adaptation de Megachile (Eutricharea) pacifica Panz. (Megachilidae) Americain en France. Apidologie 6:1–57

Tepedino VJ (1981) The pollination efficiency of the squash bee (*Peponapis pruinosa*) and the honey bee (*Apis mellifera*) on summer squash (*Cucurbita pepo*). J Kans Entomol Soc 54:359–377

Tepedino VJ (1996) A comparison of the alfalfa leafcutting bee (*Megachile rotundata*) and the honey bee (*Apis mellifera*) as pollinators for hybrid carrot seed in field cages. In VII International Symposium on Pollination 437:457–461

Tierney SM, Bernauer OM, King L, Spooner-Hart R, Cook JM (2023) Bee pollination services and the burden of biogeography. Proc R Soc B Biol Sci 290(2000):20230747

Torchio PF (1976) Use of Osmia lignaria Say (Hymenoptera: Apoidea, Megachilidae) as a pollinator in an apple and prune orchard. J Kans Entomol Soc 49:475–482

van Dulmen A (2001) Pollination and phenology of flowers in the canopy of two contrasting rain forest types in Amazonia, Colombia. Plant Ecol 153:73–85

VanEngelsdorp D, Evans JD, Saegerman C, Mullin C, Haubruge E, Nguyen BK, Frazier M, Frazier J, Cox-Foster D, Chen Y, Underwood R (2009) Colony collapse disorder: a descriptive study. PLoS One 4(8):e6481

Velayudhan R, Annadurai RS (1987) Pollen transfer and stigmatic receptivity with reference to thrips (Insecta: Thysanoptera) in *Solanum melongena* linn. Curr Sci 56(6):287–289

Velthuis HH, Van Doorn A (2006) A century of advances in bumblebee domestication and the economic and environmental aspects of its commercialization for pollination. Apidologie 37(4): 421–451

Venturieri GA (1994) Floral biology of cupuassu (Theobroma grandiflorum (Willdenow ex Sprengel) Schumann) (Doctoral dissertation, University of Reading). https://ethos.bl.uk/OrderDetails.do?uin=uk.bl.ethos.386982

Viana BF, da Encarnação Coutinho JG, Garibaldi LA, Bragança Gastagnino GL, Peres Gramacho K, Oliveira da Silva F (2014) Stingless bees further improve apple pollination and production. J Pollinat Ecol 14(25):261–269

Vicens N, Bosch J (2000) Pollinating efficacy of *Osmia cornuta* and *Apis mellifera* (Hymenoptera: Megachilidae, Apidae) on 'red delicious' apple. Environ Entomol 29(2):235–240

Vroege PW, Stelleman P (1990) Insect and wind pollination in *Salix repens* L. and *Salix caprea* L. Israel J Bot 39:125–132

Wahengbam J, Raut AM, Pal S, Banu AN (2019) Role of bumble bee in pollination. Ann Biol 35(2):290–295

Ward M, Johnson SD (2013) Generalised pollination systems for three invasive milkweeds in Australia. Plant Biol 15(3):566–572

Waser NM and Forrest JR (2014) Pollination by animals. In eLS, John Wiley & Sons. https://doi.org/10.1002/9780470015902.a0003163.pub3

Wei SG, Wang R, Smirle MJ, Xu HL (2002) Release of *Osmia excavata* and *Osmia jacoti* (Hymenoptera: Megachilidae) for apple pollination. Can Entomol 134(3):369–380

Weiblen GD (2002) How to be a fig wasp. Annu Rev Entomol 47(1):299–330

Whelan CJ, Wenny DG, Marquis RJ (2008) Ecosystem services provided by birds. Ann N Y Acad Sci 1134:25–60

Whelan CJ, Şekercioğlu ÇH, Wenny DG (2015) Why birds matter: from economic ornithology to ecosystem services. J Ornithol 156(1):227–238

Wietzke A, Westphal C, Gras P, Kraft M, Pfohl K, Karlovsky, Pawelzik E, Tscharntke T, Smit I (2018) Insect pollination as a key factor for strawberry physiology and marketable fruit quality. Agric Ecosyst Environ 258:197–204

Wilkaniec Z, Radajewska B (1996) Solitary bee *Osmia rufa* L. (Apoidea, Megachilidae) as pollinator of strawberry cultivated in an unheated plastic tunnel. In: III international strawberry symposium, vol 439, pp 489–494

Willcox BK, Howlett BG, Robson AJ, Cutting B, Evans L, Jesson L, Kirkland L, Jean-Meyzonnier-M, Potdevin V, Saunders ME, Rader R (2019) Evaluating the taxa that provide shared pollination services across multiple crops and regions. Sci Rep 9(1):13538

Wille A (1976) Las abejas jicotes del género *Melipona* (Apidae: Meliponini) de Costa Rica. Rev Biol Trop 24(1):123–147

Wille A (1983) Biology of the stingless bees. Annu Rev Entomol 28(1):41–64

Williams IH (1994) The dependence of crop production within the European Union on pollination by honey bees. Agric Zool Rev 6:229–257

Williams PH, Cameron SA, Hines HM, Cederberg B, Rasmont P (2008) A simplified subgeneric classification of the bumblebees (genus Bombus). Apidologie 39(1):46–74

Willis Chan DS, Raine NE (2021) Hoary squash bees (*Eucera pruinosa*: Hymenoptera: Apidae) provide abundant and reliable pollination services to cucurbita crops in Ontario (Canada). Environ Entomol 50(4):968–981

Willmer PG, Bataw AAM, Hughes JP (1994) The superiority of bumblebees to honeybees as pollinators: insect visits to raspberry flowers. Ecol Entomol 19:271–284

Wilson EO, Hölldobler B (2005) Eusociality: origin and consequences. Proc Natl Acad Sci USA 02(38):13367–13371

Winfree R (2008) Pollinator-dependent crops: an increasingly risky business. Curr Biol 18(20): 968–969

Winfree R, Williams NM, Dushoff J, Kremen C (2007) Native bees provide insurance against ongoing honey bee losses. Ecol Lett 10:1105–1113

Winfree R, Gross BJ, Kremen C (2011) Valuing pollination services to agriculture. Ecol Econ 71: 80–88

Wongsa K, Duangphakdee O, Rattanawannee A (2023) Pollination efficacy of stingless bees, *Tetragonula pagdeni* Schwarz (Apidae: Meliponini), on greenhouse tomatoes (*Solanum lycopersicum* Linnaeus). PeerJ 11:e15367

Wu P, Tscharntke T, Westphal C, Wang M, Olhnuud A, Xu H, Yu Z, van der Werf W, Liu Y (2021) Bee abundance and soil nitrogen availability interactively modulate apple quality and quantity in intensive agricultural landscapes of China. Agric Ecosyst Environ 305:107168

Yankit P, Rana K, Sharma HK, Thakur M, Thakur RK (2018) Effect of bumble bee pollination on quality and yield of tomato (*Solanum lycopersicum* Mill.) grown under protected conditions. Int J Curr Microbiol Appl Sci 7(1):257–263

Young HJ (1986) Beetle pollination of *Dieffenbachia longispatha* (Araceae). Am J Bot 73(6): 931–944

Yumoto T (2000) Bird-pollination of three Durio species (Bombacaceae) in a tropical rainforest in Sarawak, Malaysia. Am J Bot 87(8):1181–1188

Conservation and Augmentation of Indigenous Bumblebee *Bombus haemorrhoidalis* Smith in Cultivated Ecosystems

19

Raj Kumar Thakur, Kiran Rana, Meena Thakur, Ruchi Sharma, Priyanka Thakur, Diksha Devi, Priyanka Rani, Monika, Mangla Ram Bajiya, and Sawraj Jit Singh

Abstract

Bumblebees are well known as buzz bees and are effective natural pollinators in ecosystems. These bees contribute to pollination services to agriculture, increasing profits, producing crops and reducing labour costs. Their decline is evident from the low crop yields and quality despite necessary agronomic inputs. Conservation and management of native species and large-scale artificial rearing techniques are necessary for sustaining the natural resources and agricultural production.

Keywords

Conservation · Buzzing bees · Declining · Artificial rearing techniques

R. K. Thakur (✉) · M. Thakur · M. R. Bajiya
Department of Entomology, Dr. YSP UHF, Nauni, Solan, Himachal Pradesh, India

K. Rana · D. Devi · S. Jit Singh
Department of Entomology, Dr YS Parmar University of Horticulture and Forestry, Nauni, Himachal Pradesh, Solan, India

R. Sharma · P. Thakur
Senior Research Fellow, Department of Entomology, Dr. YSP UHF, Nauni, Solan, Himachal Pradesh, India

P. Rani · Monika
PhD Research Scholar, Department of Entomology, Dr. YSP UHF, Nauni, Solan, Himachal Pradesh, India

A. K. Chakravarthy (ed.), *The Role of Pollinators in Tropical Landscapes*,
https://doi.org/10.1007/978-981-95-9139-8_19

1 Introduction

Native pollinators play a major role in the sustenance of plant diversity, crucial to support the fauna and ecosystem. Among the different pollinators in the tropical regions of the globe, bumblebees have a significant place. They play a major role in conserving plant diversity, particularly confined to an inhospitable environment (Saini and Ghattor 2007). Their thermoregulatory abilities adapt them to cold and high elevations that fulfil an ecological and economical roles as pollinators of wild and managed flowering plants and crops. Illustrated as buzzing bees due to faster vibrations of their flight muscles, which help to burst open the pollen sacs these are the effective natural pollinators in the ecosystem (Heinrich 1979; Free 1993; Pomeroy and Fisher 2002). They have the ability to forage at low light and moderate temperature and possess longer tongue to pollinate flowers of deep corolla (Holm 1966), which makes them alternative pollinators to the honey bees (Richardson et al. 2018).

Bumblebees generally have an annual life cycle, the major obstacle that prevent bumblebees to be designated as a commercial pollinator (Sladen 1912; Velthuis and Doorn 2006). The colony has strength of up to 400 individuals, comprising three castes: queen, workers and drones. Under natural conditions, bumblebee starts life cycle on the onset of spring. Worker bees start working with queen and survive for couple of months, while drones die after mating. A daughter queen after successful mating hibernate during the winter and in spring after the emergence search for suitable nesting sites, forage and start the colony (Yadav et al. 2016). Bumblebees' species, e.g. *Bombus terrestris* Linnaeus, *B. impatiens* Cresson, *B. occidentalis* Greene, *B. lucorum* Linnaeus and *B. ignites* Smith, are being successfully reared and commercialized since 1980s in many countries viz. Belgium, Netherlands, Korea, China, USA, New Zealand, Holland, Japan, Australia and U.K. *B. terrestris* and *B. impatiens* are commercially the most reared species in Europe and North America (Velthuis and Doorn 2006).

Bumblebees are generalists and visit hundreds of native plant species. Presently, bumblebees are not being utilized as pollinators in India due to less knowledge among farmers and stakeholders of the Indian economy. The research on native species, *Bombus haemorrhoidalis* Smith and other bumblebee species, is being taken up to commercialize rearing of potential species for pollination services. Presently, it is neither being used commercially nor there is any rearing facility in India. But if successfully reared, the efforts will boost natural population in addition to their sustainable assistance in pollination services for maintenance of ecosystems for indigenous crops.

2 Bumblebees Fauna in Himachal Pradesh

Globally, 250 species of bumblebees have been identified. Among these, bumblebee species having a social worker caste are known as 'true' bumblebees, which are more or less sterile. Twenty three species of bumblebees were recorded including bees from the neighbouring countries viz. Myanmar, Bhutan, Nepal, etc. (Bingham 1897). India is home to 48 known species in high-altitude regions. These bees are occur found at 2000–15,000 feet of the entire Himalayas from Jammu and Kashmir to Nagaland. Himalayan regions are biodiverse landscapes. However, not much is documented on the abundance and distribution of plant and animal species in these regions. Bumblebees play an important role in conserving plant diversity, particularly confined to an inhospitable environment of alpine meadows and mountain tracts. The Himalayan mountain range provide suitable habitats for different bumblebees. Seven bumblebees species viz. *B. waltoni*, *B. keriensis*, *B. himalayanus*, *B. asiaticus*, *B. personatus*, *B. rufofasciatus* and *B. tunicatus* are reported from Lahaul-Spiti valley of Himachal Pradesh, India. More than 700 specimens of 21 species were found in North Eastern Himalayan states of India. In the low land tropical forest (<1000 m), *B. haemorrhoidalis*, *B. albopleuralis* and *B. breviceps* were observed, where diversity of species increased with elevation (Streinzer et al. 2019). Species diversity, evenness and species richness was more in Kashmir region compared to Ladakh and Jammu. The Kashmir Himalaya samples indicated significant differences in species diversity in altitudinal ranges of 3000–4000 m. Four species viz. *Bombus cornutus*, *B. parthenius*, *B. miniatus* and *B. morawitizianus* were documented from this region, while *B. morawitizianus* was the first record to the national list.

In Himachal Pradesh, Kashyap (2008) explored the presence of *B. haemorrhoidalis* in Kullu (Bajura) and Solan (Nauni), Kinnaur (Nichar) and *B. assamensis* Bingham in Shimla (Narkanda). Efforts have been made in search of indigenous species those can be reared in controlled conditions. Three species of bumblebees *B. tunicatius* (Keylong, Lahaul), *B. hypnorum* (Nohradhar, Sirmaur) and *B. haemorrhoidalis* (Nauni, Solan and Kullu) were reported by Rana (2019). Kalia (2020) reported three species of bumblebees *B. rufofasciatus*, *B. asiaticus* and *B. melanurus* in Chitkul valley of Kinnaur. In 2021, *B. haemorrhoidalis*, *B. tunicatus*, *B. rufofasciaticus*, *B. miniatus* and *B. melanurus* were recorded during the surveys done in district Kinnaur and Sissu, Keylong, Dharcha, Thirot, Kirting and Udaipur of district Lahaul of Himachal Pradesh, India (Sharma et al. 2022) (Table 19.1).

2.1 Life Cycle of *Bombus haemorrhoidalis* Smith

Bombus haemorrhoidalis Smith has an annual life cycle (Fig. 19.3).

There are three social castes in bumblebee:

1. Queen: The primary female to reproduce in a future colony. She is solely responsible for laying eggs after she finds a nest. She guarantees labour division and effective functioning of the colony.
2. Workers: The dominant caste is the female workers, developed from the fertilized diploid eggs which do food foraging, defend the colony and nurture larvae. They are generally sterile and do not lay eggs.

Table 19.1 Bumblebees in Himachal Pradesh

S. No.	Species	Distribution	Altitude (m amsl)	Source
1	*B. assamensis* Bingham	Narkanda, Shimla	2000–3000	Kashyap (2008)
2	*B. asiaticus*	Parts of Kinnaur (Chitkul)	350–7000	William (1991), Saini et al. (2011), Kalia (2020), Negi (2021)
4	*B. himalayanus*	Lahaul (Gramphoo, Chander tal, Losar)	4300–4980	William (1991)
5	*B. hypnorum*	Sirmaur (Nohradharc, Talangna)		Rana (2019), Sharma et al. (2022) (unpublished)
6	*B. melanurus*	Kinnaur (Chitkul, Batseri, Chitkul, Dumti), Lahaul (Sissu, Keylong, Darcha)	3000–4200	Kalia (2020), Negi (2021)
7	*Bombus miniatus*	Kinnaur (Sangla, Batseri Chitkul) Lahul Spiti (Sissu, Keylong)	3000–4200	Negi (2021)
8	*B. rufofasciatus*	Kinnaur (Reckong Peo, Sangla Chitkul, Batseri, Dumti), Lahaul (Sissu, Keylong, Darcha), Kinnaur (Chakka, sharboo, Kalpa, Kanda), Manali	3500–4200	William (1991), Kalia (2020), Saini and Ghattor (2007), Kalia (2019), Negi (2021), Sharma et al. (2022) (unpublished)
9	*B. tunicatus*	Kinnaur (Reckong Peo, Nichar, Sangla, Batseri) Lahul spiti (Keylong) Chakka, sharboo, Kalpa, Kanda	3978–4200	William (1991), Saini and Ghattor (2007), Rana (2019), Negi (2021), Sharma et al. 2022 (unpublished)
10	*B. trifasciatus*	Palampur, Mandi, Joginder nagar, Chail, Theog, Sirmaur (Nohradhar)	760–2500	Raina et al. (2017), Saini et al. (2011), Sharma et al. (2021)
11	*B. waltoni*	Kaza	3800	William (1991), Saini and Ghattor (2007)

3. Drones: Drones are raised from unfertilized haploid eggs. The sole role of drone is mating only and survive outside the colony (Figs. 19.1 and 19.2).

Indigenous bumblebees (*B. haemorrhoidalis* Smith) initiate life during the spring season when queen emerges out of hibernation. They locate suitable sites for nesting and nectar, and queen starts building nests to lay eggs. Their niche is mostly abandoned nests or cavities underground in the forests, fields and barren lands (Chauhan and Thakur 2010; Chauhan et al. 2013). They prefer darker and corner

Fig. 19.1 Dominant bumblebee species found in Himachal Pradesh

Fig. 19.2 Different castes of *Bombus haemorrhoidalis*

sides of the ground and make involucrum to protect nest from outside environment. Nests bedding materials are interwovened with dry grass, thatched leaves, pine needles, small wooden twigs and waste material, etc. (Taylor and Cameron 2003; Gonzalez et al. 2004; Thakur and Kashyap 2007). An oval or round shaped nests formed by bumblebee through rodent tunnel 80 cm long, while brood batches were arranged in a circular manner. Average length of queen, worker and honey pot cells was 20.57, 14.24 and 20.35 mm for *B. haemorrhoidalis* (Thakur 2002; Chauhan and Thakur 2010).

During summer, the colony size increases with the emergence of workers from brood cells. The mother queen starts producing the reproductive individuals, i.e. drones and gynes, in autumn season, September–October. The young gynes leave the nest in the late autumn season and mate, often more than once, with males (drones) wandering outside the nests that are forcibly driven out of the colony after mating. The young gynes after maturing forage out of the colony for mating during late autumn season and mate with drone. Some bumblebee species males patrol routes and deposit pheromones on plants along the way to attract females, while some species wait for new queens to emerge from the nest. Mating behaviour of this species still need more study for clarification as colonies with drone population were quiet few. The mated queen undergoes hibernation below the ground in the soil, until the next spring season when the environment becomes favourable for raising a new colony (Chauhan et al. 2013).

2.2 Floral Diversity of *Bombus haemorrhoidalis* Smith

Bumblebees visit flowering plants from spring to autumn for nectar/pollen. Found foraging on range of flora like fruits, vegetables, ornamental plant, medicinal plant and shrubs. Observations on different bee flora revealed that a total 43 plant species belonging to 21 different families were reported (Figs. 19.3 and 19.4). Maximum number of plant species belonged to family Fabaceae, Lamiaceae, Solanaceae and Asteraceae. Studies on the flora of native bumblebee revealed that *Caryopteris bicolor, Helichrysum bracteatum, Lupinus hartwegi, Lavendula* sp., *Scutellaria linearis, Salvia moorcroftiana, Jakranda mimmosaefolia, Adhatoda vasica* and *Punica granatum* were the important flora for sustaining queen population under mid hill conditions in Himachal Pradesh (Chauhan et al. 2014). Among the bumblebee floral plants, there is dominance of herbs (46%) with sparse distribution of shrubs (26%), climbers (19%) and tree species (9%) (Singh et al. 2017).

Singh (2015) described seasonal availability of pollen plants for *B. haemorrhoidalis* (Table 19.2). *Brassica campestris, Pisum sativum, Raphanus sativus, Prunus persica, P. armeniaca, Pyrus pashia, P. communis, Justica adhatoda, Antirrhinum majus, Glebionis coronaria, Helichrysum bracteatum, Rosmarinus officinalis, Scuutellaria* sp. and *Delphinium ajacis* were identified as spring season flora (during February–March). In the summer (April–June), bumblebee flora identified were *Bauhinia variegate, Malus domestica, Cirsium* sp., *Nicotinia tabacum, Delphinium ajacis, Papaver rhoeas, D. purpurea, D. lanata, M. laevis, Jacaranda mimosifolia, C. annuum, Cichorium intybus, H. syrianus, S. lycopersicum, S. melongena, Salvia sclarea, Phaseolus vulgaris, Gladiolus hybrida, Cucumis sativus*, etc. During rainy season (July–August), bumblebee forage on *D. primuli, B. lantana, C. annuum, O. latifolia, C. sativus, P. vulgaris, Salvia coccinea, E. purpurea, Datura stramonium, Hypericium perforatum, S. melongena, Dicliptera bupleuroides* and *Ipomoea purpurea.* Whereas during autumn season (September–November) vegetable and other crops viz. *M. annua, C. sophera, Z. elegans, Z. pinnata* and *C. ternatea, Leucas cephalotes, H. mutabilis, I. purpurea, L. cephalotes, D. bupleuroides*, Dahlia imperialis, etc. were foraged by different species of pollinators. and other crops were *M. annua, C. sophera, Z. elegans, Z. pinnata* and *C. ternatea, Leucas cephalotes, H. mutabilis, I. purpurea, L. cephalotes, D. bupleuroides, Dahlia imperialis*, etc. (Fig. 19.5)

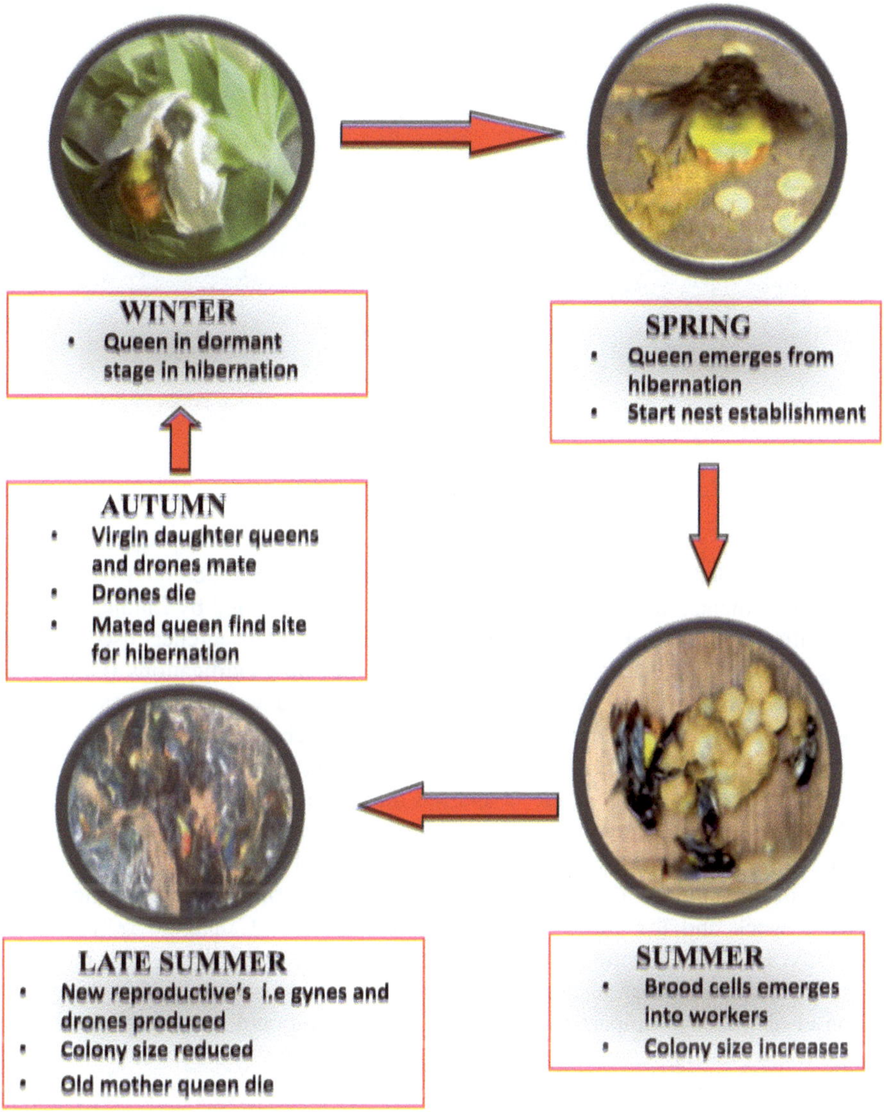

Fig. 19.3 Annual life cycle of *Bombus haemorrhoidalis*

2.3 Augmentation of *Bombus haemorrhoidalis* Colonies Domestication Under Laboratory Conditions

The bumblebee rearing has evolved since mid-1980s and now the bumblebee rearing has been evolved into a industry throughout the world worth a million of euros (Velthuis and Doorn 2006). The very first attempt for domestication of bumblebee was successfully attempted by Sladen (1912). He started rearing of bumblebees in tin

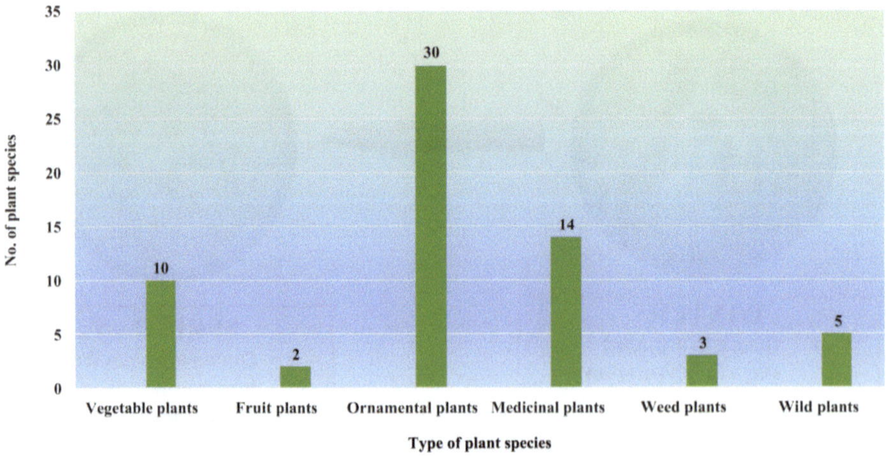

Fig. 19.4 Major flora for *B. haemorrhoidalis* Smith

domiciles and achieved 25% success. The detailed use of wooden boxes consisting two compartments, i.e., feed and brood chambers, was reported by Lindhard (1912) for in vitro bumblebee. Further for rearing practices many researchers used wooden boxes for artificial rearing of bumblebees with certain modifications (Ono et al. 1994; Jie et al. 2005). Plowright and Jay (1966) used wax paper container for artificial rearing for different species of bumblebees, i.e. *B. rufocinctus*, *B. nevadensis*, *B. fervidus*, *B. borealis*, etc., and observed successful colony initiation. They kept the boxes at 29 °C and fed the queens with sucrose solution with fresh pollen. Liebefeld cages with glass windows on front and rear sides and a feeder on the top were used by Van den Eijnde et al. (1991) for nest initiation of *B. terrestris*. They fed the queens with 50% sucrose solution and a pollen patty thrice a week and the cages were kept in dark at 29 °C and 50–60% relative humidity. Pridal and Hofleauer (1996) used the plastic boxes lined with paper for bumblebee domiciliation also by providing 50% sucrose and pollens in their feed. Yoon et al. (2004) compared the success rate of cardboard and plastic boxes for *B. ignites* kept at 27 °C and 65% relative humidity.

For the commercial bumblebee rearing in 1987, Dr. de Jonghe established the Biobest company. The Koppert Biological Systems, Dutch company Bunting Brikman Bee (BBB), also stepped in the market of commercial bumblebee rearing. Koppert uses a multiple layer secure cardboard shipping package embedded in plastic hive for pollination use. Different species being reared and commercialized successfully in many countries, e.g. Belgium, The Netherlands, Korea, China, USA, New Zealand, Holland, Japan, Australia, and UK.

Table 19.2 Host plants of Bumblebees from Indian Himalayas

S. No.	Altitudinal distribution	Host plant family/ species	Dominant Bumblebee species
1.	1500–2000 m AMSL	*Duranta plumieri* (Verbenaceae) *Citrus aurantianus* Linn. (Rutaceae) *Brassica campestris* Linn. (Cruciferae) *Helianthus annuus* Linn.,Dahlia variabilis Linn. (Asteraceae) *Callistemene linearis* (Myrtaceae) *Althaea rosea Linn.* (Malvaceae)	*B. tunicatus* *B. albopleuralis* *B. albopleuralis* *B. simillimus*
2.	2000–3500 m AMSL	*Impatiens sulcata* Wall., Impatiens scabrida Linn. (Balsaminaceae) *Carduus nutans* Linn. Echinops niveus Wall. (Asteraceae) *Trifolium pratense* Linn. (Paplionaceae) *Dipsacus inermis* Wall. (Dipsacaceae) *Buddleja paniculata* Wall. (Lamiaceae)	*B. rufofasciatus* *B. tunicatus* *B. melanurus* *B. jacobsoni* *B. breviceps* *B. friseanus* *B. avinoviellus* *B. kashmirensis*
3.	3500–5400 m AMSL	*Anaphalis busua DC, Circbita cineria Linn., Senecio chrysanthemoides DC, Taraxacum officinale, Saussurea jacea (Klotz) Cl., Saussurea lappa, Scorzonera virgata, Senecio chrysanthemoides, Saussurea costus, Lactuca orientalis, Saussurea albescens, Saussurea auriculata, Rudbeckia laciniata, Saussurea fastuosa, Leontopodium alpinum, Tanacetum dolicophyllum (Asteraceae)* *Potentilla atrosanguinea Lodd. (Rosaceae)* *Thymus serpyllum Linn., Mentha longifolia Linn., Phlomis bracteosa Royle, Stachys serecia Wall. ex Benth., Nepeta discolor Linn., Hyssopus officinalis Linn., Dracocephalum heterophyllum Benth., Nepeta laevigata (D. Don) Hand.-Mazz. (Lamiaceae)* *Polygonum amplexicaule D.Don. (Polygonaceae)* *Pedicularis pectinata Wall., Euphrasia officinalis (Scrophulariaceae)* *Gypsophil cerastioides D. Don (Caryophyllaceae)* *Myosotis sylvatica Ehrh. et Hoffm. (Boraginaceae)* *Melilotus officinalis (L.) Pall (Leguminosae)* *Cyanthus lobatus Wall. (Campanulaceae)* *Gypsophila cerastioides (Caryophyllaceae)* *Rhodiola crenulata Linn., Sedum ewersii Ladeh (Crassulaceae)* *Hyoscyamus niger Linn. (Solanaceae).*	*B. asiaticus* *B. keriensis* *B. melanurus* *B. personatus* *B. semenovianus* *B. waltoni* *B. lepidus* *B. biroi* *B. semenovi* *B. lemniscatus* *B. rufofasciatus* *B. kashmirensis* *B. hypnorum* *B. himalayanus* *B. ladakhensis* *B. miniatus* *B. subtypicus*

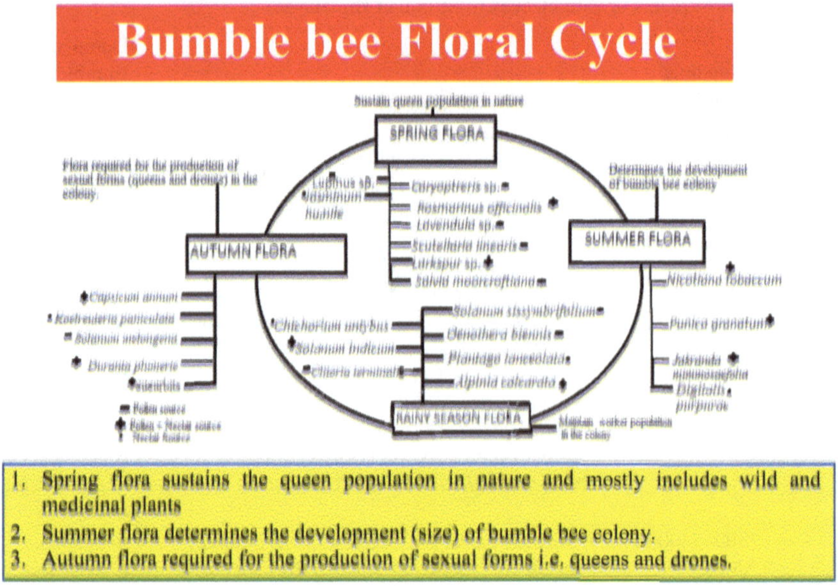

Fig. 19.5 Bumblebee floral cycle

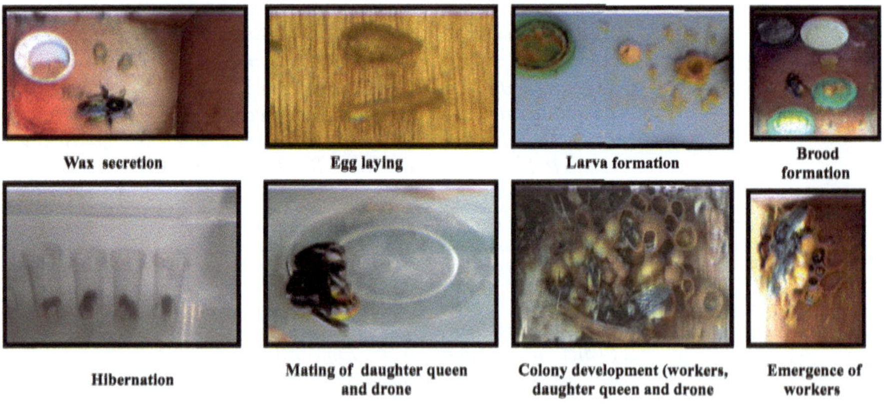

Fig. 19.6 Development biology of *Bombus haemorrhoidalis*

3 Commercially Reared Species Through Worldwide

- *B. terrestris*: Asia (Armenia, Iran, Japan Syria Turkey, Pakistan), Africa (Algeria, Libia, Spain), N. America (Mexico), S. America (Argentina, Brazil, Chile), Europe (Austria, France, Germany, Belgium, UK) (Invasive Species Compendium CABI 2019).

- *B. impatiens*: N. America (Canada, Mexico, USA), South and Central America, Europe, Asia (Robin E Owen 2016).
- *B. occidentalis*: Western United States, Canada, North West America (Hatfield et al. 2015).
- *B. ignitus*: West Asia (China, Japan, S. Korea) (www.pp-icon.eu > wild pollinators bumblebee; Ghosh et al. 2017).
- *B. lucorum*: Europe, Asia, China (Robin E Owen 2016).

In India, attempts were made from last one decade to rear largely found native bumblebee species (*B. haemorrhoidalis* Smith) in laboratory, Department of Entomology, Dr. Y.S. Parmar University of Horticulture and Forestry, Nauni, Solan, H.P. (Fig. 19.6). In India, Thakur (2002) made the first research on the nest architecture and rearing of bumblebees and used Sweden type domiciles at 25–30 °C and 60–65% relative humidity with regular feed of sucrose solutions and corbicular pollen. Kashyap (2008) used two chambered box, cardboard box, thermocole sheets and hoarding cage for rearing. In Devi 2019, Devi evaluated four different types of domiciles namely, standard wooden domiciles, cardboard, PVC and wooden nucleus for *B. haemorrhoidalis* Smith, and the standardized wooden domiciles were found effective for 80 per cent success.

During 2004 (Dayal and Rana 2004) further studied on domestication of native bumblebee species. *B. haemorrhoidalis* queens deposited wax after 8.33 days of confinement and laid 6–9 eggs. First batch of workers emerged after 28.33 days of after oviposition (Dayal and Rana 2004). However, Thakur (2006) studied the development biology of *Bombus haemorrhoidalis* Smith that found that the wax secretion in same species started in 12 days and mound building started after 18 days of confinement. One or two workers emerged daily and 10 individuals emerged after 60 days of confining the queens. The time taken for development of worker from egg to adult was found 28.40 days in *B. haemorrhoidalis* by Chauhan et al. (2013).

Small-scale rearing of *B. haemorrhoidalis* Smith was achieved up to colony development (Sharma et al. 2018) where the success rate of colony development under laboratory conditions was more in queens collected during April (60%) compared to queens collected during March (31.25%) and February (16.67%). Similarly, the per cent establishment of colonies in the field was recorded more in queens collected during April (40%) in comparison to queens collected during March (18.75%). Further investigating the bottlenecks of year round rearing success, Sharma et al. (2021) reported seasonal disease and pest incidence in laboratory reared *B. haemorrhoidalis* colonies that directed towards practical hygienic management for successful colony development.

4 Importance in Natural Ecosystems and Agriculture

Amongst pollinators, insects are acceptable for pollination and are considered the most reliable and efficient pollinators. Insects such as honey bees, bumblebees and solitary bees are exploited commercially for pollination. Bumblebees, the buzz pollinators and their communities are identified to have the potential to serve as an important indicators of ecosystems health and function (Biesmajer et al. 2006). These are large beautiful insects with specific buzzing sound and can be easily recognized and monitored for scientific analysis.

5 Importance of Bumblebees in Pollination

1. Bees pollinate plants that provide much of the food.
2. Bumblebees work well under adverse conditions of weather and have short flight range and low populations.
3. The longer tongue and wing vibrating tendencies of the bumblebees, i.e. buzzing behaviour, make them more effective pollinators for some species of plants.
4. The pollination services rendered by bumblebees are essential for the maintenance of different ecosystems.
5. Bumblebees are economically important pollinators for high economic return and production of off-season crops.

These are also considered alternate pollinators to the honey bees, serving as effective pollinators for cultivated fruits, vegetables and spices, and thus become economically valuable, as well. These are highly significant pollinators for a variety of crops in tropical and temperate regions under greenhouse and field conditions (Thornberry and Jerardo 2012). Pollination by bumblebees have positive effect on different fruit quality parameters in terms of per cent increase fruit set, fruit weight, fruit length, fruit breadth, healthy fruits, yield and reduction in misshapen fruits, especially under protected cultivation. These buzz pollinators throughout their native ranges where they developed co-evolved mutualisms. By this bees received food in return for providing pollination to plants.

6 Decline in Bumblebee Pollination

With increase in population and urbanization, a decline in pollinator's diversity and population from last 20 years (Biesmajer et al. 2006), including the buzz pollinators i.e. bumblebees, has been documented. These bees not only help in production of food but also aid in restoration of natural ecosystem and agriculture system to maintain the biological and genetic diversity in terrestrial ecosystems.

Following are the threats to wild bumblebees (Fig. 19.7):

1. Habitat loss due to agricultural intensification
2. Urbanization or pollution
3. Intense use of pesticide/chemical fertilizers

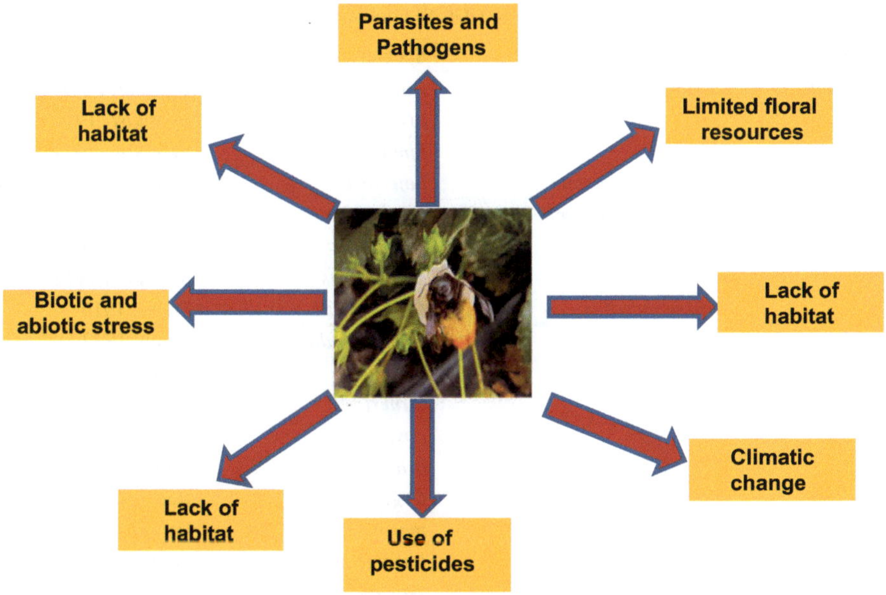

Fig. 19.7 Factors responsible for bumblebee decline

4. Pathogen spillover from managed bees
5. Climate change

In the current scenario, some of bumblebee species are declining at a very faster rate, while a few are at the verge of extinction. Some species has recently been included in the list of Critically Endangered insects by the International Union for Conservation of Nature (IUCN) (Table 19.3) that also includes species from India.

Since 1993, relative yields of crop production has either flattened or declined, while pollinator nondependent crops show no decline. This pattern of yield limitation may be several factors affect the crop yield, among which decline in pollinators population known as 'pollinator limitation' would be a major factor (Kearns et al. 1998). Pollinator decline will have serious socio-economic consequences, particularly in India where the economic value of such pollinated crops is $726 million annually (Sidhu 2005). The management and conservation of native bumblebee species is urgently needed for sustaining the available natural resources and agriculture production. Therefore, an artificial rearing technique to rear indigenous bumblebee species on large scale is required and preferred over exotic bumblebees that have possible issues of environmental and pest hazard. Thus, any threats to pollinator's community directly affect diversity and indirectly affect the natural ecosystem in addition to health of agricultural systems.

Table 19.3 List of Himalayan bumblebee (IUCN) Global Red list Assessment

Genus	Subgenus	Species	Taxonomic authority
Bombus	*Mendacibombus*	*waltoni*	Cockerell, 1910
		avinoviellus	Skorikov, 1914
		himalayanus	Skorikov, 1914
	Orientalibombus	*funerarius*	Smith, 1852
	Subterraneobombus	*personatus*	Smith, 1879
		melanurus	Lepeletier, 1835
		difficillimus	Skorikov, 1912
	Megabombus	*albopleuralis*	Friese, 1916
		haemorroidalis	Smith, 1852
	Psithyrus	*cornutus*	Frison, 1933
		turneri	Richards, 1929
		novus	Frison, 1933
		branickii	Radoszkowski, 1893
		ferganicus	Radoszkowski, 1893
		morawitzianus	Popov, 1931
		bohemicus	Seidl, 1837
		skorikovi	Popov, 1927
	Pyrobombus	*subtypicus*	Skorikov, 1914
		hypnorum	Linnaeus, 1758
		abnormis	Tkalcu, 1968
		mirus	Tkalcu, 1968
		lemniscatus	Skorikov, 1912
		lepidus	Skorikov, 1912
		pressus	Frison, 1935
		parthenius	Richards, 1934
		luteipes	Richards, 1934
		avanus	Skorikov, 1938
		biroi	Vogt, 1911
		kotzschi	Reinig, 1940
		flavescens	Smith, 1852
		rotundiceps	Friese, 1916
	Bombus	*tunicatus*	Smith, 1852
		longipennis	Friese, 1918
		jacobsoni	Skorikov, 1912
	Alpigenobombus	*genalis*	Friese, 1918
		graham	Frison, 1933
		breviceps	Smith, 1852
		kashmirensis	Friese, 1909
		nobilis	Friese, 1905
	Melanobombus	*eximius*	Smith, 1852
		festivus	Smith, 1861
		simillimus	Smith, 1852
		miniatus	Bingham, 1897

(continued)

Table 19.3 (continued)

Genus	Subgenus	Species	Taxonomic authority
		rufofasciatus	Smith, 1852
		friseanus	Skorikov, 1933
		richardsiellus	Tkalcu, 1968
		tanguticus	Morawitz, 1887
		ladakhensis	Richards, 1928
		keriensis	Morawitz, 1887
		semenovianus	Skorikov, 1914
	Sibiricobombus	*semenovi*	Morawitzi, 1887
		asiaticus	Morawitz, 1875

Acknowledgements Authors are grateful to the World Bank for funding the project on 'Managed pollination' to the state (Himachal Pradesh, India) to make this study feasible as well as all the staff which worked in unity. The preliminary work is a step in conservation that will act as strong stimulus for future investigations.

References

Biesmajer JSPM, Roberts M, Reemer R, Ohle MM, Edwards T, Peeters AP, Schaffers SG, Potte R, Kleukers CD, Thomas JS, Kunin WE (2006) Parallel decline in pollinators and insect pollinated plants in Britain and Netherlands. Science 313(5785):351–357

Bingham CT (1897) The Fauna of British India, including Ceylon and Burma, hymenoptera, Vol. I, London

Chauhan A, Thakur RK (2010) Studies on nest architecture of bumblebee, *Bombus haemorrhoidalis* Smith. Pest Manag Ecol Zoo 18(1/2):145–149

Chauhan A, Katna S, Rana BS (2013) Life cycle of bumble bee *Bombus haemorrhoidalis* Smith in Himachal Pradesh. Insect Environ 19:183–186

Chauhan A, Rana BS, Sapna K (2014) Successful rearing of bumble bee *Bombus haemorrhoidalis* Smith year round in Himachal Pradesh in India. Int J Curr Res 6(12):10891–10896

Dayal K, Rana BS (2004) Record of domestication of *Bombus* species (hymenoptera: Apidae) in India. Insect Environ 10(2):64–65

Devi D (2019) Studies on nesting material and carbon dioxide narcosis on domiciliation of bumble bee (*Bombus haemorrhoidalis* Smith). MSc thesis Dr YS Parmar University of Horticulture and Forestry, Nauni, Solan, Himachal Pradesh

Free J (1993) Insect pollination of crops. Acadamic Press, London-New York, pp 172–180

Ghosh S, Jung C, Kyoung C, Hyo YC, Hong WK, Sangwon K (2017) Body fatty and amino acid composition of a native bumblebee, *Bombus ignitus* relative to *B. terrestris* of foreign origin. Korea J Apicult 32(2):111–117. https://doi.org/10.17519/apiculture.2017.06.32.2.111

Gonzalez VH, Mejla A, Rasmussen (2004) Ecology of nesting behaviour of Bombus atratus Franklin in Andean Highlands (Hymenoptera: Apidae). J Hymenopt Res 13(2): 234–242

Hatfield R, Jepsen S, Thorp R, Richardson L, Colla S, Foltz Jordan S (2015) *Bombus occidentalis*. The IUCN Red List of Threatened Species 2015. Nature Serve, Inc. https://explorer.natureserve.org/Taxon/ELEMENT_GLOBAL.2.916920/*Bombus_occidentalis*

Heinrich B (1979) Bumblebee economics. Harvard University Press, Cambridge Mass USA

Holm SN (1966) The utilization and management of bumble bees for red clover and alfalfa seed production. Annu Rev Entomol 11: 155–182

Invasive species compendium (2019) CABI

Jie W, Wenjun P, Jiandong A, Zhanbao G, Yueming T, Jilian L (2005) Techniques for year-round rearing of *Bombus terrestris* L. (Hymenoptera: Apoidea) in China. J Apicult Sci 49:65–69

Kalia L (2020) Studies on pest incidence and effect of diets on colony development of *Bombus haemorrhoidalis* Smith. Msc Thesis, Department of Entomology, Dr YSP UHF Nauni, Solan, pp 335

Kalia S (2019) Pollination ecology and management of bumblebees in Himachal Pradesh. Ph.D. Dissertation, Dr. Yashwant Singh Parmar University of Horticulture and Forestry, Nauni, Solan, Himachal Pradesh, India

Kashyap L (2008) Domicilation of bumble bee (*Bombus* sp) and to study resource partitioning with honeybees. Msc Thesis, Dr YS Parmar University of Horticulture and Forestry, Nauni, Solan

Kearns CA, Inouye DW, Waser NM (1998) Endangered mutualisms: the conservation of plant-pollinator interactions. Ann Rev Ecol Systemat 29:83–112

Lindhard E (1912) Humlebien som Husdyr: spredte Traek af nogle danske Humlebiartes biologi. Tidsskr PlAvl 19:319–326

Mattu VK, Singh AP (2015) Pollinator diversity and relative abundance of insect visitors on apple crop in Kullu hills of Western Himalaya, India. Int J Sci Res 4(6):264–268

Negi N (2021) Diversity and altitudinal food preferences of bumblebee species in different agro-climatic zones of Himachal Pradesh. Master's Thesis, Dr. Yashwant Singh Parmar University of Horticulture and Forestry, Nauni, Solan, Himachal Pradesh

Ono M, Mitsuhata M, Sasaki M (1994) Use of introduced *Bombus terrestris* worker helpers for rapid development of Japanese native *B. hypocrita* colonies (Hymenoptera, Apidae). Appl Entomol Zool 29:413–419

Owen RE (2016) Rearing bumble bees for research and profit: practical and ethical consideration. In: Chambo ED (ed) Beekeeping and bee conservation –advances in research. Federal University of Amazonas, Nature and Culture Institute. https://doi.org/10.5772/63048

Plowright RC, Jay SC (1966) Rearing bumble bee colonies in captivity. J Apicult Res 5:155–165

Pomeroy N, Fisher RM (2002) Pollination of kiwi fruit (*Actinidia deliciosa*) by bumble bees (*Bombus terrestris*): effect of bee density and patterns of flower visitation. N Z J Entomol 25: 41–49

Pridal A, Hofleauer J (1996) Laboratory rearing and nutrition of young queens of Bumblebee (*Bombus terrestris* L.) from emergence to diapuse. Sci Papers Res Insti Fruit Grow 14:125–131

Raina RH, Saini MS, & Khan ZH (2017) Taxonomy and ecology of Bombus trifasciatus Smith (Hymenoptera: Apidae: Megabombus) from Indian Himalaya with description of new variants. J Entomol and Zool Stud 5(6):2111–2117

Rana (2019) Morphometreic and molecular characterization of bumble bee species in Himachal Pradesh. Msc Thesis,. Department of Entomology, Dr YSP UHF Nauni, Solan, p 335

Richardson LL, McFarland KP, Zahendra S, Hardy S (2018) Bumble bee (*Bombus*) distribution and diversity in Vermont, USA: a century of change. J Insect Conserv:1–29. https://doi.org/10.1007/s10841-018-0113-5

Saini MS, Ghattor HS (2007) Taxonomy and food plants of some bumble bee species of Lahaul and Spiti valley of Himachal Pradesh. Zoo's Print J 22(4):2648–2657

Saini MS, Raina RH, Khan ZH (2011) Species diversity of bumblebees (Hymenoptera: Apidae) from different mountain regions of Kashmir Himalayas. J Sci Res 4(1):263

Sharma HK, Thakur S, Rana K, Thakur M (2018) Small scale rearing of *Bombus haemorrhoidalis* Smith. Pharma Innov 7(11):41–45

Sharma HK, Kalia L, Sharma R, Thakur M, Prasad H, Devi M, Thakur P, Sharma D, Rana K (2021) Seasonal incidence, epidemiology and establishment of different pests and diseases in laboratory reared *Bombus haemorrhoidalis* Smith. Int J Trop Insect Sci 41:2555–2564

Sharma HK, Thakur S, Rana K, Thakur M, Sharma R (2022) Laboratory rearing of *Bombus haemorrhoidalis* smith. HP-HDP managed pollination (Unpublished)

Sidhu MS (2005) Fruit and vegetable processing industry in India: an appraisal of the post reform period. Econ Polit Wkly 40(28):3056–3061

Singh N, Rana K, Sharma HK, Thakur M, Thakur RK (2017) Palynological studies to determine pollen resources of *Bombus haemorrhoidalis* Smith. Int J Curr Microbiol Appl Sci 6(9): 1632–1645

Sladen FWL (1912) The bumble bee: life history and how to domesticate it. London: Mac Millan. 253 p

Streinzer M, Chakravorty J, Neumayer J, Megu K, Narah J, Schmitt T, Bharti H, Spaethe J, Brockmann A (2019) Species composition and elevational distribution of bumble bees (*Hymenoptera*, *Apidae*, *Bombus* Latreille) in the East Himalaya, Arunachal Pradesh, India. Zookeys 851:71–89

Taylor OM, Cameron SA (2003) Nest construction and architecture of the Amazonian bumble bee (Hymenoptera: Apidae). Apidologie 34:321–331

Thakur RK (2002) First attempt to study nest architecture and domicialation of bumble bee (*Bombus haemorroidalis*) in India. Ibid p 172

Thakur RK (2006) Bumble bee flora and domestication of bumble bee *(Bombus sp.)* under mid-hill conditions of Himachal Pradesh. International Workshop on integrated beekeeping development in South Asian countries, Nov-13th-14, 2006, New Delhi p 30

Thakur RK, Kashyap L (2007) Some studies on bumble bee, *Bombus haemorroidalis* Smith nests under mid hill conditions of Himachal Pradesh. Indian Bee J 69(1–4):100–102

Thornberry S, Jerardo A (2012) Vegetables and Pulses Outlook USDA, Economic Research Service, p. 52. http://www.ers.usda.gov/media/826842/vgs350.pdf

Van den Eijnde J, De Ruijter A, Van der Teen J (1991) Method of rearing *Bombus terrestris* continuously and production of bumble bee colonies for pollination purposes. Acta Hortic 288: 154–158

Velthuis HHW, Doorn VA (2006) A century of advances in bumble bee domestication and the economic and environmental aspects of its commercialization for pollination. Apidologie 37: 421–451

Williams PH (1991) The bumble bees of the Kashmir Himalaya (Hymenoptera: Apidae, Bombini). Bull Br Mus (Nat Hist) Ent 60(1):1–204

Yadav SP, Yadav S, Sharma D, Sangwan N (2016) Bumblebees, life cycle and their role in pollination - a review. Int J Enhanc Res Sci Technol Eng 5(9):60–75

Yoon HJ, Kim SE, Lee SB, Park IG, Kim NJ, Seong JH, Soo GW (2004) Comparison of the colony development between cardboard and plastic rearing box in the bumblebees, *Bombus ignitus*. Korean J Apicult 19:37–42

Part IV

Pollinators and Livelihoods

Development of Meliponine Culture in Cultivated Ecosystems

20

Abu Hassan Jalil

Abstract

The part played by pollinators in pollination of economic and wild plants is well recognized. Being native to Asia, stingless bees are more resistant to pest and diseases and sustainable and better pollinators for native plant species in comparison to foreign bees introduced in the tropics. Valuable hive by products provide additional income to the beekeepers. Researches have demonstrated sufficiently that stingless bee propolis have high clinical value and is useful in apitheraphy. Many techniques are being evolved for managing the stingless bees for crop pollination and production of honey, pollen and propolis. There are yet plenty of researchable areas to be explored, such as biology, systematics and product characterization for clinical and industrial uses.

Keywords

Stingless bees · Tropics · Pollination · Management

1 Introduction

Plant pollination is a clear example of ecosystem services providing economic benefits to nature (Millennium Ecosystem Assessment 2005; Allsopp et al. 2008). Bees and other pollinating agents provide valuable inputs to agricultural production and increasing the size and quality of the harvest (Allen-Wardell et al. 1998). Plants such as tomatoes, coffee, canola, melon and cocoa show increase production by animal pollination (NRC 2007). Meliponines bees are benefited from plant breeding efforts using pollen, nectar, gum or latex and oils (essential oils as well) from the

A. H. Jalil (✉)
Agronomist, Meliponiculturist, Stamford College, Petaling Jaya, Malaysia

A. K. Chakravarthy (ed.), *The Role of Pollinators in Tropical Landscapes*,
https://doi.org/10.1007/978-981-95-9139-8_20

flowers, as well as using phloem, resin and extra floral nectar, of floral nectar and sap which formed the plant defences against herbivores (Roubik 1993a).

In 2008, MARDI (Malaysian Agriculture Research and Development Institute) introduced a cash crop pollination business using stingless bees (Conservation and sustainable development of stingless bees for pollination in cultivated tracts in Malaysia) (Norowi et al. 2008). At the time, the monetary worth of pollination process using insects for star fruit, guava, lime, mango, watermelon, durian and coconut was estimated at USD 19 million a year (Norowi et al. 2008; Klein 2007; Ricketts et al. 2008; Vasanthkumar et al. 2017).

Among the insects identified are four species of stingless bees, i.e. *Heterotrigona itama*, *Geniotrigona thoracica*, *Tetragonilla atripes* and *Tetrigona peninsularis*. In 2010, Dr. Ibrahim Shiub with the author tried to apply this pollination service through their "Great Farm" in Hulu Selangor, Malaysia. The fruits in the fields were star fruit, guava, lemon, mango, durian and coconut. Various species of meliponines were tried including five species recommended by MARDI, *Heterotrigona itama*, *Geniotrigona thoracica*, *Lepidotrigona terminata*, *Tetragonula laeviceps* and *Tetrigona apicalis* (Slaa et al. 2006).

2 Initial Trials

Experiments and research on a pollination service business was applied with cash crops in the field. The first hindrance encountered was proper farming for the bee species diversity. (There were 11 species in the test plots at the initial stages of the venture).

Initial trials were run on the design and size of portable box hive dimensions for pollination efforts. Some obstacles faced were as follows:

1. The obstacle is to provide a universal "template" for the general species because different species will differ in size requirements as every species will have a different average population.
2. To produce durable boxes, the use of hardwood types was important for the development and sustainability of the colony.
3. The flowering and duration of flowers in bloom varied drastically. Types of stingless bees also have different affinity for flowers and colours.
4. Attempts were made to conduct pollination efforts with farmers unaware of pollination service because of the worries of damage to their products. They also do not readily accept any changes in the pesticide usage schedule in their maintenance.
5. Proposed rental rates (RM 20.00 a box of hawks hired for a duration of 7–10 days depending on the crop) are also considered expensive (Figs. 20.1 and 20.2).

Box design and meliponiculture system also change with farming system using the "Topping" box rather than the pollination portable box. With this, the pollination services in Malaysia are hard hit.

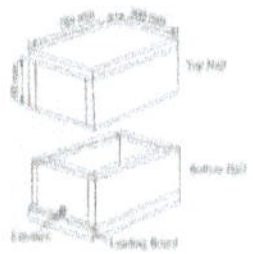

Fig. 20.1 Bee box types for pollination services

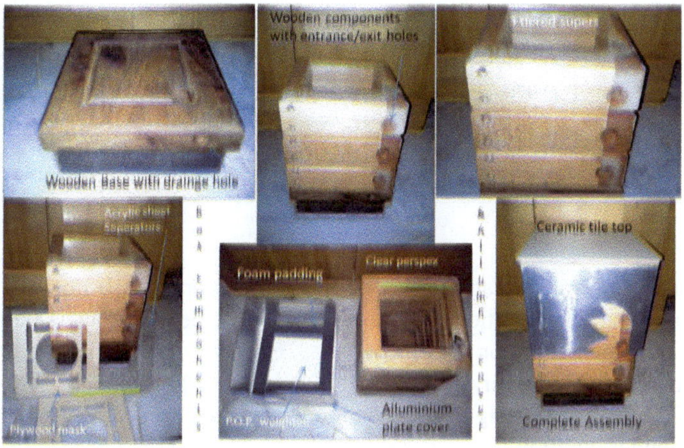

Fig. 20.2 Durable three tired box hive

In fact, in the Philippines it is still being treated in mango farm fields, coconut fields and Pili legume farm fields with a density of one nest for four mature plant mature trees. There is also a lease system of 3000 pesos per *Tetragonula biroi* box colony (Pers Comm. Cervancia and Fajardo 2012).

3 Recommendations (Table 20.1)

Based on (Momose et al. 1998; Amano 2004), social bees are 36% of insect pollinators totally. This means that IPEV (Value of pollination insect dependence) social bee is RM845M × 36% = RM 303M in 2011. Given this data is suggested, the pollination services business may be resumed by holding seminars or information dissemination sessions from time to time. This can be achieved if the representatives of the cash crop farmers are gathered together with representatives of the meliponiculture farmers for discussion and exchange of recommendations. Following factors should be considered for the successful establishment of Melipona beekeeping.

1. Extensive use of pesticides by farmers. This means crop farmers have to reschedule their maintenance regime.
2. Beekeepers do not have proper mobile pollination boxes. Fabrication of new boxes with new templates for different species will attract expenses.

Table 20.1 Economic impact of pollination services of major Malaysian fruits pollinated by insects

Commodities	Production value (USD)	Pollinator dependence ratio (USD)	Insect pollinator economic values
Starfruits	5,218,071.08	0.65	3,391,746.20
Guava	481,445.14	0.65	312,939.34
Citrus (Mandrin)	143,078.11	0.05	7,153.91
Mango	911,512.43	0.65	592,483.08
Watermelon	9,970,436.22	0.95	9,471,914.41
Durian	4,781,793.24	0.65	3,108,165.61
Coconut	9,026,651.35	0.25	2,256,662.84
Total	30,532,987.57		19,141,065.38

3. Distance and transportation from beekeeper to crop farm becomes a negative factor. Time and duration of colony movement becomes an issue with new labour and pedestals for placement.
4. Sustainability (long-term effects) for bees in "greenhouse" environment. Re-planning and redesign of "greenhouse" features needed to incorporate bees for a longer duration.
5. Frequent to and fro movement of hives from crop farm to bee farm may cause colony development issues.
6. Crop farmers are not willing to pay for hive rental without field evidence.
7. Who determines and/or monitors the rate and type of pesticide usage allowable. MyGap organic farming or EM usage be determined and adhered to.
8. The crop plants below listed benefit from insect pollination mainly insects, allspice, avocado, cashew, coconut, coffee, cotton, litchi, mango, melon, passion fruit, safflower, sunflower and several kinds of *Citrus* and *Cucurbits*.
9. Crops like safflower, mango, okra, and opium poppy give moderately increased yields when pollinated by insects particularly bees while others, for example, cashew and guava yield enhanced production.
10. In view of the inadequate manpower and facilities, investigations on tropical crop pollination are mostly preliminary. Planned investigations should be initiated to find:
    (a) Whether fruit or seed set is adequate
    (b) Whether the crops benefit from the process of pollination
    (c) The species which pollinate the flowers

4 Insects for Pollination

Having considered the demands and the needs for pollination I wish now to turn to the availability of pollinating insects. Probably humans at first encouraged wild species of bees and other pollinators as primeval forests were cleared, allowing light to penetrate and low growing plant species to multiply.

However, world over, including the tropics, the populations of wild bees and other pollinating insects have exhibited a tendency to diminish until they are less for crop pollination. There are numerous factors for this. Some of the select factors are as follows:

1. Intensive cultivation of the land, including the removal of hedgerows, banks and rough verges, has diminished the nesting and hibernation sites available for wild populations of pollinators (Gallai and Vaissière 2009; Heard 1999; Free 1999; Potts et al. 2016).
2. Regular cutting of roadside verges and around field crops and application of weedicides destroyed floral resources helping to provide beneficial insects with food; use of selective herbicides and insecticides destroy flowering weeds within crops.
3. The damage to honeybee colonies differ according to many factors including the insecticide's toxicity.
4. Steps are required to avoid destroying wild pollinators in tropical countries, so that pollination by wild honeybees and other pollinators will retreat.

5 Cash Crop Compatibility

Development of Meliponine plantation crops in following can be considered.

5.1 Oil Palm

Oil palms canopy covers does not support ground cover or shrubs for the insects' resource partitioning; also, there is practically no propolis source. However technically, oil palm is not pollinated by bees. The pollen of oil palm (*Elaeis guineensis* Jacq.) is a strong allergen and causes severe polyposis in Malaysia and Singapore (Kimura et al. 2003) and Malaysia (Thang 2005), Ministry of Agriculture, Department of Statistics, Malaysia). Of this, 50% of stingless beekeepers is estimated to practice Meliponiculture within or fringes of oil palm estates throughout Malaysia. This brings rise to a common question "How would this affect asthmatic prone individuals and / or 'hay-fever' sufferers consuming honey or 'beebread' produced from such areas of oil palm plantations?"

5.2 Pineapple

Pineapples (*Ananas comosus*) are self-pollinated in that it becomes an over matured multiflower rather than a fruit. Because hummingbirds are the principal pollinators, they are prohibited in Hawaii to avoid undesirable seeds in their pineapples (Morton 1987). *Apis* bees (having a good flight range) do gain from the "Nectar robbing"; they take nectar without any help to the pollination of pineapples. Meliponine

Fig. 20.3 Pineapples as far as the eye can see

rearing at pineapple plantations is not really a good practice because of the monoculture. Stingless bees will not have the opportunity to forage a multitude of needs because the flight range is limited especially if placed in the middle of a pineapple plantation! Meliponine in a pineapple plantation will not get the sufficient material for nest building (Fig. 20.3).

Malaysian Pineapple Industry Board (MPIB) chairman, Mr. Datuk Samsolbari Jamali, was determined to increase the pineapple production, following which the Government increased the pineapple plantation acreage. Sarawak recorded the largest pineapple plantation areas after Johor, with 1805 ha currently.

With such a vast area of pineapples, the Meliponines will have nowhere to go for the many other resources they need. As a testament to the negative effects of monoculture to meliponiculture, this is why it (melipioniculture) cannot be encouraged in a pineapple plantation. Pollination by Meliponine bees to melons can also be considered (Jaapar and Jajuli 2014; Jalil and Shuib 2014)

5.3 Coconuts

Cocos nucifera palms appear most compatible in Meliponiculture. Coconut pollen grains are monocolpate. The pollen output per anther in a flower of a healthy tree has been estimated to be between 111,000 and 221,000 grains. Besides the nectar and pollen supply for Meliponines, fallen coconuts are also a haven for them. Old fallen fruits usually pierced and foraged by squirrels leave behind a convenient nesting host for stingless bees (Jalil and Shuib 2014) (Figs. 20.4 and 20.5).

5.4 Maize, Sorghum and Other Cereals

As for corn and other cereals or even sorghum, meliponine may be ideal (http://www-saps.plantsci.cam.ac.uk/pollen/index.htm). A species with exceptionally large pollen grains with a size of 0.070 by 0.090 mm and an egg shape. *Zea mays* has the largest pollen grains among grasses. The pollen grain has pale yellowish colour. Maize pollen is too big to be travelling a long distance through the air. It falls down from the male tassels on the long female silks of the neighbouring plant (http://www.vcbio.science.ru.nl/en/virtuallessons/pollenmorphology/). Detailed studies are required on predominantly large pollen and its grain sediments in honey. This may be of concern in monoculture of large areas of maize farms. It may affect the final clarity of the resultant honey and retard marketability. Azmi et al (2016) worked on pollination of chillies in green house.

Fig. 20.4 Among the foraging favourites across all the species of Meliponini

Fig. 20.5 Meliponine nests found in neglected coconut farms in Kelanta

6 Issues in Providing Pollination Services for Meliponiculture Development

- Challenges: to provide a universal "template" for the widespread species.
- Durable boxes: hardwood types necessary for the development and sustainability of the colony (added cost).
- Pairing flower and fee: flowering and duration of fruit flowers in bloom vary. Types of stingless bees also have different affinity for a particular kind[1] of flowers and colours.
- Farmer acceptance:[2] many are unaware and less likely to try this pollination service for fear of damage to their products.
- Changes in pesticide usage: not readily accept any changes in the pesticide usage schedule in their maintenance.
- Awareness: the breeding services business could not respond because the information was not widely distributed to farmers by appropriate authorities (Fig. 20.6).

7 Meliponines and Flower Distinctiveness

The Individuality of Meliponines and Flower Distinctiveness

- Duration: The flowering and period of fruit flowers in bloom vary drastically. Types of crops have different anthesis times and a variation of flowers and colours.
- Variations: Different plants and trees have varying understory requirements, such as humidity and temperature stability.
- Predators: Trees harbour and attract varied predators or enemies of Meliponines as well (Fig. 20.7).

8 Crop Farmer Acceptance

Greenhouse operators: attempts to conduct pollination programmes with greenhouse farmers were not prepared for modifications in the structure of greenhouses.

[1] Editor's note: Some have studied a preference for a particular kind of flower or resource production in a flower patch. The rest is unknown.

[2] This is widespread and also explained and documented in the FAO book on pollination, two volumes, published in 2018. In Africa, for example, for the past 100 years or so the locals thought bee visits damaged flowers. That is 'traditional knowledge' in this part of the world. Imagine that! The real hesitancy we found in Mexico, which was justified, with growers, was that we were going to remove flowers and do harm with bagging or other manipulation, in doing our research project. Imagine that! However, since these were squash female flowers, each flower removed is one less squash produced for the farmer!

Fig. 20.6 Stingless bee pollination for corn and cereals

- Awareness: People do not try using this pollination service because of the worries of damage to their products (Cervancia 2017).
- Pesticide Regime: not prepared for modifying their pesticide usage schedule in maintenance (Figs. 20.8 and 20.9).

Fig. 20.7 Meliponines and flower distinctiveness

Fig. 20.8 Cucurbit pollination

Fig. 20.9 Honeydew melon trellis

Acceptable Fees

Agreeable Rates

- Rental rates: hive hiring for a limited duration is considered expensive by farmers.
- Records: no consistent pollination data records (effectiveness in the field) for many crops.

- Commitment: reluctant to invest upfront, and beekeepers do not want to transport their hives without initial monetary deposits.
- Binding: contract seems too complicated for short term arrangements among both parties.

9 Predicaments Faced by Beekeepers

- *Unregulated pesticides usage*: Crop farmers have to reschedule their maintenance regime.
- *Business model* (e.g. who pays whom how much concerning increase in yield) How may the beekeeper also benefit from this profit increase (if any)?
- *Fabrication of new boxes* with new templates for different species will incur new expenses as beekeepers do not have proper portable hive boxes for the pollination service.
- *Distance and transportation* Time and duration of colony movement becomes an issue with new labour and pedestals for placement.
- *The sustainability of beekeeping* in a greenhouse environment ate detrimentally insufficient in nest-building materials forage source.
- *Added costs* due to frequent movement of hives may cause colony development issues (e.g. specially customized racks for towing the boxes).

10 Predicaments Faced by Crop Farmers

Predicaments faced by crop farmers:

- *Revamping and redesign* of "greenhouse" features are needed to incorporate bees on a longer duration for economic benefits (Kasina et al. 2009; KishanTej et al. 2017).
- *The reluctance is* of farmers to pay for beehive rental without field evidence.
- *Colony Failure*: Which party bears the cost of colony fail? Colony health and pre-Q Parameters at crop farm must be suitable for colony entry.
- *Arbitration*: Who determines and monitors the rate and type of pesticide usage allowable. Require organic farming certification.

Portable Cabins and Field Hive Racks

They are constructing specially customized racks for towing box hives to and fro between farms and meliponaries (Figs. 20.10 and 20.11).

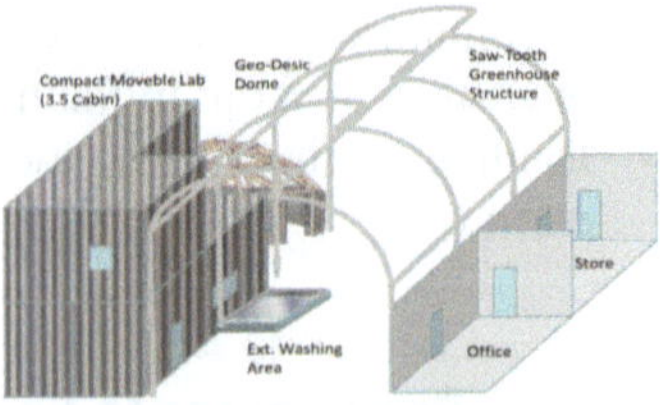

Fig. 20.10 Revamping and redesign of greenhouse features

Fig. 20.11 Portable hive boxes and field hive racks

11 Summary

Changes in the pollination community will affect the vegetation composition and the entire food chain. This is especially important for activities related to the use of forests, logging and recommendations for best practices.

It is expected that institutional research institutions will continue to focus and research on pollination efforts and ensure that this research is widely disseminated, especially among the Association of Beekeepers, Cooperatives and Government linked farmers' unions in the country. Climate conditions are unpredictable in a tropical fluctuating environment, and this may affect the survival of the colony. This subject should be considered and monitored as a long-term activity, as well as how different cultures (agriculture, horticulture and meliponiculture) adapt to it. There is a need for multiplication of hives for sustainable pollination services with the following pointers to be taken into scrutiny. Study of nest trap in different habitats/biomes would indicate which species should be prepared for the agglutination of this method. To test several trap nests and to field evaluate them before confirming that traps will be of sufficient practical use. Training of standard methods using artificial baits to validate the occurrence of local species in the environment. Salt and sugar salts have been used successfully (Roubik 1993b) in Asia.

New Honeybee Species from India

Indian subcontinent is known for cavity nesting honeybee species. Shanas and Others (2022) identified a species of cavity nesting honeybee *Apis karinjodian* n sp. (Hymenoptera, Apidae, Apinae, Apini, *Apis*). The morphological description of new species is provided by workers with molecular and behavioural aspects. Further, South India is proposed as the new centre of origin of the European honeybee, *Apis mellifera* Linnaeus, 1758.

A–C: Dorsal view of *Apis cerana*

G–I: Dorsal view of *Apis karinjodian* n sp.

(Source: Shanas et al. 2022)

References

Allen-Wardell G, Bernhardt P, Bitner R, Burquez A, Buchmann S, Cane J, Cox PA, Dalton V, Feinsinger P, Ingram M, Inouye D, Jones CE, Kennedy K, Kevan P, Koopowitz H, Medellin R, Medellin-Morales S, Nabhan GP, Pavlik B, Tepedino V, Torchio P, Walker S (1998) The potential consequences of pollinator declines on the conservation of biodiversity and stability of food crop yields. Conserv Biol 12(1):8–17

Allsopp M, Lange WJ, Veldtman R (2008) Valuing insect pollination services with cost of replacement. PLoS One 3(9):e3128. https://doi.org/10.1371/journal.pone.0003128

Amano K (2004) Attempts to introduce stingless bees for the pollination of crops under greenhouse conditions in Japan. National Institute of Livestock and Grassland Science Japan

Anonymous (2008) Tools for conservation and use of pollination services – initial survey of good pollination practices. FAO

Anonymous (n.d.-a) Bees and their role in forest livelihoods – the value of bees for crop pollination. FAO

Anonymous (n.d.-b) Pollination manual – reproductive biology and pollination in rainforest trees. In: Reproductive biology and pollination in rainforest trees

Azmi WA, Seng CT, Solihin NS (2016) Pollination efficiency of the stingless bee, Heterotrigona itama (Hymenoptera: apidae) on chili (Capsicum annuum) in greenhouse. J Trop Plant Physiol 8:1–11

Birt M (n.d.) The Trigona Species, (Stingless Bees). Their roll in higher plant pollination

Birt M (2014) The Trigona Species, (Stingless Bees): Their role in higher plant pollination. Originally published or hosted by the Blackburn Beekeepers.

Boulter SL, Kitching RL, Zalucki JM, Goodall KL (2006) Reproductive biology and pollination in rainforest trees: techniques for a community-level approach. Rainforest CRC, Cairns, Australia.

Cervancia C (2017) Asian pollination initiative. UPLB

Cervancia CR, Fajardo AC (2012) Philippine apiculture: status and RD&E agenda, 2012-2016: a roadmap for policy makers and stakeholders. University of the Philippines Los Baños Bee Program (UPLB-BP)
Free JB (n.d.) Pollination in the tropics. CMB
Free JB (1999) Pollination in the tropics. Bees for Development Journal, 50–51.
Gallai N, Vaissière BE (2009) Guidelines for the economic valuation of pollination services at a national scale. Rome: Food and Agriculture Organization of the United Nations (FAO)
Gallaia N, Sallesc J-M, Setteled J, Vaissièrea BE (2009) Economic valuation of the vulnerability of world agriculture confronted with pollinator decline. Science Direct
Heard TA (1999) The role of stingless bees in crop pollination. Annu Rev Entomol 44:183–206
Imperatriz-Fonseca VL (n.d.) Best management practices in agriculture for sustainable use and conservation of pollinators. FAO
Jaapar MF, Jajuli R (2014) Stingless bees: utilization of stingless bees pollinating rock melon under green house. MARDI
Jalil AH, Shuib I (2014) Beescape for Meliponines – conservation of indo-Malayan stingless bees. Partridge Singapore
Kasina JM, Mburu J, Kraemer M, Holm-Mueller K (2009) Economic benefit of crop pollination by bees: a case of Kakamega small-holder farming in Western Kenya. J Econ Entomol. https://doi.org/10.1603/029.102.0201
Kimura M, Chew FT, Lim SH, Lee BW (2003) Identification of the 31-kDa palm pollen glycoprotein (Ela g Bd 31 K) as a major allergen of oil palm (Elaeis guineensis) pollen. Allergy 58:284
KishanTej M, Srinivasan MR, Thakur RK (2017) Stingless bees for applied pollination in cucumber greenhouses of South India. AKM-IMS3C
Klein A-M et al (2007) Importance of pollinators in changing landscapes for world crops. Proc R Soc B. https://doi.org/10.1098/rspb.2006.3721
Momose K et al (1998) Pollination biology in a lowland dipterocarp forest in Sarawak, Malaysia. I. Characteristics of the plant- pollinator community in a lowland Dipterocarp Forest. Am J Bot 85(10):1477–1501
Morton JF (1987) Pineapple. In Fruits of warm climates. Miami, FL: Florida Flair Books, pp 18–28
Millennium Ecosystem Assessment (2005) Ecosystems and human well-being: synthesis. Washington, DC: Island Press
National Research Council (2007) Status of pollinators in North America. Washington, DC: The National Academies Press
Nicola B (2009) Bees and their role in forest livelihoods: A guide to the services provided by bees and the sustainable harvesting, processing and marketing of their products. Food and Agriculture Organization of the United Nations (FAO). Non-Wood Forest Products No. 19.
Norowi et al (2008) Conservation and sustainable utilization of stingless bees for pollination. MARDI
Potts SG, Imperatriz-Fonseca VL, Ngo HT (Eds). (2016) The assessment report of the Intergovernmental Science-Policy Platform on Biodiversity and Ecosystem Services on pollinators, pollination and food production. PBES. Secretariat of the Intergovernmental Science-Policy Platform on Biodiversity and Ecosystem Services, Bonn, German
Ricketts TH et al (2008) Landscape effects on crop pollination services: are there general patterns? Ecol Lett. https://doi.org/10.1111/j.1461-0248.2008.01157.x
Roubik DW (1993a) Direct costs of forest reproduction, bee-cycling and the efficiency of pollination modes. STRI Panama
Roubik DW (1993b) Direct costs of forest reproduction, bee-cycling and the utility of near-infrared blossoms. J Biosci 18(4):537–552
Shanas S, Krishnan GA, Mashoor K (2022) Identity of cavity nesting honey bees of the Indian subcontinent with description of a new species (Hymenoptera: Apidae: Apinae: Apini: Apis). Entomon 47(3):197–220
Slaa EJ, Chavez LA, Malagodi-Braga KS, Hofstede FE (2006) Stingless bees in applied pollination: practice and perspectives. Apidologie 37. https://doi.org/10.1051/apido:2006022

Thang KF (2005) Clinical and aerobiological studies of oil palm (Elaeis guineensis Jacq.) pollen allergy in Malaysia. Doctoral dissertation, University of Malaya

Vasanthakumar S, Srinivasan MR, Thakur RK (2017) Pollination behaviour of stingless bees Tetragonula iridipennis Smith on mango inflorescence in South India. AKM- IMS3C. https://www.thestar.com.my/business/business/news/2016/08/08/malaysia-records-pineapple-exports-worth-rm161mil-in-2015/#UQAhkOh3vCGoYG1S.99, https://www.thestar.com.my/news/community/2006/06/24/johor-still-tops-in-pineapple/#zX0mhxxZLmpApV46.99

Honey Bees in Asia: Current Status as Pollinators and Future Scope

21

Meena Yadav and Ahmad Pervez

Abstract

Asia is one of the hotspots for honey bee diversity. Apart from providing bee products, honey bees play vital role in sustaining natural- and cultivated ecosystems via pollination. Asian countries have exploited honey bees for pollination. India, China and Turkey have the largest number of beehives in the world. This chapter discusses *Apis* sp. diversity in Asia, their role in pollination, factors impacting their diversity and abundance, and their future.

Keywords

Apis sp. · Asian honey bees · Biodiversity · Pathogen · Plants · Pollination

1 Introduction

Honey bees (*Apis:* Apidae) largely known for honey production and for their societal habits. They are also major contributors of the pollination. Is it "honey bee" or "honeybee"? Kirk (2022) elucidated this issue regarding the spelling of the common name and convincingly advocated 'honey bee' as correct in usage than 'honeybees'. At least nine species of the genus *Apis* exist in Asia. Based on size and nesting style, honey bees are categorized as 'open nesting' honey bees and 'cavity nesting' honey bees. Open nesting honey bees were first recorded in Southeast Asia and include the giant honey bee (*Apis dorsata* Fabricius 1793), the Himalayan giant honey bee (*Apis*

M. Yadav (✉)
Department of Zoology, Maitreyi College, University of Delhi, Delhi, India

A. Pervez
Department of Zoology, Pt. L. M. S. Campus, Sri Dev Suman Uttarakhand University, Rishikesh, India

A. K. Chakravarthy (ed.), *The Role of Pollinators in Tropical Landscapes*,
https://doi.org/10.1007/978-981-95-9139-8_21

laboriosa Smith 1871), the red dwarf honey bee (*Apis florea* Fabricius 1787) and the black dwarf honey bee (*Apis andreniformis* Smith 1858) (Chantawannakul 2022). The giant honey bees are bigger and make extensive nests and colonies. They frequently nest in dense aggregations. The dwarf honey bees are the smallest world over and built small nests (Guerin 2020).

The 'cavity nesting' honey bees form the largest group of Asian honey bees, and their size lies between giant and dwarf honey bees. These include the red honey bee (*Apis koschevnikovi* Enderlein 1906), the mountain honey bee (*Apis nuluensis* Tingek et al. 1996), the eastern hive honey bee (*Apis cerana* Fabricius 1793) and the Sulawesi honey bee (*Apis nigrocincta* Smith 1861) (Chantawannakul 2022). Of these, *A. cerana* has been domesticated on a large scale in Asia. Apart from eight native Asian bees, Asia harbours *A. mellifera*, a native of Europe, Africa and the Middle East (Han et al. 2012). It was introduced in Asia in the 1900s and is considered one of the most economically important honey bees in Asia.

Honey bees are predominantly managed for the production of bee products and pollination worldwide. They make hives on various platforms and collect nectar and pollens for their nutrition as the former is a source of carbohydrates while the latter of protein (Winston 1987). Of all the honey bees in Asia, *A. cerana* and *A. mellifera* are managed by beekeepers on a large scale while the rest species have yet not been managed significantly (Chantawannakul et al. 2018), owing to various reasons. Native honey bees serve as for animals, such as amphibians, other insects, birds, reptiles and mammals. Thus, they are an important component of the food chain (Oldroyd and Nanork 2009).

Pollination refers to a process whereby pollens are moved from male anthers to female stigmata. This transfer may happen between flowers of the same plant (*self-pollination*) or between flowers of different plants (*cross-pollination*). Animal pollinators account for the production of 87 crops worldwide. This chapter emphasizes the honey bees of Asia with a focus on their pollination potential.

2 Benefits of Pollination

Honey bees have great ecological and economic significance, and provide multiple benefits including high-quality food products such as honey, royal jelly, and pollen, and other valuable products including beeswax, bee venom, and propolis. They serve as highly efficient pollinators in natural and agricultural ecosystems. Pollination is essential for the production of agricultural crops to meet the global food demand and cash crops to maintain the economy of the country. Crops pollinated by honey bees give better yield than self-pollinated or wind pollinated crops. Pollination is also essential for the conservation of natural ecosystems. Fruits and seeds, produced as a result of pollination, help in the regeneration of forests, which support various other animal species (Bawa 1990). In natural ecosystems, pollination by honey bees sustain herbivores and seed-eaters (Potts et al. 2006). Flowers act as microhabitats for many invertebrates and support complementary fauna to leaves. Flowers perform functions such as acting as resting or nesting sites or aggregation sites, mating sites or sites for predators. Flowers are visited by many macro- and micro-invertebrates (Wardhaugh et al. 2012). Thus, without bees, there is a serious threat to biodiversity, which is an indicator of a healthy and sustaining ecosystem. Since cultivation of many of food and cash crops depend on pollination services recently, their demand as pollinators has increased. However, managed or domesticated honey bees are unable to meet this demand, and the contribution of wild honey bees is uncertain. Given this, the scientific community is demanding to protect honey bee diversity and abundance for sustainable food security in the future (Mashilingi et al. 2022).

3 Honey Bees in Asia

Honey bees pollinate a large number of angiosperms and are considered the most efficient pollinators. One of the reasons for their success is that they are enormously hairy compared to other pollinators (Fig. 21.1). Each hair is a branched structure that makes it more efficient in catching and holding the pollen. While moving to the next flower, bees brush themselves such that some of the pollens fall in the pollen basket (made of stiff hair in hind legs), although much of it remain attached to their body hair.

Fig. 21.1 *A. mellifera* foraging on a flower. Note the enormous hairs present on the body

Fig. 21.2 *Apis mellifera* collecting pollen (Seen here with a pollen basket full of pollen

Sometimes, when the pollens are dry, they regurgitate nectar onto them to bind them so that they do not fall from the pollen basket during flight (Fig. 21.2). When such bees, carrying pollens, land on the next flower, pollens from their hair fall on the stigma of the flower. These pollens then fertilize the eggs and form seeds. Some plants require several visits by the bees so that all the eggs in the flower are fertilized. If all the eggs are not fertilized, fewer seeds develop resulting in small and poor-quality fruits (FAO 2008).

Honey bees can be put into two categories, i.e. *wild* and *managed/domesticated*. Wild honey bees occur in natural ecosystems where they pollinate various kinds of plants. The managed honey bees are maintained by beekeepers for producing bee products and pollination services in agroecosystems. While many plants are pollinated by more than one pollinator, certain crops such as berries and cherries are exclusively pollinated by honey bees. More than one species of honey bee can pollinate the same plant.

Honey bees perform two functions as a result of pollination services, *one*, in natural ecosystems they preserve and restore the diversity of wild plants and in turn support everything that depends on these plants, and *second*, in agroecosystems, they enhance crop yield and quality (Papa et al. 2022).

Wild honey bees pollinate plants in natural as well as agroecosystems (Table 21.1), but the pollination services in agroecosystems are more effectively supplemented by managed honey bees. This is because the managed honey bees are better buffered from harsh environmental conditions as they are under the care and maintenance of beekeepers (Wakgari and Yigezu 2021). Bees depend on plants for nectar and pollen, and plants need pollinators for the pollination of flowers. To ensure this, plants depend on visible cues and also secrete chemical cues to attract pollinators. For example, they secrete a rich liquid sugar from their gland that resembles nectar to attract pollinators (Shin et al. 2019).

After its success in the western world, *A. mellifera* was introduced in Asia as it gives better services in terms of bee products and pollination. The domestication of *A. mellifera* has been done on a large scale by different countries in Asia. It has become an invasive species in Asia and has dominated other native species. It has replaced 55% of *A. cerana* population, with unequal distribution; for example, it has replaced more than 95% of *A. cerana* in Afghanistan while less than 1% in Pakistan (Theisen-Jones and Bienefeld 2016).

Factors such as foraging ecology, daily behavioural patterns, blooming of flowers, flower shape (Fig. 21.3) and size, flower colour and odour, location and distribution of reproductive organs in flowers, availability of nectar and pollens and exploratory behaviour of honey bees contribute to pollination success. The fidelity of a bee to a plant species increases the efficiency of cross-pollination. Plants such as apples whose flowers have open corollas and accessible nectaries are preferred by bees as compared to plants whose flowers have narrow, deep corollas or more complex flowers such that nectaries are not easily accessible (Breed 2010).

The more time spent by honey bees foraging on flowers, the better their pollination efficiency. For example, pollination of large cardamom (*Amomum subulatum* Roxb.) is carried out by three native honey bees, i.e. *A. cerana, A. dorsata* and

Table 21.1 Select plants pollinated by honey bees

Honey bee	Plants	Country	Reference
Apis mellifera	*Malus domestica* L. (Apple[a])	Pakistan, China	Khan and Khan (2004)
	Helianthus annuus (Sunflower)	Pakistan	Abbasi et al. (2021)
	Psidium guajava L. (Guava)	India	Rajagopal et al. (2005)
	Nigella sativa L. (Black seed)	Pakistan	Munawar et al. (2009)
	Cuminum cyminum L. (Cumin)	India	Meena et al. (2018)
	Brassica juncea L. (Mustard)	India	Mandal et al. (2018)
	Brassica napus (Rapeseed)	Saudi Arabia	Khan and Ghramh (2021)
	Brassica rapa L. subspecies *chinensis*	India	Dhuria et al. (2022)
	Diospyros kaki K. (The Oriental persimmon)	East Asia	Nakamura et al. (2020)
	Paeonia ostii T. Hong et J.X. Zhang (Oil Tree Peony 'Feng Dan')	China	He et al. (2019)
	Dimocarpus longan (Longan) *Litchi chinensis* (Lychee) *Camellia sinensis* (Tea tree)	Taiwan	Lu (2018)
	Euphoria longan Stend (Longan) *Zea mays* L. (Corn) *Mimosa pigra* L. (Giant Mimosa) *Nephelium lappaceum* L. (Rambutan) *Sesamum indicum* L. (Sesame)	Thailand	Phankaew (2016)
	Mangifera indica L.	Thailand	Chuttong et al. (2022)
	Jatropha curcas L.	Thailand	Inson and Malaipan (2011)
Apis cerana	*Brassica juncia* (Mustard)	India	Stanley et al. (2017)
	Brassica rapa L. subspecies *chinensis*	India	Dhuria et al. (2022)
	Cashew *Psidium guajava* L. (Guava)	Indonesia	Göltenboth et al. (2006)
	Mangifera indica (Mango)	India	Göltenboth et al. (2006), Deuri et al. (2018)
	Capsicum annuum (Hot pepper)	Indonesia	Putra et al. (2014)
	Coriandrum sativum L. (Coriander)	India	Patil and Pastagia (2016)
	Amomum subulatum (Large cardamom)	Sikkim (India)	Partap et al. (2017)
	Fagopyrum esculentum M. (Buck wheat)	Nepal	Singh (2008)
	Raphanus sativus L. (Radish)	India	Sharma et al. (2016)
	Cocos nucifera (Coconut)	Indonesia	Göltenboth et al. (2006), Raffiudin et al. (2022)

(continued)

Table 21.1 (continued)

Honey bee	Plants	Country	Reference
	Brassica nigra L. (Black mustard)	India	Gautam et al. (2022)
	Jatropha curcas L.	Thailand	Inson and Malaipan (2011)
	Sesamum indicum (Sesame)	Sri Lanka	Ekanayake and Karunaratne (2022)
Apis dorsata	*Anisomeles indica* (Catmint); *Antigonon leptopus* (Coral vine); *Allophylus serratus* (Toothed-leaf Allophylus); *Helicteres isora* (Indian screw tree); *Hugonia mystax* (Climbing flax); *Mimosa pudica* (Humble plant); *Moringa oleifera* (Drumstick tree); *Peltophorum pterocarpum* (Copperpod); *Pongamia pinnata* (Indian beech); *Pterocarpus santalinus* (Red sandalwood); *Tamarandis indica* (Tamarind); *Tecoma stans* (Yellow trumpetbush); *Tectona grandis* (Teak); *Terminalia tomentosa* (Indian laurel); *Tribulus terrestris* (Gokharu or Gokshur); *Wrightia tinctoria* (Dyer's oleander)	India	Raju and Rao (2004)
	Cuminum cyminum L. (Cumin)	India	Meena et al. (2018)
	Coffea canephora L (Coffee)	India	Krishnan et al. (2012)
	Jatropha curcas L. (Purging nut/Physic nut)	Thailand	Inson and Malaipan (2011)
	Momordica charantia (Bitter gourd)	India	Balina et al. (2015)
Apis laboriosa	*Brassica nigra* L. (Black mustard)	India	Gautam et al. (2022)
Apis florea	*Cuminum cyminum* L. (Cumin)	India	Meena et al. (2018)
	Allium cepa L. (Onion)	India	Abrol (2015)
	Brassica juncea Czern & Coss (Raya); *Daucus carota L.* (Carrot)	India	Sihag (2021)
	Jatropha curcas L.	Thailand	Inson and Malaipan (2011)
	Mangifera indica (Mango)	India	Reddy et al. (2012)
Apis andreniformis	*Jatropha curcas* L.	Thailand	Inson and Malaipan (2011)
Apis nigrocincta	*Cocos nucifera* (Coconut)	Indonesia	Göltenboth et al. (2006), Raffiudin et al. (2022)

[a]Common names in English

Fig. 21.3 Honey bees foraging on large cardamom flower and collecting pollen and nectar. (**a**) Honey bee entering large cardamom flower, assisted due to its shape; (**b**) honey bee loaded with pollen in pollen basket

A. laboriosa in Sikkim, India (Sharma et al. 2019). Their foraging activity was the highest at 09:00–11:00 h than in early morning or evening. They entered the flower through stigma-anther column from either side and passed on the pollen to stigma during the movement. The distance of flowers from honey bee colony also matters as it impacts visitation frequency. In this case, the visitation frequency was higher in flowers that were nearer, i.e. 50 m away from the bee colony, than farther flowers, i.e. 250 m away. As a result, the pollination efficiency was also higher for flowers 50 m away from the bee colony than the farther ones. Another important factor was the availability of flowers. The cardamom flower blooms in the morning hours (05: 00–09:00 h), and the flower remains for only one day. Thus, the foraging time of honey bees coincided with the availability of flowers during day time.

Hence, pollination by honey bees increased fruit and seed set of large cardamom, i.e. increased its yield. Honey bees compete closely with bumble bees for pollination.

3.1 The European Honey Bee (*Apis mellifera*)

Apis mellifera, also known as hive bee or Western hive bee or occidental honey bee, is the cavity nesting and most widely used *managed* honey bee globally. It is a medium-sized honey bee as against 'giant honey bees'. *A. mellifera* was imported from North America, Europe and Oceania and was first introduced in Russia, in the late 1700s and later in the 1800s, it was introduced in India, Japan and Indonesia. By 1980s, *A. mellifera* was present in almost every Asian country. Currently, Asia manages the largest number of *A. mellifera* (Chantawannakul et al. 2015).

China is one of the largest honey producing countries in the world, Argentina and Mexico being the other two countries, and it utilizes services of *A. mellifera* (FAO 2008). It is a single species pollinator across the globe due to its wide availability, is economically valuable and is a good generalist pollinator but is not very effective. Although it is widely distributed in natural ecosystems, it visits only a minority of insect-pollinated plants (Hung et al. 2018).

A. mellifera shows different visitation patterns and pollination efficiencies for different plants. In some plants, it remembers the foraging areas and plants that it has visited and does not visit them again. But it has a habit of foraging plants of the same species and, thus, is consistent in pollination. However, in others, it visits the same flower multiple times (He et al. 2019). It can forage for several kilometres from the hive.

A. mellifera has established successfully bee in Asia and replaced *A. cerana* in many countries as it has better survival instinct and forage on a wider range them *A. cerana* (Koetz 2013). The colony size of *A. mellifera* is 30,000–50,000. *A. cerana* colony size is 2000–20,000. With a bigger colony size and bee size, *A. mellifera* gives higher productivity and profits. However, *A. cerana* is an excellent pollinator (Gratzer et al. 2019). On mango, *A. mellifera* starts foraging at 07:00 h and forages during sunlight. The highest foraging and pollination activity occurs between 15:00 and 17:00 hrs. However, *A. mellifera* spends less time per flower compared to *A. cerana* (Chuttong et al. 2022), and it is one of the reasons for its lower pollination efficiency.

3.2 Native Honey Bees

Asian honey bees are spread throughout the continent, but they are most notably observed in south east Asia, where they show greater diversity. They are of two groups, i.e. cavity nesting and open nesting honey bees.

3.2.1 Cavity Nesting Honey Bees

(i) *Apis cerana*

Of the four cavity-nesting species, *A. cerena* showed the largest distribution, ranging from Iran and Afghanistan in the west to Japan, China in the northeast and India, Sri Lanka, Indonesia, Philippines, in the south and southeast Asia, respectively (Chantawannakul et al. 2018). It has recently been introduced to Papua New Guinea. It builds nests within a cavity consisting of a series of parallel combs. This style is similar to that built by *A. mellifera* (FAO 2008). It is a docile species and has three major subspecies, i.e. *Apis cerana indica Apis cerana cerana* (found in mainland China, Afghanistan, east Asia up to Korea and Ussuria and Vietnam) and *Apis cerana japonica* (found only in Japan) (Chantawannakul et al. 2018).

A. cerana withstand extreme environmental conditions; for example, it can tolerate temperatures from 7 °C to 41.5 °C (Cui and Corlett 2020). It is more active during day than other species. It has a long flight duration and showed good grooming and hygienic behaviour, owing to which it is valued for high-quality by-products (Chen et al. 2022). Generally, *A. cerana* forages within a radius of 200–300 of their nest, but foraging of up to 1500–2500 m has also been observed (Koetz 2013).

In Asia, *A. cerana* is considered an excellent pollinator for a large number of fruits and vegetable crops and also pollinates nuts, oilseeds, pulses, coffee, spices, fibre and forage crops, cauliflower, okra and onion. It is considered a better pollinator than *A. mellifera. A. cerana* begins foraging quite early in the day and stops later during the day, and it employs bees for collecting pollens than those for collecting nectar. This behaviour contributes to its pollination success. Plants pollinated by *A. cerana* show better output in terms of increased quality and quantity of fruit and seed and reduced premature fruit drop (Koetz 2013).

Apis cerena pollinates plants such as *Brassica campestris* in fields. For effective pollination of one-hectare mustard field during peak flowering season, approximately 1500 bees or 2.2 colonies are required (Stanley et al. 2017). Partap et al. (2017) reported in their study that pollination of large cardamom by *A. cerana* increased the fruit set, seed set and capsule quality (fruit and seed weight) and, thus, increased the overall crop production. The foraging habits of *A. cerana* account for their effective pollination in large cardamom. They forage all day, visiting all flowers on a panicle before moving to the next panicle, visiting the same flower many times, approaching anther-stigma columns from both sides of flowers, passing over anthers and stigma and in the process brushing against anthers to collect pollen and also touching the stigma simultaneously.

(ii) *Apis koschevnikovi (The Red Honey Bee)*

Apis koscheynikovi is distributed in Indonesia, peninsular Malaysia, Sumatra, Java and Borneo (Chantawannakul 2022). The bee is slightly bigger than *A. cerana*. It builds nests similar in size and construction as *A. cerana* (FAO 2008). *A. koscheynikovi* is found at altitudes from sea level to 1600 m, but recently its range is on the decline as it is either poorly represented or not available in regions where it was previously recorded (Hadisoesilo et al. 2008). Koeniger et al. (2017) reported *A. koscheynikovi* gnawing and damaging open flowers of an orchid, *Dipodium* sp. in Malaysia. However, there are no records of their role in pollination.

(iii) *Apis nuluensis (The Mountain Honey Bee) and Apis nigrocincta (The Sulawesi Honey Bee)*

Apis nuluensis is confined to three islands, i.e. Sagihe, Sulawesi and Mindanao (Chantawannakul 2022) and Mount Kinabalu in Borneo (Guerin 2020) while *Apis nigrocincta* is distributed in Indonesia and Philippines (Guerin 2020). The nests of these bees are similar to *A. cerana* and *A. koscheynikovi* (FAO 2008). These bees have not been domesticated yet, and not much information is available about their contribution to pollination. Information on behaviour is also lacking.

3.2.2 Open Nesting Honey Bees

(i) *The Giant Honey Bees*

Apis dorsata and *Apis laboriosa* are two giant honey bees. While the former is frequently found in the Indian subcontinent and Southeast Asia, the latter is restricted to the Himalayas and mountainous regions in Myanmar, northern Laos and Vietnam.

Apis dorsata (Giant Honey Bee)

Apis dorsata, also known as the Rock bee, is the second largest honey bee and is an important pollinator of wild flowering plants and cultivated crops, in southeast Asia and the Indian subcontinent. Its distribution spreads from Afghanistan in the west to Bali in the east and in the north it is limited to the Himalayas. It makes single but large nests often suspended from a cliff, branch, or building (FAO 2008).

The visitation rate of *A. dorsata* is the highest during mid-day (11.30–14.30 hrs) as recorded by Gupta et al. (2020), although it starts foraging early morning. It tolerates temperatures between 14 °C and 41.5 °C (Cui and Corlett 2020). During the dry season, it migrates to hills in search of flowers and to escape from high temperatures. The Rock bee has never been domesticated. It collects nectar from flowers (floral reward) during daytime/sunlight hours. However, when daytime temperatures are high and unbearable, it also forages during night under moonlight illumination (Raju and Rao 2004; Young et al. 2021; Gautam et al. 2022). It often visits flowers with a shallow base and are easily accessible for floral rewards. In most of the plants visited by the Rock bee, sex organs are not concealed, which facilitates pollination. In addition to the conducive flower shape, the bee is attracted to flowers because of visual (colour of the flowers) and chemical (odour) cues. The rock bee often visits white and yellow flowers (Raju and Rao 2004).

In plants such as *T. indica*, *Anisomeles indica* (L.) Kuntze, *Helicteres isora* L. and *Tectona grandis* L., rock bee collects negligible nectar, but it visits the flowers of these trees due to exploratory nature and, hence, contributes to pollination. It visits flowers when they are in the upright position, collects floral rewards sternotribically i.e., collects pollens on its underside. The plants pollinated by *A. dorsata* are both self-pollinating as well as cross-pollinating. *A. dorsata* is a more potent pollinator compared to *A. florea* and *A. cerana indica* (Raju and Rao 2004).

Apis laboriosa (Himalayan Giant Honey Bee)

Apis laboriosa is the largest honey bee and is an important pollinator at higher altitudes (500–3500 m a.s.l.) of the Himalayas, higher than *A. dorsata*. It is restricted to the Himalayas and neighbouring mountain ranges in Asia. Its presence is recorded in Nepal, China, Bhutan, India, Vietnam, Myanmar and Laos (Kitnya et al. 2020). *A. laboriosa* is closely related to *A. dorsata* as both have similar genome composition, but number of genes are higher in *A. laboriosa* than in *A. dorsata* and the former shows the presence of more transposable elements (Lin et al. 2021). It builds a large open nest with a single comb, but these combs lie in clusters and sometimes 100 combs are suspended from a cliff close to each other (FAO 2008).

Apis laboriosa is an aggressive bee and showed comparatively less pollination efficiency. Its pollination efficiency for black mustard is lower than *A. cerana*. The lower pollination efficiency is due to less foraging time and range, as it starts foraging late during morning and stops early in the evening. Due to its aggressive nature and bigger size, *A. laboriosa* easily displaces *A. cerana* from its preoccupied flower; however, the latter is unable to do so. *A. laboriosa* stays for a long time in a flower, and due to its large size, it can carry a greater pollen load to nests (Kitnya et al. 2020; Gautam et al. 2022). The foraging duration and timings of *A. laboriosa* and *A. cerana*, when co-existing, differ. *A. cerana* forages for a longer duration (9 h) than *A. laboriosa* (8 h). *A. cerana* starts foraging early (08:00 to 08:30 hrs) and peaks at 11:00–11:30 h and stops foraging by 17:30–18:00 h. However, *A. laboriosa* starts its activity at 08:30–09:00 h, peaks at 13:00–13:30 h and stops at 16:30 h (Gautam et al. 2022).

(ii) *The Dwarf Honey bees*

Both *Apis florea* and *Apis andreniformis* are present in Southeast Asia, but *A. florea* has an extensive range, including the Indian subcontinent and the eastern region of the Arabian peninsula (Guerin 2020).

Apis florea (The Red Dwarf Honey Bee)

A. florea is a small honey bee, native to Oman and distributed in Sudan, Iraq, southeast Asia, Philippines and Indonesia (FAO 2008; Guerin 2020). It is found inhabiting steppes, semi-deserts, savannahs and rainforests, semi-arid and tropical and sub-tropical regions (Sihag 2021). It builds small nests in the open air (20–30 cm in dia) on branches of small trees and shrubs. It generally does not remain in the same location but moves with the sources of flowers. It is hunted in the wild by honey hunters. The bees are extremely quick in making nests and start producing honey on the eighth day of making their nests (Abrol 2020). Bees can tolerate temperatures between 16 °C and 46.5 °C (Cui and Corlett 2020), but their activity declines after 32 °C (Reddy et al. 2012). The visitation rate of *A. florea* is the highest during midday (11.30–14.30 hrs) (Gupta et al. 2020). Flowers with higher corolla diameter are preferred by bees for nectar collection, and *A. florea* effectively pollinates all kinds of flowers in the wild and in agroecosystems (Abrol 2020).

Apis andreniformis (The Black Dwarf Honey Bee)

Apis andreniformis was recently separated from the closely related species *A. florea* (Suwannapong et al. 2011). It is a small-sized honey bee, smaller than *A. florea*, and builds small nests (150–200 cm in diameter). It builds a single comb-nest. It is present in Philippines, Southeast Asia, Borneo and the southern Chinese peninsula (FAO 2008; Guerin 2020). In the natural environment It can tolerate temperatures between 25 °C and 41.5 °C (Cui and Corlett 2020). Little information exists for available for this species on domestication or pollination.

4 Future Scope of Pollination

The consistent rise in the global human population is putting immense pressure on the production of agricultural crops and pollinators contribute significantly to crop production as 87.5% of flowering plants are pollinated by animals (Ollerton et al. 2011). Approximately 30% of the global food is obtained from plants pollinated by animals, and approximately one-third of the total human food supply is contributed by bee-pollinated crops (Khalifa et al. 2021). However, between 1989 and 2019, there has been a global decline in the honey bee pollination services, i.e. in 1989, the global honey bee stock catered to 56.9% global pollination demand, which decreased to 44.1% in 2019 (Mashilingi et al. 2022).

The agricultural yields can be amplified using abundant and diverse honey bees. Hence, the presence of a rich and diverse honey bee population is an asset for farmers in Southeast Asia. While some crops can be pollinated by more than one species of honey bee ('generalized pollination'), others depend on specific species of honey bee for pollination ('specialized pollination'). For example, the flower of dragon fruit (*Hylocereus undatus*) blooms at night and is off-white. Its flowers are effectively pollinated in the early hours by *A. cerana*, *A. dorsata* and *A. florea* (Pushpakumara et al. 2005). However, flowers of the orchid *Cymbidium dayanum* are exclusively pollinated by *Apis cerana japonica* (Matsuda and Sugiura 2019).

The distribution of honey bees in different countries of the Asian continent varies due to different climatic conditions and vegetation patterns. While some countries manage honey bees on a large scale, others simply depend on the hunting of wild honey bees. China is one of the major providers of pollination services and the major honey producer in the world. The climatic conditions of China are cold, temperate, subtropical and tropical. So, it is resourceful in nectar plants and is home to three native species of honey bees, i.e. *A. dorsata*, *A. florea*, and *A. cerana*, and the European honey bee *A. mellifera*. Additionally, *A. andreniformis* co-exist with the native species in Yunnan (Feng-Feng 1990). The population of managed honey bees in China far exceeds than those in other countries, and bees are primarily utilized for honey production. Pollination services by honey bees are required for the cultivation of economic crops as apples, kiwi, pears, cherries and strawberries. Over the years, the planting area of these crops has increased, owing to high demand, but the pollination services are insufficient, as the bee pollination service industry is still not well developed in many regions of China and lags significantly compared to these services in other developed countries (Zhang et al. 2022). In China, for cross-pollination, farmers use artificial pollination or hand pollination system. They hire labourers for hand pollination rather than bees as they are not confident using bees for pollination. However, the labour cost of artificial pollination is higher than bee pollination, and hand pollination results are less satisfactory as the quality of fruits is affected (Zhang et al. 2022).

A recent study by Mashilingi et al. (2021), however, reported that in China, between 1961 and 2018, there was a rapid increase in dependence on insect pollinators for cultivating crops. The yield of pollinator-dependent crops increased faster than other crops. Growth in crop yield, for pollinator dependent crops, was unstable due to the insufficient availability of pollinators. In 2018, the requirement for honey bee colonies was thrice the number of available colonies. Thus, there is still a shortage of honey bees for pollination services in China.

Mongolia has harsh climatic conditions but is rich insect fauna. Several races or subspecies of European honey bee, i.e. *A. mellifera caucasia*, *A. mellifera mellifera* and *A. mellifera carpatica* (the Russian Far East bee), and Haliun bee line (local bee), have been introduced and managed. These bees are good pollinators in Mongolia (Tsevegmid et al. 2018). In Japan, beekeeping is not a big industry for bee products, but it forms a large agricultural sector due to its role in pollination (Yoshiyama and Kimura 2018). Farmers utilize honey bee pollination services for pollinating plants such as Oriental persimmon (Kamo et al. 2022). Native beekeeping was on a decline until 2005, and pollination services were negatively impacted (Donkersley et al. 2021). The two major managed honey bees in Korea include *A. cerana* F. in mountains and *A. mellifera* at the interface of mountains and open land (Jung and Lee 2018). Urban beekeeping in South Korea is encouraging, and the beekeeping industry has expanded (Kohsaka et al. 2017).

In Taiwan, *A. cerana* and *A. mellifera* are the major managed bee species and beekeeping is a well-managed and large industry (Lu 2018). Vietnam has rich honey bee diversity, and six honey bee species are reported from the country, which include *A. mellifera*, *A. cerana*, *A. laboriosa*, *A. andreniformis*, *A. dorsata* and *A. florea*. Majorly *A. mellifera*, *A. cerana* and *A. dorsata* are managed (Thai and Toan 2018). Thailand has four native honey bee species, i.e. *A. cerana*, *A. dorsata*, *A. andreniformis* and *A. florea*, while *A. mellifera* was introduced for beekeeping (Suwannapong et al. 2011). Only *A. cerana* and *A. mellifera* are managed by beekeepers (Chantawannakul et al. 2018); however, over time, *A. mellifera* has drastically replaced *A. cerana* with beekeeping with latter decreasing by 95% (Gratzer et al. 2019). The bee pollination industry is not well developed as the owners of orchards are not convinced that pollination by bees increase crop yields. Farmers believe that honey bees cause falling of the longan flowers so they don't allow beekeepers to release honey bees in orchards (Phankaew 2016).

Indonesia has a tropical climate and diverse vegetation holding rich honey bee diversity. Eight *Apis* species are reported from Indonesia, with *A. laboriosa* being absent. People in Indonesia use honey bees for products, primarily honey, and have yet not developed modern methods of beekeeping on a large scale though they manage *A. cerana* locally using traditional methods (Gratzer et al. 2019) and collect bee products by hunting *A. dorsata*. Like in Thailand, in Indonesia also, *A. mellifera* has brought down population of *A. cerana* by 45–60% over the years (Theisen-Jones and Bienefeld 2016). The contribution of *A. mellifera* is slowly decreasing due to global warming, pests and changing food sources (Kahono et al. 2018). Farmers domesticated four species, i.e. *A. cerana*, *A. mellifera*, *Apis koschevnikovi* and *A. nigrocincta*, albeit using traditional methods (Buchori et al. 2022). In the absence of modern methods, large-scale management of honey bees is not possible; hence, the future of beekeeping and using them for pollination is not bright in Indonesia.

India is home to four species of honey bees, i.e. *A. cerana*, *A. dorsata*, *A. florea* and *A. mellifera*. While *A. cerana* and *A. mellifera* are domesticated, the other two are wild. All of them have become an integral component of Indian agriculture and utilized for honey production, and pollination (Reddy et al. 2012). Several plants are pollinated by honey bees in India (Table 21.1), and farmers are rearing them for pollination services. Nepal harbours four native species of honey bees, i.e. *A. cerana*, *A. dorsata*, *A. florea* and *A. laboriosa*, and one non-native species, i.e. *A. mellifera*. *A. laboriosa* is a cliff-nesting species, and its population in on decline due to over-harvesting and deforestation. Although most of the pollination is done for wild plants, agricultural crops are pollinated by bees (Thapa et al. 2018). Bhutan has five native bee species, i.e. *A. dorsata*, *A. laboriosa*, *A. cerana*, *A. florea*, *A. andreniformis*, and one non-native species, i.e. *A. mellifera*, imported from India in 1986 (Nidup and Dorji 2016).

The native honey bee species in Pakistan include *A. cerana*, *A. florea* and *A. dorsata*. *Apis cerana* is managed primarily for honey production, but their role in pollination is being explored. Additionally, *A. mellifera* is introduced in Pakistan, but experts fear a threat to local biodiversity (Tan et al. 2021; Chantawannakul et al. 2016). Sri Lanka has three honey bee species, i.e. *A cerana*, *A. florea* and *A. dorsata* (Punchihewa 1994). There is a lack of research on honey bees in pollination or honey production in Sri Lanka. Since the country has a tropical climate and is rich in fauna and flora, there is scope for studies and identifying the potential of honey bees. Afghanistan is home to two native species of honey bee, i.e. *A. cerana* and *A. dorsata* while *A. mellifera* is non-native. Much of the work on honey bees was done in the late 1980s (Woyke 1984, 1985, 1987), and there is not much information available after that now.

The arid dry climate of Israel is not suitable for the survival of honey bees. However, the central and northern regions is inhabited by *A. mellifera*, i.e. *A. mellifera syriaca* and *A. mellifera ligustica*. These regions are urbanized and populated, i.e. more than 14 colonies per square kilometer (Soroker et al. 2018). Turkey is the second largest producer of honey in the world, after China, and *A. mellifera* is the first domesticated honey bee. There are more than 8 million beehives, and the country is third in the number of beehives after India and mainland China (Mayack and Hakanoğlu 2022).

5 Challenges for Beekeeping and Conservation

In recent decades, the population of honey bees is gradually declining in the world over (Theisen-Jones and Bienefeld 2016). Studies reported a decline in honey bee abundance in Asian countries such as Nepal (Partap 2010), India (Sihag 2021), China (Yang 2005; Liu et al. 2020), Japan (Kohsaka et al. 2017; Donkersley et al. 2021), etc.

Several studies have identified factors for the cause of their decline. These factors include parasites, pathogens, shortage of food, use of pesticides in fields, changes in land use and climate change as major causes of this decline. The factors causing honey bee decline vary among countries, due to differential geographical and climatic conditions, and beekeepers often identify these factors quickly (Fig. 21.4). The use of the insecticide ‘neonicotinoid’ has emerged as an important factor in the decline of honey bees, both wild and managed (Goulson 2013). The lack of diversity and availability of pollen and nectar add up to their declining numbers (Smith et al. 2013). This decline not only is a threat to agricultural production but also poses serious problems as the loss of important plant species, loss of diversity of honey bees and disturbances in natural ecosystems (Murray et al. 2009).

One of the major causes of honey bee decline is infection by pathogens, such as parasitic mites, viruses, bacteria, etc. (Table 21.2). Infections are responsible for

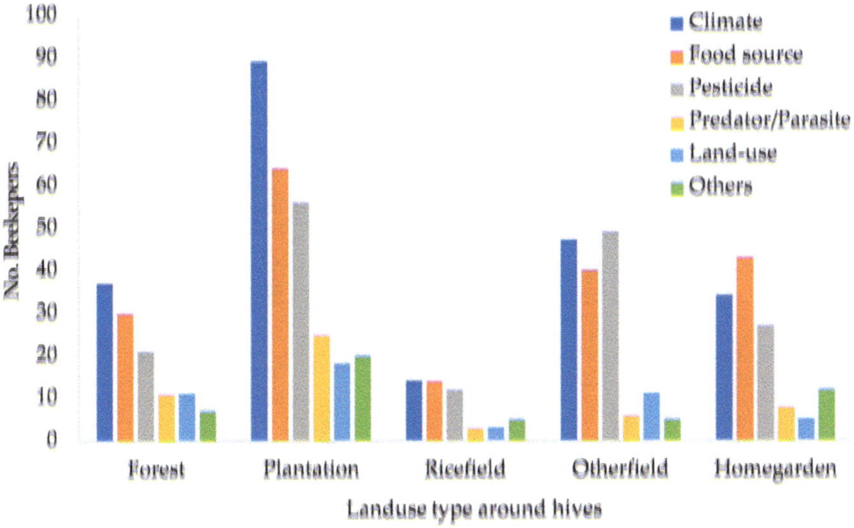

Fig. 21.4 Opinion of farmers on factors causing honey bee decline in Indonesia, the country hosting the largest honey bee diversity in Asia.

colony loss, leading to reduced pollination and loss of crop yield. The decline in honey bee populations is explicitly observed in colony collapse disorder (CCD) where there is unexplained and rapid loss of adult working bees in the colony and the colony is left with only the queen bee and few nursing bees (van Engelsdorp et al. 2009). *Varroa* spp. and *Tropilaelaps* spp. are the two major parasitic mites that are a cause of concern for beekeepers in Asia. The susceptibility of bees to these pathogens varies in different species. For example, *A. cerana* has higher resistance against most bee diseases, especially parasitic mites, but it is more sensitive to the Thai Sacbrood virus, resulting in larval deaths (Theisen-Jones and Bienefeld 2016).

The distribution and epidemiology of diseases affecting honey bee vary in different Asian countries. For example, the common diseases in Vietnam are Nosema (caused by microsporidian *Nosema ceranae*), Sacbrood disease (caused by Sacbrood virus-SBV); European foulbrood disease (EFB) (caused by the bacterium *Melissocccus plutonius*), and diseases caused by parasitic mites, *Varroa destructor* (Anderson and Trueman) and *Tropilaelaps mercedesae* Anderson and Morgan (Thai and Toan 2018). *Tropilaelaps* mites are originally parasites of Asian honey bees, but they have now become a threat to *A. mellifera* also (Gao et al. 2021). They parasitize on honey bee brood and feed on the larvae and pupae. This results in the death of bees and malformation of brood resulting in decline of colonies. The symptoms of the infection include a shorter life span, reduced body weight and wings and legs becoming deformed and shrunken (Phankaew 2016). *Varroa destructor* is eyeless and wingless and is transported to the bee colony by honey bees themselves when they visit other poorly guarded or collapsing bee colonies infected with *Varroa*. The availability of multiple bee colonies, such as in apiaries, is

Table 21.2 Pathogens and parasites of honey bees

Pathogen/Parasite	Disease	Distribution	Affected honey bee species	First Report	Reference
Bacterial					
Melissocccus plutonius	European foulbrood disease	Worldwide	*A. mellifera, A. cerana, A. laboriosa*	UK (1885)	Thai and Toan (2018), Forsgren (2010)
Paenibacillus larvae	American foulbrood disease	Nearly global	*A. mellifera, A. cerana*	US (Early 1900s)	Lu (2018), Gratzer et al. (2019), De Graaf et al. (2006), Hristov et al. (2021)
Viral					
Acute bee paralysis virus (ABPV)	–	Worldwide	*A. mellifera, A. cerana*	1963	Chanpanitkitchote et al. (2018), Bailey et al. (1963)
Deformed wing virus (DWV)	–	Worldwide	*A. mellifera, A. cerana*	UK (1991)	Chanpanitkitchote et al. (2018), Bailey and Ball (1991)
Thai sacbrood virus	Thai sacbrood disease	Asia	*A. cerana* (Thailand, India); *A. florea and A. dorsata* (India)	Thailand (1976)	Oldroyd and Nanork (2009), Theisen-Jones and Bienefeld (2016), Bailey and Collins (1982)
Sacbrood virus	Sacbrood disease	Worldwide	*A. cerana*	USA (1913)	Oldroyd and Nanork (2009), Theisen-Jones and Bienefeld (2016), White (1913)
Microsporidian					
Nosema ceranae	Nosema	Originated in southeast Asia, but now distributed worldwide	*A. mellifera, A. cerana*	Spain (2005)	National Bee Board, Government of India (n.d.), Higes et al. (2006), Grupe and Quandt (2020)

Fungal					
Ascosphaera spp. (*A. apis, A. tubingensis*)	Chalkbrood disease	Worldwide	*A. mellifera,*	Germany (Early 1900s)	Lu (2018), Gratzer et al. (2019), Cheng et al. (2022), Maassen (1913)
Parasitic mites					
Varroa spp. (Four species-*V. destructor*, *V. jacobsoni, V. underwoodi* and *V. rindereri*)	Varroosis / Varroatosis	Worldwide	*A. cerana* (Native host), *A. mellifera*	Java (1904)	Soroker et al. (2018), Traynor et al. (2020), Gao et al. (2021)
Tropilaelaps spp. (Four species—*T. koenigerum, T. clareae*, *T. thaii* and *T. mercedesae*)	-	Asia	*A. dorsata, A. laboriosa, A. mellifera,*	Philippines (1961)	Chanpanitkitchote et al. (2018), Gao et al. (2021), Aydin (2022)

a boon for the dispersal of *Varroa* mites. The foundress mite, carried by a foraging honey bee, when it lands in a fresh bee colony, settles at the bottom of the cell and feeds on the available food. When honey bee larvae start covering themselves in cocoon, i.e. at prepupae stage, foundress mite attacks it and punctures a hole in it to feed itself and its future young ones. Reproduction of *Varroa* mites is by parthenogenesis and is dependent on signals from larvae. Initially, mites attach themselves with nursing bees, but when the number of *Varroa* mites increases, they associate themselves with worker bees. When a parasitized worker bee leaves the colony, it disperses the mite (Traynor et al. 2020). *Varroa destructor* is a common ectoparasite of *A. cerana*. However, it has switched its host and now also infects *A. mellifera*. While *A. cerana* can survive *Varroa* mites on its own in the wild, *A. mellifera* requires human intervention for survival (Pattemore 2017). Similarly, the ectoparasite *Tropilaelaps mercedesae* also shifted from original host *A. dorsata* to *A. mellifera* (Chantawannakul et al. 2018).

Lately, bee viruses have also become a cause of concern for beekeepers. Many viruses are implicated in diseases affecting honey bees and thereby in their declining numbers. Viruses that infect bees are positive-sense single-stranded RNA viruses belonging to the order Picornavirales. A total of 23 viruses belonging to two families, viz. *Dicistroviridae* which includes Acute bee paralysis virus (ABPV), Israeli acute paralysis virus (IAPV), Kashmir bee virus (KBV) and Black queen cell virus (BQCV); and *Iflaviridae* which includes Deformed wing virus (DWV), Kakugo virus (KV), Sacbrood virus (SBV) and Slow bee paralysis virus (SBPV) are known to infect honey bees (McMenamin and Flenniken 2018). Sacbrood virus was the first virus that was described to be infecting honey bees and is distributed worldwide (White 1913).

DWV was earlier known to infect *A. mellifera* but there are evidences that it has made *A. cerana* its host recently, and it also infects the parasite *Tropilaelaps mercedesae*. Thus, it is notorious as it skips hosts. Similarly, ABPV also infects *A. cerana* and *Tropilaelaps mercedesae*. The phylogenetic analysis proves that DWV moved from *A. cerana* to *A. mellifera* and *T. mercedesae*, while ABPV moved from *T. mercedesae* to *A. mellifera* first and then to *A. cerana* (Chanpanitkitchote et al. 2018). Thus, *Varroa* and *Tropilaelaps* mites act as vectors for both viruses, DWV and ABPV.

In Thailand, parasitic mites have spread widely among honey bee colonies and are a cause of concern. Honey bees are also infected by the microbe *N. ceranae* (Chantawannakul et al. 2018). In Israel, *Varroa* remains the major threat to beekeepers (Soroker et al. 2018). *Varroa destructor*, *Nosema ceranae* and four viral infections affect honey bees in Mongolia. The major stress here is the climate variability, i.e. extreme cold conditions, and to add to this is the risk of predators such as bears (Tsevegmid et al. 2018). In Korea, honey bees are threatened by parasitic mites *Varroa* and *Tropilaelaps* (which infect *A. mellifera*) and Sacbrood virus (which infects *A. cerana*). Pests as yellow legged hornet (*Vespa velutina*) and small hive beetle (*Aethina tumida*) pose newer threats to beekeeping in Korea (Jung and Lee 2018). Honey bees in Taiwan are under threat of several diseases such as Nosema disease, American foulbrood disease (caused by the bacteria *Paenibacillus larvae*), European foulbrood disease, chalkbrood disease (caused by the fungus *Ascosphaera apis*), *Varroa* mite and bee viruses (Lu 2018; Gratzer et al. 2019). In Nepal, *V. underwoodi* and *V. rindereri* infect honey bees (Chantawannakul et al. 2016). *Euvarroa* mites are also pathogenic to honey bees. *Euvarroa sinhai* was first documented in India, infecting *A. florea* and *Euvarroa wongsirii* (Lekprayoon and Tangkanasing) from Thailand infecting *A. andreniformis*. *Euvarroa* spp. is documented in Malaysia, Sri Lanka, Iran and Borneo (Chantawannakul et al. 2016).

In Turkey, honey bees are affected by *Varroa* mites, stonebrood disease (caused by *Aspergillus* spp.), chalkbrood disease, American foulbrood disease, European foulbrood disease, Nosema and viruses (ABPV, BQCV, DWV, KV, and VDV1) (Mayack and Hakanoğlu 2022).

Beekeepers in India are threatened by pathogens and parasites infecting honey bees, as the Greater wax moth (*Galleria mellonella* (Linnaeus)) a major pest of *A. cerana*, the lesser wax moth (*Achroia grisella* Fabricius), ants, wasps and hornets. Bees are also infected by parasitic mites. For example, *Varroa destructor* (attacks *A. mellifera*), *Tropilaelaps* spp. (attacks *A. dorsata* is its native host, and *A. mellifera*) and Tracheal mite (*Acarapis woodi* (Rennie)) (attacks *A. mellifera* a native host, and *A. cerana*) and *Euvarroa sinhai* Delfinado and Baker (attacks *A. florea*, is its native host). Honey bees suffer from many viral and bacterial diseases. The common bee diseases include American foulbrood disease (attacks *A. mellifera*- most destructive disease), European foulbrood disease (reported from India in 1970; attacks *A. mellifera*), Chalkbrood disease, Sacbrood disease (attacks *A. mellifera*), Thai sacbrood disease (attacks *A. cerana*) and Nosema disease (National Bee Board, Government of India n.d.). A steep decline (almost 25% reduction) in the population of *A. florea* was recorded in northwest India, and this decline was not due to pathogens or parasites, loss of nesting area or dearth of flowers, but due to large-scale by pesticides (Sihag 2021). Thus, Asian beekeepers are under constant threat from various pathogens affecting bee colonies and populations resulting in economic losses.

6 Conclusion

China, India, Indonesia, Turkey, Taiwan, Vietnam and Thailand have progressed considerably in beekeeping, and farmers are using them for pollination. However, in other Asian countries, their potential has yet not been tapped as farmers either do wild hunting or use traditional methods of beekeeping. Although most of the southeast and south Asian countries have rich fauna and flora, many countries have yet not started practicing modern methods of beekeeping to manage bees on commercial scale. There is a lot of scope in this area. Many countries that practice beekeeping on large scale are facing challenges resulting in the loss of honey bee colonies. Studies have reported that the populations of different species of honey bees are on a decline owing to several factors, and cross-pollination services are hampered in many regions of the tropical world.

In future, the area of pollinator-dependent crops may increase and to meet the ever-increasing food demand, there would be a need for more pollinators. The pollination crisis has been recognized globally, and efforts are being made to address pollination management and conservation of honey bees. Encouraging beekeeping by farmers is one step adopted to increase abundance. Beekeeping is practiced for species native and abundant in a region and can be easily domesticated. The use of pesticides must be reduced in agroecosystems, which directly or indirectly affects honey bee diversity and abundance. Further, measures to contain pathogens and parasites must be adopted such that beekeeping's thrive without losses. There is a need to conserve environment so that the richness of fauna and flora is preserved, thereby providing enough food, foraging and nesting space for honey bees. Environment conservation is the most important task at hand as the global climate is rapidly changing. There is paucity of scientific studies to explore the potential of managing native honey bees for pollination.

Acknowledgements Authors are thankful to their respective institutions for the necessary support.

References

Abbasi KH, Jamal M, Ahmad S, Ghramh HA, Khanum S, Khan KA, Ullah MA, Aljedani DM, Zulfiqar B (2021) Standardization of managed honey bee (*Apis mellifera*) hives for pollination of sunflower (*Helianthus annuus*) crop. J King Saud Univ-Sci 33(8):101608. https://doi.org/10.1016/j.jksus.2021.101608

Abrol DP (2015) Foraging behaviour of *Apis florea* F., an important pollinator of *Allium cepa* L. J Apicult Res 49(4):318–325. https://doi.org/10.3896/IBRA.1.49.4.04

Abrol DP (2020) The future role of dwarf honey bees in natural and agricultural systems. CRC Press, Boca Raton. https://doi.org/10.1201/9781003033936

Aydin L (2022) *Aethina tumida* (small hive beetle; SHB) and *Tropilaelaps* spp. Mite; an emerging threat to Turkey honey bees. Ankara Univ Vet Fak Derg 69:347–354

Bailey L, Ball BV (1991) Honey bee pathology, 2nd edn. Academic, London

Bailey L, Collins MD (1982) Taxonomic studies on '*Streptococcus pluton*'. J Appl Bacteriol 53: 209–213

Bailey L, Gibbs AJ, Woods RD (1963) Two viruses from adult honey bees (*Apis mellifera* Linnaeus). Virology 21:390–395

Balina PK, Sharma SK, Rana MK (2015) Diversity, abundance and pollination efficiency of native bee pollinators of bitter gourd (*Momordica charantia* L.) in India. J Apicult Res 51(3):227–231. https://doi.org/10.3896/IBRA.1.51.3.02

Bawa KS (1990) Plant-pollinator interactions in tropical rain forests. Annu Rev Ecol Syst 21(1): 399–422

Breed MD (2010) Honey bees. Encycl Anim Behav 2010:89–96. https://doi.org/10.1016/B978-0-08-045337-8.00044-9

Buchori D, Rizali A, Priawandiputra W, Raffiudin R, Sartiami D, Pujiastuti Y, Jauharlina PMG, Meilin A, Leatemia JA, Sudiarta IP, Rustam R, Nelly N, Lestari P, Syahputra E, Hasriyanti WJF, Daud IDA, Hariani N, Jihadi A, Johannis M (2022) Beekeeping and managed bee diversity in Indonesia: perspective and preference of beekeepers. Diversity 14(1):52. https://doi.org/10.3390/d14010052

Chanpanitkitchote P, Chen Y, Evans JD, Li W, Li J, Hamilton M, Chantawannakul P (2018) Acute bee paralysis virus occurs in the Asian honey bee *Apis cerana* and parasitic mite *Tropilaelaps mercedesae*. J Invertebr Pathol 151:131–136. https://doi.org/10.1016/j.jip.2017.11.009

Chantawannakul P (2022) Conservation and management of Honey bees in Southeast Asia. In: Imperiled: the encyclopedia of conservation, pp 538–548. https://doi.org/10.1016/B978-0-12-821139-7.00038-6

Chantawannakul P, de Guzman LI, Li J, Williams GR (2015) Parasites, pathogens, and pests of honey bees in Asia. Apidologie 47:301–324. https://doi.org/10.1007/s13592-015-0407-5

Chantawannakul P, De Guzman LI, Li J, Williams GR (2016) Parasites, pathogens, and pests of honey bees in Asia. Apidologie 47:301–324. https://doi.org/10.1007/s13592-015-0407-5

Chantawannakul P, Williams G, Neumann P (2018) Asian beekeeping in the 21st century. Springer, Singapore. https://link.springer.com/book/10.1007/978-981-10-8222-1

Chen S, Wang Y, Li Y, Zhang X, Wu J (2022) Effects of honey bee (*Apis cerana*) visiting behaviour on toxic plant (*Tripterygium hypoglaucum*) reproduction. AoB Plants 14(3): plac002. https://doi.org/10.1093/aobpla/plac002

Cheng X, Zhang L, Luo J, Yang S, Deng Y, Li J, Hou C (2022) Two pathogenic fungi isolated from chalkbrood samples and honey bee viruses they carried. Front Microbiol 13:843842. https://doi.org/10.3389/fmicb.2022.843842

Chuttong B, Panyaraksa L, Tiyayon C, Kumpoun W, Chantrasri P, Lertlakkanawat P, Jung C, Burgett M (2022) Foraging behavior and pollination efficiency of honey bees (*Apis mellifera* L.) and stingless bees (*Tetragonula laeviceps* species complex) on mango (*Mangifera indica* L., cv. Nam Dokmai) in Northern Thailand. J Ecol Environ 46:15. https://doi.org/10.5141/jee.22.012

Cui Q, Corlett RT (2020) Seasonal and diurnal patterns of activity in honey bees (*Apis* spp.) on the northern edge of the Asian tropics; their implications for the climate-change resilience of pollination. Trop Conserv Sci 9(3). https://doi.org/10.1177/1940082916667142

De Graaf DC, Allipi AM, Brown M, Evans JD, Feldlaufer M, Gregor CA, Hornitzky M, Pernal SF, Schuch DMT, Titěra D, Tomkies V, Ritter W (2006) Diagnosis of American foulbrood in honey bees: a synthesis and proposed analytical protocols. Appl Microbiol 43(6):583–590. https://doi.org/10.1111/j.1472-765X.2006.02057.x

Deuri A, Rahman A, Gogoi J, Borah P, Bathari M (2018) Pollinator diversity and effect of *Apis cerana* F. pollination on yield of mango (*Mangifera indica* L.). J Entomol Zool Stud 6:957–961

Dhuria A, Thakur RK, Rani P, Rana P (2022) Pollinator diversity and foraging behavior of *Apis mellifera* Linnaeus and *Apis cerana* Fabricius on Chinese cabbage (*Brassica rapa* Linnaeus subspecies *chinensis*) grown under different farming systems. Emerg Life Sci Res 8(1): 165–174. https://doi.org/10.31783/elsr.2022.81165174

Donkersley P, Covell L, Ota T (2021) Japanese honeybees (*Apis cerana japonica* Radoszkowski, 1877) may be resilient to land use change. Insects 12(8):685. https://doi.org/10.3390/insects12080685

Ekanayake EMD, Karunaratne WAIP (2022) Native Sri Lankan bees enhance pollination of sesame, *Sesamum indicum*. Ceylon J Sci 51(1):29–36. https://doi.org/10.4038/cjs.v51i1.7976

FAO (2008). Bee species description. https://www.fao.org/3/i0842e/i0842e03.pdf

Forsgren E (2010) European foulbrood in honey bees. J Invertebr Pathol 103(2010):S5–S9. https://doi.org/10.1016/j.jip.2009.06.016

Gao J, Ma S, Wang X, Yang Y, Luo Q, Wang X, Liu F, Wang Q, Fu Z, Diao Q, Dai P (2021) *Tropilaelaps mercedesae* parasitism changes behavior and gene expression in honey bee workers. PLoS Pathog 17(7):e1009684. https://doi.org/10.1371/journal.ppat.1009684

Gautam RK, Shuyi G, Uniyal VP (2022) Comparative foraging behaviour and pollination efficiency of *Apis laboriosa* S. and *Apis cerana* F. on black mustard (*Brassica nigra* L.) in Western Himalaya, India. Curr Sci 122(7):840–846

Göltenboth F, Langenberger G, Widmann P (2006) Ecology of rice fields and other land use systems. In: Ecology of insular Southeast Asia, pp 417–463. https://doi.org/10.1016/B978-044452739-4/50020-6

Goulson D (2013) Neonicotinoids and bees: what's all the buzz? Significance 10(3):6–11. https://doi.org/10.1111/j.1740-9713.2013.00658.x

Gratzer K, Susilo F, Purnomo D, Fiedler S, Brodschneider R (2019) Challenges for beekeeping in Indonesia with autochthonous and introduced bees. Bee World 96(2):40–44. https://doi.org/10.1080/0005772X.2019.1571211

Grupe AC II, Quandt CA (2020) A growing pandemic: a review of *Nosema* parasites in globally distributed domesticated and native bees. PLoS Pathog 16(6):e1008580 https://doi.org/10.1371/journal.ppat.1008580

Guerin E (2020) Native Honey bees of Southeast Asia and Conservation Challenges. https://th.boell.org/en/2020/02/13/native-honey-bees-southeast-asia-and-conservation-challenges#:~:text=While%20some%20crops%20can%20be,dragon%20fruit%20(Hylocereus%20undatus

Gupta D, Mohammad MN, Jangid BL (2020) Influence of annual ornamental flowers (Asteraceae) on the relative abundance of honey bee species in the hot semi-arid environment. Ann Arid Zone 59(1&2):1–7

Hadisoesilo S, Raffiudin R, Susanti W, Atmowidi T, Hepburn C, Radloff SE, Fuchs S, Hepburn HR (2008) Morphometric analysis and biogeography of *Apis koschevnikovi* Enderlein (1906). Apidologie 39:495–503. https://doi.org/10.1051/apido:2008029

Han F, Wallberg A, Webster MT (2012) From where did the Western honey bee (*Apis mellifera*) originate? Ecol Evol 2(8):1949–1957

He C, Zhang K, Hou X, Han D, Wang S (2019) Foraging behavior and pollination efficiency of *Apis mellifera* L. on the oil tree Peony 'Feng Dan' (*Paeonia ostii* T. Hong et JX Zhang). Insects 10(4):116. https://doi.org/10.3390/insects10040116

Higes M, Martín-Hernández R, Meana A, Higes M, Martin R, Meana A (2006) *Nosema ceranae*, a new microsporidian parasite in honey bees in Europe. J Invertebr Pathol 92:93–95. https://doi.org/10.1016/j.jip.2006.02.005

Hristov YV, Le Roux JJ, Allsopp MH, Wossler TC (2021) Identity and distribution of American foulbrood (*Paenibacillus larvae*) in South Africa. J Apicult Res. https://doi.org/10.1080/00218839.2021.1887635

Hung K-LJ, Kingston JM, Albrecht M, Holway DA, Kohn JR (2018) The worldwide importance of honey bees as pollinators in natural habitats. Proc R Soc B Biol Sci 285(1870). https://doi.org/10.1098/rspb.2017.2140

Inson C, Malaipan S (2011) Diversity of bees (Hymenoptera: Apoidea) as insect pollinators on physic nuts (Euphorbiaceae: *Jatropha curcas* L.) in Thailand. Thai J Agric Sci 44(4):263–269

Jung C, Lee MI (2018) Beekeeping in Korea: past, present, and future challenges. In: Chantawannakul P, Williams G, Neumann P (eds) Asian beekeeping in the 21st century. Springer, Singapore. https://doi.org/10.1007/978-981-10-8222-1_8

Kahono S, Chantawannakul P, Engel MS (2018) Social bees and the current status of beekeeping in Indonesia. In: Chantawannakul P, Williams G, Neumann P (eds) Book: Asian beekeeping in the 21st century. Springer, Singapore, pp 287–306

Kamo T, Nikkeshi A, Inoue H, Yamamoto S, Sawamura N, Nakamura S, Kishi S (2022) Pollinators of oriental persimmon in Japan. Appl Entomol Zool 57:237–248. https://doi.org/10.1007/s13355-022-00784-8

Khalifa SA, Elshafiey EH, Shetaia AA, El-Wahed AAA, Algethami AF, Musharraf SG, El-Seedi HR (2021) Overview of bee pollination and its economic value for crop production. Insects 12(8):688. https://doi.org/10.3390/insects12080688

Khan KA, Ghramh HA (2021) Pollen source preferences and pollination efficacy of honey bee, *Apis mellifera* (Apidae: Hymenoptera) on *Brassica napus* crop. J King Saud Univ 33(6):101487. https://doi.org/10.1016/j.jksus.2021.101487

Khan MR, Khan MR (2004) The role of honey bees *Apis mellifera* L. (Hymenoptera: Apidae) in pollination of apple. Pak J Biol Sci 7:359–362

Kirk WDJ (2022) Is it honey bee or honeybee? Bumble bee or bumblebee? Who decides the common names of bees? Bee World 99(2):38–39. https://doi.org/10.1080/0005772X.2021.1982315

Kitnya N, Prabhudev MV, Bhatta CP, Pham TH, Nidup T, Megu K, Chakravorty J, Brockmann A, Otis GW (2020) Geographical distribution of the giant honey bee *Apis laboriosa* Smith, 1871 (Hymenoptera, Apidae). ZooKeys 951:67–81. https://doi.org/10.3897/zookeys.951.49855

Koeniger G, Koeniger N, O'Byrne P, Lamb A, Phiancharoen M, Lim H, Tihoi H, Bosuang S (2017) Scientific note on *Apis koschevnikovi* chewing petals of an orchid flower (*Dipodium* species O'Byrne KIP1272) in Borneo. Apidologie 48:259–261. https://doi.org/10.1007/s13592-016-0470-6

Koetz AH (2013) Ecology, behaviour and control of *Apis cerana* with a focus on relevance to the Australian incursion. Insects 4(4):558–592. https://doi.org/10.3390/insects4040558

Kohsaka R, Park MS, Uchiyama Y (2017) Beekeeping and honey production in Japan and South Korea: past and present. J Ethn Foods 4(2):72–79. https://doi.org/10.1016/j.jef.2017.05.002

Krishnan S, Kushalappa CG, Shaanker RU, Ghazoul J (2012) Status of pollinators and their efficiency in coffee fruit set in a fragmented landscape mosaic in South India. Basic Appl Ecol 13:277–285. https://doi.org/10.1016/j.baae.2012.03.007

Lin D, Lan L, Zheng T, Shi P (2021) Comparative genomics reveals recent adaptive evolution in Himalayan giant honey bee *Apis laboriosa*. Genome Biol Evol 13(10):evab227. https://doi.org/10.1093/gbe/evab227

Liu F, Zhang G, Qin J, Jiang W, Luo Q, Xu X, Xi F, Yuan F, Yi Z, He H, Ye W, Dai P (2020) Lowering forest habitat does not reduce forage availability to the eastern honey bees in summer and fall. Apidologie 51:609–619. https://doi.org/10.1007/s13592-020-00746-5

Lu MC (2018) Beekeeping on Taiwan Island. In: Chantawannakul P, Williams G, Neumann P (eds) Asian beekeeping in the 21st century. Springer, Singapore. https://doi.org/10.1007/978-981-10-8222-1_7

Maassen A (1913) Weitere Mitteilungen uber der seuchenhaften Brutkrankheiten der Bienen [Further communication on the epidemic brood disease of bees]. Mitteilungen aus der Kaiserlichen Biologischen Anstalt fur Land- und Forstwirtscshaft 14:48–58

Mandal E, Amin MR, Rahman H, Akanda AM (2018) Abundance and foraging behavior of native insect pollinators and their effect on mustard (*Brassica juncea* L.). Bangladesh J Zool 46:117–123. https://doi.org/10.3329/bjz.v46i2.39045

Mashilingi SK, Zhang H, Chen W, Vaissiere BE, Garibaldi LA, An J (2021) Temporal trends in pollination deficits and its potential impacts on Chinese agriculture. J Econ Entomol 114(4): 1431–1440. https://doi.org/10.1093/jee/toab100

Mashilingi SK, Zhang H, Garibaldi LA, An J (2022) Honey bees are far too insufficient to supply optimum pollination services in agricultural systems worldwide. Agric Ecosyst Environ 335: 108003. https://doi.org/10.1016/j.agee.2022.108003

Matsuda Y, Sugiura N (2019) Specialized pollination by honey bees in *Cymbidium dayanum*, a fall–winter flowering orchid. Plant Species Biol 34(1):19–26. https://doi.org/10.1111/1442-1984.12231

Mayack C, Hakanoğlu H (2022) Honey bee pathogen prevalence and interactions within the Marmara region of Turkey. Vet Sci 9:573. https://doi.org/10.3390/vetsci9100573

McMenamin AJ, Flenniken ML (2018) Recently identified bee viruses and their impact on bee pollinators. Curr Opin Insect Sci 26:120–129

Meena NK, Lal G, Meena RS, Meena BM, Meena RD (2018) Pollinator's diversity and abundance on cumin (*Cuminum cyminum* L.) and their impact on yield enhancement at semi-arid regions. J Entomol Zool Stud 6:1017–1021

Munawar MS, Sarwar G, Raja S, Waghchoure ES, Iftikhar F, Mahmood R (2009) Pollination by honey bee (*Apis mellifera*) increases seed setting and yield in black seed (*Nigella sativa*). Int J Agric Biol 11:611–615

Murray TE, Kuhlmann M, Potts SG (2009) Conservation ecology of bees: populations, species and communities. Apidologie 40:211–236. https://doi.org/10.1051/apido/2009015

Nakamura S, Yamamoto S, Sawamura N, Nikkeshi A, Kishi S, Kamo T (2020) Pollination effectiveness of European honey bee, *Apis mellifera* (Hymenoptera: Apidae), in an oriental persimmon, *Diospyros kaki* (Ericales: Ebenaceae), orchard. Appl Entomol Zool 55:405–412. https://doi.org/10.1007/s13355-020-00696-5

National Bee Board, Government of India (n.d.) Pests and diseases of honey bees and their management. https://nbb.gov.in/pdf/Pests&DiseasesHoneybees&Management.pdf

Nidup T, Dorji P (2016) The honey bees (hymenoptera: Apidae) of Bhutan with a key to the Apis species. Biol Bull 2(2):01–07

Oldroyd BP, Nanork P (2009) Conservation of Asian honey bees. Apidologie 40:296–312

Ollerton J, Winfree R, Tarrant S (2011) How many flowering plants are pollinated by animals? Oikos 120:321–326. https://doi.org/10.1111/j.1600-0706.2010.18644.x

Papa G, Maier R, Durazzo A, Lucarini M, Karabagias IK, Plutino M, Bianchetto E, Aromolo R, Pignatti G, Ambrogio A, Pellecchia M, Negri I (2022) The honey bee *Apis mellifera*: an insect at the interface between human and ecosystem health. Biology (Basel) 11(2):233. https://doi.org/10.3390/biology11020233

Partap U (2010) Innovations in revival strategies for declining pollinators with particular reference to the indigenous honey bees: experiences of ICIMOD'S initiatives in the Hindu Kush-Himalayan region. Pest Manag Econ Zoo 18:85–95

Partap U, Sharma DP, Sharma G, Gaira KS, Rasul G, Sharma E (2017) The indigenous honey bee, *Apis cerana*– a pollen robber or pollinator of large cardamom? Studies on the pollinators and pollination of large cardamom with a particular focus on the role of *Apis cerana* in Sikkim (ICIMOD Working Paper 2017/8). ICIMOD, Kathmandu

Patil PN, Pastagia JJ (2016) Effect of bee pollination on yield of coriander. *Coriandrum sativum* Linnaeus. Int J Plant Prot 9:79–83

Pattemore DE (2017) Pollination. In: Thomas B (ed) Encyclopedia of applied plant sciences, vol 2, 2nd edn. Elsevier, pp 309–320. https://doi.org/10.1016/B978-0-12-394807-6.00044-7

Phankaew C (2016) Apiculture and pollinator industry survey in Thailand. Int J Agric Ext 4(02): 95–103

Potts SG, Petanidou T, Roberts S, O'Toole C, Hulbert A, Willmer P (2006) Plant-pollinator biodiversity and pollination services in a complex Mediterranean landscape. Biol Conserv 129:519–529. https://doi.org/10.1016/j.biocon.2005.11.019

Prof Feng-Feng (1990) Apiculture in China. Bee World 71(3):104–106. https://doi.org/10.1080/0005772X.1990.11099046

Punchihewa RWK (1994) Beekeeping for honey production in Sri Lanka: management of Asiatic hive honey bee *Apis cerana* in its natural tropical monsoonal environment. Sri Lanka Department of Agriculture and Canadian International Development Agency, Peradeniya, p 232

Pushpakumara DKNG, Gunasena HPM, Kariayawasam M (2005) Flowering and fruiting phenology, pollination vectors and breeding system of dragon fruit (*Hylocereus* spp.). Sri Lankan J Agric Sci 42:81–91

Putra RE, Permana AD, Kinasih I (2014) Application of Asiatic honey bees (*Apis cerana*) and stingless bees (*Trigona laeviceps*) as pollinator agents of hot pepper (*Capsicum annuum* L.) at local Indonesia farm system. Psyche:687979. https://doi.org/10.1155/2014/687979

Raffiudin R, Ariyanti NS, Aprilianingrum I, Anwar H, Shullia NI, Bening S, Wiyati SY, Priawandiputra W, Saleh S, Suardi, Fahri F, Putra RE, Soesilohadi RCH, Purnobasuki H (2022) Flight activity and pollen resources of *Apis nigrocincta* and *Apis cerana* in Central Sulawesi, Indonesia. Agric Nat Resour 56(3):463–472. https://doi.org/10.34044/j.anres.2022.56.3.03

Rajagopal D, Eswarappa G, Raju AJS (2005) Changing trends in pollen spore research. Today and tomorrow's Printers & Publishers; New Delhi, India: 2005. Pollination potentiality of honey bees in increasing productivity of guava in Karnataka, pp 131–141

Raju AJS, Rao SP (2004) The rock honey bee *Apis dorsata* and its pollination potential in some plants. J Nat Conser 16(1):59–65

Reddy PVR, Verghese A, Rajan VV (2012) Potential impact of climate change on honey bees (*Apis* spp.) and their pollination services. Pest Manage Hortic Ecosyst 18(2):121–127

Sharma HK, Katna S, Rana BS, Rana K (2016) *Apis cerana* F. As an important natural pollinator of radish (*Raphanus sativus* L.) under mid-hill conditions of Himachal Pradesh. Int J Bio-resour Stress Manag 7(5):1156–1160

Sharma G, Partap U, Sharma DP (2019) Pollination biology of large cardamom (*Amomum subulatum* Roxb.) with special emphasis on honey bees (*Apis* spp.) and bumble bees (*Bombus* spp.) pollinators. Trop Ecol 60:507–517. https://doi.org/10.1007/s42965-020-00056-y

Shin D, Choi WT, Lin H, Qu Z, Breedveld V, Meredith JC (2019) Humidity-tolerant rate-dependent capillary viscous adhesion of bee-collected pollen fluids. Nat Commun 10:1–9. https://doi.org/10.1038/s41467-019-09372-x

Sihag RC (2021) The red dwarf honey bee (*Apis florea* F.) faces the threat of extirpation in Northwest India. Ukr J Ecol 11(2):1–11

Singh MM (2008) Foraging behaviour of the Himalayan honey bee (*Apis cerana* F.) on flowers of *Fagopyrum esculentum* M and its impact on grain quality and yield. Ecoprint 15:37–46

Smith KM, Loh EH, Rostal MK, Zambrana-Torrelio CM, Mendiola L, Daszak P (2013) Pathogens, pests, and economics: drivers of honey bee colony declines and losses. EcoHealth 10(4): 434–445. https://doi.org/10.1007/s10393-013-0870-2

Soroker V, Yossi S, Chejanovsky N (2018) Apiculture in Israel. In: Chantawannakul P, Williams G, Neumann P (eds) Asian beekeeping in the 21st century. Springer, Singapore. https://doi.org/10.1007/978-981-10-8222-1_4

Stanley J, Sah K, Subbanna ARNS (2017) How efficient is the Asian honey bee, *Apis cerana* in pollinating mustard, *Brassica campestris* var. toria? Pollination behavior, pollinator efficiency, pollinator requirements and impact of pollination. J Apicult Res 56(4):439–451. https://doi.org/10.1080/00218839.2017.1329796

Suwannapong G, Benbow ME, Nieh JC (2011) Biology of Thai honey bees: natural history and threats. In: Florio RM (ed) Bees: Biology, threats and colonies. Nova Science Publishers Inc, pp 1–103. https://labs.biology.ucsd.edu/nieh/papers/ThaiHoneybeesPageProofs.pdf

Tan H, Naeem M, Ali H, Shakeel M, Kuang H, Zhang Z, Sun C (2021) Genome sequence of the Asian honey bee in Pakistan sheds light on its phylogenetic relationship with other honey bees. Insects 12(7):652. https://doi.org/10.3390/insects12070652

Thai PH, Toan TV (2018) Beekeeping in Vietnam. In: Chantawannakul P, Williams G, Neumann P (eds) Asian beekeeping in the 21st century. Springer, Singapore. https://doi.org/10.1007/978-981-10-8222-1_11

Thapa R, Aryal S, Jung C (2018) Beekeeping and honey hunting in Nepal: current status and future perspectives. In: Chantawannakul P, Williams G, Neumann P (eds) Asian beekeeping in the 21st century. Springer, Singapore. https://doi.org/10.1007/978-981-10-8222-1_5

Theisen-Jones H, Bienefeld K (2016) The Asian honey bee (*Apis cerana*) is significantly in decline. Bee World 93(4):90–97. https://doi.org/10.1080/0005772X.2017.1284973

Traynor KS, Mondet F, de Miranda JR, Techer M, Kowallik V, Oddie MAY, Chantawannakul P, McAfee A (2020) *Varroa destructor*: a complex parasite, crippling honey bees worldwide. Trends Parasitol 36(7):592–606

Tsevegmid K, Dooshin S, Ramsey S, Chantawannakul P (2018) Beekeeping in Mongolia. In: Chantawannakul P, Williams G, Neumann P (eds) Asian beekeeping in the 21st century. Springer, Singapore. https://doi.org/10.1007/978-981-10-8222-1_9

van Engelsdorp D, Evans JD, Saegerman C, Mullin C, Haubruge E, Nguyen BK, Frazier M, Frazier J, Cox-Foster D, Chen Y, Underwood R, Tarpy DR, Pettis JS (2009) Colony collapse disorder: a descriptive study. PLoS One 4:e6481. https://doi.org/10.1371/journal.pone.0006481

Wakgari M, Yigezu G (2021) Honey bee keeping constraints and future prospects. Cogent Food Agric 7(1):1872192. https://doi.org/10.1080/23311932.2021.1872192

Wardhaugh CW, Stork NE, Edwards W, Grimbacher PS (2012) The overlooked biodiversity of flower-visiting invertebrates. PLoS One 7:e45796. https://doi.org/10.1371/journal.pone.0045796

White GF (1913) Sacbrood, a disease of bees. US Department of Agriculture, Bureau of Entomology; Washington, DC

Winston ML (1987) The biology of the honey bee. Harvard University Press, Cambridge, MA
Woyke J (1984) Survival and prophylactic control of *Tropilaelaps clareae* infesting *Apis mellifera* colonies in Afghanistan. Apidologie 15(4):421–434
Woyke J (1985) Further investigations into control of the parasite bee mite *Tropilaelaps clarae* without medication. J Apicult Res 24(4):250–254
Woyke J (1987) Comparative population dynamics of *Tropilaelaps clareae* and *Varroa jacobsoni* mites on honey bees. J Apic Res 26(3):196–202
Yang GH (2005) Harm of introducing the western honey bee *Apis mellifera* L. to the Chinese honey bee *Apis cerana* F. and its ecological impact. Acta Entomol Sinica 48:401–406
Yoshiyama M, Kimura K (2018) Bee diversity and current status of beekeeping in Japan. In: Chantawannakul P, Williams G, Neumann P (eds) Asian beekeeping in the 21st century. Springer, Singapore. https://doi.org/10.1007/978-981-10-8222-1_10
Young AM, Kodabalagi S, Brockmann A, Dyer FC (2021) A hard day's night: patterns in the diurnal and nocturnal foraging behavior of *Apis dorsata* across lunar cycles and seasons. PLoS One 16(10):e0258604. https://doi.org/10.1371/journal.pone.0258604
Zhang S, Ma J, Zhang L, Sun Z, Zhao Z, Khan N (2022) Does adoption of honey bee pollination promote the economic value of kiwifruit farmers? Evidence from China. Int J Environ Res Public Health 19:8305. https://doi.org/10.3390/ijerph19148305

MIX
Papier aus verantwortungsvollen Quellen
Paper from responsible sources
FSC® C105338

If you have any concerns about our products,
you can contact us on
ProductSafety@springernature.com

In case Publisher is established outside the EU,
the EU authorized representative is:
Springer Nature Customer Service Center GmbH
Europaplatz 3, 69115 Heidelberg, Germany

Printed by Libri Plureos GmbH
in Hamburg, Germany